W9-BNZ-166

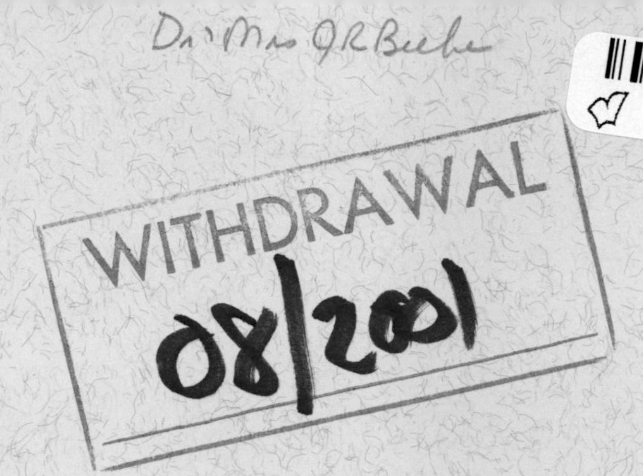
WITHDRAWAL
08/2001

Encyclopedia of Drugs and Alcohol

Editorial Board

Encyclopedia of Drugs and Alcohol

VOLUME 4

Jerome H. Jaffe, M.D.
Editor in Chief
University of Maryland, Baltimore

MACMILLAN LIBRARY REFERENCE USA
SIMON & SCHUSTER MACMILLAN
NEW YORK

SIMON & SCHUSTER AND PRENTICE HALL INTERNATIONAL
LONDON MEXICO CITY NEW DELHI SINGAPORE SYDNEY TORONTO

Macmillan Library Reference
Simon & Schuster Macmillan
1633 Broadway
New York, NY 10019-6785

Library of Congress Catalog Card Number: 94-21458

Printed in the United States of America

printing number
2 3 4 5 6 7 8 9 10

Library of Congress Cataloging-in-Publication Data
Encyclopedia of drugs and alcohol / Jerome H. Jaffe, editor-in-chief.
p. cm.
Includes bibliographical references and index.
ISBN 0-02-897185-X (set)
1. Drug abuse—Encyclopedias. 2. Substance abuse—Encyclopedias.
3. Alcoholism—Encyclopedias. 4. Drinking of alcoholic beverages—
Encyclopedias. I. Jaffe, Jerome.

HV5804.E53 1995
362.29'03—dc20 95-2321
CIP

This paper meets the requirements of ANSI-NISO Z39.48-1992 (Permanence of Paper). ∞™

Contents

Encyclopedia of Drugs and Alcohol

APPENDIX I

Poison Control Centers for Drug Overdoses (ODs) and Emergencies

INTRODUCTION

The list of Poison Control Centers has been compiled from information furnished by the American Association of Poison Control Centers (AAPCC), 3201 New Mexico Avenue NW, Suite 310, Washington, DC, 20016.

The list includes facilities that provide information on the treatment and prevention of accidents involving ingestion of poisonous (toxic) and potentially poisonous substances, including alcohol and drugs. Alcohol and drug overdoses (ODs) often cause "blackouts," coma, and death. A call to poison control in your area, with symptoms described, can mean immediate help, first-aid suggestions, and the swift response of emergency medical services (EMS). If in doubt, call Poison Control before 911.

Household products, garden supplies, and hobby materials may be inhaled or swallowed, accidentally or on purpose. Prescription drugs and over-the-counter (OTC) medications are sometimes taken in larger doses than may be safe. "Kiddie dope"—drugs sold legally without prescription, by mail and in shops, which mimic the effects of amphetamines (speed) may be taken in great quantities; these products usually contain a combination of caffeine, phenylpropanolamine, phenylephrine, ephedrine, or pseudoephedrine. Kiddie dope is taken by youngsters who expect increased energy, weight loss, and a pleasant high—but handfuls of such pills often lead to seizures, heart failure, and cerebral bleeding (stroke). Poison control units are available to answer your questions and help out in any suspected poisoning emergency—and any chemical substance can be toxic if inhaled or taken in inappropriate quantities.

The list below is current through the end of 1994. Call the phone number or 800-number closest to you for help. Try any other poison control number if you cannot get through. Some units will call an ambulance for you and/or suggest procedure for you to follow until medical help arrives.

CERTIFIED REGIONAL POISON CENTERS

ALABAMA

Alabama Poison Center, Tuscaloosa
408-A Paul Bryant Drive
Tuscaloosa, AL 35401
(800) 462–0800 (AL only)
(205) 345–0600

Regional Poison Control Center The Children's Hospital of Alabama
1600 7th Ave. South
Birmingham, AL 35233–1711
(205) 939–9201
(800) 292–6678 (AL only)
(205) 933–4050

ARIZONA

Arizona Poison and Drug Information Center
Arizona Health Sciences Center, Rm. #3204-K
1501 North Campbell Ave.
Tucson, AZ 85724
(800) 362–0101 (AZ only)
(602) 626–6016

Samaritan Regional Poison Center
Teleservices Department
1441 North 12th Street
Phoenix, AZ 85006
(602) 253–3334

CALIFORNIA

Central California Regional Poison Control Center
Valley Children's Hospital
3151 North Millbrook, IN31
Fresno, CA 93703
(800) 346–5922 (Central CA only)
(209) 445–1222

San Diego Regional Poison Center
UCSD Medical Center
200 West Arbor Drive
San Diego, CA 92103–8925
(619) 543–6000
(800) 876–4766 (in 619 area code only)

San Francisco Bay Area Regional Poison Control Center
San Francisco General Hospital
1001 Potrero Ave., Building 80, Room 230
San Francisco, CA 94110
(800) 523–2222

Santa Clara Valley Regional Poison Center
Valley Health Center, Suite 310
750 South Bascom Ave.
San Jose, CA 95128
(408) 885–6000
(800) 662–9886 (CA only)

University of California, Davis, Medical Center Regional Poison Control Center
2315 Stockton Blvd.
Sacramento, CA 95817
(916) 734–3692
(800) 342–9293 (Northern California only)

COLORADO

Rocky Mountain Poison and Drug Center
645 Bannock St.
Denver, CO 80204
(303) 629-1123

DISTRICT OF COLUMBIA

National Capital Poison Center
3201 New Mexico Avenue, NW, Suite 310
Washington, DC 20016
(202) 625-3333
(202) 362–8563 (TTY)

FLORIDA

Florida Poison Information Center–Jacksonville
University Medical Center
University of Florida Health Science Center–Jacksonville
655 West 8th Street
Jacksonville, FL 32209
(904) 549–4480
(800) 282–3171 (FL only)

The Florida Poison Information Center and Toxicology Resource Center
Tampa General Hospital
Post Office Box 1289
Tampa, FL 33601
(813) 253–4444 (Tampa)
(800) 282–3171 (Florida)

GEORGIA

Georgia Poison Center
Grady Memorial Hospital
80 Butler Street S.E.
P.O. Box 26066
Atlanta, GA 30335–3801
(800) 282–5846 (GA only)
(404) 616–9000

INDIANA

Indiana Poison Center
Methodist Hospital of Indiana
1701 N. Senate Boulevard
P.O. Box 1367
Indianapolis, IN 46206–1367
(800) 382–9097 (IN only)
(317) 929–2323

KENTUCKY

Kentucky Regional Poison Center of Kosair Children's Hospital
P.O. Box 35070
Louisville, KY 40232–5070
(502) 629–7275
(800) 722–5725 (KY only)

MARYLAND

Maryland Poison Center
20 N. Pine St.
Baltimore, MD 21201
(410) 528–7701
(800) 492–2414 (MD only)

National Capital Poison Center (DC suburbs only)
3201 New Mexico Avenue, NW, Suite 310
Washington, DC 20016
(202) 625–3333
(202) 362–8563 (TTY)

MASSACHUSETTS

Massachusetts Poison Control System
300 Longwood Ave.
Boston, MA 02115
(617) 232–2120
(800) 682–9211

MICHIGAN

Poison Control Center
Children's Hospital of Michigan
3901 Beaubien Blvd.
Detroit, MI 48201
(313) 745–5711

MINNESOTA

Hennepin Regional Poison Center
Hennepin County Medical Center
701 Park Ave.
Minneapolis, MN 55415
(612) 347–3141
Petline: (612) 337–7387
(612) 337–7474

Minnesota Regional Poison Center
St. Paul-Ramsey Medical Center
640 Jackson Street
St. Paul, MN 55101
(612) 221–2113

MISSOURI

Cardinal Glennon Children's Hospital Regional Poison Center
1465 S. Grand Blvd.
St. Louis, MO 63104
(314) 772–5200
(800) 366–8888

MONTANA

Rocky Mountain Poison and Drug Center
645 Bannock St.
Denver, CO 80204
(303) 629–1123

NEBRASKA

The Poison Center
8301 Dodge St.
Omaha, NE 68114
(402) 390–5555 (Omaha)
(800) 955–9119 (NE & WY)

NEW JERSEY

New Jersey Poison Information and Education System
201 Lyons Ave.
Newark, NJ 07112
(800) 962–1253

NEW MEXICO

New Mexico Poison and Drug Information Center
University of New Mexico
Albuquerque, NM 87131–1076
(505) 843–2551
(800) 432–6866 (NM only)

NEW YORK

Hudson Valley Poison Center
Nyack Hospital
160 N. Midland Ave.
Nyack, NY 10960
(800) 336–6997
(914) 353–1000
Long Island Regional Poison Control Center
Winthrop University Hospital
259 First Street
Mineola, NY 11501
(516) 542–2323, 2324, 2325, 3813
New York City Poison Control Center
N.Y.C. Department of Health
455 First Ave., Room 123
New York, NY 10016
(212) 340–4494
(212) P-O-I-S-O-N-S
(212) 689–9014

OHIO

Central Ohio Poison Center
700 Children's Drive
Columbus, OH 43205–2696
(614) 228–1323
(800) 682–7625
(614) 228–2272 (TTY)
(614) 461–2012
Cincinnati Drug & Poison Information Center and Regional Poison Control System
231 Bethesda Avenue, M.L. 144
Cincinnati, OH 45267–0144
(513) 558–5111
(800) 872–5111 (OH only)

OREGON

Oregon Poison Center
Oregon Health Sciences University
3181 S.W. Sam Jackson Park Road
Portland, OR 97201
(503) 494–8968
(800) 452–7165 (OR only)

PENNSYLVANIA

Central Pennsylvania Poison Center
University Hospital
Milton S. Hershey Medical Center
Hershey, PA 17033
(800) 521–6110
The Poison Control Center serving the greater Philadelphia metropolitan area
One Children's Center
Philadelphia, PA 19104–4303
(215) 386–2100
Pittsburgh Poison Center
3705 Fifth Avenue
Pittsburgh, PA 15213
(412) 681–6669

RHODE ISLAND

Rhode Island Poison Center
583 Eddy St.
Providence, RI 02903
(401) 277–5727

TEXAS

North Texas Poison Center
5201 Harry Hines Blvd.
P.O. Box 35926
Dallas, TX 75235
(214) 590-5000
Texas Watts (800) 441–0040
Texas State Poison Center
The University of Texas Medical Branch
Galveston, TX 77550–2780
(409) 765–1420 (Galveston)
(713) 654–1701 (Houston)

UTAH

Utah Poison Control Center
410 Chipeta Way, Suite 230
Salt Lake City, UT 84108
(801) 581–2151
(800) 456–7707 (UT only)

VIRGINIA

Blue Ridge Poison Center
Box 67
Blue Ridge Hospital
Charlottesville, VA 22901
(804) 924–5543
(800) 451–1428
National Capital Poison Center (Northern VA only)
3201 New Mexico Avenue, NW, Suite 310
Washington, DC 20016
(202) 625–3333
(202) 362–8563 (TTY)

WEST VIRGINIA

West Virginia Poison Center
3110 MacCorkle Ave. S.E.
Charleston, WV 25304
(800) 642–3625 (WV only)
(304) 348–4211

WYOMING

The Poison Center
8301 Dodge St.
Omaha, NE 68114
(402) 390–5555 (Omaha)
(800) 955–9119 (NE & WY)

APPENDIX II

U.S. and State Government Drug Resources Directory

INTRODUCTION

This is a guide to state agencies that address substance abuse concerns. It begins with listings of federal agencies and goes on to state and local listings. The listings were originally compiled for a report by the U.S. Department of Justice, Bureau of Justice Statistics. Six groupings are presented here: Federal Information Centers and Clearinghouses; Other Federal Sources; Drug Abuse Resistance Education (DARE) Regional Training Centers; National Prevention Network; Treatment Alternatives to Street Crime (TASC) Programs; and the State Listings.

In Appendix III, which follows, an extensive State-by-State Directory is presented for drug abuse and alcoholism treatment and prevention programs, both public and private.

FEDERAL INFORMATION CENTERS AND CLEARINGHOUSES

CRIMINAL JUSTICE

Drugs & Crime Data Center & Clearinghouse
1600 Research Boulevard
Rockville, MD 20850
(800) 666–3332
Sponsored by: Bureau of Justice Statistics

National Institute of Justice/ National Criminal Justice Reference Service
PO Box 6000
Rockville, MD 20850
(800) 851–3420
Sponsored by: National Institute of Justice

Bureau of Justice Assistance Clearinghouse
PO Box 6000
Rockville, MD 20850
(800) 688–4BJA/4252
Sponsored by: Bureau of Justice Assistance

Juvenile Justice Clearinghouse
PO Box 6000
Rockville, MD 20850
(800) 638–8736
Sponsored by: Office of Juvenile Justice and Delinquency Prevention

Justice Statistics Clearinghouse
PO Box 6000
Rockville, MD 20850
(800) 732–3277
Sponsored by: Bureau of Justice Statistics

NIJ/AIDS Clearinghouse
PO Box 6000
Rockville, MD 20850
(301) 251–5500
Sponsored by: National Institute of Justice

Technology Assessment Program Information Center
PO Box 6000
Rockville, MD 20850
(800) 248–2742
Sponsored by: National Institute of Justice

National Victims Resource Center
PO Box 6000
Rockville, MD 20850
(301) 251–5525
Sponsored by: Office for Victims of Crime

National Institute of Corrections Information Center
1790 30th Street
Suite 130
Boulder, CO 80301
(303) 444–1101
Sponsored by: National Institute of Corrections

HEALTH

National Clearinghouse for Alcohol and Drug Information
PO Box 2345
Rockville, MD 20852
(301) 468–2600
Sponsored by: Office of Substance Abuse Prevention

National Drug Information and Treatment Referral Hotline
5600 Fishers Lane
Rockville, MD 20857
(800) 662–HELP
Sponsored by: National Institute on Drug Abuse

Drug-Free Workplace Helpline
5600 Fishers Lane
Rockville, MD 20857
(800) 843–4971
Sponsored by: National Institute on Drug Abuse

National AIDS Information Clearinghouse
PO Box 6003
Rockville, MD 20850
(800) 458–5231
Sponsored by: Centers for Disease Control

PUBLIC HOUSING

Drug Information and Strategy Clearinghouse
PO Box 6424
Rockville, MD 20850
(800) 245–2691
Sponsored by: Housing and Urban Development

EDUCATION

ACCESS ERIC
Department CCE
1600 Research Boulevard
Rockville, MD 20850
(800) USE–ERIC
Sponsored by: United States Department of Education

OTHER FEDERAL SOURCES

Executive Office of the President
Office of National Drug Control Policy
Executive Office of the President
Washington, DC 20500
(202) 673–2823

U.S. Department of Justice
Office of Justice Programs
633 Indiana Avenue NW
Washington, DC 20531
(202) 724–5933

Bureau of Justice Assistance
633 Indiana Avenue NW
Washington, DC 20531
(202) 272–6121

Bureau of Justice Statistics
633 Indiana Avenue NW
Washington, DC 20531
(202) 724–7765

National Institute of Justice
633 Indiana Avenue NW
Washington, DC 20531
(202) 724–2942

Office of Juvenile Justice and Delinquency Prevention
633 Indiana Avenue NW
Washington, DC 20531
(202) 724–7751

Office for Victims of Crime
633 Indiana Avenue NW
Washington, DC 20531
(202) 724–5983

Executive Office for United States Attorneys
10th and Constitution Avenue NW
Washington, DC 20009
(202) 633–2000

Drug Enforcement Administration
600 Army Navy Drive
Arlington, VA 20537
(202) 307–7977 Public Affairs
(202) 307–8932 Library

U.S. Courts
Administrative Office of the United States Courts
811 Vermont Avenue NW
Washington, DC 20544
(202) 633–6040

Federal Judicial Center
1520 H Street NW
Washington, DC 20005
(202) 633–6011

U.S. Department of Health and Human Services
National Institute on Drug Abuse
Parklawn Building
5600 Fishers Lane
Rockville, MD 20857
(301) 443–6245

Office of Substance Abuse Prevention
Parklawn Building
5600 Fishers Lane
Rockville, MD 20857
(301) 443–0365

U.S. Department of State
Bureau of International Narcotics Matters
Room 7331
2201 C Street NW
Washington, DC 20520
(202) 647–6642

U.S. Department of the Treasury
Bureau of Alcohol, Tobacco, and Firearms
Public Affairs
Room 4402
1200 Pennsylvania Avenue NW
Washington, DC 20226
(202) 566–7135

United States Customs Service
Room 611
1301 Constitution Avenue NW
Washington, DC 20229
(202) 566–8195

U.S. Department of Transportation
United States Coast Guard
2100 2nd Street SW
Washington, DC 20593
(202) 267–2229

U.S. Department of Education
Drug-Free Schools Staff
400 Maryland Avenue SW
Washington, DC 20202
(202) 732–4599

U.S. Department of Housing and Urban Development
Office for Drug-Free Neighborhoods
451 7th Street SW
Washington, DC 20241
(202) 755–7197

DRUG ABUSE RESISTANCE EDUCATION (DARE) REGIONAL TRAINING CENTERS

Bureau of Justice Assistance
DARE Program Manager
Bureau of Justice Assistance
633 Indiana Avenue NW
Washington, DC 20531
(202) 272–4604

Project DARE
Coordinator
3110 North 19th Avenue
Suite 290
Phoenix, AZ 85015
(602) 262–8111

Los Angeles Police Department
DARE Coordinator
Los Angeles Police Department
Juvenile Division—DARE
150 North Los Angeles Street
Los Angeles, CA 90012
(213) 485–4856

Strategic Development Bureau
Division of Administration
Illinois State Police
201 East Adams Street
Suite 300
Springfield, IL 62707
(217) 782–5227

Project DARE
Education Development Center
55 Chapel Street
Newton, MA 02160
(617) 969–7100

Project DARE
Virginia State Police Department
Box 27472
Richmond, VA 23261–7472
(804) 323–2059

NATIONAL PREVENTION NETWORK

State Substance Abuse Coordinator
Alabama Department of Mental Health
PO Box 3710
Montgomery, AL 36193
(205) 271–9243

Executive Director
Alaska Council on Prevention of Alcohol and Drug Abuse
7521 Old Seward Highway
Suite A
Anchorage, AK 99518
(907) 349–6602

Director of the Governor's Partnership in Substance Abuse Prevention
DADAP
PO Box 1437
Little Rock, AR 72203–1437
(501) 682–6664

Program Representative
Office of Community Behavioral Health
411 North 24th Street
Phoenix, AZ 85008
(602) 220–6478

Chief
Drug Prevention Programs
Department of Alcohol and Drug Programs
111 Capitol Mall
Room 420
Sacramento, CA 95814
(916) 323–2088

Director
Prevention/Intervention
Colorado Department of Health
Alcohol and Drug Abuse Division
4210 East 11th Avenue
Denver, CO 80220
(303) 331–8201

Prevention Director
Connecticut Alcohol and Drug Abuse Commission
999 Asylum Avenue
Hartford, CT 06105
(203) 566–7458

Public Health Analyst
Commission of Public Health
Health Planning and Development
1660 L Street NW
7th Floor
Washington, DC 20036
(202) 724–5637

Administrator
Office of Prevention
Delaware Youth and Family Center
1825 Faulkland Road
Wilmington, DE 19805
(302) 633–2678

Director of Training
Division of Alcoholism, Drug Abuse and Mental Health
1901 North DuPont Highway
New Castle, DE 19720
(302) 421–8251

Bureau Chief
Florida Department of Health and Rehabilitation Services
Alcohol and Drug Abuse Programs
1317 Winewood Boulevard
Tallahassee, FL 32301
(904) 488–0090

Alcohol and Drug Abuse Services
Division of Mental Health and Mental Retardation
878 Peachtree Street NE
Room 319
Atlanta, GA 30309
(404) 894–4749

Prevention Coordinator
Alcohol and Drug Abuse Division
PO Box 3378
Honolulu, HI 96801
(808) 548–4280

Bureau Chief
Public Health Department
Iowa Division of Substance Abuse and Health Promotion
Lucas State Office Building
4th Floor
321 East 12th Street
Des Moines, IA 50319–0075
(515) 281–3641

Bureau of Substance Abuse
Department of Health and Welfare
450 West State Street
Boise, ID 83720
(208) 334–5934

Department of Alcoholism and Substance Abuse
State of Illinois Center
Suite 5–600
100 West Randolph Street
Chicago, IL 60601
(312) 917–6332

Director
Prevention and Planning
Department of Mental Health
Addiction Services Division
117 East Washington Street
Indianapolis, IN 46204
(317) 232–7919

Alcohol and Drug Abuse Services
Biddle Building
300 Southwest Oakley
Topeka, KS 66606
(913) 296–3925

Substance Abuse Branch
275 East Main Street
Frankfort, KY 40621
(502) 564–2880

Division of Prevention for Alcohol and Drug Abuse
2744 B Wooddale Boulevard
Baton Rouge, LA 70805
(504) 342–9351

Director
Prevention Services
Massachusetts Department of Public Health
Division of Substance Abuse Services
150 Tremont Street
Boston, MA 02111
(617) 727–1960

Alcohol and Drug Abuse Administration
201 West Preston Street
Room 410
Baltimore, MD 21201
(301) 225–6543

Office of Alcohol and Drug Abuse Prevention
State House
Station 11
Augusta, ME 04333
(207) 289–2781

Office of Substance Abuse Services
PO Box 30195
Lansing, MI 48909
(517) 373–4700

Chemical Dependency Program
Department of Human Services
444 Lafayette Road
St. Paul, MN 55101
(612) 296–4711

Division of Alcohol and Drug Abuse
1915 Southridge
PO Box 687
Jefferson City, MO 65102
(314) 751–4942

Division of Alcohol and Drug Abuse
Department of Mental Health
1101 Robert E. Lee Building
239 North Lamar Street
Jackson, MS 39201
(601) 359–1288

Alcohol and Drug Abuse Division
1539–11th Avenue
Helena, MT 59620
(406) 444–2878

Division of Mental Health, Mental Retardation, and Substance Abuse
Department of Human Resources
325 North Salisbury Street
Suite 628
Raleigh, NC 27611
(919) 733–4555

Research and Evaluation
Division of Alcohol and Drug Abuse
1839 East Capitol Avenue
Bismarck, ND 58501
(701) 224–2769

Counselor Certification Coordinator
Division on Alcoholism and Drug Abuse
Department of Public Institutions
PO Box 94728
Lincoln, NE 68509
(402) 471–2851

Special Projects
Office of Alcohol and Drug Abuse Prevention
Health and Human Services
6 Hazen Drive
Concord, NH 03301–6525
(603) 271–4629

Special Projects
Division of Narcotic and Drug Abuse Control
New Jersey State Department of Health
CN–360–Room 100
Trenton, NJ 08625–0360
(609) 292–4346

Public Information/Special Projects
Prevention and Education Unit
New Jersey Division of Alcoholism
129 East Hanover Street
Trenton, NJ 08625
(609) 984–3313

Chief
Research and Evaluation
Substance Abuse Bureau
PO Box 968
Santa Fe, NM 87503
(505) 827–2587

Representative for Drug Abuse Prevention
Bureau of Alcohol and Drug Abuse
Capitol Complex
Room 500
505 East King Street
Carson City, NV 89710
(702) 885–4790

Representative-Alcohol Abuse Prevention
Division of Alcoholism and Alcohol Abuse
194 Washington Avenue
Albany, NY 12210
(518) 473–0887

Representative for Drug Abuse Prevention
Deputy Director for Substance Abuse Prevention
Executive Park South
Albany, NY 12210
(518) 457–5840

Special Projects
B.A.A.A.R.
Ohio Department of Health
170 North High Street
3rd Floor
Columbus, OH 43215
(614) 466–3445

State Prevention Coordinator
Bureau of Drug Abuse
170 North High Street
3rd Floor
Columbus, OH 43215
(614) 466–7893

Director of Prevention Services
Department of Mental Health and Substance Abuse Services
1200 N.E. 13th Street
Oklahoma City, OK 73117
(405) 271–7474

Prevention Manager
Office of Alcohol and Drug Abuse Programs
1178 Chemeketa Street NE
Salem, OR 97310
(503) 378–2163

Director of Prevention and Intervention
Pennsylvania Department of Health
Office of Drug and Alcohol Programs
H & W Building
Room 929
Harrisburg, PA 17108
(717) 783–8200

Policy and Planning
Division of Substance Abuse
PO Box 20363
Cranston, RI 02920
(401) 464–2191

South Carolina Commission on Alcohol and Drug Abuse
3700 Forest Drive
Suite 300
Columbia, SC 29204
(803) 734–9552

State Prevention Coordinator
South Dakota Human Services Department
Division of Alcohol and Drug Abuse
523 East Capitol Street
Room 125
Pierre, SD 57501
(605) 773–3123

Division of Alcohol and Drug Abuse
Department of Mental Health
Doctor's Building
4th Floor
706 Church Street
Nashville, TN 37219
(615) 741–3862

Director
Program Development
Texas Commission on Alcohol and Drug Abuse
1705 Guadalupe Street
Austin, TX 78701–1214
(512) 463–5510

Executive Committee
Alcohol and Drug Abuse Clinic
50 North Medical Drive
Salt Lake City, UT 84132
(801) 581–6228

Research and Evaluation
Prevention, Promotion and Library
PO Box 1797
Richmond, VA 23214
(804) 786–1530

Executive Committee
Office of Alcohol and Drug Abuse Program
103 South Main Street
Waterbury, VT 05676
(802) 241–2170

Executive Committee
Community Mobilization Against Substance Abuse
Mail Stop OB–44E
Olympia, WA 98504
(206) 586–6135

Special Projects
Office of Alcohol and Other Drug Abuse
PO Box 7851
Madison, WI 53707–7851
(608) 266–9485

Division on Alcoholism and Drug Abuse
West Virginia Department of Health
1800 Washington Street East
Room 451
Charleston, WV 25305
(304) 348–2276

Alcohol and Drug Program
Department of Human Resources
Government of American Samoa
Pago Pago, AS 96799

Department of Mental Health and Substance Abuse
PO Box 9400
Tamuning, GU 96911
(671) 646–9261–69

Supervisor
Prevention Branch
Department of Mental Health and Substance Abuse
PO Box 9400
Tamuning, GU 96911
(671) 646–9261–69

Puerto Rico Department of Anti-Addiction Services
Box B–Y, Rio Piedras Station
Rio Piedras, PR 00928
(809) 751–6915

Department of Health
Division of Mental Health, Alcoholism and Drug Dependency
PO Box 520
Christiansted
St. Croix, VI 00820
(809) 773–8443

TREATMENT ALTERNATIVES TO STREET CRIME (TASC) PROGRAMS

TASC Project Director
University of Alabama at
Birmingham TASC
718 30th Street South
Birmingham, AL 35233
(205) 934–7430

TASC Project Director
Treatment Assessment Screening
Center, Incorporated
2234 North 7th Street
Phoenix, AZ 85006
(602) 254–7328

TASC Program Manager
Yuma Behavioral Health Services
Department
1073 West 23rd Street
Yuma, AZ 85364
(602) 783–0197

TASC Treatment Study
University of Arkansas
Medical Sciences
Department of Pathology
Mail Slot 502
Little Rock, AR
(501) 686–7148

TASC Program Coordinator
Sonoma County Drug Abuse
Services
837 5th Street
Santa Rosa, CA 95404
(707) 527–2543

TASC Coordinator
Denver TASC
1626 High Street
Denver, CO 80206
(303) 321–6563

Program Director
Northeast Denver TASC
7255 Irving Street
Suite 106
Westminster, CO 80030
(303) 428–5264

Director
TASC/Continuing Care Unit
1501-C North Broom Street
Wilmington, DE 19806
(302) 429–5928

Statewide TASC Coordinator
Alcohol and Drug Programs
Department of Health and
Rehabilitative Services
1317 Winewood Boulevard
Tallahassee, FL 32301
(904) 488–0900

Levy
451 Capital Street
PO Box 516
Bronson, FL 32621
(904) 495–2726

Volusia County TASC
440 1/2 South Beach Street
Daytona Beach, FL 32114
(904) 252–8026

Assistant Director
Stewart Treatment Center/TASC
124 Michigan Avenue
Daytona Beach, FL 32014
(904) 255–0447

Director
Program to Aid Drug Abusers
TASC
2920 Franklin Street
PO Box 1593
Eaton Park, FL 33840
(813) 665–2211

Spectrum TASC
2801 East Oakland Park Boulevard
Room 210
Ft. Lauderdale, FL 33306
(305) 564–2266

Indian River Community Mental
Center
800 Avenue H
Ft. Pierce, FL 33450
(407) 464–8111

Lee Mental Health Center-20th
Judicial
TASC Program
PO Box 06137
Ft. Myers, FL 33906
(813) 275–3222

Addictions and Family Health
Services/TASC
4300 SW 13th Street
Gainesville, FL 32608
(904) 374–5690

Starting Place
2057 Coolidge Street
Hollywood, FL 33020
(407) 925–2225

Gateway Community Services,
Incorporated
555 Stockton Street
Jacksonville, FL 32204
(904) 387–4661

River Region Human Services
Center
421 West Church Street
Suite 702
Jacksonville, FL 32202
(904) 359–6571

MH Services of Osceola County
917 Emmett Street
Kissimmee, FL 32741
(407) 846-0023

North Florida Mental Health
Centers/TASC
PO Box 2818
Lake City, FL 32056–2818
(904) 752–1045

Operation PAR, Incorporated
Adult TASC
13800 66th Street North
Largo, FL 34641
(813) 536–9493

Program Director
Mental Health Care Center of the
Lower Keys
PO Box 488
Key West, FL 33041
(305) 464–6723

Director
Metro-Dade Office of Rehabilitative
Services/TASC
Juvenile TASC
3300 NW 27th Avenue
Miami, FL 33137
(305) 638–6005

Program Supervisor
David Lawrence Center
Court Related Services
6075 Golden Gate Parkway
Naples, FL 33999
(813) 455–1031 ext. 292

Program Administrator
Human Development Center
PASCO/TASC
8251 Arevee Drive
New Port Richey, FL 34653
(813) 847–4700

Marion/Citrus Mental Health
Tri-County TASC
717 SW Martin Luther King Jr. Avenue
Ocala, FL 32671
(904) 629–8893

Director
The Center for Drug Free Living/ TASC
100 West Columbia Street
Orlando, FL 32806
(407) 423–6618

Putnam County TASC
PO Drawer 1355
Palatka, FL 32077
(904) 328–3461

Escambia County TASC
1190 West Leonard Street
Pensacola, FL 32501
(904) 436–9855

Director
Brevard County TASC/Circles of Care
1770 Cedar Street
Rockledge, FL 32955
(407) 632–9480

Director
St. Johns County Mental Health Services
PO Drawer 1209
St. Augustine, FL 32085–1209
(904) 824–4454

Executive Director
Storefront Centers for Counseling, Incorporated
1670 Main Street
Sarasota, FL 34236
(813) 953–3595

Brafford/Union/Putnam Guidance Clinic, Incorporated/TASC
PO Box 399
Starke, FL 32091
(904) 964–8382

Director
Disc Village/TASC
3333 West Pensacola Street
Suite 100
Tallahassee, FL 32304
(904) 488–6520

Director
Dacco TASC
3200 Henderson Boulevard
Tampa, FL 32304
(904) 488–6520

Director
Lake Sumpter TASC
112 Sinclair Avenue
Tavares, FL 32778
(904) 343–8102

Director
The Grove Counselling Center
218 South Oak Avenue
Sanford, FL 32771
(407) 327–2686

Hawaii Department of Corrections
Drug Screening Project/TASC
Gold Bond Building
677 Ala Moana Boulevard
Suite 700
Honolulu, HI 96813
(808) 548–3630

Executive Director
TASC, Incorporated
1500 North Halsted
Chicago, IL 60622
(312) 787–0208

Coordinator
Youth Services
TASC, Incorporated
1100 South Hamilton
Room 21
Chicago, IL 60612
(312) 666–7339

Court Services Coordinator
Court Services
TASC, Incorporated
2600 South California Avenue
Room 107
Chicago, IL 60608
(312) 376–0950

Area Representative Coordinator
Court Outposts
TASC, Incorporated
1500 North Halsted
2nd Floor
Chicago, IL 60622
(312) 787–0208 ext. 54

TASC, Incorporated-DuPage
201 Reber Street
Room 203B
Wheaton, IL 60187
(708) 260–0891

TASC, Incorporated-Waukegan
415 Washington Street
Waukegan, IL 60085
(708) 249–2200

TASC, Incorporated-Joliet
58 North Chicago Office
Suite 508
Joliet, IL 60431
(815) 727–6397

TASC, Incorporated
119 North Church Street
Suite 200
Rockford, IL 61101
(815) 965–1106

Area Coordinator
TASC, Incorporated
Plaza Building
1705 2nd Avenue
Suite 401
Rock Island, IL 61201
(309) 788–0816

Area Coordinator
TASC, Incorporated
909 1st National Bank Building
Peoria, IL 61602
(309) 673–3769

TASC, Incorporated
Three Old Capitol Plaza West
Springfield, IL 62701
(217) 544–0842

TASC, Incorporated
104 West University Avenue
Urbana, IL 61801
(217) 344–4546

Area Coordinator
TASC, Incorporated
100 West Main Street
Belleville, IL 61801
(618) 277–0410

Area Coordinator
TASC, Incorporated
1009 Chestnut Street
Murphysboro, IL 62966
(618) 687–2321

Area Coordinator
TASC, Incorporated
103 Plaza Court
Edwardsville, IL 62025
(618) 656–7672

TASC, Incorporated
5600 West Old Orchard Road
Skokie, IL 60077
(708) 470–7427

TASC Director
Lawrence Circuit Court TASC Project
1502 I Street
Room 208
Bedford, IN 47421
(812) 275–1980

Assistant Director
Department of Correctional Services/TASC
1035 3rd Avenue SE
Cedar Rapids, IA 52403
(319) 398-3672

Executive Director
National Council on Alcoholism and Other Drug Dependencies
Fleming Building
Suite 606
218 6th Avenue
Des Moines, IA 50309
(515) 244–2297

Department of Correctional Services
PO Box 2596
527 East 5th Street
Waterloo, IA 50704
(319) 291–2091

Assistant Director
Division of Probation and Parole
State House
Station 111
Augusta, ME 04333
(207) 289–4381

TASC/Early Intervention
Kennebec County Jail
Augusta, ME 04330
(207) 623–3591

TASC/Early Intervention
Somerset County Jail
High Street
Skowhegan, ME 04976
(207) 474–9591

TASC Project Director
Baltimore County Alternative Sentencing/TASC
201 West Chesapeake
Towson, MD 21204
(301) 887–2056

TASC Project
105 Fleet Street
Rockville, MD 20850
(301) 279-1332

Director of Probation
Main Drug Intake and Referral/ TASC Unit
Detroit Recorder's Court
1441 St. Antoine Street
Detroit, MI 48226
(313) 224–5187

State TASC Coordinator
Administrative Offices of the Court
CN 982
Trenton, NJ 08625
(609) 292–3488

Senior Court Consultant
Burlington TASC-Court Liaison Program
3rd Floor
Room 307
County Office Building
Mount Holly, NJ 08060
(609) 265–5335

TASC Coordinator
Hudson County Administrative Building
595 Newark Avenue
Jersey City, NJ 07306
(201) 795–6857

TASC Coordinator
Middlesex County Probation Department/TASC Project
John F. Kennedy Square
PO Box 789
New Brunswick, NJ 08846
(908) 745–3649

Project Director
TASC of the Capitol District, Incorporated
87 Columbia Street
Albany, NY 12210
(518) 465–1455

Associate Executive Director
TASC/EAC, Incorporated
100 East Old Country Road
Mineola, NY 12210
(516) 741–5580

Project Director
Queens TASC/EAC, Incorporated
North County Complex
Building 16
Hauppauge, NY 11787
(516) 360–5777

Project Director
Staten Island TASC/EAC, Incorporated
25 Hyatt Street
5th Floor
Staten Island, NY 10301
(718) 727–9722

Project Director
Nassau TASC/EAC, Incorporated
250 Fulton Avenue
Hempstead, NY 11550
(516) 486–8944

Westchester County Department of Mental Health/TASC
85 Court Street
Room 103
White Plains, NY 10601
(914) 285–5832

Program Assistant
Westchester County TASC
112 East Post Road
White Plains, NY 10601
(914) 285–5265

Criminal Justice Specialist
Alcohol and Drug Abuse Services
North Carolina Department of Mental Health, Mental Retardation and Substance Abuse Services
325 North Salisbury Street
Raleigh, NC 27611
(919) 733–4555

Blue Ridge Area Mental Health, Mental Retardation, and Substance Abuse Services
283 Biltmore Avenue
Asheville, NC 28801
(704) 252–8748

Alamance-Caswell Area Mental Health, Mental Retardation, and Substance Abuse Services
1946 Martin Street
Burlington, NC 27215
(919) 228–0580

Open House/TASC
145 Remount Road
Charlotte, NC 28203
(704) 332–9001

Durham County Substance Abuse Services
705 South Mangum Street
Durham, NC 27701
(919) 560–7500

Cumberland County Mental Health Center/TASC
PO Box 2068
Fayetteville, NC 28302
(919) 433–2712

Sycamore Center for Substance Abuse
101 West Sycamore Street
Suite 140
Greensboro, NC 27401
(919) 275–9341

High Point Drug Action Council
119 Chestnut Drive
PO Box 2714
High Point, NC 27261
(919) 882–2125

Pitt Area Mental Health, Mental Retardation, and Substance Abuse Services
2310 Stantonsburg Road
Greenville, NC 27834
(919) 752–7151

Director
Drug Action of Wake County
2809 Industrial Drive
Raleigh, NC 27526
(919) 832–4453

TASC Director
Cape Fear Substance Abuse Center
419 Chestnut Street
Wilmington, NC 28401
(919) 762–5333

TASC Director
Step One, Incorporated-TASC Program
310 East 3rd Street
PO Box 2110
Winston-Salem, NC 27102
(919) 725–8389

Oklahoma Department of Corrections
3400 Martin Luther King Boulevard
PO Box 1140
Oklahoma City, OK 73136
(405) 425–2555

TASC Coordinator
District I TASC
201 Court Street
Suite 201
Muskogee, OK 74401
(918) 683–0243

TASC Coordinator
District II TASC
1328 South Denver
Tulsa, OK 74129
(918) 587–8269

TASC Coordinator
District III TASC
PO Box 669
McAlester, OK 74502
(918) 423–1668

District IV TASC
415 11th Street
PO Box 2649
Lawton, OK 73502
(405) 248–1444

District V TASC
808 West Maine
Enid, OK 73701
(405) 237–3396

District VI TASC
4640 South May
Oklahoma City, OK 73119
(405) 681–4663

Executive Director
TASC of Oregon, Incorporated
1727 NE 13th Avenue
Portland, OR 97212
(503) 281–0037

TASC Coordinator
Marion County Department of Correction
3060 Center Street NE
Salem, OR 97301
(503) 588–5289

State TASC Coordinator
Pennsylvania Department of Health
Office of Drug and Alcohol Programs
Health and Welfare Building
Room 923
PO Box 90
Harrisburg, PA 17102
(717) 783–8200

TASC Director
Lehigh County TASC
521 Court Street
Allentown, PA 11801
(215) 432–6760

Executive Director
Bucks County TASC/TODAY, Incorporated
PO Box 98
Newtown, PA 18940
(215) 343–7800

TASC Coordinator
Greater Erie Community Action Committee TASC
809 Peach Street
Erie, PA 16501
(814) 870–5424

Executive Director
Chester County Council on Addictive Disease/TASC
313 East Lancaster Avenue
Exton, PA 19341
(215) 363–7709

Executive Director
Adams County TASC Program
108 North Stratton Street
Gettysburg, PA 17325
(717) 334–8154

Administrative Director
Mon-Yough/Westmoreland Drug and Alcohol Services
Miller Square
104 West 4th Street
Suite 1
Greensburg, PA 15601
(412) 832–5880

Program Director
Dauphin County TASC
Department of Drug and Alcohol Services
25 South Front Street
Harrisburg, PA 17101
(717) 255–2985

Director
Montgomery County TASC
319 Swede Street
Norristown, PA 19401
(215) 279–4262

Director
Allegheny County TASC Program
232 1st Avenue
Pittsburgh, PA 15222
(412) 261–2817

Director
Berks County TASC Program
36 North 6th Street
2nd Floor
Reading, PA 19601
(215) 375–4426

Associate Director
Luzerene/Wyoming County TASC
Court Advocate Program
33 East Northhampton Street
Wilkes-Barre, PA 18701
(717) 822–7118

Executive Director
York County TASC
York Alcohol and Drug Services
211 South George Street
York, PA 17401
(717) 854–9591

TASC Project
Substance Abuse Administration Building
Rhode Island Medical Center
Cranston, RI 02920
(401) 464–2381

Capital Area Planning Council/ Regional TASC
2520 Interstate Highway 35 South
Suite 100
Austin, TX 78704
(512) 443–7653

Case Management Supervisor
Richmond TASC
Virginia Department of Probation
804 West Main Street
Richmond, VA 23220
(804) 649–7673

Executive Director
Snohomish TASC
Drug Abuse Council
2720 Rucker
Everett, WA 98201
(206) 259–7142

TASC Program Supervisor
King County TASC
Alternative Intervention Resources
710 2nd Avenue
Suite 1111
Seattle, WA 98104
(206) 467–0338

Executive Director
Spokane County TASC
1320 North Ash Street
Spokane, WA 99201
(509) 326–7740

Executive Director
Tacoma TASC/Pierce County Alliance
710 South Fawcett
Tacoma, WA 98402
(206) 572–4750

Executive Director
Clark County TASC
1209 Jefferson Street
Vancouver, WA 98660
(206) 693–2243

TASC Coordinator
Yakima County TASC
Yakima County Courthouse
Room B–6
128 North 2nd Street
Yakima, WA 98901
(509) 575–4472

Coordinator
Treatment Alternatives Program
Wisconsin Office of Alcohol and Other Drug Abuse
PO Box 7851
Madison, WI 53707–7851
(608) 266–0907

Dane County Treatment Alternatives Program
16 North Carroll Street
Suite 700
Madison, WI 53704
(608) 256–4502

Project Director
Eau Claire Treatment Alternatives Program
Triniteam, Incorporated
305 1/2 Barstow Street
Eau Claire, WI 54701
(715) 836–8106

Rock Valley Treatment Alternatives Program
431 Olympian Boulevard
Beloit, WI 53511
(608) 362–4690

Wisconsin Correctional Service
436 West Wisconsin Avenue
Milwaukee, WI 53203
(414) 271–2512

Programa TASC DSCA
Apartado 1190
Arecibo, PR 00613
(809) 879–2021

Director
TASC
Departmento de Servicios
Contra La Addiccion
414 Barbosa Avenue
Hato Rey, PR 00928
(809) 763–7575

STATE LISTINGS

State Office Functions

STATE POLICY OFFICES

Office of the Governor
Establishes policy priorities and issues executive orders; responsible for the implementation of legislation; responsible for designating the State agency that applies for Federal drug law enforcement, education, treatment, and prevention funds.

State Legislature
Enacts enabling legislation and provides oversight of executive agency activities; sets funding levels for statewide drug law enforcement, treatment, and prevention.

State Drug Program Coordinator
Establishes a statewide drug abuse action plan and coordinates the activities of executive branch agencies; helps to establish program priorities.

STATE CRIMINAL JUSTICE OFFICES

Attorney General's Office
Establishes legal guidelines for the implementation of legislation and the prosecution of offenders; helps coordinate statewide drug task force activities.

Law Enforcement Planning Office
Executive branch agency responsible for coordinating statewide criminal justice initiatives.

Statistical Analysis Centers
Assembles statewide criminal justice statistics and issues periodic reports; acts as a clearinghouse for statewide crime information and statistics.

Uniform Crime Reports
Assembles statewide UCR offense and arrest data and produces annual report; submits statewide arrest statistics to the FBI's National Uniform Crime Reports for inclusion in the annual *Crime in the United States*.

BJA Strategy Preparation Agency
Prepares and submits to the Bureau of Justice Assistance (BJA) a State drug strategy; distributes BJA grant funds in accordance with the strategy; performs other analyses of statewide drug problems and appropriate interventions.

Judicial Agency
The administrative office of the State court system coordinates the activities of the various judicial districts; gathers State court data and issues periodic reports.

Corrections Agency
Operates the State prison system; establishes in-prison programs; collects statistics on correctional populations.

STATE HEALTH OFFICES

RADAR (Regional Alcohol and Drug Awareness Resource) Network Agency
State office responsible for distributing alcohol and drug abuse prevention and education materials. Established by the U.S. Department of Health and Human Services' Office of Substance Abuse Prevention, these activities are coordinated by the National Clearinghouse for Alcohol and Drug Information.

HIV-Prevention Program
Coordinates State AIDS prevention activity and oversees State AIDS prevention funding.

Drugs and Alcohol Agency
Sets prevention and treatment priorities and administers State and Federal funds, particularly those from the U.S. Department of Health and Human Services' Office of Substance Abuse Prevention.

STATE EDUCATION OFFICE

State Coordinator For Drug-Free Schools
Establishes school-based drug and alcohol prevention/education programs and administers Federal Drug-Free Schools and Communities funds.

Statistical Analysis Center
Information Analysis Section
Department of Public Safety
PO Box 6638
Phoenix, AZ 85005
(602) 223–2630

Uniform Crime Reports Contact
Arizona Department of Public Safety
PO Box 6638
Phoenix, AZ 85005
(602) 223–2232

BJA Strategy Preparation Agency
Arizona Criminal Justice Commission
Suite 250
1700 North 7th Avenue
Phoenix, AZ 85007
(602) 255–1928

Judicial Agency
Supreme Court
State Capitol
West Wing
Room 209
1700 West Washington Street
Phoenix, AZ 85007
(602) 255–4359

Corrections Agency
Department of Corrections
1601 West Jefferson
Phoenix, AZ 85007
(602) 255–5536

STATE HEALTH OFFICES

HIV-Prevention Program
Department of Health Services
Division of Disease Prevention
3008 North Third Street
Phoenix, AZ 85012
(602) 250–5843

Drug and Alcohol Agency
Office of Community Behavioral Health Services
Department of Health Services
701 East Jefferson Street
Phoenix, AZ 85304
(602) 255–1152

STATE EDUCATION OFFICE

State Coordinator For Drug-Free Schools
Chemical Abuse Prevention Specialist
Arizona Department of Education
1535 West Jefferson
Phoenix, AZ 85007
(602) 255–3847

Arkansas

STATE POLICY OFFICES

Governor's Office
Office of the Governor
State Capitol
Room 250
Little Rock, AR 72201
(501) 371–2345

State Legislative Contact
Bureau of Legislative Research
Legislative Council
State Capitol
Room 315
5th and Woodlane
Little Rock, AR 72201
(501) 371–1937

State Drug Program Coordinator
Arkansas Alcohol and Drug Abuse Coordinating Council
State Capitol
Suite 250
Little Rock, AR 72201
(501) 682–2345

STATE CRIMINAL JUSTICE OFFICES

Attorney General's Office
Office of the Attorney General
Heritage West Building
201 East Markham Street
Little Rock, AR 72201
(501) 371–2007

Crime Prevention Offices
Arkansas Crime Information Center
Office of Crime Prevention
One Capitol Mall
Little Rock, AR 72201
(501) 371–2221

Arkansas Crime Prevention Council
201 North Spring Street
Springdale, AR 72764
(501) 756–8200

Statistical Analysis Center
Special Services Section
Arkansas Crime Information Center
One Capitol Mall
Little Rock, AR 72201
(501) 682–2222

Uniform Crime Reports Contact
Arkansas Crime Information Center
One Capitol Mall
4D-200
Little Rock, AR 72201
(501) 682–2222

BJA Strategy Preparation Agency
Department of Finance and Administration
Office of Intergovernmental Services
401 DFA Building
PO Box 3278
Little Rock, AR 72203
(501) 682–1074

Judicial Agency
Judicial Department
Justice Building
Capitol Grounds
Little Rock, AR 72201
(501) 371–2295

Corrections Agency
Department of Corrections
Administration Building
Princeton Pike
PO Box 8707
Pine Bluff, AR 71611
(501) 247–1800

STATE HEALTH OFFICES

RADAR Network Agency
Office on Alcohol and Drug Abuse Prevention
PO Box 1437
400 Donaghey Plaza North
7th and Main Streets
Little Rock, AR 72203–1437
(501) 682–6653

HIV-Prevention Program
Arkansas Department of Health
Sexually Transmitted Diseases
4815 West Markham
Room 455
Little Rock, AR 72205
(501) 661–2133

Drug and Alcohol Agency
Office on Alcohol and Drug Abuse Prevention
Department of Human Services
1515 West 7th Street
Suite 310
Little Rock, AR 72201
(501) 371–2604

STATE EDUCATION OFFICE

State Coordinator For Drug-Free Schools
Arkansas Department of Education
Health, Physical Education and Drug Education
#4 Capitol Mall, 405B
Little Rock, AR 72201–1071
(501) 682–4472

California

STATE POLICY OFFICES

Governor's Office
Office of the Governor
State Capitol
1st Floor
Sacramento, CA 95814
(916) 445–2841

State Legislative Contact
Assembly Office of Research
Committee on Policy Research Legislature
1100 J Street
Room 535
Sacramento, CA 95814
(916) 445–1638

State Drug Program Coordinator
Chairman
Governor's Policy Council
111 Capitol Mall
Sacramento, CA 95814
(916) 445–1943

STATE CRIMINAL JUSTICE OFFICES

Attorney General's Office
Attorney General
Department of Justice
1515 K Street
Suite 511
Sacramento, CA 95814
(916) 445–9555

Law Enforcement Planning
Office of Criminal Justice Planning
Office of the Governor
1130 K Street
Suite 300
Sacramento, CA 95814
(916) 324–9100

Crime Prevention Offices
Crime Prevention Center
Office of the Attorney General
PO Box 944255
Sacramento, CA 94244–2550
(916) 324–7863

Governor's Office of Criminal Justice Planning
Crime Prevention Division
1130 K Street
Suite 300
Sacramento, CA 95814
(916) 323–7722

California Crime Prevention Officers Association
100 Presidio
San Clemente, CA 92672
(714) 361–8213

Statistical Analysis Center
Bureau of Criminal Statistics and Special Services
PO Box 903427
Sacramento, CA 94203–4270
(916) 739–5568

Uniform Crime Reports Contact
Bureau of Criminal Statistics
Department of Justice
PO Box 903427
Sacramento, CA 94203
(916) 739–5173

BJA Strategy Preparation Agency
Office of Criminal Justice Planning
Suite 300
1130 K Street
Sacramento, CA 95814
(916) 324 9140

Judicial Agency
Administrative Office of the Courts
Judicial Council
State Building
Room 3154
350 McAllister Street
San Francisco, CA 94102
(415) 557–3203

Corrections Agency
Department of Corrections
630 K Street
PO Box 942883
Sacramento, CA 94283–0001
(916) 445–7688

STATE HEALTH OFFICES

RADAR Network Agency
State of California Department of Alcohol and Drug Programs
111 Capitol Mall
Room 250
Sacramento, CA 95814–3229
(916) 324–7234

HIV-Prevention Program
Department of Health Services
AIDS Program Office
313 North Figueroa Street
Room 1014
Los Angeles, CA 90012
(213) 974–7803

Drug and Alcohol Agency
Department of Alcohol and Drug Programs
111 Capitol Mall
Sacramento, CA 95814
(916) 445–0834

STATE EDUCATION OFFICE

State Coordinator For Drug-Free Schools
Manager
State Department of Education
Critical Health Initiatives
P.O. Box 944272–2720
Sacramento, CA 94244–2720
(916) 322–4018

Colorado

STATE POLICY OFFICES

Governor's Office
Office of the Governor
State Capitol
Room 136
Denver, CO 80203
(303) 866–2471

State Legislative Contact
Legislative Council
State Capitol
Room 029
200 East Colfax Avenue
Denver, CO 80203
(303) 866–3521

State Drug Program Coordinator
Project Director
Communities for a Drug-Free Colorado
140 East 19th Avenue
Suite 100
Denver, CO 80203
(303) 894–2750

STATE CRIMINAL JUSTICE OFFICES

Attorney General's Office
Department of Law
3rd Floor
State Services Building
1525 Sherman Street
Denver, CO 80203
(303) 866–3611

Law Enforcement Planning
Division of Criminal Justice
Department of Local Affairs
700 Kipling Street
Suite 3000
Denver, CO 80215
(303) 239–4442

Crime Prevention Office
Colorado Crime Prevention Association
1777 6th Street
Boulder, CO 80302
(303) 441–3322

Statistical Analysis Center
Division of Criminal Justice
Department of Public Safety
700 Kipling Street
Suite 3000
Denver, CO 80215
(303) 239–4442

Uniform Crime Reports Contact
Colorado Bureau of Investigation
690 Kipling Street
Denver, CO 80215
(303) 239–4300

BJA Strategy Preparation Agency
Division of Criminal Justice
Suite 3000
700 Kipling Street
Denver, CO 80215
(303) 239–4442

Judicial Agency
Judicial Department
2 East 14th Avenue
Denver, CO 80203
(303) 861–1111 ext. 125

Corrections Agency
Department of Corrections
Springs Office Park
Suite 400
2860 South Circle Drive
Colorado Springs, CO 80906
(303) 579–9580

STATE HEALTH OFFICES

RADAR Network Agency
Resource Department
Colorado Alcohol and Drug Abuse Division
4210 East 11th Avenue
Denver, CO 80220
(303) 331–8201

HIV-Prevention Program
Department of Health
STD Control Program
4210 East 11th Avenue
Denver, CO 80220
(303) 331–8320

Drug and Alcohol Agency
Alcohol and Drug Abuse Division
Department of Health
4210 East 11th Avenue
Denver, CO 80220
(303) 331–8201

STATE EDUCATION OFFICE

State Coordinator For Drug-Free Schools
Colorado Department of Education
High Risk Intervention
201 East Colfax Avenue
Denver, CO 80203
(303) 866–6766

Connecticut

STATE POLICY OFFICES

Governor's Office
Office of the Governor
State Capitol
210 Capitol Avenue
Hartford, CT 06106
(203) 566–4840

State Legislative Contact
Office of Legislative Research
Legislative Office Building
18-20 Trinity Street
Hartford, CT 06106
(203) 566–8400

State Drug Program Coordinator
Chairman
Connecticut Alcohol and Drug Abuse Commission
999 Asylum Avenue
Hartford, CT 06105
(203) 566–4145

STATE CRIMINAL JUSTICE OFFICES

Attorney General's Office
30 Trinity Street
Hartford, CT 06106
(203) 566–2026

Law Enforcement Planning
Justice Planning Division
Office of Policy and Management
80 Washington Street
Hartford, CT 06106
(203) 566–3020

Crime Prevention Office
Connecticut Law Enforcement and Crime Prevention Association
120 Main Street
Danbury, CT 06810
(203) 797–4577

Statistical Analysis Center
Justice Planning Division
Office of Policy and Management
80 Washington Street
Hartford, CT 06106
(203) 566–3522

Uniform Crime Reports Contact
Uniform Crime Reporting Program
294 Colony Street
Meriden, CT 06450
(203) 238–6594

BJA Strategy Preparation Agency
Office of Policy and Management
Justice Planning Division
80 Washington Street
Hartford, CT 06106
(203) 566–3020

Judicial Agency
Judicial Department
State Library and Supreme Court Building
231 Capitol Avenue
PO Drawer N, Station A
Hartford, CT 06106
(203) 566–4461

Corrections Agency
Department of Corrections
340 Capitol Avenue
Hartford, CT 06106
(203) 566–5710

STATE HEALTH OFFICES

RADAR Network Agency
Connecticut Clearinghouse
334 Farmington Avenue
Plainville, CT 06062
(203) 793-9791

HIV-Prevention Program
Department of Health Services
AIDS Program
150 Washington Street
Hartford, CT 06106
(203) 566–2048

Drug and Alcohol Agency
Alcohol and Drug Abuse Commission
999 Asylum Avenue
Hartford, CT 06105
(203) 566–4145

STATE EDUCATION OFFICE

State Coordinator For Drug-Free Schools
Connecticut Department of Education
PO Box 2219
Hartford, CT 06145
(203) 566–2931

Delaware

STATE POLICY OFFICES

Governor's Office
Office of the Governor
Legislative Hall
Legislative Avenue
Dover, DE 19901
(302) 736–4101

State Legislative Contact
Legislative Council
Legislative Hall
Legislative Avenue
PO Box 1401
Dover, DE 19901
(302) 736–4114

State Drug Program Coordinator
Chairman
Drug Abuse Coordinating Council
Elbert N. Carvel State Office Building
820 North French Street
Wilmington, DE 19801
(302) 571–3017

STATE CRIMINAL JUSTICE OFFICES

Attorney General's Office
Department of Justice
Elbert N. Carvel State Office Building
820 North French Street
Wilmington, DE 19801
(302) 571–3838

Law Enforcement Planning
Criminal Justice Council
Elbert N. Carvel State Office Building
820 North French Street
Wilmington, DE 19801
(302) 571–3430

Statistical Analysis Center
60 The Plaza
Dover, DE 19901
(302) 736–4846

Uniform Crime Reports Contact
State Bureau of Identification
PO Box 430
Dover, DE 19901
(302) 736–5875

BJA Strategy Preparation Agency
Criminal Justice Council
Elbert N. Carvel State Office Building
4th Floor
820 North French Street
Wilmington, DE 19801
(302) 571–3430

Judicial Agency
Administrative Office of the Courts
Elbert N. Carvel State Office Building
11th Floor
820 North French Street
PO Box 8911
Wilmington, DE 19801
(302) 571–2480

Corrections Agency
Department of Corrections
80 Monrovia Avenue
Smyrna, DE 19977
(302) 736–5601

STATE HEALTH OFFICES

RADAR Network Agency
Educational Services
The Resource Center of the YMCA of Delaware
11th and Washington Streets
Wilmington, DE 19801
(302) 571–6975

HIV-Prevention Program
AIDS Program Office
3000 Newport Gap Pike
Building G
Wilmington, DE 19808
(302) 995–8422

Drug and Alcohol Agency
Bureau of Alcoholism and Drug Abuse
C T Building
Delaware State Hospital
1901 North DuPont Highway
New Castle, DE 19720
(302) 421–6101

STATE EDUCATION OFFICE

State Coordinator For Drug-Free Schools
State Supervisor
Department of Public Instruction
Health Education and Services
Townsend Building
PO Box 1402
Dover, DE 19903
(302) 736–4886

District of Columbia

POLICY OFFICES

Mayor's Office
Executive Office of the Mayor
District Building
Room 520
1350 Pennsylvania Avenue NW
Washington, DC 20004
(202) 727–6319

Legislative Contact
Office of Intergovernmental Relations
Executive Office of the Mayor
District Building
Room 416
1350 Pennsylvania Avenue NW
Washington, DC 20004
(202) 727–6265

Drug Program Coordinator
Drug Czar
District of Columbia
1111 E Street NW
Suite 500E
Washington, DC 20004
(202) 727–9472

CRIMINAL JUSTICE OFFICES

Attorney General's Office
Office of the Corporation Counsel
Room 329
District Building
1350 Pennsylvania Avenue NW
Washington, DC 20004
(202) 727–6248

Law Enforcement Planning
Office of Criminal Justice Plans and Analysis
Executive Office of the Mayor
717 14th Street NW
Suite 500
Washington, DC 20005
(202) 727–6537

Statistical Analysis Center
Office of Criminal Justice Plans and Analysis
717 14th Street NW
Suite 500
Washington, DC 20005
(202) 727–6554

Uniform Crime Reports Contact
Data Processing Division
Metropolitan Police Department
300 Indiana Avenue NW
Washington, DC 20001
(202) 727–4301

BJA Strategy Preparation Agency
Office of Criminal Justice Plans and Analysis
Suite 500C
1111 E Street NW
Washington, DC 20004
(202) 727–6537

Judicial Agency
District of Columbia Courts
District of Columbia Courthouse
Room 1500
500 Indiana Avenue NW
Washington, DC 20001
(202) 879–1700

Corrections Agency
Department of Corrections
Grimke Building
Room N–203
1923 Vermont Avenue NW
Washington, DC 20001
(202) 673–7316

HEALTH OFFICES

RADAR Network Agency
Coordinator of Information and Referral
Washington Area Council on Alcoholism and Drug Abuse (WACADA)
1232 M Street NW
Washington, DC 20005
(202) 682–1716

HIV-Prevention Program
Commission of Public Health
AIDS Program
1975 Connecticut Avenue NW
Washington, DC 20009
(202) 637–7700

Drug and Alcohol Agency
Alcohol and Drug Abuse Services Administration
Universal North Building
Room 837
1875 Connecticut Avenue NW
Washington, DC 20009
(202) 673–6759

EDUCATION OFFICE

Coordinator For Drug-Free Schools
Director
Substance Abuse Program
D.C. Public Schools
Lovejoy Administrative Unit
12th and D Streets NE
Room 102
Washington, DC 20002
(202) 724–3610

Florida

STATE POLICY OFFICES

Governor's Office
Office of the Governor
The Capitol
Tallahassee, FL 32399
(904) 488–4441

State Legislative Contact
Division of Legislative Library Services
Joint Legislative Management Committee
State Legislature
The Capitol
Room 701
Tallahassee, FL 32399
(904) 488–2812

State Drug Program Coordinator
Chairman
Drug Policy Task Force
Executive Office of the Governor
2106 Capitol
Tallahassee, FL 32399–0001
(904) 488–1363

STATE CRIMINAL JUSTICE OFFICES

Attorney General's Office
Department of Legal Affairs
Plaza Level 01
The Capitol
Tallahassee, FL 32301
(904) 487–1963

Law Enforcement Agency
Florida Department of Law Enforcement
Public Information Officer
PO Box 1489
Tallahassee, FL 32302
(904) 488–8771

Law Enforcement Planning
Office of Planning and Budgeting
Carlton Building
Room 415
Calhoun Street
Tallahassee, FL 32301
(904) 488–0090

Crime Prevention Offices
Help Stop Crime!
Florida Attorney General's Office
The Capitol
Tallahassee, FL 32399–1050
(904) 487–3712

Florida Crime Prevention Association
251 West Plant
Winter Garden, FL 32787
(305) 656–3636

Statistical Analysis Center
Division of Criminal Justice Information Systems
Florida Department of Law Enforcement
PO Box 1489
Tallahassee, FL 32302
(904) 487–4808

Uniform Crime Reports Contact
Uniform Crime Reports Section
Special Services Bureau
Florida Department of Law Enforcement
PO Box 1489
Tallahassee, FL 32302
(904) 487–1179

BJA Strategy Preparation Agency
Bureau of Public Safety Management
The Rhyne Building
2740 Centerview Drive
Tallahassee, FL 32399–2100
(904) 488–8016

Judicial Agency
State Courts Administrator
Supreme Court Building
Tallahassee, FL 32399–1900
(904) 488–8621

Corrections Agency
Bureau of Planning and Research
1311 Winewood Boulevard
Tallahassee, FL 32399–2500
(904) 488–5021

STATE HEALTH OFFICES

RADAR Network Agency
Florida Alcohol and Drug Abuse Association Clearinghouse
1286 Paul Russell Road
Tallahassee, FL 32301
(904) 878–6922

HIV-Prevention Program
Health and Rehabilitative Services
1309 Winewood Boulevard
Building 6
Tallahassee, FL 32399
(904) 488–2905

Drug and Alcohol Agency
Alcohol and Drug Abuse Program
Building 6
Room 156
1317 Winewood Boulevard
Tallahassee, FL 32399–0700
(904) 488–0900

STATE EDUCATION OFFICE

State Coordinator For Drug-Free Schools
Director
Prevention Center
Florida Department of Education
Suite 414
325 West Gaines Street
Tallahassee, FL 32399–0400
(904) 488–6304

Georgia

STATE POLICY OFFICES

Governor's Office
Office of the Governor
State Capitol
Room 203
Atlanta, GA 30334
(404) 656–1776

State Legislative Contact
Legislative Counsel
State Capitol
Room 316
Atlanta, GA 30334
(404) 656–5000

State Drug Program Coordinator
Deputy Director for Substance Abuse Services
Department of Human Resources
878 Peachtree Street NE
Suite 319
Atlanta, GA 30309
(404) 894–4200

STATE CRIMINAL JUSTICE OFFICES

Attorney General's Office
Law Department
Judicial Building
Room 132
40 Capitol Square, SW
Atlanta, GA 30334
(404) 656–4586

Law Enforcement Planning
Criminal Justice Coordinating Council
Office of the Governor
Floyd Memorial Building
Balcony Level, Suite 470
205 Butler Street SE
Atlanta, GA 30334
(404) 656–1721

Crime Prevention Offices
Georgia Crime Prevention Program
40 Marietta Street NW
Suite 800
Atlanta, GA 30303
(404) 656–3851

Georgia Crime Prevention Association
4400 Memorial Drive
Decatur, GA 30032
(404) 294–2574

Statistical Analysis Center
Georgia Crime Information Center
PO Box 370748
Decatur, GA 30037
(404) 244–2601

Uniform Crime Reports Contact
Georgia Crime Information Center
Georgia Bureau of Investigation
PO Box 370748
Decatur, GA 30037
(404) 244–2606

BJA Strategy Preparation Agency
Criminal Justice Coordinating Council
Suite 470
1 West Court Square
Decatur, GA 30330
(404) 370–5080

Judicial Agency
Administrative Office of the Courts
State Office Building Annex
Room 550
224 Washington Street SW
Atlanta, GA 30334
(404) 656–5171

Corrections Agency
Department of Corrections
Floyd Memorial Building
East Tower, Suite 866
205 Butler Street SE
Atlanta, GA 30334
(404) 656–4605

STATE HEALTH OFFICES

RADAR Network Agency
Department of Human Resources
Division of Mental Health
878 Peachtree Street NE
Room 319
Atlanta, GA 30309
(404) 894–4204

HIV-Prevention Program
Department of Human Resources
Office of Infectious Diseases
878 Peachtree Street NE
Suite 109
Atlanta, GA 30309
(404) 894–5304

Drug and Alcohol Agency
Division of Mental Health, Mental Retardation and Substance Abuse
Department of Human Resources
878 Peachtree Street NE
Suite 319
Atlanta, GA 30309
(404) 894–4785

STATE EDUCATION OFFICE

State Coordinator For Drug-Free Schools
Coordinator
Georgia State Board of Education
Health and Physical Education
1952 Twin Towers East
Atlanta, GA 30334–5040
(404) 656–2414

Hawaii

STATE POLICY OFFICES

Governor's Office
Office of the Governor
State Capitol Building
5th Floor
415 Beretania Street
Honolulu, HI 96813
(808) 548–5420

State Legislative Contact
Office of the Legislative Reference Bureau
State Capitol
415 South Beretania Street
Honolulu, HI 96813
(808) 548–6237

State Drug Program Coordinator
State Drug Program Coordinator
PO Box 3044
Honolulu, HI 96802
(808) 548–2272

STATE CRIMINAL JUSTICE OFFICES

Attorney General's Office
Department of the Attorney General
State Capitol
Room 405
415 South Beretania Street
Honolulu, HI 96813
(808) 548–4740

Law Enforcement Planning
State Law Enforcement Planning Agency
Department of the Attorney General
Kamamalu Building
Room 412
250 South King Street
Honolulu, HI 96813
(808) 548–3800

Crime Prevention Office
Hawaii Criminal Justice Commission
Education and Information Division
222 South Vineyard Street
Suite 703
Honolulu, HI 96813
(808) 548–6714

Statistical Analysis Center
Hawaii Criminal Justice Data Center
Department of the Attorney General
465 South King Street
Honolulu, HI 96813
(808) 548–2090

Uniform Crime Reports Contact
Hawaii Criminal Justice Information Data Center
Department of the Attorney General
First Floor
Kekuanao'a Building
465 South King Street
Honolulu, HI 96813
(808) 548–2090

BJA Strategy Preparation Agency
Attorney General
State of Hawaii
Resource Coordination Division
Room 201
426 Queen Street
Honolulu, HI 96813
(808) 548–3800

Judicial Agency
Administrative Director of Courts
Hawaii State Judiciary
417 South King Street
PO Box 2560
Honolulu, HI 96804
(808) 548–4605

Corrections Agency
Corrections Division
Department of Social Services and Housing
Queen Liliuokalani Building
1390 Miller Street
PO Box 339
Honolulu, HI 96809
(808) 548–6440

STATE HEALTH OFFICES

RADAR Network Agency
Hawaii Substance Abuse Information Center
200 North Vineyard Boulevard
Suite 603
Honolulu, HI 96817
(808) 536–7234

HIV-Prevention Program
State of Hawaii
Department of Health
3627 Kilauea Avenue
Suite 304
Honolulu, HI 96816
(808) 735–5303

Drug and Alcohol Agency
Drug Control Program
Office of Narcotics Enforcement
1100 Ward Ave.
Suite 875
Honolulu, HI 96814
(808) 548–7186

STATE EDUCATION OFFICE

State Coordinator For Drug-Free Schools
Assistant Superintendent
Department of Education
PO Box 2360
Honolulu, HI 96804
(808) 548–2360

Idaho

STATE POLICY OFFICES

Governor's Office
Office of the Governor
State Capitol Building
2nd Floor
Boise, ID 83720
(208) 334-2100

State Legislative Contact
Legislative Council
State Capitol Building
700 West Jefferson Street
Boise, ID 83720
(208) 334-2475

STATE CRIMINAL JUSTICE OFFICES

Attorney General's Office
Attorney General
State Capitol Building Room 210
700 West Jefferson Street
Boise, ID 83720
(208) 334-2400

Law Enforcement Planning
Police Services Division
Department of Law Enforcement
6083 Clinton Street
Boise, ID 83704
(208) 334-3889

Crime Prevention Office
Idaho Crime Prevention Association
7200 Barrister Driver
Boise, ID 83704
(208) 377-6622

Statistical Analysis Center
Support Services Bureau
Department of Law Enforcement
6111 Clinton Street
Boise, ID 83704
(208) 334-2162

Uniform Crime Reports Contact
Criminal Identification Bureau
Department of Law Enforcement
6062 Corporal Lane
Boise, ID 83704
(208) 334-3889

BJA Strategy Preparation Agency
Idaho Department of Law Enforcement
Support Services Bureau
6111 Clinton Street
Boise, ID 83704
(208) 327-7170

Judicial Agency
Administrative Director of the Courts
Supreme Court
Supreme Court Building
451 West State Street
Boise, ID 83720
(208) 334-2246

Corrections Agency
Department of Corrections
State Office Building
700 West State Street
Statehouse Mail
Boise, ID 83720
(208) 334-2318

STATE HEALTH OFFICES

RADAR Network Agency
Health Watch Foundation
Suite 270
1101 West River Road
Boise, ID 83702
(208) 377-0068

HIV-Prevention Program
Department of Health and Welfare
Bureau of Preventive Medicine
450 West State Street
4th Floor
Boise, ID 83720
(208) 334-4309

Drug and Alcohol Agency
Bureau of Social Services
Department of Health and Welfare
Towers Building
7th Floor
450 West State Street
Boise, ID 83720
(208) 334-4085

STATE EDUCATION OFFICE

State Coordinator For Drug-Free Schools
Drug Education Consultant
Idaho Department of Education
Len B. Jordan Building
Boise, ID 83720
(208) 334-2165

Illinois

STATE POLICY OFFICES

Governor's Office
Office of the Governor
State Capitol
Room 207
Springfield, IL 62706
(217) 782-6830

State Legislative Contact
Legislative Research Unit
222 South College Street
3rd Floor, Suite A
Springfield, IL 62704
(217) 782-6851

State Drug Program Coordinator
Department of Alcohol and Substance Abuse
222 South College Street
2nd Floor
Springfield, IL 62704
(217) 782-0685

STATE CRIMINAL JUSTICE OFFICES

Attorney General's Office
Office of the Attorney General
500 South 2nd Street
Springfield, IL 62706
(217) 782-1090

Law Enforcement Planning
Criminal Justice Authority
120 South Riverside Plaza
Chicago, IL 60606
(312) 793-8550

Crime Prevention Offices
Illinois Criminal Justice Information Authority
Illinois State McGruff Public Information Campaign
120 South Riverside Plaza
Suite 1016
Chicago, IL 60606
(312) 793–8550

Illinois Attorney General's Office
Crime Prevention Unit
174 West Randolph
Chicago, IL 60601
(312) 793–7575

Illinois Crime Prevention Association
PO Box 426
Tinley Park, IL 60477
(708) 377–4435

Statistical Analysis Center
Information Resource Center
Criminal Justice Information Authority
120 South Riverside Plaza
Chicago, IL 60606
(312) 793–8550

Uniform Crime Reports Contact
Bureau of Identification
Illinois Department of State Police
726 South College Street
Springfield, IL 62704
(217) 782–8263

BJA Strategy Preparation Agency
Illinois Criminal Justice Information Authority
Suite 1016
120 South Riverside Plaza
Chicago, IL 60606–3997
(312) 793–8550

Judicial Agency
Administrative Office of the Courts
Supreme Court Building
118 West Edwards Street
Springfield, IL 62706
(217) 782–7770

Corrections Agency
Department of Corrections
1301 Concordia Court
PO Box 19277
Springfield, IL 62794–9277
(217) 782–7777

STATE HEALTH OFFICES

RADAR Network Agency
Prevention Resource Center Library
901 South 2nd Street
Springfield, IL 62704
(217) 525–3456

HIV-Prevention Program
Illinois Department of Public Health
AIDS Activity Section
100 West Randolph
Suite 6–600
Chicago, IL 60601
(312) 917–4846

Drug and Alcohol Agency
Department of Alcoholism and Substance Abuse
State of Illinois Center
Room 5–600
100 West Randolph Street
Chicago, IL 60601
(312) 917–3840

STATE EDUCATION OFFICE

State Coordinator For Drug-Free Schools
Illinois State Board of Education
Program Support Office
100 North 1st Street
Springfield, IL 62777
(217) 782–3810

Indiana

STATE POLICY OFFICES

Governor's Office
Office of the Governor
State House
Room 206
Indianapolis, IN 46204
(317) 232–4567

State Legislative Contact
Legislative Services Agency
State House
Room 302
200 West Washington Street
Indianapolis, IN 46204
(317) 232–9550

State Drug Program Coordinator
Chairman
Governor's Commission on a Drug-Free Indiana
c/o Office of the Governor
State House
Indianapolis, IN 46204
(317) 232–2588

STATE CRIMINAL JUSTICE OFFICES

Attorney General's Office
Office of the Attorney General
State House
Room 219
200 West Washington Street
Indianapolis, IN 46204
(317) 232–6201

Law Enforcement Planning
Criminal Justice Institute
150 West Market Street
Suite 200
Indianapolis, IN 46204
(317) 232–1233

Statistical Analysis Center
Indiana Criminal Justice Institute
150 West Market Street
Suite 200
Indianapolis, IN 46204
(317) 232–1619

BJA Strategy Preparation Agency
Indiana Criminal Justice Institute
Suite 1030
101 West Ohio
Indianapolis, IN 46204
(317) 232–1230

Judicial Agency
Supreme Court
State House
Room 312
200 West Washington Street
Indianapolis, IN 46204
(317) 232–2540

Corrections Agency
Department of Corrections
State Office Building
Room 804
100 North Senate Avenue
Indianapolis, IN 46204
(317) 232–5766

STATE HEALTH OFFICES

RADAR Network Agency
Indiana Prevention Resource Center for Substance Abuse
840 State Road
46 Bypass
Room 110
Indiana University
Bloomington, IN 47405
(812) 855–1237

HIV-Prevention Program
State Board of Health
AIDS Program
PO Box 1964
Indianapolis, IN 46202
(317) 633–0851

Drug and Alcohol Agency
Division of Addiction Services
Department of Mental Health
117 East Washington Street
Indianapolis, IN 46204–3647
(317) 232–7816

STATE EDUCATION OFFICE

State Coordinator For Drug-Free Schools
Department of Education
Center for School Improvement
State House, Room 229
Indianapolis, IN 46204–2798
(317) 232–6984

Iowa

STATE POLICY OFFICES

Governor's Office
Office of the Governor
State Capitol Building
Des Moines, IA 50319
(515) 281–5211

State Legislative Contact
Legislative Service Bureau
State Capitol Building
Des Moines, IA 50319
(515) 281–3566

State Drug Program Coordinator
Coordinator
Governor's Alliance on Substance Abuse
Lucas State Office Building
Second Floor
Des Moines, IA 50319
(515) 281–3784

STATE CRIMINAL JUSTICE OFFICES

Attorney General's Office
Attorney General
Hoover State Office Building
1300 East Walnut Street
Des Moines, IA 50319
(515) 281–8760

Crime Prevention Office
Iowa Crime Prevention Association
420 Harrison Street
Davenport, IA 52801
(319) 326–7960

Statistical Analysis Center
Division of Criminal Justice and Juvenile Planning
Executive Hills East
Des Moines, IA 50319
(515) 281–5816

Uniform Crime Reports Contact
Iowa Department of Public Safety
Wallace State Office Building
Des Moines, IA 50319
(515) 281–7962

BJA Strategy Preparation Agency
Department of Public Health
Lucas State Office Building
321 East 12th Street
Des Moines, IA 50309
(515) 281–3788

Judicial Agency
Judicial Department
State Capitol Building
10th Street and Grand Avenue
Des Moines, IA 50319
(515) 281–5241

Corrections Agency
Department of Corrections
Jewett Building
Room 250
914 Grand Avenue
Des Moines, IA 50309
(515) 281–4811

STATE HEALTH OFFICES

RADAR Network Agency
Iowa Substance Abuse Information Center
Cedar Rapids Public Library
500 First Street SE
Cedar Rapids, IA 52401
(319) 398–5133

HIV-Prevention Program
Department of Health
Division of Disease Prevention
Lucas State Office Building
Des Moines, IA 50319
(515) 281–4936

Drug and Alcohol Agency
Substance Abuse Division
Lucas State Office Building
4th Floor
321 East 12th Street
Des Moines, IA 50319
(515) 281–3641

STATE EDUCATION OFFICE

State Coordinator For Drug-Free Schools
Substance Education Consultant
Iowa Department of Education
Grimes State Office Building
Des Moines, IA 50319
(515) 281–3021

Kansas

STATE POLICY OFFICES

Governor's Office
Office of the Governor
State Capitol
2nd Floor
Topeka, KS 66612
(913) 296–3232

State Legislative Contact
Legislative Research Department
State House
Room 545–N
Topeka, KS 66612
(913) 296–3181

State Drug Program Coordinator
Special Assistant on Drug Abuse
Governor's Office of Drug Abuse Programs
State House
Room 265E
Topeka, KS 66612–1572
(913) 296–2584

STATE CRIMINAL JUSTICE OFFICES

Attorney General's Office
Attorney General
Kansas Judicial Center
2nd Floor
301 West 6th Street
Topeka, KS 66612
(913) 296–2215

Crime Prevention Office
Kansas Bureau of Investigation
Crime Prevention Unit
1620 Southwest Tyler
Topeka, KS 66612
(913) 232–6000

Kansas Crime Prevention Association
200 East 7th Street
Topeka, KS 66606
(913) 295–4391

Statistical Analysis Center
Kansas Bureau of Investigation
1620 Southwest Tyler
Topeka, KS 66612
(913) 232–6000

Uniform Crime Reports Contact
Kansas Bureau of Investigation
1620 Southwest Tyler
Topeka, KS 66612
(913) 232–6000

BJA Strategy Preparation Agency
Department of Administration
Room 215-E
Statehouse
Topeka, KS 66612–1572
(913) 296–3011

Judicial Agency
Judicial Branch
Kansas Judicial Center
301 West 10th Street
Topeka, KS 66612
(913) 296–4873

Corrections Agency
Department of Corrections
Jayhawk Towers
700 Jackson Street
Topeka, KS 66603
(913) 296–3317

STATE HEALTH OFFICES

RADAR Network Agency
Public Information Officer
Kansas Alcohol and Drug Abuse Services
Department of Social and Rehabilitation Services
300 Southwest Oakley
Topeka, KS 66606
(913) 296–3925

HIV-Prevention Program
Kansas Department of Health and Environment
Landon State Office Building
900 Southwest Jackson
Topeka, KS 66612
(913) 296–5585

Drug and Alcohol Agency
Alcohol and Drug Abuse Services
Department of Social and Rehabilitation Services
Biddle Building
2nd Floor
2700 West 6th Street
Topeka, KS 66606
(913) 296–3925

STATE EDUCATION OFFICE

State Coordinator For Drug-Free Schools
Program Specialist
Substance Abuse
Kansas State Department of Education
120 East 10th Street
Topeka, KS 66612
(913) 296–4946

Kentucky

STATE POLICY OFFICES

Governor's Office
Office of the Governor
State Capitol
Room 100
Frankfort, KY 40601
(502) 564–2611

State Legislative Contact
Legislative Research Commission
State Capitol Building
Frankfort, KY 40601
(502) 564–8100

State Drug Program Coordinator
Executive Director
Champions Against Drugs
612 B Shelby Street
Frankfort, KY 40601
(502) 564–7889

STATE CRIMINAL JUSTICE OFFICES

Attorney General's Office
Office of the Attorney General
State Capitol Building
Room 116
Frankfort, KY 40601
(502) 564–7600

Law Enforcement Planning
Justice Cabinet
Commonwealth Credit Union Building
3rd Floor
417 High Street
Frankfort, KY 40601
(502) 564–7554

Crime Prevention Office
Kentucky Crime Prevention Association
c/o NCPI
University of Louisville
Shelby Campus
#142 Burhans Hall
Louisville, KY 40292
(502) 425–0644

Statistical Analysis Center
Statistical Analysis Center
Office of the Attorney General
Capitol Building
Frankfort, KY 40601
(502) 564–4002

Uniform Crime Reports Contact
Records Section
Kentucky State Police
New State Office Building
Frankfort, KY 40601
(502) 227–8717

BJA Strategy Preparation Agency
Kentucky Justice Cabinet
Division of Grants Management
3rd Floor
417 High Street
Frankfort, KY 40601
(502) 564–7554

Judicial Agency
Administrative Office of the Courts
Court of Justice
Bush Building
403 Wapping Street
Frankfort, KY 40601
(502) 564–7486

Corrections Agency
Corrections Cabinet
State Office Building
5th Floor
Holmes and High Streets
Frankfort, KY 40601
(502) 564–4726

STATE HEALTH OFFICES

RADAR Network Agency
Drug Information Service for Kentucky
Division of Substance Abuse
275 East Main Street
Frankfort, KY 40621
(502) 564–2880

HIV-Prevention Program
Cabinet for Human Resources
STD Control (CTS)
275 East Main Street
Frankfort, KY 40621
(502) 564–4804

Drug and Alcohol Agency
Division of Substance Abuse
Department of Mental Health and Mental Retardation
Cabinet for Human Resources
Health Services Building
275 East Main Street
Frankfort, KY 40621
(502) 564–2880

STATE EDUCATION OFFICE

State Coordinator For Drug-Free Schools
Branch Manager
State Department of Education
Alcohol/Drug Unit
1720 Capitol Plaza Tower
Frankfort, KY 40601
(502) 564–6720

Louisiana

STATE POLICY OFFICES

Governor's Office
Office of the Governor
State Capitol
900 Riverside North
PO Box 94004
Baton Rouge, LA 70804–9004
(504) 342–7015

State Legislative Contact
Legislative Services
House of Representatives
State Capitol
900 Riverside North
PO Box 44486
Baton Rouge, LA 70804
(504) 342–7393

State Drug Program Coordinator
Liaison to Coordinating Council
Office of the Governor
PO Box 94004
Baton Rouge, LA 70804–9004
(504) 342–7015

STATE CRIMINAL JUSTICE OFFICES

Attorney General's Office
Attorney General
Department of Justice
State Capitol
900 Riverside North
PO Box 94005
Baton Rouge, LA 70804
(504) 342–7013

Crime Prevention Office
Louisiana Crime Resistance Association
PO Box 1581
Monroe, LA 71210–1581
(318) 322–1925

Statistical Analysis Center
Louisiana Commission on Law Enforcement
2121 Wooddale Boulevard
Baton Rouge, LA 70806
(504) 925–4440

BJA Strategy Preparation Agency
Louisiana Commission on Law Enforcement and Administration of Criminal Justice
2121 Wooddale Boulevard
Baton Rouge, LA 70806–1442
(504) 925–4418

Judicial Agency
Judicial Administrator
Supreme Court Building
Room 109
301 Loyola Avenue
New Orleans, LA 70112
(504) 568–5747

Corrections Agency
Department of Public Safety and Corrections
654 Main Street
PO Box 94304
Baton Rouge, LA 70804–9304
(504) 342–6740

STATE HEALTH OFFICES

RADAR Network Agency
Division of Alcohol and Drug Abuse
PO Box 3868
Baton Rouge, LA 70821–3868
(504) 342–9352

HIV-Prevention Program
Health and Human Resources
VD Control Section
PO Box 60630
New Orleans, LA 70160
(504) 568–5275

Drug and Alcohol Agency
Prevention and Recovery from Alcohol and Drug Abuse Office
Department of Health and Human Resources
2744-B Wooddale Boulevard
PO Box 53219
Baton Rouge, LA 70892
(504) 922–0730

STATE EDUCATION OFFICE

State Coordinator For Drug-Free Schools
Director
Bureau of Student Services
Louisiana Department of Education
PO Box 94064
Baton Rouge, LA 70804–9064
(504) 342–3388

Maine

STATE POLICY OFFICES

Governor's Office
Office of the Governor
State House
Station One
Augusta, ME 04333
(207) 289–3531

State Legislative Contact
Revisor of Statutes Office
State House
Room 108
Station 7
Augusta, ME 04333
(207) 289–1650

State Drug Program Coordinator
Director
Office of Alcohol and Drug Abuse Prevention
State House Station Number 11
Augusta, ME 04333
(207) 289–2781

STATE CRIMINAL JUSTICE OFFICES

Attorney General's Office
Department of the Attorney General
State Office Building
6th Floor
State House
Station 6
Augusta, ME 04333
(207) 289–3661

Crime Prevention Office
Maine Criminal Justice Academy
Maine Department of Public Safety
93 Silver Street
Waterville, ME 04901
(207) 289–2789

Statistical Analysis Center
Maine Criminal Justice Data Center
Department of Corrections
State House
Station 111
Augusta, ME 04333
(207) 289–2711

Uniform Crime Reports Contact
Uniform Crime Reporting Division
Maine State Police
Station 42
36 Hospital Street
Augusta, ME 04333
(207) 289–2025

BJA Strategy Preparation Agency
Department of Public Safety
State House Station Number 42
Augusta, ME 04333
(207) 289–3801

Judicial Agency
Administrative Office of the Courts
Judicial Department
70 Center Street
PO Box 4820
DTS
Portland, ME 04112
(207) 879–4792

Corrections Agency
Department of Corrections
State Office Building
4th Floor
State House
Station 111
Augusta, ME 04333
(207) 289–2711

STATE HEALTH OFFICES

RADAR Network Agency
Clearinghouse Coordinator
Maine Alcohol and Drug Abuse Clearinghouse
Office of Alcoholism and Drug Abuse Prevention
State House Station Number 11
Augusta, ME 04333
(207) 289–2781

HIV-Prevention Program
Department of Human Services
State House Station
Augusta, ME 04333
(207) 289–3747

Drug and Alcohol Agency
Office of Alcoholism and Drug Abuse Prevention
Department of Human Services
Station 11
State House
Augusta, ME 04333
(207) 289–2781

STATE EDUCATION OFFICE

State Coordinator For Drug-Free Schools
Program Resources Coordinator
Department of Education and Cultural Services
Stevens School Complex
State House Station Number 57
Augusta, ME 04333
(207) 289–3876

Maryland

STATE POLICY OFFICES

Governor's Office
Office of the Governor
State House
Annapolis, MD 21404
(301) 269–3901

State Legislative Contact
Department of Legislative Reference
Legislative Services Building
90 State Circle
Annapolis, MD 21401
(301) 841–3865

State Drug Program Coordinator
Chairman
Governor's Drug and Alcohol Abuse Commission
State House
Annapolis, MD 21401
(301) 974–3077

STATE CRIMINAL JUSTICE OFFICES

Attorney General's Office
Office of the Attorney General
Munsey Building
7 North Culvert Street
Baltimore, MD 21202
(301) 576–6300

Crime Prevention Offices
Maryland Department of Public Safety and Correctional Services
Maryland Crime Watch
6776 Reisterstown Road
Suite 310
Baltimore, MD 21215
(301) 764–4035

Maryland Crime Prevention Association
PO Box 20397
Baltimore, MD 21284–0397
(301) 665–8647

Statistical Analysis Center
Maryland Justice Analysis Center
Institute of Criminal Justice and Criminology
University of Maryland
College Park, MD 20742
(301) 454–4538

Uniform Crime Reports Contact
Criminal Records Central Repository
Maryland State Police Headquarters
Pikesville, MD 21202
(301) 653–4462

BJA Strategy Preparation Agency
Governor's Office of Justice Assistance
Suite 1105
300 East Joppa Road
Baltimore, MD 21204
(301) 321–3521

Judicial Agency
Administrative Office of the Courts
Courts of Appeal Building
361 Rowe Boulevard
Annapolis, MD 21401
(301) 269–2141

Corrections Agency
Division of Corrections
Department of Public Safety and Correctional Services
6776 Reisterstown Road
Suite 309
Baltimore, MD 21215
(301) 764–4100

STATE HEALTH OFFICES

RADAR Network Agency
Addictions Program Advisor
Alcohol and Drug Abuse Administration
Department of Health and Mental Hygiene
201 West Preston Street
4th Floor
Baltimore, MD 21201
(301) 225–6543

HIV-Prevention Program
Center for AIDS Education
Health and Mental Hygiene
201 West Preston Street
Baltimore, MD 21201
(301) 225–6707

Drug and Alcohol Agency
Drug Abuse Administration
Department of Health and Mental Hygiene
Herbert R. O'Conor State Office Building
201 West Preston Street
Baltimore, MD 21201
(301) 225–6925

STATE EDUCATION OFFICE

State Coordinator For Drug-Free Schools
Director
State Department of Education
Drug-Free Schools Program
200 West Baltimore Street
Baltimore, MD 21201
(301) 333-2324

Massachusetts

STATE POLICY OFFICES

Governor's Office
Executive Office
State House
Room 360
Boston, MA 02133
(617) 727-3600

State Legislative Contact
Legislative Research Bureau
30 Winter Street
11th Floor
Boston, MA 02108
(617) 722-2345

State Drug Program Coordinator
Executive Director
Governor's Alliance Against Drugs
One Ashburton Place
Room 2131
Boston, MA 02108
(617) 727-0786

STATE CRIMINAL JUSTICE OFFICES

Attorney General's Office
Department of the Attorney General
John W. McCormack State Office Building
One Ashburton Place
Boston, MA 02108
(617) 727-2200

Law Enforcement Planning
Committee on Criminal Justice
Leverett Saltonstall State Office Building
100 Cambridge Street
Boston, MA 02202
(617) 727-6300

Crime Prevention Offices
Massachusetts Criminal Justice Training Council
Massachusetts Crime Watch
1155 Central Avenue
Needham, MA 02192
(617) 727-1907

Massachusetts Crime Prevention Officers Association
MIT Police
120 Massachusetts Avenue
Cambridge, MA 02139
(617) 455-7570

Statistical Analysis Center
Massachusetts Committee on Criminal Justice
100 Cambridge Street
Room 2100
Boston, MA 02202
(617) 727-0237

Uniform Crime Reports Contact
Criminal History Systems Board
1010 Commonwealth Avenue
Boston, MA 02215
(617) 727-0090

BJA Strategy Preparation Agency
Massachusetts Committee on Criminal Justice
Room 2100
100 Cambridge Street
Boston, MA 02202
(617) 727-6300

Judicial Agency
Supreme Judicial Court
Courthouse
Room 1300
Boston, MA 02108
(617) 725-8083

Corrections Agency
Department of Corrections
Leverett Saltonstall State Office Building
100 Cambridge Street
Boston, MA 02202
(617) 727-3301

STATE HEALTH OFFICES

RADAR Network Agency
Massachusetts Information and Referral Service
675 Massachusetts Avenue
Cambridge, MA 02139
(617) 445-6999

HIV-Prevention Program
Department of Public Health
State Laboratory Institute
305 South Street
Jamaica Plain, MA 02130
(617) 522-3700

Drug and Alcohol Agency
Division of Alcoholism and Drug Rehabilitation
Department of Public Health
150 Tremont Street
Boston, MA 02111
(617) 727-8614

STATE EDUCATION OFFICE

State Coordinator For Drug-Free Schools
Executive Director
Governor's Alliance
Office of the Governor
One Ashburton Place
Room 2131
Boston, MA 02108
(617) 727-0786

Michigan

STATE POLICY OFFICES

Governor's Office
Office of the Governor
State Capitol Building
Lansing, MI 48913
(517) 373-3423

State Legislative Contact
Legislative Service Bureau
Legislative Council
Billie Farnum Building
3rd Floor
125 West Allegany Street
PO Box 30036
Lansing, MI 48909
(517) 373–0170

State Drug Program Coordinator
Director
Office of Drug Agencies
Grandview Plaza
206 East Michigan Avenue
PO Box 30026
Lansing, MI 48909
(517) 373–4700

STATE CRIMINAL JUSTICE OFFICES

Attorney General's Office
Department of the Attorney General
Law Building
7th Floor
525 West Ottawa Street
Lansing, MI 48913
(517) 373–1110

Law Enforcement Planning
Office of Criminal Justice
Department of Management and Budget
Lewis Cass Building
320 West Walnut Street
PO Box 30026
Lansing, MI 48909
(517) 373–6655

Crime Prevention Office
Crime Prevention Association of Michigan
2110 Park Avenue
Suite 332
Detroit, MI 48201
(313) 224–4479

Statistical Analysis Center
Office of Criminal Justice
Lewis Cass Building
PO Box 30026
Lansing, MI 48909
(517) 373–6510

Uniform Crime Reports Contact
Uniform Crime Reporting Section
Michigan State Police
7150 Harris Drive
Lansing, MI 48913
(517) 322–5542

BJA Strategy Preparation Agency
Office of Criminal Justice
2nd Floor
Lewis Cass Building
PO Box 30026
Lansing, MI 48909
(517) 373–6655

Judicial Agency
State Court Administrative Office
Judicial Department
Ottawa Building North
611 West Ottawa Street
PO Box 48909
Lansing, MI 48913
(517) 373–0130

Corrections Agency
Department of Corrections
Stevens T. Mason Building
PO Box 30003
Lansing, MI 48909
(517) 373–0720

STATE HEALTH OFFICES

RADAR Network Agency
Michigan Substance Abuse and Traffic Safety Information Center
925 East Kalamazoo
Lansing, MI 48912
(517) 482–9902

HIV-Prevention Program
Special Office on AIDS
Center for Health Promotion
PO Box 30035
Lansing, MI 48906
(517) 335–8399

Drug and Alcohol Agency
Office of Substance Abuse Services
Department of Public Health
3500 North Logan Street
PO Box 30035
Lansing, MI 48909
(517) 335–8810

STATE EDUCATION OFFICE

State Coordinator For Drug-Free Schools
Health Education Specialist
Comprehensive School Health Unit
Department of Education
PO Box 30008
Lansing, MI 48909
(517) 373–2589

Minnesota

STATE POLICY OFFICES

Governor's Office
Office of the Governor
State Capitol
Aurora Avenue
Room 130
St. Paul, MN 55155
(612) 296–3391

State Legislative Contact
Legislative Reference Library
State Office Building
Room 645
550 Cedar Street
St. Paul. MN 55155
(612) 296–3398

State Drug Program Coordinator
Director
Office of Drug Policy
Department of Public Safety
Transportation Building
Room 211
John Ireland Boulevard
St. Paul, MN 55155
(612) 296–1057

STATE CRIMINAL JUSTICE OFFICES

Attorney General's Office
Office of the Attorney General
State Capitol
Aurora Avenue and Park Street
Room 102
St. Paul, MN 55155
(612) 296–6196

Law Enforcement Planning
Criminal Justice Program
State Planning Agency
Capitol Square Building
Room 100
550 Cedar Street
St. Paul, MN 55101
(612) 297–2436

Crime Prevention Offices
Minnesota Crime Prevention Officers Association
318 Transportation Building
St. Paul, MN 55155
(612) 296–7541

Minnesota Department of Public Safety
Public Information Office
318 Transportation Building
St. Paul, MN 55155
(612) 296 7541

Statistical Analysis Center
Criminal Justice Statistical Analysis Center
State Planning Agency
658 Cedar Street
St. Paul, MN 55155
(612) 296–7819

Uniform Crime Reports Contact
Criminal Justice Information Systems
1246 University Avenue
St. Paul, MN 55104
(612) 642–0670

BJA Strategy Preparation Agency
Minnesota Department of Public Safety
Office of Drug Policy
316 Transportation Building
St. Paul, MN 55155
(612) 297–4749

Judicial Agency
Supreme Court
State Capitol
Room 230
Aurora Avenue and Park Street
St. Paul, MN 55155
(612) 296–2474

Corrections Agency
Department of Corrections
Bigelow Building
Room 300
450 North Syndicate Street
St. Paul, MN 55104
(612) 642–0282

STATE HEALTH OFFICES

RADAR Network Agency
Coordinator of Information Services
Minnesota Prevention Resource Center
2829 Verndale Avenue
Anoka, MN 55303
(800) 233–9513

HIV-Prevention Program
Department of Health
Acute Disease EPI Section
717 Southeast Delaware Street
Minneapolis, MN 55440
(612) 623–5363

Drug and Alcohol Agency
Chemical Dependency Program Division
Department of Human Services
Space Center Building
444 Lafayette Road
St. Paul, MN 55155
(612) 296–3991

STATE EDUCATION OFFICE

State Coordinator For Drug-Free Schools
Supervisor
Drug Abuse Program
State Department of Education
Learner Support Systems
994 Capitol Square Building
St. Paul, MN 55101
(612) 296–3925

Mississippi

STATE POLICY OFFICES

Governor's Office
Office of the Governor
State Capitol
PO Box 139
Jackson, MS 39205
(601) 359–3150

State Legislative Contact
Legislative Reference Bureau
PO Box 1018
Jackson, MS 39205
(601) 359–3135

State Drug Program Coordinator
Executive Director
Substance Abuse Policy Council
PO Box 220
Jackson, MS 39205–0220
(601) 359–3692

STATE CRIMINAL JUSTICE OFFICES

Attorney General's Office
Attorney General
Carroll Gartin Justice Building
450 High Street
PO Box 220
Jackson, MS 39205
(601) 359–3680

Law Enforcement Planning
Criminal Justice Planning
Federal-State Building
301 West Pearl Street
Jackson, MS 39203–3088
(601) 949–2225

Statistical Analysis Center
Department of Criminal Justice Planning
301 West Pearl Street
Jackson, MS 39203
(601) 949–2006

BJA Strategy Preparation Agency
Division of Public Safety Planning
Office of Justice Programs
301 West Pearl Street
Jackson, MS 39203–3088
(601) 949–2225

Judicial Agency
Supreme Court
Carroll Gartin Justice Building
450 High Street
PO Box 117
Jackson, MS 39205
(601) 359–3697

Corrections Agency
Department of Corrections
723 North President Street
Jackson, MS 39202
(601) 354–6454

STATE HEALTH OFFICES

RADAR Network Agency
Mississippi Department of Mental Health
Division of Alcoholism and Drug Abuse
1101 Robert E. Lee Building
9th Floor
239 North Lamar Street
Jackson, MS 39207
(601) 359–1288

HIV-Prevention Program
Department of Public Health
AIDS Program Director
PO Box 1700
Jackson, MS 39215
(601) 960–7725

Drug and Alcohol Agency
Division of Alcohol and Drug Abuse
Department of Mental Health
1500 Woolfolk Road
Jackson, MS 39201
(601) 359–1297

STATE EDUCATION OFFICE

State Coordinator For Drug-Free Schools
Coordinator
State Department of Education
550 High Street
Jackson, MS 39205
(601) 359–3598

Missouri

STATE POLICY OFFICES

Governor's Office
Office of the Governor
State Capitol
Room 216
PO Box 720
Jefferson City, MO 65102
(314) 751–3222

State Legislative Contact
Committee on Legislative Research
State Capitol
Room 117A
Jefferson City, MO 65101
(314) 751–4223

State Drug Program Coordinator
Interagency Working Group for Drug and Alcohol Abuse
PO Box 720
Jefferson City, MO 65102
(314) 751–3222

STATE CRIMINAL JUSTICE OFFICES

Attorney General's Office
Attorney General
Supreme Court Building
PO Box 899
Jefferson City, MO 65102
(314) 751–3321

Law Enforcement Planning
Department of Public Safety
Truman State Office Building
8th Floor
301 West High Street
PO Box 749
Jefferson City, MO 65102–0749
(314) 751–4905

Crime Prevention Offices
Missouri Crime Prevention Association
321 East Chestnut Expressway
Springfield, MO 65802
(417) 864–1730

Missouri Department of Public Safety
Statewide Crime Prevention Resource Center
Truman State Office Building
Room 870
Jefferson City, MO 65102
(314) 751–4905

Statistical Analysis Center
Department of Public Safety
Missouri Highway Patrol
1510 East Elm
Jefferson City, MO 65101
(314) 751–4026

BJA Strategy Preparation Agency
Missouri Department of Public Safety
Truman State Office Building
PO Box 749
Jefferson City, MO 65102–0749
(314) 751–4905

Judicial Agency
Supreme Court
1105R Southwest Boulevard
Jefferson City, MO 65101
(314) 751–4377

Corrections Agency
Board of Probation and Parole
Department of Corrections and Human Resources
117 Commerce Street
PO Box 1157
Jefferson City, MO 65102
(314) 751–2389

STATE HEALTH OFFICES

RADAR Network Agency
Prevention and Education
Missouri Division of Alcohol and Drug Abuse
1915 Southridge Drive
Jefferson City, MO 65109
(314) 751–4942

HIV-Prevention Program
Division of Health
AIDS Program
PO Box 570
Jefferson City, MO 65102
(314) 751–6438

Drug and Alcohol Agency
Division of Alcohol and Drug Abuse
Department of Mental Health
1915 Southridge
PO Box 687
Jefferson City, MO 65102
(314) 751–4942

STATE EDUCATION OFFICE

State Coordinator For Drug-Free Schools
Commissioner
State Department of Elementary and Secondary Education
PO Box 480
Jefferson City, MO 65102
(314) 751–4234

Montana

STATE POLICY OFFICES

Governor's Office
Office of the Governor
State Capitol
Room 204
Helena, MT 59620
(406) 444–3111

State Legislative Contact
Legislative Council
Capitol Building
Room 138
Helena, MT 59620
(406) 444–3064

State Drug Program Coordinator
Administrator of Crime Control
Scott Hart Building
Room 463
303 North Roberts
Helena, MT 59620
(406) 444–3604

STATE CRIMINAL JUSTICE OFFICES

Attorney General's Office
Department of Justice
Justice Building
Room 317
215 North Sanders Street
Helena, MT 59620
(406) 444–2026

Law Enforcement Planning
Crime Control Division
Department of Justice
Scott Hart Building
Room 462
303 North Roberts Street
Helena, MT 59620
(406) 444–3604

Crime Prevention Office
Montana Crime Prevention Association
414 East Callender
Livingston, MT 59047
(406) 222–6120

Statistical Analysis Center
Research Planning Bureau
Board of Crime Control
303 North Roberts Street
Helena, MT 59620
(406) 444–3604

Uniform Crime Reports Contact
Montana Board of Crime Control
303 North Roberts
Helena, MT 59620
(406) 444–3604

BJA Strategy Preparation Agency
Montana Board of Crime Control
Scott Hart Building
303 North Roberts
Helena, MT 59620
(406) 444–3604

Judicial Agency
Judiciary Division
Supreme Court
Justice Building
Room 315
215 North Sanders Street
Helena, MT 59620
(406) 444–2621

Corrections Agency
Corrections Division
Department of Institutions
1539 11th Avenue
Helena, MT 59620
(406) 444–5671

STATE HEALTH OFFICES

RADAR Network Agency
Department of Institutions
Chemical Dependency Bureau
1539 11th Avenue
Helena, MT 59620
(406) 444–2878

HIV-Prevention Program
Department of Health
Health Education
Cogswell Building
Helena, MT 59620
(406) 444–4740

Drug and Alcohol Agency
Alcohol and Drug Abuse Division
Department of Institutions
1539 11th Avenue
Helena, MT 59620
(406) 444–2827

STATE EDUCATION OFFICE

State Coordinator For Drug-Free Schools
Drug-Free Coordinator
State Department of Education
Office of Public Instruction
Capitol Building
Helena, MT 59620
(406) 444–4434

Nebraska

STATE POLICY OFFICES

Governor's Office
Office of the Governor
State Capitol
Room 2316
Lincoln, NE 68509
(402) 471–2244

State Legislative Contact
Legislative Council
State Capitol
7th Floor
1445 K Street
Lincoln, NE 68509
(402) 471–2221

State Drug Program Coordinator
Toward a Drug-Free Nebraska
PO Box 94601
Lincoln, NE 68509
(402) 471–2414

STATE CRIMINAL JUSTICE OFFICES

Attorney General's Office
Department of Justice
State Capitol
Room 2115
1445 K Street
PO Box 94906
Lincoln, NE 68509
(402) 471–2682

Law Enforcement Planning
Commission on Law Enforcement and Criminal Justice
State Office Building
301 Centennial Mall South
Lincoln, NE 68509
(402) 471–2194

Crime Prevention Office
Nebraska Crime Prevention Association
233 South 10th Street
Lincoln, NE 68508
(402) 471–7261

Statistical Analysis Center
Commission on Law Enforcement and Criminal Justice
PO Box 94946
Lincoln, NE 68509
(402) 471–2194

Uniform Crime Reports Contact
Uniform Crime Reporting Section
The Nebraska Commission on Law Enforcement and Criminal Justice
PO Box 94946
Lincoln, NE 68509
(402) 471–3982

BJA Strategy Preparation Agency
Nebraska Commission on Law Enforcement and Criminal Justice
PO Box 94946
Lincoln, NE 68509
(402) 471–2194

Judicial Agency
Supreme Court
State Capitol
Room 1220
1445 K Street
Lincoln, NE 68509
(402) 471–2643

Corrections Agency
Department of Correctional Services
Lincoln Regional Center Campus
West Van Dorn and Folsom Streets
Lincoln, NE 68509
(402) 471–2654

STATE HEALTH OFFICES

RADAR Network Agency
Public Information Director
Alcohol and Drug Information Clearinghouse
Alcoholism Council of Nebraska
215 Centennial Mall South
Room 412
Lincoln, NE 68508
(402) 474–0930

HIV-Prevention Program
Department of Health
AIDS Program
PO Box 95007
Lincoln, NE 68509
(402) 471–4091

Drug and Alcohol Agency
Division on Alcoholism and Drug Abuse
Department of Public Institutions
Lincoln Regional Center Campus
West Van Dorn and Folsom Streets
Lincoln, NE 68509
(402) 471–2851

STATE EDUCATION OFFICE

State Coordinator For Drug-Free Schools
Administrator of Curriculum
Nebraska State Department of Education
PO Box 94987
301 Centennial Mall South
Lincoln, NE 68509–4987
(402) 471–4332

Nevada

STATE POLICY OFFICES

Governor's Office
Office of the Governor
Capitol Complex
Carson City, NV 89710
(702) 885–5670

State Legislative Contact
Legislative Counsel Bureau
Legislative Building
Room 148
401 South Carson Street
Carson City, NV 89710
(702) 885–5668

State Drug Program Coordinator
Coordinator of Drug and Alcohol Programs
State of Nevada
Las Vegas, NV 89158
(702) 486–4181

STATE CRIMINAL JUSTICE OFFICES

Attorney General's Office
Hero's Memorial Building
198 South Carson Street
Carson City, NV 89710
(702) 885–4170

Law Enforcement Planning
Department of Motor Vehicles and Public Safety
555 Wright Way
Carson City, NV 89711–0900
(702) 885–5375

Crime Prevention Offices
Nevada Crime Prevention Association
400 East Stewart Street
Las Vegas, NV 89101
(702) 386–3507

Nevada Attorney General's Office
Community Crime Prevention
Hero's Memorial Building
Capitol Complex
Carson City, NV 89710
(702) 855–4170

Statistical Analysis Center
Crime Information Services Bureau
Nevada Highway Patrol
555 Wright Way
Carson City, NV 89701
(702) 885–5713

BJA Strategy Preparation Agency
Department of Motor Vehicles and Public Safety
555 Wright Way
Carson City, NV 89711–0900
(702) 885–5375

Judicial Agency
Administrative Office of the Courts
400 West King Street
Room 406
Carson City, NV 89710
(702) 885–5076

Corrections Agency
Department of Prisons
PO Box 7000
Carson City, NV 89702
(702) 887–3285

STATE HEALTH OFFICES

RADAR Network Agency
Bureau of Alcohol and Drug Abuse
505 East King Street
Suite 500
Carson City, NV 89710
(702) 885–4790

HIV-Prevention Program
Nevada Health Division
Communicable Disease Section
505 East King Street
Room 200
Carson City, NV 89710
(702) 885–4800

Drug and Alcohol Agency
Bureau of Alcohol and Drug Abuse
Department of Human Resources
Kinkead Building
505 East King Street
Carson City, NV 89710
(702) 885–4790

STATE EDUCATION OFFICE

State Coordinator For Drug-Free Schools
Deputy Superintendent
State Department of Education
Office of Public Instruction
Capitol Complex
Carson City, NV 89710
(702) 885–3100

New Hampshire

STATE POLICY OFFICES

Governor's Office
Office of the Governor
214 State House
Room 208
Concord, NH 03301
(603) 271–2121

State Legislative Contact
Office of Legislative Services
State House
Room 109
107 North Main Street
Concord, NH 03301
(603) 271–3435

State Drug Program Coordinator
Director
Alcohol and Drug Abuse Prevention
Hazen Drive
Concord, NH 03301
(603) 271–4627

STATE CRIMINAL JUSTICE OFFICES

Attorney General's Office
State House Annex
Room 208
25 Capitol Street
Concord, NH 03301
(603) 271–3655

Statistical Analysis Center
Statistical Analysis Center
Office of the Attorney General
State House Annex
Concord, NH 03301
(603) 271–3658

Uniform Crime Reports Contact
Division of State Police
10 Hazen Drive
Concord, NH 03301
(603) 271–2535

BJA Strategy Preparation Agency
Office of the Attorney General
State House Annex
Concord, NH 03301
(603) 271–3658

Judicial Agency
Supreme Court Building
Noble Drive
Concord, NH 03301
(603) 271–2647

Corrections Agency
Department of Corrections
PO Box 769
Concord, NH 03301
(603) 224–3500

STATE HEALTH OFFICES

RADAR Network Agency
Prevention and Education
New Hampshire Office of Alcohol and Drug Abuse Prevention
6 Hazen Drive
Concord, NH 03301
(603) 271–4638

HIV-Prevention Program
Division of Public Health Services
Bureau of Disease Control
6 Hazen Drive
Concord, NH 03301
(603) 271–4477

Drug and Alcohol Agency
Office of Alcohol and Drug Abuse Prevention
Department of Health and Human Services
Health and Human Services Building
6 Hazen Drive
Concord, NH 03301–6525
(800) 852–3345

STATE EDUCATION OFFICE

State Coordinator For Drug-Free Schools
Department of Education
State Office Park, South
101 Pleasant Street
Concord, NH 03301
(603) 271–2632

New Jersey

STATE POLICY OFFICES

Governor's Office
Office of the Governor
State House
Trenton, NJ 08625
(609) 292–6000

State Legislative Contact
Office of Legislative Services
State House Annex
CN 068
Trenton, NJ 08625
(609) 292–4661

State Drug Program Coordinator
Chairman
Governor's Coordinating Council
c/o Department of Alcoholism
129 East Hanover Street
Trenton, NJ 08608
(609) 292–8949

STATE CRIMINAL JUSTICE OFFICES

Attorney General's Office
Department of Law and Public Safety
CN 080
Trenton, NJ 08625
(609) 292–4925

Law Enforcement Planning
State Law Enforcement Planning Agency
CN 083
Trenton, NJ 08625–0083
(609) 588–3920

Crime Prevention Offices
New Jersey Crime Prevention Officers Association
PO Box 580
Howell, NJ 07731
(908) 938–4575 ext. 308

New Jersey Department of Community Affairs
Crime Prevention Program
363 West State Street
Trenton, NJ 08625
(609) 292–6110

Statistical Analysis Center
Research and Evaluation
Department of Law and Public Safety
Hughes Justice Complex CN-085
Trenton, NJ 08625
(609) 984-2814

Uniform Crime Reports Contact
Division of State Police
PO Box 7068
West Trenton, NJ 08625
(609) 882–2000

BJA Strategy Preparation Agency
New Jersey Department of Law and Public Safety
CN 081
Trenton, NJ 08625
(609) 292–4478

Judicial Agency
Administrative Office of the Courts
Richard J. Hughes Justice Complex
CN 037
Trenton, NJ 08625
(609) 984–0275

Corrections Agency
Department of Corrections
Whittlesey Road
CN 863
Trenton, NJ 08625
(609) 292–9860

STATE HEALTH OFFICES

RADAR Network Agency
Community Information and Education
129 East Hanover Street
Trenton, NJ 08625
(609) 292–0729

HIV-Prevention Program
Department of Health
AIDS Program
363 West State Street
CN 360
Trenton, NJ 08625
(609) 984–6000

Drug and Alcohol Agency
Department of Health
129 East Hanover Street
Trenton, NJ 08608
(609) 292–5760

STATE EDUCATION OFFICE

State Coordinator For Drug-Free Schools
Manager
New Jersey State Department of Education
General Academic Education
225 West State Street, CN500
Trenton, NJ 08625
(609) 292–5780

New Mexico

STATE POLICY OFFICES

Governor's Office
Office of the Governor
State Capitol
4th Floor
Santa Fe, NM 87503
(505) 827–3000

State Legislative Contact
Legislative Council Service
State Capitol
Room 334
Santa Fe, NM 87503
(505) 984–9600

State Drug Program Coordinator
Cabinet Secretary
Department of Public Safety
Drug Policy Board
PO Box 1628
Santa Fe, NM 87504–1628
(505) 827–3370

STATE CRIMINAL JUSTICE OFFICES

Attorney General's Office
Office of the Attorney General
Bataan Memorial Building
Room 260
Santa Fe, NM 87504–1508
(505) 827–6000

Law Enforcement Planning
Administrative Services Division
Corrections Department
113 Washington Avenue
Santa Fe, NM 87501
(505) 827–8631

Crime Prevention Offices
New Mexico Crime Prevention Association
401 Marquette NW
Albuquerque, NM 76102
(505) 768–2180

New Mexico Criminal Justice Resource Center
Community Crime Prevention Programs
PO Box 2740
Santa Fe, NM 87504
(505) 827–9970

Statistical Analysis Center
Institute for Criminal Justice Studies
Onate Hall
University of New Mexico
Albuquerque, NM 87131
(505) 277–4257

BJA Strategy Preparation Agency
Department of Public Safety
4491 Cerillos Road
PO Box 1628
Santa Fe, NM 87504–1628
(505) 827–3370

Judicial Agency
Administrative Office of the Courts
Supreme Court Building
Room 25
237 Don Gaspar Avenue
Santa Fe, NM 87503
(505) 827–4800

Corrections Agency
Correction Department
Peralta Compound
1422 Paseo de Peralta
Santa Fe, NM 87501
(505) 827–8645

STATE HEALTH OFFICES

RADAR Network Agency
Health and Environment Department
Substance Abuse Bureau
Harold Runnels Building
Room 3350
1190 St. Francis Drive
Santa Fe, NM 87504–0968
(505) 827–2587

HIV-Prevention Program
Health and Environment
AIDS Prevention Program
PO Box 968
Santa Fe, NM 87504
(505) 827–0086

Drug and Alcohol Agency
Drug Abuse Bureau
Health and Environment Department
Harold Runnels Building
1190 St. Francis Drive
Santa Fe, NM 87504–0968
(505) 827–2589

STATE EDUCATION OFFICE

State Coordinator For Drug-Free Schools
Drug-Free Schools Coordinator
State Department of Education
300 Don Gaspar Avenue
Santa Fe, NM 87501
(505) 827–6648

New York

STATE POLICY OFFICES

Governor's Office
Office of the Governor
State Capitol
Albany, NY 12224
(518) 474–8390

State Legislative Contact
Legislative Library
State Capitol
Room 337
Albany, NY 12224
(518) 463–5683

State Drug Program Coordinator
Chairman
Governor's Anti-Drug Abuse Council
State Capitol
Room 326
Albany, NY 12224
(518) 474–4623

STATE CRIMINAL JUSTICE OFFICES

Attorney General's Office
Department of Law
State Capitol
Room 221
Albany, NY 12224
(518) 474–7330

Law Enforcement Planning
Director of Criminal Justice
Executive Department
State Capitol
Room 245
Albany, NY 12224
(518) 474–3334

Crime Prevention Offices
New York State Division of Criminal Justice Services
Crime Prevention
Executive Park Tower
Stuyvesant Plaza
Albany, NY 12203
(518) 457–3670

New York State Crime Prevention Coalition
407 South State Street
Syracuse, NY 13202
(315) 425–3006

Statistical Analysis Center
Bureau of Statistical Services
Division of Criminal Justice Services
Executive Park Tower
Stuyvesant Plaza
Albany, NY 12203
(518) 457–8393

Uniform Crime Reports Contact
Statistical Services
New York State Division of Criminal Justice Services
Executive Park Tower
Stuyvesant Plaza
Albany, NY 12203
(518) 457–8381

BJA Strategy Preparation Agency
New York State Division of Criminal Justice Services
Office of Funding and Program Assistance
Executive Park Tower
Stuyvesant Plaza
Albany, NY 12203–3764
(518) 485–7911

Judicial Agency
Office of Court Administration
270 Broadway
Room 1400
New York, NY 10007
(212) 587–2004

Corrections Agency
Department of Correctional Services
Building 2
State Campus
Albany, NY 12226
(518) 457–8134

STATE HEALTH OFFICES

RADAR Network Agency
Prevention/Intervention Group
194 Washington Avenue
Albany, NY 12210
(518) 473–3460

HIV-Prevention Program
AIDS Institute
Corning Tower
1315 Empire State Plaza
Albany, NY 12237
(518) 486–1320

Drug and Alcohol Agency
Division of Substance Abuse Services
Office of Alcoholism and Substance Abuse
Executive Park South
2nd Floor
Stuyvesant Plaza
Albany, NY 12203
(518) 457–2061

STATE EDUCATION OFFICE

State Coordinator For Drug-Free Schools
Chief
State Education Department
Bureau of Health & Drug Education
Washington Avenue
Albany, NY 12234
(518) 474–1491

North Carolina

STATE POLICY OFFICES

Governor's Office
Office of the Governor
State Capitol
Raleigh, NC 27611
(919) 733–4240

State Legislative Contact
Legislative Administration Officer
General Assembly of North Carolina
Dobbs Building
Room 1072
430 North Salisbury Street
Raleigh, NC 27687
(919) 733–4000

State Drug Program Coordinator
North Carolina Drug Cabinet
116 West Jones Street
Room G068
Raleigh, NC 27603–8006
(919) 733–5002

STATE CRIMINAL JUSTICE OFFICES

Attorney General's Office
Department of Justice
PO Box 629
Raleigh, NC 27602
(919) 733–3377

Law Enforcement Planning
Governor's Crime Commission
Department of Crime Control and Public Safety
Dobbs Building
Room 1072
430 North Salisbury Street
Raleigh, NC 27611
(919) 733–4000

Crime Prevention Offices
North Carolina Crime Prevention Division
PO Box 27687
Raleigh, NC 27611
(919) 733–5522

North Carolina Crime Prevention
Officers Association
PO Box 11324
Raleigh, NC 27604
(919) 733–5522

Statistical Analysis Center
Criminal Justice Analysis Center
Governor's Crime Commission
Department of Crime Control
PO Box 27687
Raleigh, NC 27611
(919) 733–5013

Uniform Crime Reports Contact
State Bureau of Investigation
Division of Criminal Information
407 North Blount Street
Raleigh, NC 27601
(919) 733–3171

BJA Strategy Preparation Agency
Governor's Crime Commission
North Carolina Department of
Crime Control and Public Safety
PO Box 27687
Raleigh, NC 27611
(919) 733–4000

Judicial Agency
Administrative Office of the Courts
Department of Justice
Justice Building
2 East Morgan Street
Raleigh, NC 27602
(919) 733–7107

Corrections Agency
Department of Corrections
840 West Morgan Street
Raleigh, NC 27603
(919) 733–4926

STATE HEALTH OFFICES

RADAR Network Agency
North Carolina Alcohol/Drug
Resource Center
G5 1200 Broad Street
Durham, NC 27705
(919) 286–5118

HIV-Prevention Program
Communicable Disease Control
AIDS Program
PO Box 2091
Raleigh, NC 27602
(919) 733–3419

Drug and Alcohol Agency
Alcohol and Drug Abuse Services
Section
Division of Mental Health, Mental
Retardation, and Substance Abuse
Services
Department of Human Resources
Albemarle Building
Room 1124
325 North Salisbury Street
Raleigh, NC 27611
(919) 733–4670

STATE EDUCATION OFFICE

State Coordinator For Drug-Free Schools
Director
Department of Public Instruction
Division of Alcohol & Drug Defense
210 North Dawson Street
Raleigh, NC 27603–1712
(919) 733–6615

North Dakota

STATE POLICY OFFICES

Governor's Office
Office of the Governor
Capitol Building
1st Floor
Bismarck, ND 58505
(701) 224–2200

State Legislative Contact
Legislative Council
State Capitol
Bismarck, ND 58505
(701) 224–2916

State Drug Program Coordinator
Administrative Assistant
Office of the Governor
600 East Boulevard
State Capitol
Bismarck, ND 58505
(701) 224–2200

STATE CRIMINAL JUSTICE OFFICES

Attorney General's Office
Attorney General
State Capitol
Bismarck, ND 58505
(701) 224–2210

Law Enforcement Planning
Criminal Justice Training and
Statistics
Office of the Attorney General
State Capitol
Bismarck, ND 58505
(701) 224–2594

Statistical Analysis Center
Criminal Justice Research
Office of the Attorney General
State Capitol
Bismarck, ND 58505
(701) 221–6108

Uniform Crime Reports Contact
Criminal Justice Training and
Statistics Division
Attorney General's Office
State Capitol Building
Bismarck, ND 58505
(701) 224–2594

BJA Strategy Preparation Agency
Attorney General's Office
Bureau of Criminal Investigation
PO Box 1054
Bismarck, ND 58502
(701) 224–2594

Judicial Agency
Supreme Court
State Capitol
Bismarck, ND 58505
(701) 224–4216

Corrections Agency
Institutions Office
State Capitol
Bismarck, ND 58505
(701) 224–2471

STATE HEALTH OFFICES

RADAR Network Agency
School Prevention Specialist
North Dakota Prevention Resource Center
1839 East Capitol Avenue
Bismarck, ND 58501
(701) 224–3603

HIV-Prevention Program
Department of Health
State Capitol Building
Bismarck, ND 58505
(701) 224–2378

Drug and Alcohol Agency
Division of Alcoholism and Drug Abuse
Department of Human Services
State Capitol
Judicial Wing
Bismarck, ND 58505
(701) 224–2769

STATE EDUCATION OFFICE

State Coordinator For Drug-Free Schools
Director
Department of Public Instruction
Guidance/Drug-Free Schools
State Capitol -9th Floor
Bismarck, ND 58505–0440
(701) 224–2269

Ohio

STATE POLICY OFFICES

Governor's Office
Office of the Governor
State Capitol
Columbus, OH 43215
(614) 466–3555

State Legislative Contact
Legislative Service Commission
State House
Broad and High Streets
Columbus, OH 43215
(614) 466–3615

State Drug Program Coordinator
Director
Governor's Office of Alcohol and Drug Recovery Systems
170 North High Street
3rd Floor
Columbus, OH 43215
(614) 644–7231

STATE CRIMINAL JUSTICE OFFICES

Attorney General's Office
State Office Tower
17th Floor
30 East Broad Street
Columbus, OH 43215
(614) 466–3376

Law Enforcement Planning
Governor's Office of Criminal Justice Services
65 East State Street
Suite 312
Columbus, OH 43266–0527
(614) 466–7782

Crime Prevention Office
Ohio Crime Prevention Association
1560 Fishinger Road
Columbus, OH 43221
(614) 459–0580

Statistical Analysis Center
Office of Criminal Justice Services
Ohio Department of Development
65 East State Street
Suite 312
Columbus, OH 43216
(614) 466–0310

BJA Strategy Preparation Agency
Governor's Office of Criminal Justice Service
400 East Town Street
Suite 120
Columbus, OH 43215
(614) 466–7782

Judicial Agency
Administrative Director of the Courts
Supreme Courts
State Office Tower
30 East Broad Street
Columbus, OH 43266–0419
(614) 466–2653

Corrections Agency
Department of Rehabilitation and Correction
1050 Freeway Drive North
Columbus, OH 43229
(614) 431–2762

STATE HEALTH OFFICES

RADAR Network Agency
Prevention Specialist
Bureau of Drug Abuse
Bureau on Alcohol Abuse and Alcoholism Recovery
170 North High Street
3rd Floor
Columbus, OH 43266–0586
(614) 466–7893

HIV-Prevention Program
Department of Health
Epidemiology Division
246 North High Street
Eighth Floor
Columbus, OH 43266
(614) 466–5480

Drug and Alcohol Agency
Bureau of Drug Abuse
Division of Mental Health
170 North High Street
3rd Floor
Columbus, OH 43215
(614) 466–7893

STATE EDUCATION OFFICE

State Coordinator For Drug-Free Schools
Assistant Director
Ohio Department of Education
Division of Education Services
65 South Front Street
Room 719
Columbus, OH 43266–0308
(614) 466–3708

Oklahoma

STATE POLICY OFFICES

Governor's Office
Office of the Governor
State Capitol
Room 212
Oklahoma City, OK 73105
(405) 521–2342

State Drug Program Coordinator
Chairman
Drug Policy Board
State Capitol
Room 112
Oklahoma City, OK 73105
(405) 521–3921

STATE CRIMINAL JUSTICE OFFICES

Attorney General's Office
State Capitol
Room 112
Lincoln Boulevard
Oklahoma City, OK 73105
(405) 521–3921

Crime Prevention Office
Oklahoma Crime Prevention Association
3901 Northwest 62
Oklahoma City, OK 73112
(405) 943–9198

Statistical Analysis Center
Planning and Research
Oklahoma Department of Corrections
PO Box 11400
Oklahoma City, OK 73136
(405) 425–2590

Uniform Crime Reports Contact
Uniform Crime Reporting Section
Oklahoma Bureau of Investigation
6600 North Harvey
Suite 300
Oklahoma City, OK 73116
(405) 848–6724

BJA Strategy Preparation Agency
District Attorney's Council
2200 Classen Boulevard
Suite 1800
Oklahoma City, OK 73106–5811
(405) 521–2349

Judicial Agency
Administrative Office of the Courts
Supreme Court
Denver Davison Building
Room 305
1915 North Stiles
Oklahoma City, OK 73105
(405) 521–2450

Corrections Agency
Department of Corrections
2400 Martin Luther King Avenue
Oklahoma City, OK 73136
(405) 427–6511

STATE HEALTH OFFICES

RADAR Network Agency
Oklahoma State Department of Mental Health
PO Box 53277
Oklahoma City, OK 73152
(405) 271–7474

HIV-Prevention Program
Department of Health
AIDS Division
PO Box 53551
Oklahoma City, OK 73152
(405) 271–4636

Drug and Alcohol Agency
Alcohol and Drug Abuse Programs
Programs Division
PO Box 53277
Oklahoma City, OK 73152
(405) 521–0044

STATE EDUCATION OFFICE

State Coordinator For Drug-Free Schools
Director
Office of Federal Financially Assisted Programs
State Department of Education
2500 North Lincoln Blvd.
Oklahoma City, OK 73105–4599
(405) 521–2106

Oregon

STATE POLICY OFFICES

Governor's Office
Office of the Governor
State Capitol
Room 254
Salem, OR 97310
(503) 378–3111

State Legislative Contact
Legislative Research
Legislative Administration Committee
State Capitol
Room S420
Salem, OR 97310
(503) 378–8871

State Drug Program Coordinator
Assistant Director
Office of Alcohol and Drug Abuse Programs
Department of Human Service
1178 Cherneketa Street NE
Salem, OR 97310
(503) 378–2163

STATE CRIMINAL JUSTICE OFFICES

Attorney General's Office
Department of Justice
Justice Building
Court and 12th Streets
Salem, OR 97310
(503) 378–6002

Crime Prevention Offices
Oregon Board on Police Standards and Training
Oregon Crime Watch
550 North Monmouth Avenue
Monmouth, OR 97361–0070
(503) 378–2100

Crime Prevention Association of Oregon
PO Box 19148
Portland, OR 97219
(503) 248–4592

Statistical Analysis Center
Crime Analysis Center
Department of Justice
Justice Building
Salem, OR 97310
(503) 378–8056

Uniform Crime Reports Contact
Law Enforcement Data Systems Division
Oregon Executive Department
155 Cottage Street NE
Salem, OR 97310
(503) 378–3057

BJA Strategy Preparation Agency
Criminal Justice Services Division
Executive Department
155 Cottage Street NE
Salem, OR 97310–0310
(503) 378–4123

Judicial Agency
State Court Administrator
Supreme Court
Supreme Court Building
Salem, OR 97310
(503) 378–6046

Corrections Agency
Corrections Division
Department of Human Resources
2575 Center Street NE
Salem, OR 97310
(503) 378–2467

STATE HEALTH OFFICES

RADAR Network Agency
Oregon Drug and Alcohol Information Center
100 North Cook
Portland, OR 97227
(503) 280–3673

HIV-Prevention Program
AIDS Coordinator
Department of Human Resources
1400 Southwest Fifth Avenue
Portland, OR 97201
(503) 229–5792

Drug and Alcohol Agency
Office of Alcohol and Drug Abuse Programs
Department of Human Resources
Public Service Building
Room 301
Capitol Mall
Salem, OR 97310
(503) 378–2163

STATE EDUCATION OFFICE

State Coordinator For Drug-Free Schools
Associate Superintendent
State Department of Education
Division of Special Student Service
700 Pringle Parkway SE
Salem, OR 97310
(503) 378–2677

Pennsylvania

STATE POLICY OFFICES

Governor's Office
Office of the Governor
Main Capitol Building
Room 225
Harrisburg, PA 17120
(717) 787–2500

State Legislative Contact
Legislative Reference Bureau
Main Capitol Building
Room 641
Harrisburg, PA 17120
(717) 787–5323

State Drug Program Coordinator
Executive Director
Drug Policy Council
Executive Office of the Governor
Finance Building
Room 310
Harrisburg, PA 17120
(717) 783–8626

STATE CRIMINAL JUSTICE OFFICES

Attorney General's Office
Office of the Attorney General
Strawberry Square
4th and Walnut Streets
16th Floor
Harrisburg, PA 17120
(717) 787–3391

Law Enforcement Planning
Commission on Crime and Delinquency
Executive House
101 South 2nd Street
Harrisburg, PA 17108
(800) 692–7292

Crime Prevention Offices
Pennsylvania Commission on Crime and Delinquency
Pennsylvania Bureau of Crime Prevention
PO Box 1167
Federal Square Station
Harrisburg, PA 17108–1167
(717) 787–1777

Pennsylvania Crime Prevention Officers Association
PO Box 15086
Reading, PA 19612–5086
(215) 250–6660

Statistical Analysis Center
Bureau of Statistics and Policy Research
Pennsylvania Commission on Crime and Delinquency
PO Box 1167
Harrisburg, PA 17108
(717) 787–5152

Uniform Crime Reports Contact
Bureau of Research and Development
Pennsylvania State Police
1800 Elmerton Avenue
Harrisburg, PA 17120
(717) 783–5536

BJA Strategy Preparation Agency
Pennsylvania Commission on Crime and Delinquency
PO Box 1167
Federal Square Station
Harrisburg, PA 17108–1167
(717) 787–2040

Judicial Agency
Administrative Office of the Courts
Supreme Court
1515 Market Street
Suite 1414
Philadelphia, PA 19102
(215) 496–4500

Corrections Agency
Department of Corrections
Central Office Building
Camp Hill, PA 17011
(717) 975–4860

STATE HEALTH OFFICES

RADAR Network Agency
ENCORE
Pennsylvania Department of Health
Department of Health Programs
PO Box 2773
Harrisburg, PA 17105
(717) 787–2606

HIV-Prevention Program
Department of Health
AIDS Education and Risk Education
PO Box 90
Harrisburg, PA 17108
(717) 787–5900

Drug and Alcohol Agency
Office of Drug and Alcohol Programs
Department of Health
Health and Welfare Building
Room 809
Forster Street and Commonwealth Avenue
Harrisburg, PA 17108
(717) 787–9857

STATE EDUCATION OFFICE

State Coordinator For Drug-Free Schools
Drug-Free Schools Coordinator
Division of Student Services
State Department of Education
333 Market Street
Harrisburg, PA 17126–0333
(717) 783–9294

Rhode Island

STATE POLICY OFFICES

Governor's Office
Office of the Governor
State House
Providence, RI 02903
(401) 277–2080

State Legislative Contact
Legislative Council
State House
Room 101
82 Smith Street
Providence, RI 02909
(401) 277–3757

State Drug Program Coordinator
Director of Drug Programs
Office of the Governor
State House
Providence, RI 02903
(401) 277–1290

STATE CRIMINAL JUSTICE OFFICES

Attorney General's Office
Department of the Attorney General
72 Pine Street
Providence, RI 02903
(401) 274–4400

Law Enforcement Planning
Governor's Justice Commission
222 Quaker Lane
West Warwick, RI 02893
(401) 277–2620

Crime Prevention Office
Rhode Island Crime Prevention Association
99 Veterans Memorial Drive
Warwick, RI 02886
(401) 737–2244

Statistical Analysis Center
Statistical Analysis Center
Governor's Commmission on Justice
222 Quaker Lane
Suite 100
West Warwick, RI 02893
(401) 277–2620

Uniform Crime Reports Contact
Rhode Island State Police
PO Box 185
North Scituate, RI 02857
(401) 647–3311

BJA Strategy Preparation Agency
Rhode Island Governor's Justice Commission
222 Quaker Lane
Suite 100
West Warwick, RI 02893
(401) 277–2620

Judicial Agency
Office of the State Court Administrator
Providence County Courthouse
250 Benefit Street
Providence, RI 02903
(401) 277–3263

Corrections Agency
Department of Corrections
Staff House
75 Howard Avenue
Cranston, RI 02920
(401) 464–2611

STATE HEALTH OFFICES

RADAR Network Agency
Division of Substance Abuse
Louis Pasteur Building
Howard Avenue
Cranston, RI 02920
(401) 464–2140

HIV-Prevention Program
Department of Health
Disease Control
75 Davis Street
Providence, RI 02908
(401) 277–2362

Drug and Alcohol Agency
Division of Substance Abuse
Department of Mental Health, Retardation and Hospitals
Substance Abuse Administration Building
Cranston, RI 02920
(401) 464–2091

STATE EDUCATION OFFICE

State Coordinator For Drug-Free Schools
Director
State Department of Education
School Support Services
22 Hayes Street
Providence, RI 02908
(401) 277–2638

South Carolina

STATE POLICY OFFICES

Governor's Office
Office of the Governor
State House
1st Floor
Columbia, SC 29211
(803) 734–9818

State Legislative Contact
Code Commissioner and Director
Legislative Council
State House
Columbia, SC 29211
(803) 734–2145

State Drug Program Coordinator
State Drug Program Coordinator
South Carolina Law Enforcement Division
PO Box 21398
Columbia, SC 29221
(803) 737–9051

STATE CRIMINAL JUSTICE OFFICES

Attorney General's Office
Office of the Attorney General
Rembert C. Dennis Office Building
Room 729
1000 Assembly Street
Columbia, SC 29211
(803) 734–3970

Law Enforcement Planning
Division of Public Safety Programs
Office of the Governor
Edgar A. Brown Building
1205 Pendleton Street
Columbia, SC 29201
(803) 734–0425

Crime Prevention Offices
South Carolina Governor's Office
State Crime Prevention Office
1205 Pendleton Street
Columbia, SC 29201
(803) 734–0427

South Carolina State Association of Crime Prevention Officers
PO Box 210-831
Columbia, SC 29221–0831
(803) 271–5359

Statistical Analysis Center
Office of Criminal Justice Programs
Office of the Governor
1205 Pendleton Street
Columbia, SC 29201
(803) 734–0423

Uniform Crime Reports Contact
South Carolina Law Enforcement Division
PO Box 21398
Columbia, SC 29221
(803) 737–9061

BJA Strategy Preparation Agency
Division of Public Safety Programs
1205 Pendleton Street
Columbia, SC 29201
(803) 734–0423

Judicial Agency
Department of Court Administration
Five Points Executive Building
2221 Devine Street
Columbia, SC 29250
(803) 734–9300

Corrections Agency
Department of Corrections
4444 Broad River Road
Columbia, SC 29221
(803) 737–8555

STATE HEALTH OFFICES

RADAR Network Agency
Programs and Services
South Carolina Commission on Alcohol and Drug Abuse
The Drug Store Information Clearinghouse
3700 Forest Drive
Suite 300
Columbia, SC 29204
(803) 734–9559

HIV-Prevention Program
Health and Environmental Control
2600 Bull Street
Columbia, SC 29201
(803) 734–5482

Drug and Alcohol Agency
Commission on Alcohol and Drug Abuse
3700 Forest Drive
Suite 300
Columbia, SC 29204
(803) 734–9542

STATE EDUCATION OFFICE

State Coordinator For Drug-Free Schools
Department of Education
At-Risk Program
1429 Senate Street
Room 1206
Columbia, SC 29201
(803) 734-8097

South Dakota

STATE POLICY OFFICES

Governor's Office
Office of the Governor
500 East Capitol Avenue
Pierre, SD 57501
(605) 773-3212

State Legislative Contact
Legislative Research Council
State Capitol Annex
500 East Capitol Avenue
Pierre, SD 57501
(605) 773-3251

State Drug Program Coordinator
Special Assistant for Human Resources
Office of the Governor
Pierre, SD 57501
(605) 773-3212

STATE CRIMINAL JUSTICE OFFICES

Attorney General's Office
Office of the Attorney General
State Capitol Building
500 East Capitol Avenue
Pierre, SD 57501
(605) 773-3215

Statistical Analysis Center
State Statistical Center
Criminal Justice Training Center
Division of the Attorney General
Pierre, SD 57501
(605) 773-3331

BJA Strategy Preparation Agency
Office of Operations
State Capitol Building
Pierre, SD 57501
(605) 773-3212

Judicial Agency
Unified Judicial System of South Dakota
State Capitol
500 East Capitol Avenue
Pierre, SD 57501
(605) 773-3474

Corrections Agency
Board of Charities and Corrections
Joe Foss Building
523 East Capitol Avenue
Pierre, SD 57501
(605) 773-3478

STATE HEALTH OFFICES

RADAR Network Agency
State Prevention Coordinator
Department of Health
Division of Alcohol and Drug Abuse
Joe Foss Building
Room 125
523 East Capitol
Pierre, SD 57501
(605) 773-3123

HIV-Prevention Program
Department of Health
Communicable Disease
523 East Capitol
Pierre, SD 57501
(605) 773-3357

Drug and Alcohol Agency
Division of Alcohol and Drug Abuse
Department of Health
Joe Foss Building
523 East Capitol Avenue
Pierre, SD 57501-3182
(605) 773-3123

STATE EDUCATION OFFICE

State Coordinator For Drug-Free Schools
Drug-Free Schools Director
State Department of Education
Division of Education
700 Governors Drive
Pierre, SD 57501-2291
(605) 773-4670

Tennessee

STATE POLICY OFFICES

Governor's Office
Office of the Governor
State Capitol
1st Floor
Nashville, TN 37219-5081
(615) 741-2001

State Legislative Contact
Office of Legislative Services
General Assembly
State Capitol
Room G3
Nashville, TN 37219
(615) 741-3511

State Drug Program Coordinator
Coordinator
Drug-Free Tennessee
c/o Governor's Planning Office
309 John Sevier Building
Nashville, TN 37219
(615) 741-1676

STATE CRIMINAL JUSTICE OFFICES

Attorney General's Office
Attorney General
450 James Robertson Parkway
Nashville, TN 37219-5025
(615) 741-3491

Crime Prevention Office
Tennessee Crime Prevention Association
201 Poplar Avenue
Suite 11-26
Memphis, TN 38103
(901) 576-5380

Statistical Analysis Center
State Planning Office
307 John Sevier Building
500 Charlotte Avenue
Nashville, TN 37219
(615) 741–1676

BJA Strategy Preparation Agency
State Planning Office
307 John Sevier Building
500 Charlotte Avenue
Nashville, TN 37219–5082
(615) 741–1676

Judicial Agency
Supreme Court
Supreme Court Building
Room 422
401 7th Avenue North
Nashville, TN 37219
(615) 741–2687

Corrections Agency
Department of Correction
Rachel Jackson State Office Building
4th Floor
320 6th Avenue N
Nashville, TN 37219–5252
(615) 741–2071

STATE HEALTH OFFICES

RADAR Network Agency
Tennessee Alcohol and Drug Association
545 Mainstream Drive
Suite 404
Nashville, TN 37228
(615) 244–7066

HIV-Prevention Program
Department of Health
AIDS Education Coordinator
100 Ninth Avenue N
Nashville, TN 37219
(615) 741–7387

Drug and Alcohol Agency
Division of Alcohol and Drug Abuse Services
Department of Mental Health and Mental Retardation
James K. Polk State Office Building
4th Floor
505 Deaderick Street
Nashville, TN 37219
(615) 741–1921

STATE EDUCATION OFFICE

State Coordinator For Drug-Free Schools
Coordinator
Tennessee Department of Education
Drug-Free Schools Program
140 Cordell Hull Building
Nashville, TN 37219
(615) 741–6055

Texas

STATE POLICY OFFICES

Governor's Office
Office of the Governor
PO Box 12428
Capitol Station
Austin, TX 78711
(512) 463–2000

State Legislative Contact
Legislative Council
State Capitol
Room 155
Congress Avenue
Austin, TX 78711
(512) 463–1151

State Drug Program Coordinator
General Counsel
State of Texas
PO Box 12428
Austin, TX 78711
(512) 463–1988

STATE CRIMINAL JUSTICE OFFICES

Attorney General's Office
PO Box 12548
Capitol Station
Austin, TX 78711
(512) 463–2100

Law Enforcement Planning
Criminal Justice Division
Office of the Governor
Sam Houston State Office Building
Room 300
201 East 14th Street
Austin, TX 78711
(512) 463–1919

Crime Prevention Office
Texas Crime Prevention Association
PO Box 924
San Marcos, TX 78667–0924
(817) 485–1450

Statistical Analysis Center
Criminal Justice Policy Council
PO Box 13332
Capitol Station
Austin, TX 78711
(512) 463–1810

Uniform Crime Reports Contact
Uniform Crime Reporting Bureau
Crime Records Division
Texas Department of Public Safety
PO Box 4143
Austin, TX 78765
(512) 465–2091

BJA Strategy Preparation Agency
Criminal Justice Division
Office of the Governor
PO Box 12428
Austin, TX 78711
(512) 463–1788

Judicial Agency
Office of Court Administration of the Texas Judicial System
Texas Law Center
Room 602
1414 Colorado Street
Austin, TX 78711
(512) 463–1625

Corrections Agency
Department of Correction
PO Box 99
Huntsville, TX 77340
(409) 295–6371

STATE HEALTH OFFICES

RADAR Network Agency
Texas Commission on Alcohol and Drug Abuse Resource Center
1705 Guadalupe
Austin, TX 78701–1214
(512) 463–5510

HIV-Prevention Program
Texas Department of Health
AIDS Division
1100 West 49th Street
Austin, TX 78756
(512) 458–7207

Drug and Alcohol Agency
Commission on Alcohol and Drug Abuse
1705 Guadalupe Street
Austin, TX 78701
(512) 463–5510

STATE EDUCATION OFFICE

State Coordinator For Drug-Free Schools
Drug-Free Schools Coordinator
Texas Education Agency
Drug Abuse Prevention Program
1701 North Congress Avenue
Room 5–123
Austin, TX 78701–1494
(512) 463–9501

Utah

STATE POLICY OFFICES

Governor's Office
Office of the Governor
State Capitol
Room 210
Salt Lake City, UT 84114
(801) 533–5231

State Legislative Contact
Office of Legislative Research and General Counsel
State Capitol
Room 436
Salt Lake City, UT 84114
(801) 533–5481

State Drug Program Coordinator
Commission on Criminal and Juvenile Justice
101 State Capitol
Salt Lake City, UT 84114
(801) 538–1031

STATE CRIMINAL JUSTICE OFFICES

Attorney General's Office
State Capitol
Room 236
Salt Lake City, UT 84114
(801) 533–5261

Law Enforcement Planning
Council for Crime Prevention
Department of Public Safety
DOT/Public Safety Building
4501 South 2700 W
Salt Lake City, UT 84119
(801) 965–4587

Crime Prevention Office
Utah Department of Public Safety
Utah Council for Crime Prevention
4501 South 2700 W
Salt Lake City, UT 84010
(801) 965–4587

Statistical Analysis Center
Commission on Criminal and Juvenile Justice
Utah State Capitol
Room 101
Salt Lake City, UT 84114
(801) 538–1031

Uniform Crime Reports Contact
Utah Department of Public Safety
4501 South 2700 W
Salt Lake City, UT 84119
(801) 965–4577

BJA Strategy Preparation Agency
Commission on Criminal and Juvenile Justice
Room 101
State Capitol
Salt Lake City, UT 84114
(801) 538–1031

Judicial Agency
Office of Court Administrator
230 South 500 E
Suite 300
Salt Lake City, UT 84102
(801) 533–6371

Corrections Agency
Department of Corrections
6065 South 300 E
Salt Lake City, UT 84107
(801) 261–2817

STATE HEALTH OFFICES

RADAR Network Agency
Utah Federation of Parents
120 North 200 W
4th Floor
Salt Lake City, UT 84103
(801) 538–3949

HIV-Prevention Program
Utah Department of Health
Bureau of Epidemiology
PO Box 16660
Salt Lake City, UT 84116
(801) 538–6191

Drug and Alcohol Agency
Division of Alcoholism and Drugs
Department of Social Services
Social Services Building
150 West North Temple Street
Salt Lake City, UT 84145–0500
(801) 533–6532

STATE EDUCATION OFFICE

State Coordinator For Drug-Free Schools
Educational Specialist
Utah State Office of Education
Drug-Free Schools Program
250 East 500 S
Salt Lake City, UT 84111
(801) 538–7713

Vermont

STATE POLICY OFFICES

Governor's Office
Office of the Governor
Pavilion Building
5th Floor
Montpelier, VT 05602
(802) 828–3333

State Legislative Contact
Legislative Council
State House
115 State Street
Montpelier, VT 05602
(802) 828–2231

State Drug Program Coordinator
State Drug Program Coordinator
Director of State Police
Waterbury, VT 05676
(802) 244–7345

STATE CRIMINAL JUSTICE OFFICES

Attorney General's Office
Pavilion Office Building
109 State Street
Montpelier, VT 05602
(802) 828–3171

Law Enforcement Planning
Department of Public Safety
103 South Main Street
Waterbury, VT 05676
(802) 244–8718

Statistical Analysis Center
Vermont Criminal Justice Center
State Office Building
Montpelier, VT 05602
(802) 828–3897

Uniform Crime Reports Contact
Support Services
Vermont Department of Public Safety
PO Box 189
Waterbury, VT 05676
(802) 244–8786

BJA Strategy Preparation Agency
Vermont Department of Public Safety
103 South Main Street
Waterbury, VT 05676
(802) 244–8781

Judicial Agency
Supreme Court
Supreme Court Building
111 State Street
Montpelier, VT 05602
(802) 828–3281

Corrections Agency
Department of Corrections
Agency of Human Services
Osgood Building
Waterbury Office Complex
103 South Main Street
Waterbury, VT 05676
(802) 241–2263

STATE HEALTH OFFICES

RADAR Network Agency
Clearinghouse Manager
Office of Alcohol and Drug Programs
103 South Main Street
Waterbury, VT 05676
(802) 241–2178

HIV-Prevention Program
Vermont Department of Health
VD Control Program
PO Box 70
60 Main Street
Burlington, VT 05402
(802) 863–7245

Drug and Alcohol Agency
Office of Alcohol and Drug Programs
Agency of Human Services
Waterbury Office Complex
Building 1 North
103 South Main Street
Waterbury, VT 05676
(802) 241–2170

STATE EDUCATION OFFICE

State Coordinator For Drug-Free Schools
Vermont Department of Education
Drug-Free School Program
120 State Street
Montpelier, VT 05602–2703
(802) 828–3111

Virginia

STATE POLICY OFFICES

Governor's Office
Office of the Governor
State Building
Richmond, VA 23212
(804) 786–2211

State Legislative Contact
Division of Legislative Services
General Assembly Building
910 Capitol Street
Richmond, VA 23208
(804) 786–3591

State Drug Program Coordinator
State Drug Program Coordinator
Secretary of Health and Human Resources
Box 1475
Richmond, VA 23212
(804) 786–7765

STATE CRIMINAL JUSTICE OFFICES

Attorney General's Office
Department of Law
101 North Eighth Street
Richmond, VA 23219
(804) 786–2071

Law Enforcement Planning
Department of Criminal Justice Services
Eighth Street Office Building
10th Floor
805 East Broad Street
Richmond, VA 23219
(804) 786–4000

Crime Prevention Offices
Virginia Department of Criminal Justice Services
Virginia Crime Prevention Center
805 East Broad Street
Richmond, VA 23219
(804) 786–8467

Virginia Crime Prevention Association, Inc.
PO Box 6942
Richmond, VA 23230
(804) 747–9193

Statistical Analysis Center
Department of Criminal Justice Services
805 East Broad Street
Richmond, VA 23219
(804) 786–4000

Uniform Crime Reports Contact
Records and Statistics Division
Department of State Police
PO Box 27472
Richmond, VA 23261
(804) 674–2023

BJA Strategy Preparation Agency
Department of Criminal Justice Services
805 East Broad Street
Richmond, VA 23219
(804) 786–4000

Judicial Agency
Supreme Court
Supreme Court Building
100 North 9th Street
Richmond, VA 23219
(804) 786–6455

Corrections Agency
Department of Corrections
4615 West Broad Street
Room 314
Richmond, VA 23261
(804) 257–6172

STATE HEALTH OFFICES

RADAR Network Agency
Virginia Department of Mental Health, Mental Retardation and Substance Abuse
109 Governor Street
Richmond, VA 23219
(804) 786–3909

HIV-Prevention Program
VD Control Section
109 Governor Street
Room 722
Richmond, VA 23219
(804) 786–6267

Drug and Alcohol Agency
Office of Substance Abuse Services
Department of Mental Health and Mental Retardation
Madison Building
109 Governor Street
Richmond, VA 23214
(804) 786–3906

STATE EDUCATION OFFICE

State Coordinator For Drug-Free Schools
Supervisor
Department of Education
Health and Physical Education
PO Box 6-Q
Richmond, VA 23216–2060
(804) 225–2733

Washington

STATE POLICY OFFICES

Governor's Office
Office of the Governor
Legislative Building
Room AS–13
Olympia, WA 98504
(206) 753–6780

State Legislative Contact
Office of Program Research
House of Representatives
House Office Building
Room 230
Olympia, WA 98504
(206) 786–7102

State Drug Program Coordinator
Special Assistant on Drug Issues
Insurance Building
4th Floor
Mail Stop AQ-44
Olympia, WA 98504
(206) 586–0827

STATE CRIMINAL JUSTICE OFFICES

Attorney General's Office
Temple of Justice
Mail Stop AV–21
Olympia, WA 98504
(206) 753–2550

Crime Prevention Offices
Washington State Criminal Justice Commission
Washington Crime Watch
2450 South 142nd
Seattle, WA 98168
(206) 764–4301

Washington State Crime Prevention Association
1920 West Dry Creek Road
Ellensburg, WA 98926
(509) 925–2280

Statistical Analysis Center
Office of Financial Management
Forecasting and Estimation Division
Insurance Building
MS AQ–44
Olympia, WA 98504
(206) 586–2501

Uniform Crime Reports Contact
Uniform Crime Reporting Program
Washington Association of Sheriffs and Police Chiefs
PO Box 826
Olympia, WA 98507
(206) 586–3221

BJA Strategy Preparation Agency
Washington State Department of Community Development
MS/GH–51
9th and Columbia Building
Olympia, WA 98504–4151
(206) 586–0487

Judicial Agency
Office of Administrator for the Courts
1206 South Quince
Olympia, WA 98504
(206) 753–5780

Corrections Agency
Department of Corrections
Capital Center Building
410 West 5th Street
Olympia, WA 98504
(206) 753–2500

STATE HEALTH OFFICES

RADAR Network Agency
Clearinghouse Coordinator
Washington State Substance Abuse Coalition (WSSAC)
14700 Main Street
Bellevue, WA 98007
(206) 747–9111

HIV-Prevention Program
Office on HIV-AIDS
Airdustrial Park
Building 9
Mail Stop LJ–17
Olympia, WA 98504
(206) 753–5810

Drug and Alcohol Agency
Bureau of Alcohol and Substance Abuse
Office Building 2
12th Avenue and Franklin Street
Olympia, WA 98504
(206) 753–5866

STATE EDUCATION OFFICE

State Coordinator For Drug-Free Schools
Supervisor
Department of Public Instruction
Substance Abuse Education
Old Capitol Building, FG–11
Olympia, WA 98504
(206) 753–5595

West Virginia

STATE POLICY OFFICES

Governor's Office
Office of the Governor
State Capitol
Charleston, WV 25305
(304) 348–2000

State Legislative Contact
Legislative Services
State Capitol
Room E–132
Charleston, WV 25305
(304) 348–2040

State Drug Program Coordinator
State Drug Program Coordinator
Secretary
Department of Public Safety
PO Box 2930
State Capitol Complex
Charleston, WV 25305
(304) 348–2930

STATE CRIMINAL JUSTICE OFFICES

Attorney General's Office
Room E–26
State Capitol
Charleston, WV 25305
(304) 348–2021

Law Enforcement Planning
Criminal Justice and Highway Safety Office
5790A MacCorkle Avenue SE
Charleston, WV 25304
(304) 348–8814

Statistical Analysis Center
West Virginia State Police
725 Jefferson Road
South Charleston, WV 25309
(304) 746–2124

Uniform Crime Reports Contact
Uniform Crime Reporting Program
725 Jefferson Road
South Charleston, WV 25309
(304) 746–2159

BJA Strategy Preparation Agency
Criminal Justice and Highway Safety Office
5790–A MacCorkle Avenue SE
Charleston, WV 25304
(304) 348–8814

Judicial Agency
State Court Administrator
State Capitol
Room E–402
Charleston, WV 25305
(304) 348–0145

Corrections Agency
Department of Corrections
State Office Building 4
Room 300
112 California Avenue
Charleston, WV 25305
(304) 348–2037

STATE HEALTH OFFICES

RADAR Network Agency
Field Consultant
West Virginia Library Commission
Cultural Center
Charleston, WV 25305
(304) 348–2041

HIV-Prevention Program
Department of Health
VD Control Section
151 11th Avenue
Charleston, WV 25303
(304) 348–2950

Drug and Alcohol Agency
Division on Alcoholism and Drug Abuse
State Office Building 3
Room 451
1800 Washington Street E
Charleston, WV 25305
(304) 348–2276

STATE EDUCATION OFFICE

State Coordinator For Drug-Free Schools
Coordinator of Drug Education
State Department of Education
Student Support Services
Capitol Complex, B–309
Charleston, WV 25305
(304) 348–8830

Wisconsin

STATE POLICY OFFICES

Governor's Office
Office of the Governor
East State Capitol
Room 115
Madison, WI 53707–7863
(608) 266–1212

State Legislative Contact
Legislative Reference Bureau
State Capitol
Room 201N
Madison, WI 53702
(608) 266–0361

State Drug Program Coordinator
Coordinator
State Alliance for a Drug-Free Wisconsin
Madison, WI 53702
(608) 266–9923

STATE CRIMINAL JUSTICE OFFICES

Attorney General's Office
Department of Justice
State Capitol
Room 114 East
Madison, WI 53702
(608) 266–1221

Law Enforcement Planning
Council on Criminal Justice
30 West Mifflin Street
Madison, WI 53702
(608) 266–7488

Crime Prevention Office
Wisconsin Crime Prevention Officers Association, Inc.
211 South Carroll
Madison, WI 53710
(608) 266–4238

Statistical Analysis Center
Office of Justice Assistance
30 West Mifflin Street
Suite 330
Madison, WI 53702
(608) 266–7646

Uniform Crime Reports Contact
Office of Justice Assistance
30 West Mifflin Street
Suite 330
Madison, WI 53703
(608) 266–3323

BJA Strategy Preparation Agency
Wisconsin Office of Justice Assistance
30 West Mifflin Street
Suite 330
Madison, WI 53702
(608) 266–7282

Judicial Agency
Director of State Courts
State Capitol
Room 213 NE
Madison, WI 53701–1688
(608) 266–6828

Corrections Agency
Division of Corrections
Wilson Street State Office Building
Room 1050
One West Wilson Street
Madison, WI 53707
(608) 266–2471

STATE HEALTH OFFICES

RADAR Network Agency
Wisconsin Clearinghouse
315 North Henry Street
Madison, WI 53703
(608) 263–2797

HIV-Prevention Program
Wisconsin Department of Health
One West Wilson Street
Madison, WI 53703
(608) 263–2797

Drug and Alcohol Agency
Office of Alcohol and Drug Abuse
Wilson Street State Office Building
Room 434
One West Wilson Street
Madison, WI 53707
(608) 266–2717

STATE EDUCATION OFFICE

State Coordinator For Drug-Free Schools
Chief, Programs Development
Department of Public Instruction
Bureau for Pupil Services
125 South Webster Street
Madison, WI 53707
(608) 266–0963

Wyoming

STATE POLICY OFFICES

Governor's Office
Office of the Governor
State Capitol
Cheyenne, WY 82002–0010
(307) 777–7434

State Legislative Contact
Legislative Service Office
State Capitol
Room 213
200 West 24th Street
Cheyenne, WY 82002
(307) 777–7881

State Drug Program Coordinator
Chairman
Governor's State Drug Policy Board
316 West 22nd Street
Cheyenne, WY 82002–0001
(307) 777–7181

STATE CRIMINAL JUSTICE OFFICES

Attorney General's Office
Office of the Attorney General
State Capitol
Room 123
200 West 24th Street
Cheyenne, WY 82002
(307) 777–7841

Law Enforcement Planning
Office of the Attorney General
State Capitol
Room 123
200 West 24th Street
Cheyenne, WY 82002
(307) 777–7841

Crime Prevention Office
Wyoming Crime Prevention Coalition
45 West 12th Street
Sheridan, WY 82801
(307) 672–2413

Statistical Analysis Center
Identification Unit
Division of the Attorney General
316 West 22nd Street
Cheyenne, WY 82002
(307) 777–7523

Uniform Crime Reports Contact
Criminal Justice Information Section
Division of Criminal Investigation
316 West 22nd Street
Cheyenne, WY 82002
(307) 777–7625

BJA Strategy Preparation Agency
Division of Criminal Investigation
316 West 22nd Street
Cheyenne, WY 82002
(307) 777–7181

Judicial Agency
Court Coordinator
Supreme Court Building
2301 Capitol Avenue
Cheyenne, WY 82002
(307) 777–7581

Corrections Agency
Herschler Building
122 West 25th Street
1st Floor
Cheyenne, WY 82002
(307) 777–7405

STATE HEALTH OFFICES

RADAR Network Agency
WY CARE Program
University of Wyoming
PO Box 3425
Laramie, WY 82071
(307) 766–4119

HIV-Prevention Program
Wyoming Health and Human Services
AIDS Prevention Program
Hathaway Building
4th Floor
Cheyenne, WY 82002
(307) 777–5800

Drug and Alcohol Agency
Office of Substance Abuse Programs
Hathaway Building
Room 350
2300 Capitol Avenue
Cheyenne, WY 82002
(307) 777–7115

STATE EDUCATION OFFICE

State Coordinator For Drug-Free Schools
Coordinator
State Department of Education
Office of Public Instruction
Hathaway Building
2300 Capitol Avenue
Cheyenne, WY 82002
(307) 777–6202

American Samoa

POLICY OFFICES

Governor's Office
Office of the Governor
Pago Pago, AS 96799
011–684–633–4116

CRIMINAL JUSTICE OFFICES

Attorney General's Office
Department of Legal Affairs
Fagatogo
Pago Pago, AS 96799
011–684–633–4163

Law Enforcement Planning
Criminal Justice Planning Agency
Utulei
Pago Pago, AS 96799
011–684–633–5221

Uniform Crime Reports Contact
Commissioner
Department of Public Safety
PO Box 1086
Pago Pago, AS 96799
011–684–633–5221

BJA Strategy Preparation Agency
Government of American Samoa
Department of Public Safety
PO Box 1086
Pago Pago, AS 96799
011–684–633–1111

Judicial Agency
High Court and District Court
Court House
Fagatogo
Pago Pago, AS 96799
011–684–633–4131

Corrections Agency
Department of Public Safety
PO Box 1086
Pago Pago, AS 96799
011–684–633–1111

HEALTH OFFICES

RADAR Network Agency
Department of Human Resources
Social Services Division
Government of American Samoa
Pago Pago, AS 96799
011–684–633–2696

HIV-Prevention Program
Director of Health
Government of American Samoa
LBJ Tropical Medical Center
Pago Pago, AS 96799
011–684–633–2732

Drug and Alcohol Agency
Alcohol and Drug Abuse
Department of Health Services
LBJ Tropical Medical Center
Pago Pago, AS 96799
011–684–633–5139

EDUCATION OFFICE

Coordinator For Drug-Free Schools
Coordinator for Federal Programs
Department of Education
Office of Pupil Services
American Samoa Government
Pago Pago, AS 96799
011–684–633–1246

Guam

POLICY OFFICES

Governor's Office
Office of the Governor
Agana, GU 96910
011–671–472–8931

CRIMINAL JUSTICE OFFICES

Attorney General's Office
Department of Law
PO Box 2950
Agana, GU 96910
011–671–472–6841

Uniform Crime Reports Contact
Guam Police Department
Planning, Research and Development
Pedro's Plaza
287 West O'Brien Drive
Agana, GU 96910
011–670–332–6311

BJA Strategy Preparation Agency
Bureau of Planning
Governor's Office
PO Box 2950
Agana, GU 96910
011–671–472–4201 ext. 3

Judicial Agency
Superior Court of Guam
110 West O'Brien Drive
Agana, GU 96910
011–671–472–8961

Corrections Agency
Department of Corrections
Maimai Road
Mangilao
Agana, GU 96910
011–671–734–3980

HEALTH OFFICES

RADAR Network Agency
Prevention Branch
Department of Mental Health and Substance Abuse
PO Box 9400
Tamuning, GU 96911
011–671–642–9261

HIV-Prevention Program
Department of Public Health
PO Box 2816
Agana, GU 96910
011–671–734–2964

Drug and Alcohol Agency
Department of Mental Health and Substance Abuse
PO Box 8896
Tamuning, GU 96911
011–671–646–9261

EDUCATION OFFICE

Coordinator For Drug-Free Schools
Director of Education
Government of Guam
Office of the Director
PO Box DE
Agana, GU 96910
011–671–472–8901 ext. 307

Puerto Rico

POLICY OFFICES

Governor's Office
Office of the Governor
PO Box 82
San Juan, PR 00901
(809) 721–7000

Legislative Contact
Office of Legislative Services
Old School of Tropical Medicine
2nd Floor
Stop 3
Munoz Rivera Avenue
San Juan, PR 00904
(809) 723–4112

CRIMINAL JUSTICE OFFICES

Attorney General's Office
Department of Justice
PO Box 192
San Juan, PR 00902
(809) 721–2900 ext. 205

Law Enforcement Planning
Crime Commission
Office of the Governor
La Fortaleza
Fortaleza Street
San Juan, PR 00901
(809) 721–7000

Statistical Analysis Center
Statistical Analysis Center
Office of the Attorney General
PO Box 192
San Juan, PR 00902
(809) 783–3382

Uniform Crime Reports Contact
Puerto Rico Police
Roosevelt Avenue 101
Puerto Neuvo Hato Rey
San Juan, PR 00918
(809) 782–1540

BJA Strategy Preparation Agency
Attorney General
Department of Justice
Commonwealth of Puerto Rico
PO Box 192
San Juan, PR 00902
(809) 721–0335

Judicial Agency
Office of Court Administration
10 Vela Street
Stop 35 1/2
Hato Rey, PR 00919
(809) 763–3049

Corrections Agency
Administrator of Corrections
Barbosa Plaza Building
306 Barbosa Avenue
San Juan, PR 00917
(809) 751–2670

HEALTH OFFICES

RADAR Network Agency
Department of Anti-Addiction Services
Apartado 21414
Rio Piedras Station
Rio Piedras, PR 00928–1414
(809) 763–3133

HIV-Prevention Program
STD Control Program
Call Box STD
Caparra Heights Station
San Juan, PR 00922
(809) 754–8118

Drug and Alcohol Agency
Department of Addiction Services
Lincoln Building
414 Barbosa Avenue
Rio Piedras, PR 00928–1474
(809) 764–3670

EDUCATION OFFICE

Coordinator For Drug-Free Schools
Director
Department of Education
Office of Federal Affairs
PO Box 759
Hato Rey, PR 00919
(809) 758–4949 ext. 6047

Virgin Islands

POLICY OFFICES

Governor's Office
Office of the Governor
21–22 Kongens Gade
St. Thomas, VI 00801
(809) 774–0001

Legislative Contact
Legislative Counsel's Office
Veterans Drive
Charlotte Amalie
St. Thomas, VI 00801
(809) 774–0880

Drug Program Coordinator
Drug Policy Advisor
Office of the Governor
116 and 164 Sub Base
Estate Nisky #6
St. Thomas, VI 00802
(809) 774–6400

CRIMINAL JUSTICE OFFICES

Attorney General's Office
Department of Law
Charlotte Amalie
St. Thomas, VI 00801
(809) 774–5666

Law Enforcement Planning
Law Enforcement Planning Commission
Office of the Governor
8 Crown Bay
Charlotte Amalie
St. Thomas, VI 00801
(809) 774–6400

Crime Prevention Office
Virgin Islands Office of the Governor
Crime Prevention Project
8 Crown Bay
Sub Base
PO Box 3807
St. Thomas, VI 00801
(809) 774–6400

Statistical Analysis Center
Office of Justice Research Services
Law Enforcement Planning Commission
Office of the Governor
8 Crown Bay
Sub Base
St. Thomas, VI 00801
(809) 774–6400

Uniform Crime Reports Contact
Records Bureau
Department of Public Safety
PO Box 210
Charlotte Amalie
St. Thomas, VI 00801
(809) 774–2211

BJA Strategy Preparation Agency
Virgin Islands Law Enforcement Planning Commission
116 and 164 Sub Base
Estate Nisky Number 6
St. Thomas, VI 00802
(809) 774–6400

Judicial Agency
Territorial Court of the Virgin Islands
Barbel Plaza South
Charlotte Amalie
St. Thomas, VI 00801
(809) 774–6680

Corrections Agency
Bureau of Corrections
Office of the Governor
4A La Grande Princess
Kingshill
St. Croix, VI 00820
(809) 773–6309

HEALTH OFFICES

RADAR Network Agency
Division of Mental Health,
Alcoholism and Drug Dependency
PO Box 1117
St. Croix, VI 00821
(809) 773–8443

HIV-Prevention Program
Department of Health
PO Box 1026
Christiansted
St. Croix, VI 00820
(809) 773–1059

Drug and Alcohol Agency
Substance Abuse Services
Department of Health
Charlotte Amalie
St. Thomas, VI 00801
(809) 774–7265

EDUCATION OFFICE

Coordinator For Drug-Free Schools
State Coordinator
Department of Education
44-56 Kongens Gade
St. Thomas, VI 00802
(809) 774–4976

APPENDIX III

State-by-State Directory of Drug Abuse and Alcoholism Treatment and Prevention Programs

INTRODUCTION

This directory is a compilation of U.S. public and private facilities responsible for providing alcoholism and drug-abuse treatment and prevention services. The information was collected in 1992 and published in 1993 by the Substance Abuse and Mental Health Services Administration (SAMHSA), Office of Applied Studies (OAS).

The directory is provided as a resource for program managers, treatment personnel, researchers, education officials, parents, and students interested in the location of such facilities. Phone numbers have not been provided since they may change—check your local telephone directory for new listings or contact your information operator service for current listings.

ALABAMA

ALEXANDER CITY

Lighthouse of Tallapoosa County, Inc.
201 Franklin Street
Alexander City, AL 35010

ANNISTON

Agency for Substance Abuse Prevention of Calhoun and Cleburne Counties
1302 Noble Street
Lyric Square Suite 3-B
Anniston, AL 36201

Anniston Fellowship House, Inc.
1219 Leighton Avenue
Anniston, AL 36201

Calhoun/Cleburne Mental Health Center New Directions
331 East 8 Avenue
Anniston, AL 36202

ATMORE

Poarch Band of Creeks Substance Abuse Prevention Program
Route 3
Atmore, AL 36502

BIRMINGHAM

Alcohol and Drug Abuse Council
1923 14 Avenue South
Birmingham, AL 35205

Alcoholism Recovery Services, Inc.
2701 Jefferson Avenue
Birmingham, AL 35211

Aletheia House, Inc.
3600 8th Avenue South
Suite W-106
Birmingham, AL 35222

Birmingham TASC Program
718 30th Street South
Birmingham, AL 35233

Family and Child Services Prevention Outreach Programs
3600 8th Avenue South
Suite W-102
Birmingham, AL 35222

Fellowship House, Inc.
1625 12th Avenue South
Birmingham, AL 35205

Jefferson County Economic Opportunity Alcoholism Outreach/Aftercare Program
3040 Ensley Avenue
Birmingham, AL 35208

Oakmont Center
1915 Avenue H
Ensley
Birmingham, AL 35218

Parkside Lodge of Birmingham Outpatient Services
631 Beacon Parkway West
Suite 211
Birmingham, AL 35209

Saint Anne's Home, Inc.
2772 Hanover Circle
Birmingham, AL 35205

Salvation Army Halfway House Program for Males
1401 F L Shuttlesworth Drive
Birmingham, AL 35234

University of Alabama in Birmingham UAB Substance Abuse Programs
3015 7 Avenue South
Birmingham, AL 35233

Veterans Affairs Medical Center Substance Abuse Treatment Program
700 South 19th Street
Birmingham, AL 35233

CALERA

Chilton/Shelby Mental Health Center Substance Abuse Division
Calera, AL 35040

CENTRE

Alcoholism/Substance Abuse Council of Cherokee County
Cherokee County Courthouse Annex
Centre, AL 35960

CLAYTON

Ventress Correctional Facility Substance Abuse Services
State Road 239
Clayton, AL 36016

CULLMAN

Cullman Area Mental Health Center Substance Abuse Services
1909 Commerce Avenue NW
Cullman, AL 35055

Lighthouse, Inc.
925 Convent Road NE
Cullman, AL 35055

North Central Alabama Substance Abuse Council
Cullman County Courthouse
Room 22
Cullman, AL 35055

DECATUR

North Central Alabama Quest Rec Center Substance Abuse Treatment
Highway 31 South
Decatur, AL 35601

DEMOPOLIS

West Alabama Mental Health Center Substance Abuse Program
1215 South Walnut Avenue
Demopolis, AL 36732

DOTHAN

Charter Woods Hospital Addictive Disease Program
700 Cottonwood Road
Dothan, AL 36301

Spectra Care
1903 West Main Street
Suite 8A
Dothan, AL 36303

FAIRHOPE

Baldwin County MH/MR Center Substance Abuse Program
372 South Greeno Road
Fairhope, AL 36532

Drug Education Council/Baldwin County
372 South Greeno Road
Fairhope, AL 36532

FLORENCE

Riverbend Center for Mental Health Stratford Clinic
635 West College Street
Florence, AL 35630

GADSDEN

Cherokee/Etowah/De Kalb Fellowship House
312 South 5 Street
Gadsden, AL 35901

Mental Health Center Substance Abuse Services
901 Goodyear Avenue
Gadsden, AL 35903

Substance Abuse Council of Etowah County
943 3 Avenue
Gadsden, AL 35901

The Bridge, Inc.
3232 Lay Springs Road
Route 3
Gadsden, AL 35901

GUNTERSVILLE

Marshall/Jackson Mental Health Authority Cedar Lodge
22165 U.S. Highway 431
Guntersville, AL 35976

HUNTSVILLE

Madison County Mental Health Center New Horizons Recovery Center
600 Saint Clair Street
Number 9 Suite 23
Huntsville, AL 35801

The Pathfinder, Inc. Recovery Home for Male Alcoholics
3104 Ivy Avenue SW
Huntsville, AL 35805

JASPER

Northwest Alabama Mental Health Center Substance Abuse Program
1100 7 Avenue
Jasper, AL 35501

MOBILE

Dauphin Way Lodge Treatment Services Quarterway/Halfway/Int Outpatient
1009 Dauphin Street
Mobile, AL 36604

Drug Education Council, Inc.
954 Government Street
Mobile, AL 36604

Franklin Memorial Primary Health Center Prevention Services
1303 Dr. Martin Luther King Jr. Avenue
Mobile, AL 36603

Mobile Bay Area Partnership For Youth
305A Glenwood Street
Mobile, AL 36606

Mobile Mental Health Center, Inc. Substance Abuse Services
2400 Gordon Smith Drive
Mobile, AL 36617

Parkside Lodge of Mobile Outpatient Services
2864 Dauphin Street
Mobile, AL 36606

USA Doctors Hospital Renaissance Substance Abuse Program
1700 Center Street
Mobile, AL 36604

MONTGOMERY

Administrative Office of the Court Alabama Court Referral Program
817 South Court Street
Montgomery, AL 36130

Baptist Medical Center Addictive Disease Program
2105 East South Boulevard
Montgomery, AL 36116

Chemical Addictions Program, Inc.
1155 Air Base Boulevard
Montgomery, AL 36108

Council on Substance Abuse/ NCADD
415 South McDonough Street
Montgomery, AL 36104

Federal Prison Camp Drug Abuse Program
Maxwell Air Force Base
Montgomery, AL 36112

Jackson Hospital Psychiatric Unit
1235 Forest Avenue
Montgomery, AL 36106

Lighthouse Counseling Center, Inc. Intensive Outpatient Unit
1415 East South Boulevard
Montgomery, AL 36116

NORMAL

Alabama A and M University Center for Drug Abuse Prevention and Education
Normal, AL 35762

PELHAM

Bradford at Birmingham Adolescent Services
2280 Highway 35
Pelham, AL 35124

PHENIX CITY

Phenix Medical Park Hospital Genesis Center
1707 21 Avenue
Phenix City, AL 36868

RED LEVEL

First Step Substance Abuse Treatment
Smiley Street
Red Level, AL 36474

RUSSELLVILLE

Sunrise Lodge Substance Abuse Treatment Center
Route 9
Russellville, AL 35653

SELMA

Cahaba Center for MH/MR Substance Abuse Services
1017 Medical Center Parkway
Selma, AL 36701

SYLACAUGA

Cheaha Mental Health Center Adapt
1628 Old Birmingham Highway
Sylacauga, AL 35150

TALLADEGA

Federal Correctional Center Substance Abuse Services
Talladega, AL 35160

TUSCALOOSA

Indian Rivers Mental Health Center Alcohol and Drug Abuse Program
505 19 Avenue
Tuscaloosa, AL 35401

Phoenix Houses, Inc. Men and Women
2008 University Boulevard
Tuscaloosa, AL 35401

Veterans Affairs Medical Center Addictions Treatment Unit
3701 Loop Road East
Tuscaloosa, AL 35404

WARRIOR

Parkside Lodge of Birmingham Inpatient Residential Services
1189 Allbritton Road
Warrior, AL 35180

WETUMPKA

Alternatives, Inc.
Wetumpka, AL 36092

ALASKA

ANCHORAGE

Alaska Division of Alcoholism and Drug Abuse Anchorage Alcohol Safety Action Program
941 West 4 Avenue
3 Floor
Anchorage, AK 99501

Alaska Council on Prevention of Alcohol and Drug Abuse, Inc.
3333 Denali Street
Suite 201
Anchorage, AK 99503

Alaska Human Services Outpatient Alcohol/Substance Abuse Treatment Program
4050 Lake Otis Parkway
Suite 111
Anchorage, AK 99508

Alaska Native Alcoholism Recovery Center
120 North Hoyt Street
Anchorage, AK 99508

Alaska Women's Resource Center New Dawn
111 West 9 Avenue
Anchorage, AK 99501

Aleutian Pribilof Islands Association
401 East Fireweed Lane
Suite 201
Anchorage, AK 99503

Center for Alcohol and Addiction Studies
3211 Providence Drive
UAA
Anchorage, AK 99508

Charter North Hospital and Counseling Centers
2530 Debarr Road
Anchorage, AK 99508

Elmendorf AFB Social Actions Substance Abuse Office
2900 9 Street
Elmendorf Air Force Base
Anchorage, AK 99506

Humana Hospital/Alaska Addictive Behavior Center
2801 De Barr Road
Anchorage, AK 99508

Narcotic Drug Treatment Center, Inc. Center for Drug Problems
520 East 4 Avenue
Suite 102
Anchorage, AK 99501

North Star Adolescent Hospital Substance Abuse Services
1650 South Bragaw Street
Anchorage, AK 99508

Rural Cap, Inc. Alcohol and Drug Abuse Prevention
731 East 8 Street
Anchorage, AK 99501

Salvation Army Clitheroe Center
2207 Spenard Road
Anchorage, AK 99503

Spectrum Counseling
4325 Laurel Street
Anchorage, AK 99508

Veterans Affairs Addictions Treatment Program
2925 Debarr Road
Anchorage, AK 99508

Volunteers of America of Alaska, Inc.
911 West 8 Avenue
Suite 100
Anchorage, AK 99501

ANIAK

Kuskokwim Native Association Community Counseling Program
Aniak, AK 99557

COPPER CENTER

Copper River Mental Health and Substance Abuse Services
Mile 104 Old Richardson Highway
Copper Center, AK 99573

CORDOVA

Cordova Community Hospital Mental Health and Alcohol Clinic
602 Chase Avenue
Cordova, AK 99574

CRAIG

Communities Organized for Health Options (COHO) Substance Abuse Program
210 Cold Storage Road
Craig, AK 99921

DILLINGHAM

Bristol Bay Area Health Corporation Alcohol/Drug Abuse Program
Dillingham, AK 99576

FAIRBANKS

Alaska Counseling
232 2 Street
Graehl
Fairbanks, AK 99701

Cultural Heritage and Education Institution
201 Cushman Street
Old Courthouse Building Suite 3-B
Fairbanks, AK 99701

Fairbanks Memorial Hospital Family Recovery Center
1650 Cowles Street
Fairbanks, AK 99701

Fairbanks Native Association, Inc. Regional Center for Alcoholism and Other Addictions
3100 South Cushman Street
Fairbanks, AK 99701

Fairbanks Substance Abuse Center
3098 Airport Way
Fairbanks, AK 99709

Knopf Counseling
748 8 Avenue
Fairbanks, AK 99701

Ianana Chiefs Conference, Inc. Ukon/Tanana CMH/Substance Abuse Center
1302 21 Avenue
Fairbanks, AK 99701

FORT WAINWRIGHT

Fort Wainwright Alcohol and Drug Abuse Prevention Control Program (ADAPCP)
Building 3401
Fort Wainwright, AK 99703

GALENA

Yukon Koyukuk Mental Health Program Alcohol and Drug Abuse Services
Galena, AK 99741

JUNEAU

City and Borough of Juneau Health and Social Services Chemical Dependency Division
3406 Glacier Highway
Juneau, AK 99801

Gastineau Manor Recovery Home
5597 Aisek Street
Juneau, AK 99801

Lakeside Recovery Center Inc. Juneau
9097 Glacier Highway
Suite 205
Juneau, AK 99801

National Council on Alcoholism/ Juneau
211 4 Street
Suite 102
Juneau, AK 99801

KENAI

Cook Inlet Council on Alcohol and Drug Abuse
10200 Kenai Spur Highway
Kenai, AK 99611

KETCHIKAN

Gateway Center for Human Services Alcohol and Drug Abuse Division
3050 5 Avenue
Ketchikan, AK 99901

Ketchikan Youth Services
Ketchikan, AK 99901

KODIAK

Kodiak Council on Alcoholism, Inc.
Kodiak, AK 99615

NENANA

Railbelt Mental Health and Addictions
2 Street
Nenana, AK 99760

NOME

Norton Sound Health Corporation Northern Lights Recovery Center
5 Avenue and Division Streets
Community Health Services Building
Nome, AK 99762

Turning Point
Nome, AK 99762

PETERSBURG

Petersburg Council on Alcoholism and Other Drug Abuse
102 A Haugen Drive
Petersburg, AK 99833

SEWARD

Seward Life Action Council
504 Adams Street
Seward, AK 99664

SITKA

Sitka Council on Alcoholism and Other Drug Abuse, Inc.
207 Moller Drive
Sitka, AK 99835

Sitka Teen Resource Center
201 Katlian Street
Sitka, AK 99835

SOLDOTNA

Central Penn General Hospital Family Recovery Center
250 Hospital Circle
Soldotna, AK 99669

WASILLA

Alaska Addiction Rehab Services, Inc. Nugens Ranch
3701 Palmer-Wasilla Highway
Wasilla, AK 99687

Matanuska/Susitna Council on Alcoholism
2801 Bogard Road
Wasilla, AK 99654

WRANGELL

Wrangell Council on Alcoholism and Related Drug Dependencies (WCA/D)
406 Alaska Avenue
Wrangell, AK 99929

AMERICAN SAMOA

PAGO PAGO

LBJ Tropical Medical Center Alcohol and Drug Program Human Services Clinic
Pago Pago, AS 96799

ARIZONA

APACHE JUNCTION

PGBHA Superstition Mountain Mental Health Center/ Substance Abuse Services
564 North Idaho Road
Apache Junction, AZ 85219

AVONDALE

Codama Catholic Social Services
501 West Van Buren Street
Suite T
Avondale, AZ 85323

BISBEE

Cochise Community Counseling Services Bisbee Outpatient
1100 Highway 92
El Rancho Plaza Room 8
Bisbee, AZ 85603

BULLHEAD CITY

Mohave Mental Health Clinic Outpatient and Day Treatment
2135 Highway 95
Suites 125 and 241
Bullhead City, AZ 86442

CASA GRANDE

PGBHA A Y Smith and Associates
222 East Cottonwood Lane
Casa Grande, AZ 85222

RGBHA Behavioral Health Agency of Central Arizona (BHACA)
Administrative Unit
Outpatient
120 West Main Street
Casa Grande, AZ 85222

CHAMBERS

Navajo Nation Behavioral Health Services
Newlands Office
Chambers, AZ 86502

CHANDLER

Sunburst Treatment Services, Inc.
1300 West Queen Creek Road
Chandler, AZ 85248

Valley Hope Alcoholism Treatment Center
501 North Washington Street
Chandler, AZ 85225

COTTONWOOD

Verde Valley Guidance Clinic, Inc.
19 East Beech Street
Cottonwood, AZ 86326

DOUGLAS

Cochise Community Counseling Services Douglas Outpatient
640 10 Street
Douglas, AZ 85607

ELOY

PGBHA Pinal County Hispanic Council
311 North Main Street
Eloy, AZ 85231

FLAGSTAFF

Native Americans for Community Action Substance Abuse Services
2717 North Steves Boulevard
Suite 11
Flagstaff, AZ 86004

Northern Arizona Methadone Maintenance and Detoxification Clinic
425 South San Francisco Street
Flagstaff, AZ 86001

The Guidance Center, Inc.
2309 North Center Street
Flagstaff, AZ 86004

519 North Leroux Street
Flagstaff, AZ 86001

FOUNTAIN HILLS

Fort McDowell Alcohol Program
Fountain Hills, AZ 86268

GLENDALE

Community Care Network (CCN) Glendale Youth Center
5401 West Ocotillo Road
Glendale, AZ 85301

GLOBE

PGBHA La Questa Alcohol and Drug Treatment Center
Ice House Canyon
Globe, AZ 85501

GREEN VALLEY

La Frontera Center Green Valley
75 Calle de las Tiendas
Suite B105
Green Valley, AZ 85614

HOLBROOK

Community Counseling Centers, Inc.
105 North 5 Avenue
Holbrook, AZ 86025

KEARNY

PGBHA Copper Basin Behavioral Health Services (CBBHS)
1116 Tilbury Street
Kearny, AZ 85237

KINGMAN

Mohave Mental Health Clinic Halfway House
1968 Atlantic Avenue
Kingman, AZ 86401

Substance Abuse Residential Services
617 Oak Street
Kingman, AZ 86401

Substance Abuse Services
1750 Beverly Street
Kingman, AZ 86401

LAKE HAVASU CITY

Mohave Mental Health Clinic
Substance Abuse Services
2178 McCulloch Boulevard
Suites 2 and 12
Lake Havasu City, AZ 86403

MARICOPA

Behavioral Health Assoc. of Central Arizona (BHACA) West Pinal Human Services
Maricopa, AZ 85239

MESA

Centro de Amistad, Inc. Mesa Office
734 East Broadway
Suite 5
Mesa, AZ 85204

Codama Family Service Agency
925 South Gilbert Road
Suite 207
Mesa, AZ 85204

East Valley Alcoholism Council
554 South Bellview Street
Mesa, AZ 85204

East Valley Catholic Social Services
610 East Southern Avenue
Mesa, AZ 85204

Jewish Family/Children's Services
1930 South Alma School Road
Room A114
Mesa, AZ 85210

Mesa YMCA
207 North Mesa Drive
Mesa, AZ 85201

New Hope Behavioral Health Center, Inc.
215 South Power Road
Suite 113
Mesa, AZ 85208

Prehab of Arizona, Inc.
Center for Family Enrichment
868 East University Drive
Mesa, AZ 85203

Helaman House
2613 South Power Road
Mesa, AZ 85216

Homestead Residence
1131 East University Drive
Mesa, AZ 85203

Pic Program
1025 East University Street
Mesa, AZ 85203

Tri City Behavioral Services, Inc.
1255 West Baseline Road
Suite 296
Mesa, AZ 85202

MORENCI

Graham/Greenlee Counseling Center, Inc. Morenci
Burro Alley and Coronado Boulevard
Morenci, AZ 85540

NOGALES

Santa Cruz Family Guidance Center, Inc.
Outpatient Unit
489 North Arroyo Boulevard
Nogales, AZ 85621

Substance Abuse Residential
Ruby Road
Fast J Ranch
Nogales, AZ 85621

Southeastern Arizona Behavioral Health Services (SEABHS)/ Administration Unit
2935 North Grand Avenue
Nogales, AZ 85621

ORACLE

PGBHA Tri Community Counseling Services
Dodge Road and American Avenue
Oracle, AZ 85623

PAGE

The Guidance Center, Inc.
112 West 6 Avenue
Page, AZ 86040

PARKER

Behavioral Health Services, Inc. (BHS) New Life Guidance Center
1713 Kofa Avenue
Suite L
Parker, AZ 85344

PAYSON

PGBHA Rim Guidance Center
404 West Aero Drive
Payson, AZ 85547

PHOENIX

Calvary Rehabilitation Center
329 North 3 Avenue
Phoenix, AZ 85003

Chicanos por la Causa, Inc.
Centro de la Familia
8225 West Indian School Road
Phoenix, AZ 85033

Corazon/Vida Nueva
3639 West Lincoln Street
Phoenix, AZ 85009

Codama Intensive Treatment Systems Services Program
550 West Indian School Road
Room 100
Phoenix, AZ 85013

Codama Phoenix South CMHC (PSCMHC) Southminister Substance Abuse Services
1923 East Broadway Road
Phoenix, AZ 85040

Crossroads
1845 East Ocotillo Road
Phoenix, AZ 85016

Drug and Alcohol Treatment Institute Clarence Lawson Foundation
2230 North 24 Street
Phoenix, AZ 85008

Ebony House, Inc.
6222 South 13 Street
Phoenix, AZ 85040

Federal Correctional Institution Substance Abuse Services
Black Canyon Stage 1
Phoenix, AZ 85027

Guide Post, Inc.
5850 South 7 Avenue
Phoenix, AZ 85041

Hohokam Room
525 North 18 Street
Suite 406
Phoenix, AZ 85006

Indian Rehabilitation, Inc.
650 North 2 Avenue
Phoenix, AZ 85003

Menninger Phoenix Saint Joseph's Hospital and Medical Center
300 West Clarendon Street
Suite 275
Phoenix, AZ 85013

National Council on Alcohol and Drug Dependency/Central and Northern Arizona
2701 North 16 Street
Suite 103
Phoenix, AZ 85006

Phoenix Adolescent Recovery Center
Parc Place
5116 East Thomas Road
Phoenix, AZ 85018

Phoenix Indian Center Alcohol Treatment Program
2601 North 3 Street
Suite 100
Phoenix, AZ 85004

Phoenix Larc Public Inebriate Program
3101 East Watkins Road
Phoenix, AZ 85034

Progress Valley III
931 East Devonshire Avenue
Phoenix, AZ 85014

Progress Valley IV
4430 North 23 Avenue
Phoenix, AZ 85015

Saint Luke's Behavioral Health Center Chemical Dependency Treatment
1800 East Van Buren Street
Phoenix, AZ 85006

Salvation Army Harbor Light Center
2707 East Van Buren Street
Treatment Center
Phoenix, AZ 85008

Terros, Inc.
711 East Missouri Street
Suite 317
Phoenix, AZ 85014

The New Casa de Amigas
303 West Portland Street
Phoenix, AZ 85003

The New Foundation
6401 South 8 Place
Phoenix. AZ 85040

Valle Del Sol, Inc. Behavioral Health Program
1209 South First Avenue
Phoenix, AZ 85003

Veterans' Affairs Medical Center Substance Abuse Treatment Program
650 East Indian School Road
Phoenix, AZ 85012

44th Street Drug Abuse Program
Larkspur Medical Center
12426 North 28 Drive
Phoenix, AZ 85029

44th Street Unit
4909 North 44 Street
Suite A
Phoenix, AZ 85018

PRESCOTT

Veterans' Affairs Medical Center Substance Abuse Treatment Program
Room 116A3
Prescott, AZ 86313

West Yavapai Guidance Clinic
Cortez Street Unit
505 South Cortez Street
Prescott, AZ 86301

Hillside Center
642 Dameron Drive
Prescott, AZ 86301

Yavapai/Prescott Substance Abuse Program
530 East Merritt Street
Prescott, AZ 86301

SAFFORD

Federal Correctional Institution Substance Abuse Services
Safford, AZ 85546

Graham/Greenlee Counseling Center, Inc. Highway 70 Unit
1456 East Highway 70
Safford, AZ 85546

SAINT JOHNS

Little Colo Behavioral Health Center, Inc.
4 and Washington Streets
Saint Johns, AZ 85936

SCOTTSDALE

Community Care Network (CCN) Jewish Family and Children's Service
8040 East Morgan Trail
Suite 4
Scottsdale, AZ 85258

SELLS

Tohono Oodham Nation of Arizona Alcoholism and Substance Abuse Branch
Sells, AZ 85634

SHOW LOW

Community Counseling Centers
Outpatient Unit
2350 Show Low Lake Road
Show Low, AZ 85901

Pineview Center
2550 Show Low Lake Road
Show Low, AZ 85901

SIERRA VISTA

Cochise Community Counseling Services Sierra Vista Outpatient
185 South Moorman Street
Sierra Vista, AZ 85635

SOMERTON

Cocopah Indian Tribe Cocopah Alcohol/Drug Abuse Prevention Program
Somerton, AZ 85350

SPRINGERVILLE

Little Colo Behavioral Health Center, Inc.
130 South Mountain Avenue
Springerville, AZ 85938

STANFIELD

PGBHA Stanfield Community Center
208 East Cooper Drive
Stanfield, AZ 85272

TEMPE

Center for Behavioral Health
2123 East Southern Avenue
Suite 2
Tempe, AZ 85282

Codama Family Service Agency
3030 South Rural Road
Suite 109
Tempe, AZ 85282

Saint Luke's Hospital/Tempe Chemical Dependency Services
1500 South Mill Avenue
Tempe, AZ 85281

Valle Del Sol, Inc. East Clinic
509 South Rockford Drive
Tempe, AZ 85281

TUCSON

Adapt, Inc. La Frontera Central Office
502 West 29 Street
Tucson, AZ 85713

Amity, Inc.
Desert Willow Program
10755 East Tanque Verde Road
Tucson, AZ 85749

Firehouse
1030 North 4 Avenue
Tucson, AZ 85701

Arizona Center for Clinical Management, Inc.
4575 East Broadway
Tucson, AZ 85711

CODAC Behavioral Health Services of Pima County, Inc.
CODAC Counseling Center
Stratford Center
2530 East Broadway
Suite D
Tucson, AZ 85716

Prevention Unit
101 South Stone Avenue
Tucson, AZ 85701

Wildflowers
700 North 7 Avenue
Tucson, AZ 85705

Cottonwood de Tucson
4110 Sweetwater Drive
Tucson, AZ 85745

Davis Monthan Air Force Base Drug/Alcohol Abuse Control
Social Actions Office
Building 2300 2nd Floor
Tucson, AZ 85707

Federal Correctional Institution Substance Abuse Services
8901 South Wilmot Road
Tucson, AZ 85706

Haven, Inc. Weigel Haus Transitional Living
1015 East Adelaide Drive
Tucson, AZ 85719

La Frontera Center
Casa de Vida
410 South 6 Avenue
Tucson, AZ 85701

East Clinic
7820 East Broadway Boulevard
Suite 120
Tucson, AZ 85710

Hope Center
260 South Scott Street
Tucson, AZ 85701

Mountain View Ranch
4747 East Fort Lowell Street
Tucson, AZ 85705

Tucson Alcoholic Recovery Home, Inc.
1809 East 23 Street
Tucson, AZ 85713

3HO Foundation of Arizona, Inc. 3HO Super Health
2545 North Woodland Road
Tucson, AZ 85749

WILLCOX

Chochise Community Counseling Services Willcox Treatment Center
418 North Austin Boulevard
Willcox, AZ 85643

WILLIAMS

The Guidance Center, Inc.
301 South 7 Street
Williams, AZ 86046

WINSLOW

Community Counseling Centers Outpatient Clinic
209 East 3 Street
Winslow, AZ 86047

YUMA

Behavioral Health Services, Inc. (BHS) The Family Center
2549 Arizona Avenue
Suite H
Yuma, AZ 85364

Prevention and Intervention Associates
242 West 28 Street
Suite J
Yuma, AZ 85364

ARKANSAS

ALEXANDER

Alexander Youth Services Center
1501 Woody Drive
Alexander, AR 72002

BATESVILLE

Northcentral Arkansas Development Council Alcohol Treatment Program
133 Broad Street
Batesville, AR 72501

BENTON

Benton Detoxification Services Center Division of Alcohol/ Drug Abuse Prevention
6701 Highway 67
Benton, AR 72015

Counseling Clinic, Inc.
307 East Servier Street
Benton, AR 72015

CAMDEN

Ouachita County Hospital Chemical Dependency Unit
638 California Street
Camden, AR 71701

EL DORADO

South Arkansas Regional Health Center Recovery Center
710 West Grove Street
El Dorado, AR 71730

FAYETTEVILLE

Charter Vista Hospital Addictive Diseases Program
4253 Crossover Road
Fayetteville, AR 72702

FORREST CITY

Baptist Memorial Hospital Alcohol and Drug Treatment Program
Interstate 40 and Highway 284
Forrest City, AR 72335

East Central Arkansas Economic Opportunity Corp. Opportunity House
145 West Broadway
Forrest City, AR 72335

FORT SMITH

Gateway House, Inc.
1715 Grand Avenue
Fort Smith, AR 72901

Harbor House, Inc.
615 North 19 Street
Fort Smith, AR 72901

Harbor View Mercy Hospital Chemical Dependency Program
10301 Mayo Drive
Fort Smith, AR 72913

Sparks Regional Medical Center Sparks Recovery Center
1311 South I Street
Fort Smith, AR 72901

Western Arkansas Counseling and Guidance Center/Horizon Adolescent Program
3111 South 70 Street
Fort Smith, AR 72903

GASSVILLE

Ozark Mountain Alcoholism Residential Treatment, Inc. (OMART)
Highway 62
Gassville, AR 72635

GRADY

Dept. of Correction/Cummins Unit Substance Abuse Treatment Program (SATP)
Grady, AR 71644

HELENA

East Arkansas Regional Mental Health Center Helena Service Center
305 Valley Drive
Helena, AR 72342

HOT SPRINGS

Quapaw House, Inc.
115 Market Street
4th Floor
Hot Springs, AR 71902

JONESBORO

Crowley's Ridge Development Council
Critical Populations Ethnic Minority
1500 East Washington Street
Suite C
Jonesboro, AR 72401

Northeast Arkansas Women's Recovery
417 West Jefferson Street
Jonesboro, AR 72401

Substance Abuse Program
520 West Monroe Street
Jonesboro, AR 72401

George W. Jackson CMHC Alcohol and Drug Abuse Program
2920 McClellan Drive
Jonesboro, AR 72401

Greenleaf Center Inc./Arkansas Substance Abuse Treatment Unit
2712 East Johnson Avenue
Jonesboro, AR 72401

LITTLE ROCK

Baptist Rehabilitation Institute Recover
9601 Interstate 630 Exit 7
Little Rock, AR 72205

Central Arkansas Substance Abuse Program (CASAP)
7107 West 12 Street
Suite 203B
Little Rock, AR 72204

CPC Pinnacle Pointe Hospital Substance Abuse Services
11501 Financial Centre Parkway
Little Rock, AR 72211

Gyst House
4201 Barrow Road
Little Rock, AR 72204

Saint Vincent Medical Center Restore Unit
2 Saint Vincent Circle
Little Rock, AR 72205

Serenity Park, Inc.
2801 Roosevelt Road
Little Rock, AR 72204

Twenty Four Hour Center, Inc.
Men's Rehabilitation Center
2021 Main Street
Little Rock, AR 72206

Women's Chemical Dependency Treatment
3900 Affolter Lane
Route 5
Little Rock, AR 72212

MAUMELLE

Charter Hospital of Little Rock
1601 Murphy Drive
Maumelle, AR 72113

MONTICELLO

Delta Counseling Associates Outpatient
790 Roberts Drive
Monticello, AR 71655

MOUNTAIN HOME

Ozark Counseling Services
8 Medical Plaza
Mountain Home, AR 72653

NORTH LITTLE ROCK

Family Service Agency of Central Arkansas
2700 North Willow Street
North Little Rock, AR 72115

Riverbend Recovery Center
1201 River Road
North Little Rock, AR 72114

The Bridgeway
21 Bridgeway Road
North Little Rock, AR 72113

Veterans' Affairs Medical Center Drug Dependency Treatment Program/116E
2200 Fort Roots Drive
North Little Rock Division
North Little Rock, AR 72114

PARAGOULD

Crowley's Ridge Development Council Northeast Arkansas Regional Recovery Center
Route 6
Paragould, AR 72401

PINE BLUFF

Dept. of Correction/Women's Unit Substance Abuse Treatment Program (SATP)
8000 West 7 Street
Pine Bluff, AR 71603

Human Development and Research Services, Inc.
6841 West 13 Street
Pine Bluff, AR 71602

Jefferson Regional Medical Center First Step Chemical Dependency Unit
1515 West 42 Avenue
Pine Bluff, AR 71603

Pine Bluff Youth Services Center
7301 West 13 Street
Pine Bluff, AR 71602

Southeast Arkansas Mental Health Center Substance Abuse Treatment Program
2500 Rike Drive
Pine Bluff, AR 71613

POCAHONTAS

Black River Area Development Corp. Alcoholism Treatment Program
1405 Hospital Drive
Pocahontas, AR 72455

RUSSELLVILLE

Arkansas River Valley Area Council Freedom House
900 Dike Road
Russellville, AR 72801

Counseling Associates/ Russellville
110 Skyline Drive
Russellville, AR 72801

SEARCY

Wilbur D. Mills Alcoholism Treatment Center
3204 East Moore Avenue
Searcy, AR 72143

SPRINGDALE

215 Club, Inc. DBA Decision Point
301 Holcomb Street
Springdale, AR 72764

TEXARKANA

Red River Regional Council Alcohol/Drug Abuse Dowd House
2101 Dudley Street
Texarkana, AR 75502

TUCKER

Dept. of Correction/Tucker Unit Substance Abuse Treatment Program (SATP)
Star Route
Tucker, AR 72168

WRIGHTSVILLE

Dept. of Correction/Wrightsville Unit Substance Abuse Treatment Program (SATP)
Building 6
Wrightsville, AR 72183

CALIFORNIA

ACTON

Department of Health Services Acton Rehabilitation Center
30500 Arrastre Canyon Road
Acton, CA 93510

ALAMEDA

Catholic Counseling Services
1109 Chestnut Street
Alameda, CA 94501

Navy Counseling and Assistance Center Naval Air Station
Building 116
Alameda, CA 94501

ALPINE

McAlister Institute for Treatment and Education (MITE) Rural East County Rec. Center
2751-B Alpine Boulevard
Alpine, CA 91901

ALTURAS

Alcoholics Anonymous
100 South Main Street
Alturas, CA 96101

ANAHEIM

Casa Elena
832 South Anaheim Boulevard
Anaheim, CA 92805

Hope House
707 North Anaheim Boulevard
Anaheim, CA 92805

Hotline Help Center
Anaheim, CA 92815

Orange County Center for Health Counseling Center/Hispanic Drug Abuse Outpatient
3170 East La Palma Avenue
Anaheim, CA 92806

Orange County Health Care Agency Anaheim Drug Abuse Services
1133 North Homer Street
Anaheim, CA 92801

School Ten
2384 Orangethorpe Avenue
Anaheim, CA 92806

Spencer Recovery Centers/ Anaheim
222 West Ball Road
Anaheim, CA 92805

Vista Recovery Center
1830 West Romneya Drive
Anaheim, CA 92801

ANGELS CAMP

Changing Echoes
7632 Pool Station Road
Angels Camp, CA 95222

ANTIOCH

Alcohol Intervention/Recovery Services Post Conviction Drinking Driver Program
2400 Sycamore Drive
Suite 36
Antioch, CA 94509

Family and Community Services of Contra Costa
3700 Delta Fair Boulevard
Suite 210
Antioch, CA 94509

Reach Project
1915 D Street
Antioch, CA 94509

APPLE VALLEY

Jackson/Bibby Awareness Group
19031 Outer Highway 18
Suite 150-170
Apple Valley, CA 92307

Saint John of God Health Care Services Alpha House/Alpha Tot Women's Program
11726 Deep Creek Road
Apple Valley, CA 92308

APTOS

Acacia Associates
9057 Soquel Drive
Suite C-B
Aptos, CA 95003

ARCADIA

Choices/Transformations
630 West Duarte Road
Suite 202
Arcadia, CA 91006

Cocaine Anonymous Hotline San Gabriel Valley
114 Saint Joseph Street
Arcadia, CA 91006

ARLETA

El Proyecto Del Barrio
8902 Woodman Avenue
Arleta, CA 91331

ARROWBEAR LAKE

Fourth Dimension Life Center
2312 Blue Jay Lane
Arrowbear Lake, CA 92382

ARROYO GRANDE

San Luis Obispo County Mariposa Community Recovery Center
1106 Grand Avenue
Suite A
Arroyo Grande, CA 93420

ATASCADERO

Community Health Projects/ Atascadero
6500 Morro Road
Suite D
Atascadero, CA 93422

San Luis Obispo County North County Connection/Comm. Rec. Center
1350 El Camino Real
Atascadero, CA 93424

AUBURN

Chapa de Indian Health Substance Abuse Services
1240 High Street
Suite 2
Auburn, CA 95603

Pacific Educational Services
11812 Kemper Road
Auburn, CA 95603

Placer County Alcohol Drug and Tobacco Programs
11533 C Avenue
De Witt Center
Auburn, CA 95603

Sierra Council on Alcohol and Drug Dependency Auburn Service Center
610 Auburn Ravine Road
Suite A
Auburn, CA 95603

South Placer Residential Treatment Program
11417 D Avenue
Auburn, CA 95603

Sierra Family Services/Auburn
1141 High Street
Auburn, CA 95603

AZUSA

ABC Traffic Safety Programs Covina Valley Traffic Safety Programs
453 Arrow Highway
Suite J
Azusa, CA 91702

Social Model Recovery Systems River Community
23701 East Fork Road
Azusa, CA 91702

BAKERSFIELD

Addictive Drugs Educational Foundation
2639 Belle Terrace
Bakersfield, CA 93304

Alcoholics Anonymous Central Office of Kern County
930 Truxten Avenue
Room 110
Bakersfield, CA 93301

Aware Teenage and Family Counseling For Drugs and Alcohol Program
1631 30 Street
Bakersfield, CA 93301

Community Health Projects Medical Group
1018 21 Street
Bakersfield, CA 93301

Community Service Organization (CSO) Brotherhood
500 Baker Street
Bakersfield, CA 93305

Ebony Counseling Center
1301 California Avenue
Bakersfield, CA 93304

Family and Substance Abuse Counseling Agency
1680 20 Street
Bakersfield, CA 93301

Kern County Hispanic Com. on Alcohol/Drug Abuse El Camino Nuevo
1211 S Street
Bakersfield, CA 93301

Kern County Alcohol Center
Cedar Women's Center
2429 19 Street
Bakersfield, CA 93301

Kern Recovery Center
2105 F Street
Bakersfield, CA 93301

Kern County Com. on Alcohol/ Drug Abuse Services Casa Serena
Bakersfield, CA 93301

Kern County Dept. of Mental Health Services Substance Abuse Program
1401 L Street
Bakersfield, CA 93301

Kern County Economic Opportunity Corp. Substance Abuse Program
699 East Brundage Lane
Bakersfield, CA 93307

Narcotics Anonymous Hotline Kern County
4909 Stockdale Highway
Suite 343
Bakersfield, CA 93309

Special Treatment Education and Prevention Services, Inc. (STEPS)
3533 Mount Vernon Avenue
Bakersfield, CA 93306

Teen Challenges of Southern California Teen Challenge of Kern County
301 East Roberts Lane
Bakersfield, CA 93308

Traffic and Alcohol Awareness School of Kern (TAASK)
2020 Anita Lane
Bakersfield, CA 93304

BALDWIN PARK

Community Health Projects Baldwin Park
14418 East Pacific Avenue
Baldwin Park, CA 91706

Crossroads
3143 Puente Street
Baldwin Park, CA 91706

BANNING

Desert Dawn Centers Drug and Alcohol Prevention
569 East King Street
Banning, CA 92220

House of Hope/Banning
628 8 Street
Banning, CA 92220

Koalacare of California Ace Program
60 West Hays Street
Banning, CA 92220

Riverside County Drug Abuse Program Second Chance
1626 Hargrave Street
Banning, CA 92220

BARSTOW

Hi Desert Mental Health Center Substance Abuse Services
500 Melissa Street
Barstow, CA 92311

Jackson/Bibby Awareness Group
222 Main Street
Suite 218
Barstow, CA 92311

BELL GARDENS

Bell Gardens Youth Services Bureau
5840 Florence Avenue
Bell Gardens, CA 90201

BELLFLOWER

Little House
9718 Harvard Street
Bellflower, CA 90706

BELMONT

Community Psychiatric Center (CPC) Belmont Hills Hospital/ Dual Diagnosis
1301 Ralston Avenue
Belmont, CA 94002

Center for Independence of the Disabled Alcohol and Drug Program
875 Oneill Avenue
Belmont, CA 94002

BENICIA

Youth and Family Services Benicia
150 East K Street
Benicia, CA 94510

BERKELEY

Berkeley Addiction Treatment Services Outpatient Methadone Maintenance Services
2975 Sacramento Street
Berkeley, CA 94702

East Bay Asian Youth Center
1950 Carleton Street
Room D-6
Berkeley, CA 94704

Mandana North Community Recovery Center
606 Bancroft Way
Berkeley, CA 94710

New Bridge Foundation
1820 Scenic Avenue
Berkeley, CA 94709

VA Northern California System of Clinics Berkeley Outpatient Clinic
841 Folger Avenue
Berkeley, CA 94710

BEVERLY HILLS

Matrix Center/Beverly Hills
9033 Wilshire Boulevard
Suite 201
Beverly Hills, CA 90211

BIG BEAR LAKE

Operation Breakthrough
40585 Lakeview Drive
Big Bear Lake, CA 92315

BISHOP

Inyo County Substance Abuse Services
162 H Grove Street
Suite H
Bishop, CA 93514

BLOOMINGTON

Social Science Services, Inc.
Cedar House Rehabilitation Center
18612 Santa Ana Avenue
Bloomington, CA 92316

Cedar House Sober Living Environment
18717 6 Street
Bloomington, CA 92316

Drinking Driver Programs
11608 Cedar Avenue
Bloomington, CA 92316

BLUE JAY

Rim Communities Recovery Center
27186 Highway 189
Blue Jay, CA 92317

BLYTHE

Metcalf Recovery Ranch
9826 18 Street
Blythe, CA 92225

Riverside County Dept. of Mental Health Alcohol/Drug Abuse Control Services/Blythe
134 East Hobson Way
Suite 5
Blythe, CA 92225

BORON

Federal Prison Camp Drug Abuse Program
Boron, CA 93516

BOULEVARD

Kumeyaay Lodge Native American Alcohol Recovery Center
La Posta Reservation
Boulevard, CA 91905

BRISBANE

Latino Commission on Alcohol and Drug Abuse Services of San Mateo County/Casa Maria
105 McLain Avenue
Brisbane, CA 94005

BUENA PARK

City of Buena Park Employee Assistance Program
6650 Beach Boulevard
Buena Park, CA 90620

Orange County Community Hospital of Buena Park/Focus on Recovery
6850 Lincoln Avenue
Buena Park, CA 90620

BURBANK

Bridge Focus Community Resource Managemenet
2829 North Glenoaks Boulevard
Suite 206
Burbank, CA 91504

New Way Foundation
Avalon House
126 East Olive Avenue
Burbank, CA 91502

Aware Program
207 North Victory Boulevard
Burbank, CA 91502

Padre, Inc.
2410 West Olive Avenue
Burbank, CA 91506

Safety Education Center
134 South San Fernando Boulevard
Burbank, CA 91502

Saint Joseph Medical Center Chemical Dependency Services/Outpatient
3413 Pacific Avenue
Burbank, CA 91505

Valley Lodge
446 North Varney Street
Burbank, CA 91502

BURLINGAME

Alcohol and Drug Helpline
1811 Trousdale Drive
Burlingame, CA 94010

Mills Peninsula Hospital Chemical Dependency Center
1783 El Camino Real
Burlingame, CA 94010

Women's Recovery Association Residential and Outpatient
1450 Chapin Street
1st Floor
Burlingame, CA 94010

BURNEY

Crossroads Clinic
20597 Commerce Way
Burney, CA 96013

Pit River Health Services Substance Abuse Services
36977 Park Avenue
Burney, CA 96013

CALABASAS

Client Appraisal Resources/ Calabasas
5026 North Parkway
Calabasas, CA 91302

CALEXICO

Imperial Valley Methadone Clinic
535 Railroad Boulevard
Calexico, CA 92231

CALISTOGA

Alcoholics Anonymous Up Valley Intergroup
Calistoga, CA 94515

Duffy's Myrtledale, Inc. Alcohol Recovery Facility
3076 Myrtledale Road
Calistoga, CA 94515

CAMARILLO

Alcoholics Anonymous Central Service Office
321 North Aviador Street
Suite 115
Camarillo, CA 93010

Gateway Recovery and Intervention Program
200 Horizon Circle
Camarillo, CA 93010

Palmer Drug Abuse Program of Ventura County
1840 East Ventura Boulevard
Camarillo, CA 93010

CAMBRIA

San Luis Obispo Cambria Connection/Community Rec. Center
2380-F Main Street
Cambria, CA 93428

CAMPBELL

Alanon Family Groups Santa Clara Valley
1 West Campbell Avenue
Campbell Community Ctr Bldg Q
Room 83
Campbell, CA 95008

Alert Driving, Inc. (ADI)
1475 South Bascom Avenue
Suite 108
Campbell, CA 95008

Good Samaritan Hospital Recovery Center
3333 South Bascom Avenue
Campbell, CA 95008

Santa Clara County Foundation for Alcohol Awareness
1550 South Bascom Avenue
Suite 300
Campbell, CA 95008

CAMPO

San Diego Freedom Ranch
1777 Buckman Springs Road
Campo, CA 91906

CANOGA PARK

Canoga Park Safety Programs
21054 Sherman Way
Suite 220
Canoga Park, CA 91303

Dignity Counseling Center
20944 Sherman Way
Suite 104
Canoga Park, CA 91303

Humana Hospital/West Hills Careunit
7300 Medical Center Drive
Canoga Park, CA 91307

Pine Grove Hospital and Mental Health Center Chemical Dependency Services
7011 Shoup Avenue
Canoga Park, CA 91307

CARMICHAEL

Associated Family Therapy Effective in Recovery (AFTER)
5120 Manzanita Street
Suite 120
Carmichael, CA 95608

Associated Rehab Program for Women
8400 Fair Oaks Boulevard
Carmichael, CA 95608

People Reaching Out
5433 El Camino Avenue
Suite 700
Carmichael, CA 95608

CARPINTERIA

Klein Bottle Youth Programs Carpinteria
1056 Eugenia Place
Suite B
Carpinteria, CA 93013

Salvation Army Adult Rehabilitation Center
6410 Cindy Lane
Carpinteria, CA 93013

CARSON

Abstinent Living Centers
218 Street Services
329 West 218 Street
Carson, CA 90745

Fresh Start Training Center
1167 East 215 Place
Carson, CA 90745

Kaiser Permanente Hospital Chemical Dependency Recovery Program/Carson
23621 South Main Street
Carson, CA 90745

Palm House Alcoholism Recovery Home
2515 East Jefferson Street

Carson, CA 90810

CASTAIC

Antelope Valley Rehabilitation Centers Warm Springs Rehabilitation Center
38200 North Lake Hughes Road
Castaic, CA 91310

CATHEDRAL CITY

Alcoholics Anonymous
32150 Candlewood Street
Suite 2
Cathedral City, CA 92234

CERES

Memorial Hospital Association Recovery Resources Alcohol and Drug Center
1905 Memorial Drive
Ceres, CA 95307

CERRITOS

Southeast Council on Alcoholism and Drug Programs, Inc.
13205 South Street
Cerritos, CA 90701

CHICO

Alcoholics Anonymous
234 West 3 Street
Suite E
Chico, CA 95926

Butte County Alcohol and Drug Services
Unit 578
578 Rio Lindo Avenue
Suite 1
Chico, CA 95926

Youth Services Division
492 Rio Lindo Avenue
Chico, CA 95926

Prevention Education Program
500 Cohasset Road
Suite 28
Chico, CA 95926

Touch Stone
1802 Esplanade Street
Chico, CA 95926

CHINO

City of Chino Human Services Division
13271 Central Avenue
Chino, CA 91710

Jericho Outreach
Men's Home
5151 F Street
Chino, CA 91710

Women's Home
12591 Benson Avenue
Chino, CA 91710

San Bernardino County Chino Multiple Diagnosis Clinic
6180 Riverside Drive
Suite H
Chino, CA 91710

W and G Enterprises Alcohol Education and Recovery Center
12560 Central Avenue
Chino, CA 91710

CHULA VISTA

Episcopal Community Services (ECS)
South Bay Drug Abuse Services/ Naltrexone
314 Park Way
Suite A
Chula Vista, CA 91910

South Bay Alcohol Recovery Services
314 Park Way
Suite B
Chula Vista, CA 91910

MAAC Project
Drinking Driver Programs
45 3 Avenue
Suite 101
Chula Vista, CA 91910

Nosotros
73 North 2 Avenue
Building B
Chula Vista, CA 91910

McAlister Institute for Treatment and Education (MITE) Options for Recovery/South Bay
251 Palomar Street
Suite A
Chula Vista, CA 91911

San Diego Treatment Services Third Avenue Clinic
1161 3 Avenue
Chula Vista, CA 91911

Southwood Hospital and Residential Treatment Center
330 Moss Street
Chula Vista, CA 91911

CITRUS HEIGHTS

Oak House Corporation
Oak House I
7987 Oak Avenue
Citrus Heights, CA 95610

Oak House II
7919 Oak Avenue
Citrus Heights, CA 95610

CITY OF INDUSTRY

Twin Palms Recovery Center
16054 Amar Road
City of Industry, CA 91744

CLAYTON

Bi Bett Corporation Diablo Valley Ranch Male Recovery Community
11540 Marsh Creek Road
Clayton, CA 94517

CLOVIS

Central Valley Indian Health Program Substance Abuse Services
20 North Dewitt Street
Suite 9B
Clovis, CA 93612

Clovis Community Hospital Recovery Center for Chemical Dependency
88 North Dewitt Avenue
Clovis, CA 93612

Clovis Community Hospital Renaissance Adolescent Center
2755 North Herndon Street
Clovis, CA 93612

COLEVILLE

Toiyabe Indian Health Project Family Service Department
Camp Antelope Road and Eastside Rd
Coleville, CA 96107

COLOMA

Progress House/Men's Facility
838 Beach Court Road
Coloma, CA 95613

COLTON

Alcoholics Anonymous Central Office
1265 North Mount Vernon Avenue
Colton, CA 92324

San Bernardino County Medical Society
952 South Mount Vernon Avenue
Colton, CA 92324

COLUSA

Department of Substance Abuse Services Counseling Center
642 5 Street
Colusa, CA 95932

COMPTON

American Indian Free Clinic Main Artery Alcoholism Recovery Home
1330 South Long Beach Boulevard
Compton, CA 90221

Charles R. Drew First Offender King Drew Substance Abuse Program
2715 North Wilmington Avenue
Compton, CA 90222

Compton Special Services Center
404 North Alameda Street
Compton, CA 90221

Get Off Drugs/Dan Anderson Women's Home
1416 South Tamarind Street
Compton, CA 90220

Kazi House Residential Drug Program
930 West Compton Boulevard
Compton, CA 90220

Mini Twelve Step House The Solution Drop-In Center
347 West Compton Boulevard
Compton, CA 90220

CONCORD

Bi Bett Corporation
Diablo Valley Ranch Annex
1860 Belmont Street
Concord, CA 94520

Frederic Ozanam Center
2931 Prospect Street
Concord, CA 94518

Shennum Center
2090 Commerce Avenue
Concord, CA 94520

Catholic Charities Catholic Counseling Services
2350 Pacheco Street
Concord, CA 94520

Center for Human Development NEAT Family
391 Taylor Boulevard
Suite 120
Concord, CA 94523

Family Stress Center
2086 Commerce Street
Concord, CA 94520

Mount Diablo Hospital Medical Center Center for Recovery
2540 East Street
Concord, CA 94520

New Connections The Keller House
1760 Clayton Road
Concord, CA 94520

Occupational Health Services Driving Under the Influence Program
2401 Stanwell Drive
Suite 460
Concord, CA 94520

Sunrise House
135 Mason Circle
Unit M
Concord, CA 94520

CORNING

Tehama Alcohol Recovery Center
DBA Right Road
275 Solano Street
Corning, CA 96021

Right Road Counseling Center/II
670 Edith Avenue
Corning, CA 96021

CORONA

ACT Family Treatment Program Chemical Dependency Unit
2214 Vesper Circle
Suite E-2
Corona, CA 91719

California Institution for Women Methadone Program
16756 Chino-Corona Road
Corona, CA 91720

Charter Hospital of Corona
2055 Kellogg Avenue
Corona, CA 91719

Koalacare of California Ace Program
223 East 3 Street
Corona, CA 91720

Riverside County Alcohol Control Program/Corona
212 East Grand Boulevard
Suite A
Corona, CA 91719

CORONADO

Coronado Recovery Center
250 Prospect Place
Coronado, CA 92118

COSTA MESA

Academy of Defensive Driving Orange County Fairgrounds
88 Fair Drive
Costa Mesa, CA 92626

Addiction Institute
3151 Airway Building
Suite C-1
Costa Mesa, CA 92626

Association Renaissance Creators (ARC)
3142 Cork Lane
Costa Mesa, CA 92626

Pine Creek
1300 Adams Lane
Suite 27C
Costa Mesa, CA 92626

The Dove Cottage
2826 Portola Drive
Costa Mesa, CA 92626

Cope Center
440 Fair Drive
Suite K
Costa Mesa, CA 92626

First Step House of Orange County
2015 Charle Street
Costa Mesa, CA 92627

Heritage House
2212–2218 Placentia Street
Costa Mesa, CA 92626

New Directions for Women
2601 Willo Lane
Costa Mesa, CA 92627

Orange Avenue Recovery Home
1976 Orange Avenue
Costa Mesa, CA 92627

Orange County Alcohol Program South Region
3115 Redhill Avenue
Costa Mesa, CA 92626

Orange County Health Care Agency Newport Mesa Drug Abuse Service
3115 Redhill Avenue
Costa Mesa, CA 92626

Rap Institute
666 West Baker Street
Suite 421
Costa Mesa, CA 92626

South Coast Counseling Center
2950 Airway Street
Suite B-3
Costa Mesa, CA 92626

693 Plumer Street
Costa Mesa, CA 92627

Starting Point of Orange County
350 West Bay Street
Costa Mesa, CA 92627

The Way Back
254 15 Street
Suite B
Costa Mesa, CA 92627

COTATI

A Step Up
420 East Cotati Avenue
Cotati, CA 94931

COVINA

Crossroads
4525 Grenfinan Street
Covina, CA 91723

Crossroads of Grand Avenue
3634 North Grand Avenue
Covina, CA 91723

MCC Managed Behavioral Care of California
545 Eremland Street
Covina, CA 91723

National Council on Alcohol/ Drug Dependency of East San Gabriel and Pomona Valleys
754 East Arrow Highway
Suite F
Covina, CA 91722

Spencer Recovery Center/Covina
444 West Badillo Street
Covina, CA 91723

Stepping Stones Recovery Home
17727 East Cypress Street
Covina, CA 91722

CRESCENT CITY

Alcoholics Anonymous
Crescent City, CA 95531

Del Norte County Drug and Alcohol Services
384 Elk Valley Road
Crescent City, CA 95531

Humboldt Addictions Services Programs (HASP) Del Norte County
200 Marine Way
Crescent City, CA 95531

CUDAHY

East Los Angeles Health Task Force Aztec Driving School
7810 Otis Avenue
Cudahy, CA 90201

CULVER CITY

Brotman Hospital and Medical Center Choices at Brotman
3828 Delmas Terrace
Suite T-4
Culver City, CA 90231

Century Services
5427 Sepulveda Boulevard
Suite 1
Culver City, CA 90230

Cocaine Anonymous Central Office
6125 Washington Boulevard
Suite 202
Culver city, CA 90232

CYPRESS

Straight Talk Clinic
5712 Camp Street
Cypress, CA 90630

DANA POINT

Community Counseling Center of San Juan Capistrano/Casa Del Cerro I
26882–26884 Avenida Las Palmas Street
Dana Point, CA 92624

Straight Ahead
34185 Coast Highway
Dana Point, CA 92629

DANVILLE

San Ramon Valley Discovery Center
530 La Gonda Way
Suite A
Danville, CA 94526

DAVIS

Alcoholics Anonymous
Davis, CA 95617

Yolo County Department of Alcohol and Drug Programs, Davis Branch
Family Treatment Services
Youth Outreach Program
600 A Street
Davis, CA 95616

DEER PARK

Crutchers Serenity House
50 Hillcrest Street
Deer Park, CA 94576

Saint Helena Hospital Alcohol and Chemical Recovery
650 Sanitarium Road
Deer Park, CA 94576

DELANO

Catholic Social Services Kaleidoscope/Delano
908 Main Street
Delano, CA 93215

Kern County Alcohol Centers Pathways Family Recovery Center
1224 Jefferson Street
Suite 21
Delano, CA 93215

DESCANSO

Phoenix House San Diego Residential Drug Free Program
23981 Sherilton Valley Road
Descanso, CA 91916

DESERT HOT SPRINGS

Desert Rehabilitation Services
Cayton House
66800 First Street
Desert Hot Springs, CA 92240

Hacienda Valdez
12890 Quinta Way
Desert Hot Springs, CA 92240

The Ranch
7885 Annandale Avenue
Desert Hot Springs, CA 92240

Soroptimist House of Hope
13525 Cielo Azul Way
Desert Hot Springs, CA 92240

DINSMORE

Center for Individual Recovery Services Dinsmore Lodge
Highway 36
Dinsmore, CA 95526

DIXON

Dixon Family Services
155 North 2 Street
Dixon, CA 95620

DOWNEY

Awakenings Program for Deaf and Hard of Hearing Persons/ Outpatient
8515 East Florence Avenue
Suite 200–203
Downey, CA 90240

DUBLIN

Catholic Counseling Services
11750 Dublin Boulevard
Suite 115
Dublin, CA 94568

Family Service of the East Bay
7080 Donlon Way
Dublin, CA 94568

DULZURA

Rancho L Abri
18091 Bee Canyon Road
Dulzura, CA 91917

EAST PALO ALTO

Catholic Charities of the Archdiocese of San Francisco/ Thea Bowman House
2109 Dumbarton Street
East Palo Alto, CA 94303

Daytop Adult
2560 Pulgas Avenue
East Palo Alto, CA 94303

Links to Positive People Center for Community Change
2524 Pulgas Avenue
East Palo Alto, CA 94303

Ravenswood Parent Child Intervention Program
1286 Runnymede Street
East Palo Alto, CA 94303

EL CAJON

Episcopal Community Services (ECS)
East County Accord
900 North Cuyamaca Street
El Cajon, CA 92020

East County Neighborhood Recovery Center
1089 El Cajon Boulevard
El Cajon, CA 92020

McAlister Institute for Treatment and Education (MITE)
East County Center
700 North Johnson Street
Suite G
El Cajon, CA 92020

Pregnant Inmates Program
810 Arnele Avenue
El Cajon, CA 92020

Professional Community Services
900 North Cuyamaca Street
Suite 201
El Cajon, CA 92020

San Diego Health Alliance East Office
234 North Magnolia Avenue
El Cajon, CA 92020

EL CENTRO

Imperial County MH Alcohol and Drug Programs
Adolescent Treatment Program
1331 Clark Road
Building 2
El Centro, CA 92243

Outpatient Clinic
1073 West Ross Avenue
Suite F
El Centro, CA 92243

Healthy New Life/Perinatal Treatment Program
1331 Clark Road
Building 3
El Centro, CA 92243

Imperial Valley Safety Services First Offender Program
480 Olive Avenue
Suite 3A
El Centro, CA 92243

Sober Roads
1030 Broadway
Suite 103
El Centro, CA 92243

Sure Helpline Center
120 North 6 Street
El Centro, CA 92243

Volunteers of America Imperial Alcohol and Drug Program
1331 Clark Road
El Centro, CA 92243

EL MONTE

California Hispanic Commission on Alcohol/Drug Abuse Casa Blanca
12042 Ramona Boulevard
El Monte, CA 91732

Community Health Projects Medical Group El Monte
11041 Valley Boulevard
El Monte, CA 91731

Community Service Organization (CSO) El Monte
4549 Peck Road
El Monte, CA 91731

East Los Angeles Health Task Force Aztec Driving School
10816 Ramona Boulevard
El Monte, CA 91731

Mid Valley Alcohol Education Centers
3430 Cogswell Road
El Monte, CA 91732

Women's Resource Center
9961 Valley Boulevard
Unit C
El Monte, CA 91732

Project Info Community Prevention and Recovery Program
9870 Garvey Avenue
El Monte, CA 91733

Twin Palms Recovery Center
3574 Lexington Avenue
El Monte, CA 91731

Community Prevention and Recovery Program
11025 Lower Azusa Road
El Monte, CA 91731

Whitney Group Whitney House
4524 Whitney Drive
El Monte, CA 91731

EL TORO

National Council on Alcoholism of Orange County
22471 Aspan Street
Suite 103
El Toro, CA 92630

ELK GROVE

Maguire Ranch
8121 Sloughouse Road
Elk Grove, CA 95624

ENCINITAS

Recovery Lifestyles of the William Beck Institute
535 Encinitas Boulevard
Suite 115
Encinitas, CA 92024

ESCONDIDO

Fellowship Center Alcohol and Other Drug Services
736 East 2 Avenue
Escondido, CA 92025

Mental Health Systems North Inland Drug Abuse Program
1299 East Pennsylvania Street
Escondido, CA 92027

North County Serenity House
123 South Elm Street
Escondido, CA 92025

Serenity Too
760 East 2 Avenue
Escondido, CA 92025

EUREKA

Alcohol/Drug Care Services
1335 C Street
Eureka, CA 95501

County of Humboldt Alcohol and Other Drug Programs
734 Russ Street
Eureka, CA 95501

Crossroads
1205 Myrtle Avenue
Eureka, CA 95501

Healthy Moms Program
2944 D Street
Eureka, CA 95501

Humboldt Addictions Services Programs (HASP) Humboldt County
924 4 Street
2nd Floor
Eureka, CA 95501

Humboldt County Perinatal Clinic
529 I Street
Eureka, CA 95501

Humboldt Recovery Center
1303 11 Street
Eureka, CA 95501

Saint Joseph Hospital Family Recovery Services
2700 Dolbeer Street
Eureka, CA 95501

United Indian Lodge
1116 9 Street
Eureka, CA 95501

FAIR OAKS

Fair Oaks Hospital and Counseling Centers Outpt. Int. CD Program Dual Diagnosis Program
11228 Fair Oaks Boulevard
Fair Oaks, CA 95628

FAIRFIELD

A K Bean Foundation Drinking Driver Program
623 Great Jones Street
Fairfield, CA 94533

Solano County Economic Opportunity Council
Solano Recovery Projects
2130 West Texas Street
Fairfield, CA 94533

Vose House
710 Beck Street
Fairfield, CA 94533

Solano County Substance Abuse Services Freedom Outreach
1735 Enterprise Drive
Building 3
Fairfield, CA 94533

FARMERSVILLE

Alcoholics Anonymous Central Office
628 North Farmersville Boulevard
Farmersville, CA 93223

FONTANA

Kaiser Permanente Hospital Chemical Dependency Recovery Program/Fontana
17046 Marygold Avenue
Marygold Annex
Fontana, CA 92335

Merrill Community Services, Inc.
16846 Merrill Avenue
Suite 202
Fontana, CA 92335

Rehabilitation Alcohol Program (RAP)
8253 Sierra Street
Suite 105
Fontana, CA 92335

FOREST KNOLLS

Serenity Knolls Chemical Dependency Recovery Program
145 Tamal Road
Forest Knolls, CA 94933

FORT ORD

Fort Ord Alcohol and Drug Abuse Division
Building 3012
Fort Ord, CA 93941

FOUNTAIN VALLEY

Pathways to Discovery
18350 Mount Langley Street
Suite 205
Fountain Valley, CA 92708

School Ten Garfield Street Unit
9555 Garfield Street
Fountain Valley, CA 92708

Straight Talk
11770 East Warner Avenue
Suite 119
Fountain Valley, CA 92708

FREMONT

Carnales Unidos Reformando Adictos (CURA) Therapeutic Community
37437 Glenmore Drive
Fremont, CA 94536

Catholic Counseling Services
3767 Decoto Road
Fremont, CA 94555

43148 Mission Boulevard
Fremont, CA 94539

Community Counseling and Education Center
39355 California Street
Suite 100
Fremont, CA 94538

Family Service of the East Bay
4537 Mattos Drive
Fremont, CA 94536

Second Chance Fremont Recovery Center
37957 Fremont Boulevard
Fremont, CA 94536

Solidarity Fellowship
34413 Blackstone Way
Fremont, CA 94555

FRENCH CAMP

San Joaquin County
Methadone Maintenance Clinic
Office of Substance Abuse Recovery House
Outpatient Methadone Detox Clinic
Residential Treatment Center
Women's Program
500 West Hospital Road
French Camp, CA 95231

FRESNO

Alcoholism and Drug Abuse Council
1651 L Street
Fresno, CA 93721

Alpha House
Non-Residential
3233 East Illinois Street
Fresno, CA 93702

Residential Program
645 South Minnewawa Avenue
Fresno, CA 93727

Transitional
Tulare
4670 East Tulare Street
Fresno, CA 93702

Bay Area Addiction Research/ Treatment, Inc.
South Orange Clinic
2851 South Orange Avenue
Fresno, CA 93725

Van Ness Clinic
539 North Van Ness Street
Fresno, CA 93728

Community Health Projects, Inc. Medical Group/Pinedale
34 East Minarets Avenue
Fresno, CA 93650

Community Hospital of Central California Avanti Center
1279 North Wishon Avenue
Fresno, CA 93728

Comprehensive Alcohol Program (CAP) Residential
2445 West Whitesbridge Road
Fresno, CA 93706

Department of Rehabilitation
2550 Mariposa Street
Room 2000
Fresno, CA 93721

DWI Services
1219 E Street
Fresno, CA 93706

Eleventh Hour Residential and Outpatient Programs
4925 East Turner Street
Fresno, CA 93727

Family Communication Center Fresno Youth Advocates
1039 U Street
Fresno, CA 93721

Family Service Center
3030 North Fresno Street
Suite 106
Fresno, CA 93703

Focus to Life
535 Tulare Street
Fresno, CA 93706

Fresno Community Hospital
Olive Street Bridge
1279 North Wishon Avenue
Fresno, CA 93728

Fresno County Substance Abuse Probation Team
744 South 10th Street
Fresno, CA 93702

Fresno County Hispanic Commission on Alcohol and Drug Abuse Services
1444 Fulton Street
Fresno, CA 93721

King of Kings
Aftercare/Non-Residential Program
1126 East Tower Street
Fresno, CA 93706

Alcohol Recovery Home
2267 South Geneva Street
Fresno, CA 93706

Pregnant Post-Partum Women's Program
1350 East Annadale Avenue
Fresno, CA 93706

West Fresno DF HIV/AIDS Intervention Program
2385 South Fairview Avenue
Suite 17
Fresno, CA 93706

Women in Transition Residential Facility
4065 North Cecelia Avenue
Fresno, CA 93722

New Directions Woman's Program Comprehensive Alcohol Program
1530 West Whitesbridge Road
Fresno, CA 93706

Nuestra Casa Recovery Home
1414 West Kearny Boulevard
Fresno, CA 93706

The Third Floor
North Howard
120 North Howard Street
Fresno, CA 93702

West Olive
2731 West Olive Avenue
Fresno, CA 93728

Turning Point of Central California United States Probation Aftercare
283 North Fresno Street
Fresno, CA 93701

FULLERTON

KC Services, Inc.
905 South Euclid Street
Suite 201
Fullerton, CA 92632

Orange County Health Care Agency North Orange County Alcohol Services
211 West Commonwealth Avenue
Suite 204
Fullerton, CA 92632

Western Pacific Fullerton Program Outpatient Detox and Methadone Maintenance
218 East Commonwealth Avenue
Fullerton, CA 92632

Woodglen Recovery Junction
771 West Orangethorpe Avenue
Fullerton, CA 92632

GALT

Altua Village
12490 Alta Mesa Road
Galt, CA 95632

GARBERVILLE

Singing Trees Recovery Center
2061 Highway 101 South
Garberville, CA 95440

GARDEN GROVE

Alcohol Services for Homosexuals (ASH, Inc.)
11918 Garden Grove Boulevard
Garden Grove, CA 92643

Garden Grove Stephouse
13472 Gilbert Street
Garden Grove, CA 92644

Hispanic Alcoholism Services Center of Orange County
9842 West 13 Street
Suite B
Garden Grove, CA 92644

Larson House
10111 Larson Street
Garden Grove, CA 92643

Bowen Street Unit
13521 Bowen Street
Garden Grove, CA 92643

Trask Avenue Unit
10172 Trask Avenue
Garden Grove, CA 92643

Roque Center Residential
9842 West 13 Street
Garden Grove, CA 92644

GARDEN VALLEY

Progress House II Women's Facility
5607 Mount Murphy Road
Garden Valley, CA 95633

GARDENA

City of Gardena Recreation and Human Services Dept.
1651 West 162 Street
Gardena, CA 90247

South Bay Chemical Dependency and South Bay Family Recovery Center
15519 Crenshaw Boulevard
Gardena, CA 90249

GEORGE AIR FORCE BASE

U.S. Air Force Social Actions Drug and Alcohol Services
35th MSSQ/MSL
George Air Force Base, CA 92394

GILROY

South County Alternative Drug Abuse
7700 Monterey Street
Gilroy, CA 95020

South Valley Counseling Center, Inc. Outpatient Alcoholism Service
8475 Forest Street
Suite A2
Gilroy, CA 95020

GLEN ELLEN

Mountain Vista Farm
3020 Warm Springs Road
Glen Ellen, CA 95442

GLENDALE

Glendale Adventist Alcohol and Drug Services
335 Mission Road
Glendale, CA 91205

Glendale Memorial Hospital and Health Center Alpha Addiction Center
Central and Los Feliz Streets
Glendale, CA 91225

New Insights
431 North Brand Boulevard
Suite 303
Glendale, CA 91203

Right on Programs
102 East Broadway
Glendale, CA 91205

Verdugo Mental Health Center
Substance Abuse Program
417 Arden Avenue
Glendale, CA 91203

Positive Directions
225-D North Maryland Avenue
Glendale, CA 91206

Western Pacific Glendale Program Outpatient Detox and Methadone Maintenance
4628 San Fernando Road
Glendale, CA 91204

GLENDORA

Project Info Community Prevention and Recovery Programs
1505 South Sunflower Avenue
Glendora, CA 91740

GOLETA

Zona Seca Goleta
269 South Magnolia Street
Goleta, CA 93117

GRAND TERRACE

Drug Alternative Program Recovery House
11786 Kingston Street
Grand Terrace, CA 92324

GRASS VALLEY

Nevada County Council on Alcoholism Grass Valley Service Center
139 1/2 Mill Street
Grass Valley, CA 95945

Team 3 Prevention Activities
11745 Maltman Drive
Grass Valley, CA 95945

GROVER CITY

Casa Solana
383 South 13 Street
Grover City, CA 93433

GUERNEVILLE

West County Community Services
15999 River Road
Guerneville, CA 95446

HANFORD

Alcohol/Drug Education and Counseling Center
1393 Bailey Drive
Hanford, CA 93230

Cornerstone Community Alcohol/ Drug Recovery Systems
Heart-to-Heart Perinatal Services
819 West 7 Street
Hanford, CA 93230

Men's Recovery
801–807 West 7 Street
Hanford, CA 93230

Women's Program
817 West 7 Street
Hanford, CA 93230

HAPPY CAMP

Karuk Tribal Health Program Substance Abuse Services
101 Indian Creek Road
Happy Camp, CA 96039

Siskiyou County Mental Health Happy Camp Health Services
38 Park Way
Happy Camp, CA 96039

HARBOR CITY

Western Health Harbor City Clinic
1647 West Anaheim Street
Harbor City, CA 90710

HAWAIIAN GARDENS

ABC Traffic Safety Programs
12018 East Centralia Road
Suite 200
Hawaiian Gardens, CA 90716

HAWTHORNE

Behavioral Health Services
Pacifica House
2501 West El Segundo Boulevard
Hawthorne, CA 90250

The Wayback Inn
12917 Cerise Avenue
Hawthorne, CA 90250

Omni Outpatient Methadone Maintenance/Detox
12954 Hawthorne Boulevard
Suite 103
Hawthorne, CA 90250

Rickman Recovery Center/ Hawthorne
4238 West 130 Street
Hawthorne, CA 90250

HAYWARD

Bi Bett Corporation Drinking and Driving Program
21192 Hesperian Boulevard
Hayward, CA 94541

Catholic Counseling Services
1544 B Street
Suite 1
Hayward, CA 94541

Family Service of the East Bay
24301 Southland Drive
Suite 605
Hayward, CA 94545

Horizon Services Cronin House
2595 Depot Road
Hayward, CA 94545

Kaiser Permanente Medical Center Chemical Dependency Service
27400 Hesperian Boulevard
Hayward, CA 94545

Project Eden Outpatient Services
680 West Tennyson Road
Hayward, CA 94544

Second Chance/Hayward Recovery Center
22297 Mission Boulevard
Hayward, CA 94541

Terra Firma Diversion/Education Services
26785 Mission Boulevard
Hayward, CA 94544

HEMET

Double Check Retreat
47552 East Florida Avenue
Hemet, CA 92544

I Am New Life Ministries
38400 San Ignacio Road
Hemet, CA 92543

Koalacare of California, Inc. Ace Program
736 North State Street
Suite 108
Hemet, CA 92543

Riverside Recovery Resources
Our House
41040 Acacia Avenue
Hemet, CA 92544

First Step House
40329 Stetson Avenue
Hemet, CA 92544

Solutions Family Treatment Center
25845 Cowston Avenue
Suite 255
Hemet, CA 92545

HIGHLAND

San Manuel Indian Health Clinic Riverside/San Bernardino Indian Health
5771 North Victoria Street
Highland, CA 92346

HOLLISTER

San Benito County Substance Abuse Program
321 San Felipe Road
Suite 9
Hollister, CA 95023

HUNTINGTON PARK

Diversion Safety Program, Inc.
6612 Pacific Boulevard
Victorian Plaza Suite D
Huntington Park, CA 90255

INDIO

ABC Recovery Center, Inc.
44-374 Palm Street
Indio, CA 92201

Awareness Program Drinking Driver
45-561 Oasis Street
Indio, CA 92201

Riverside County Dept. of Mental Health Drug Abuse Program
83-912 Avenue 45
Suite 9
Indio, CA 92201

INGLEWOOD

Bridge for the Needy
601 Venice Way
Inglewood, CA 90302

Centinela Hospital Medical Center Life Starts
555 East Hardy Street
Inglewood, CA 90301

Community Health Projects Inglewood Clinic
614 West Manchester Boulevard
Suite 104
Inglewood, CA 90301

Inglewood Mental Health Service Drug Abuse Treatment Program
4450 West Century Boulevard
Inglewood, CA 90304

IONE

Arbor and Manzanita Substance Abuse Programs
201 Waterman Road
Ione, CA 95640

ISLA VISTA

Isla Vista Medical Clinic
970 Embarcadero Del Mar
Isla Vista, CA 93117

JACKSON

Amador County Alcohol and Drug Services
1001 Broadway
Suite 103
Jackson, CA 95642

Amador County Crisis Hotline Operation Care
114 Main Street
Suite 212
Jackson, CA 95642

JOSHUA TREE

Morongo Basin Mental Health Services Associates
6393 Sunset Road
Joshua Tree, CA 92252

Panorama Ranch Morongo Basin Mental Health
65675 Sullivan Road
Joshua Tree, CA 92252

KINGS BEACH

Sierra Council on Alcohol and Drug Dependency Kings Beach Womens Residential Facility
8677 Golden Avenue
Kings Beach, CA 96143

LA JOLLA

New Horizons/San Diego
8950 Villa La Jolla Drive
Suite 1130
La Jolla, CA 92037

Scripps Clinic Alcohol/Chemical Dependency Center
10666 North Torrey Pines Road
TMU 3
La Jolla, CA 92037

LA MESA

Alvarado Parkway Institute Chemical Dependency Program
7050 Parkway Boulevard
La Mesa, CA 91942

LA PUENTE

Bay Area Addiction Research/ Treatment, Inc. La Puente
15229 East Amar Road
La Puente, CA 91744

Crown of Glory Men's Home
305 Roundabout Street
La Puente, CA 91744

LAGUNA BEACH

First Step Treatment Centers
21095 Raquel Road
Laguna Beach, CA 92651

Orange County Alcohol Program
Allso Viejo
5 Mareblu Street
Laguna Beach, CA 92656

Laguna Beach
30818 South Coast Highway
Laguna Beach, CA 92677

Orange County Health Care Agency South County Drug Abuse Services
5 Mareblu Street
Suite 200
Laguna Beach, CA 92656

South Coast Counseling Center
28052 Camino Capistrano
Suite 213
Laguna Beach, CA 92677

South Coast Medical Center Genesis Chemical Dependency Services
31872 Coast Highway
Laguna Beach, CA 92677

LAKE FOREST

Chapman Counseling Adolescent Program
22772 Centre Drive
Suite 160-E
Lake Forest, CA 92630

LAKE ISABELLA

Council on Substance Abuse Awareness Lake Isabella
5220 Lake Isabella Boulevard
Suite 3
Lake Isabella, CA 93240

LAKESIDE

McAlister Institute for Treatment and Education (MITE)
Kiva House
10123 Los Ranchitos Road
Lakeside, CA 92040

LAKEWOOD

Lakewood Regional Medical Center New Beginnings
5300 North Clark Avenue
Lakewood, CA 90712

LAMONT

Community Service Organization (CSO) de Colores
10910 Main Street
Lamont, CA 93241

LANCASTER

ABC Traffic Safety Programs
44742 North Beech Avenue
Lancaster, CA 93534

Alcoholics Anonymous Antelope Valley Central Office
44751 North Beech Avenue
Suite 2
Lancaster, CA 93534

Antelope Valley Council on Alcoholism and Drug Dependency
44815 Fig Avenue
Suite 206
Lancaster, CA 93534

Antelope Valley Hospital Medical Center Chemical Dependency Outpatient Treatment
1600 West Avenue J
Lancaster, CA 93534

High Road Program
706 West Avenue J
Lancaster, CA 93534

Keystone Recovery Center, Inc.
43423 Division Street
Suite 108
Lancaster, CA 93535

Western Pacific Lancaster Program Outpatient Detox and Methadone Maintenance
45335 Sierra Highway
Lancaster, CA 93534

LARKSPUR

Marin Services for Women Outpatient Unit
444 Magnolia Avenue
Larkspur, CA 94939

Residential Unit
127 King Street
Larkspur, CA 94939

New Perspectives
375 Doherty Drive
Larkspur, CA 94939

LEMON GROVE

McAlister Institute for Treatment and Education (MITE)
Options Recovery East
2045 Skyline Drive
Lemon Grove, CA 91945

LEMOORE

Naval Air Station/Lemoore Counseling and Assistance Center
Barracks 1
Lemoore, CA 93246

LIVERMORE

Catholic Counseling Services
458 Mapel Street
Livermore, CA 94550

Family Service of the East Bay
3311 Pacific Avenue
Livermore, CA 94550

LODI

San Joaquin County Alcohol Recovery Center
1209 West Tokay Street
Suite 12
Lodi, CA 95240

LOMA LINDA

Jerry L. Peddis Memorial VA Hospital Alcohol and Drug Treatment Program
11201 Benton Street
Room 116A1
Loma Linda, CA 92357

LOMPOC

Casa Floral Counseling and Education Center
4010 Jupiter Street
Lompoc, CA 93436

Central Coast Headway/Lompoc Drug and Alcohol Awareness Program
1100 West Laurel Avenue
Lompoc, CA 93436

Farm House, Inc.
Lompoc, CA 93436

Klein Bottle Youth Programs
110 South C Street
Suite A
Lompoc, CA 93436

United States Penitentiary Drug Abuse Treatment Program
3901 Klein Boulevard
Lompoc, CA 93436

LONG BEACH

Behavioral Health Services Redgate Memorial Hospital
1775 Chestnut Avenue
Long Beach, CA 90813

Cambodian Association of America Community Prevention and Recovery Program
2501 Atlantic Avenue
Long Beach, CA 90806

Charter Hospital of Long Beach Chemical Dependency Program
6060 Paramount Boulevard
Long Beach, CA 90805

Family Services of Long Beach
1133 Rhea Street
Suite 104
Long Beach, CA 90806

Flossie Lewis Alcoholism Recovery Center
Traditional Living Center
351 East 6 Street
Long Beach, CA 90802

New Life Center
615 Elm Street
Long Beach, CA 90802

Get Off Drugs
Outpatient
525 East 55 Street
Long Beach, CA 90805

Men's Home
Rehabilitation 1
515 East 55 Street
Long Beach, CA 90805

Harbor Area High Gain Program, Inc.
330 East 3 Street
Long Beach, CA 90802

Industry Community Interface Projects (ICI Projects)
1850 Pacific Avenue
Suite 100
Long Beach, CA 90806

Long Beach Alcohol and Drug Rehab Program
Central Clinic
1133 East Rhea Street
Long Beach, CA 90806

Grand Avenue Clinic
2525 Grand Avenue
Long Beach, CA 90815

North Clinic
6335 Myrtle Avenue
Long Beach, CA 90805

Long Beach Community Services Outpatient Drug Free
2452 Pacific Avenue
Long Beach, CA 90806

Men's Twelfth-Step House Harbor Area
1005 East 6 Street
Long Beach, CA 90802

National Council on Alcoholism and Other Drug Dependencies/ Woman to Woman
836 Atlantic Avenue
Long Beach, CA 90813

Phoenix Family Treatment Center
3620 Long Beach Boulevard
Suite C1
Long Beach, CA 90807

Safety Consultant Services
5518 North Long Beach Boulevard
Long Beach, CA 90805

Salsido Recovery Center Freedom House
268 East Louise Street
Long Beach, CA 90805

Sober Living House
1110 Saint Louis Avenue
Long Beach, CA 90804

Sobriety House of Long Beach
3125 East 7 Street
Long Beach, CA 90804

Southeast Council on Alcohol/ Drug Problems Baby Step Inn
1755 Freeman Avenue
Long Beach, CA 90804

Tarzana Treatment Center/Long Beach
2101–45 Magnolia Avenue
Long Beach, CA 90806

Veterans' Affairs Medical Center Substance Abuse Treatment Program
5901 East 7 Street
Ward K2
Long Beach, CA 90822

Watkins and Gilmore Alcohol Education and Recovery Center
3580 East Pacific Coast Highway
Suite 4
Long Beach, CA 90804

West County Medical Clinic Substance Abuse Program
100 East Market Street
Long Beach, CA 90805

Western Health Long Beach Clinic
2933 East Anaheim Street
Long Beach, CA 90804

LOS ALAMITOS

Aldine Corporation Twin Town Treatment Center
10741 Los Alamitos Boulevard
Los Alamitos, CA 90720

MCC Managed Behavioral Care of California
10861 Cherry Street
Suite 202
Los Alamitos, CA 90720

LOS ANGELES

A Los Angeles Driver Education Center
Eagle Rock
2607 Colorado Boulevard
Suite 104
Los Angeles, CA 90041

Santa Monica Boulevard Unit
8350 Santa Monica Boulevard
Suite 107
Los Angeles, CA 90069

A Step to Freedom
1665 South Kingsley Drive
Los Angeles, CA 90016

Adapt Foundation
1644 Wilshire Boulevard
Suite 303
Los Angeles, CA 90026

Alcohol and Drug Program Administration Information and Referral Unit
714 West Olympic Boulevard
9th Floor
Los Angeles, CA 90015

Alcoholics Anonymous
Central Office
767 South Harvard Boulevard
Los Angeles, CA 90005

Alcoholism Center for Women, Inc.
Units 1 and 2
1147 South Alvarado Street
Los Angeles, CA 90006

Alpha Omega Services
4167 West Washington Boulevard
Los Angeles, CA 90018

Alta Med Health Services Buena Care
1701 Zonal Avenue
Los Angeles, CA 90033

Alternative Action Programs
2511 South Barrington Avenue
Los Angeles, CA 90064

Asian American Drug Abuse Program, Inc.
Therapeutic Community Residential
5318 South Crenshaw Boulevard
Los Angeles, CA 90043

Special Deliveries/Perinatal Services
3850 Martin Luther King Boulevard
Suite 201
Los Angeles, CA 90008

Avalon/Carver Community Center Drug Abuse Program
4920 South Avalon Boulevard
Los Angeles, CA 90011

Bay Area Addiction Research/ Treatment, Inc.
Southeast Clinic
4920 South Avalon Boulevard
Los Angeles, CA 90011

West Olympic
1020 West Olympic Boulevard
Los Angeles, CA 90015

Bea Attitude Center for Women
1650 Rockwood Street
Los Angeles, CA 90026

Behavioral Health Services Hollywood Community Prevention/Recovery Center
6838 Sunset Boulevard
Los Angeles, CA 90028

Unit 1
3421 East Olympic Boulevard
Los Angeles, CA 90023

Unit 2
4099 North Mission Road
Building A
Los Angeles, CA 90032

Behavioral Systems Southwest
6051 Hollywood Boulevard
Suite 205
Los Angeles, CA 90028

California Hispanic Comm. on Alcohol and Drug Abuse
Aguila Recovery Home
6157 North Figueroa Street
Los Angeles, CA 90042

Hermanos/Consejos/CPRP
5838 East Beverly Boulevard
Los Angeles, CA 90022

Latinas Recovery Home/Saint Louis
327 North Saint Louis Street
Los Angeles, CA 90033

Hispanic Alcohol Recovery Community Center
4754 East Brooklyn Avenue
Los Angeles, CA 90022

Latinos Recovery Home/Wabash
2436 Wabash Avenue
Los Angeles, CA 90033

Mujeres Recovery Home
530 North Avenue 54
Los Angeles, CA 90042

Paloma Recovery Home
328 North Avenue 59
Los Angeles, CA 90042

Vista CPRP/Sol Youth Resources Project
109 and 111 North Avenue 56
Los Angeles, CA 90042

Canon Human Services Center
9705 South Holmes Avenue
Los Angeles, CA 90002

Casa de Hermandad
Community Prevention Recovery Program
133 North Sunol Drive
Suite 249
Los Angeles, CA 90063

West Area Opportunity Center
11821 West Pico Boulevard
Los Angeles, CA 90064

Cedars/Sinai Medical Center Center for Chemical Dependency
8700 Beverly Boulevard
Shuman Building 7th Floor
Los Angeles, CA 90048

Chabad Residential Treatment Center for Men
1952 South Robertson Boulevard
Los Angeles, CA 90034

Children's Hospital of Los Angeles Division of Adolescent Medicine/Substance Abuse Services
4650 Sunset Boulevard
Los Angeles, CA 90027

Clare Foundation Culver Vista Family Center
11325 Washington Boulevard
Los Angeles, CA 90066

Covenant House
1325 North Western Avenue
Los Angeles, CA 90027

CPC Hospital
2112 South Barrington Avenue
Los Angeles, CA 90025

Creative Neighbors Always Sharing Recovery Shelter
8224 South Broadway
Los Angeles, CA 90003

CRI Help Socorro RDF
5110 South Huntington Drive
Los Angeles, CA 90032

Dial Alcohol and Drug Education Center
3921 Wilshire Boulevard
Suite 301
Los Angeles, CA 90010

Central Los Angeles
1212 North Vermont Avenue
Suite 101
Los Angeles, CA 90029

Do It Now Foundation of Southern California, Inc.
7080 Hollywood Boulevard
Suite 906
Los Angeles, CA 90028

East Los Angeles Alcoholism Council
916 South Atlantic Boulevard
Los Angeles, CA 90022

East Los Angeles Health Task Force Comprehensive Substance Abuse Program
630 South Saint Louis Street
Los Angeles, CA 90023

El Centro Human Services Corporation El Centro Substance Abuse Treatment Center
1972 East Brooklyn Avenue
Los Angeles, CA 90033

Escuela Latina de Traffico and Alcohol of East Los Angeles
4532 East Whittier Boulevard
Suite 210
Los Angeles, CA 90022

Fairwood Villa Recovery Program
127 East 25 Street
Los Angeles, CA 90011

Family Services of Los Angeles Dignity Center
626 South Kingsley Drive
Los Angeles, CA 90005

Father Figure Residential Continuing Care Center
4410 South Main Street
Los Angeles, CA 90037

Felicity House
3701 Cardiff Avenue
Los Angeles, CA 90034

Friendly House Hand Foundation Friendly House
347 South Normandie Avenue
Los Angeles, CA 90020

Gay and Lesbian Community Services Center Addiction Recovery Services Los Angeles
1625 North Hudson Avenue
Los Angeles, CA 90028

Good News Shields for Families Project
11534 Croesus Avenue
Apartment 413
Los Angeles, CA 90059

Health Care Delivery Services The Resource Center
2311 Pontius Avenue
Los Angeles, CA 90064

His Sheltering Arms
Family Services Center
112 West 111 Street
Los Angeles, CA 90061

Recovery Home
10615 South Avalon Boulevard
Los Angeles, CA 90003

Homeless Health Care Los Angeles
1010 South Flower Street
Suite 500
Los Angeles, CA 90015

Indian Alcoholism Commission of California
225 West 8 Street
Suite 910
Los Angeles, CA 90014

Industry Community Interface Projects
La Brea
2126 South La Brea Street
Suite 203
Los Angeles, CA 90016

Manchester
519 West Manchester Street
Suite 204
Los Angeles, CA 90044

Jeff Grand Medical Group Outpatient Methadone Maintenance and Detox
3130 South Hill Street
Los Angeles, CA 90007

Jewish Family Services of Los Angeles Alcohol and Drug Action Program
8838 West Pico Boulevard
Los Angeles, CA 90035

Kaiser Permanente
Chemical Dependency Recovery Program
4700 Sunset Boulevard
Los Angeles, CA 90027

Culver Marina
1201 West Washington Boulevard
Los Angeles, CA 90066

Kedren Community Mental Health Center Drug Abuse Prevention
4211 South Avalon Boulevard
Los Angeles, CA 90011

King/Drew Drinking Driver Program Community Prevention Recovery Program
9307 South Central Avenue
Los Angeles, CA 90002

Korean Community Services
650 North Berendo Street
Los Angeles, CA 90004

LAC/USC Medical Center Professional Staff Association Infants of Substance Abuse/ Mothers' Clinic
1129 North State Street
Room 1D35
Los Angeles, CA 90033

Los Angeles Centers for Alcohol/ Drug Abuse (LACADA)
2854 Lancaster Avenue
Suite 359
Los Angeles, CA 90033

Los Angeles Treatment Services Quit for Good
11427 South Avalon Boulevard
Los Angeles, CA 90061

Mary Lind Foundation
Bimini Recovery Home
155 South Bimini Place
Los Angeles, CA 90004

Rena B Recovery Home
4445 Burns Avenue
Los Angeles, CA 90029

Royal Palms Recovery Home
360 South Westlake Avenue
Los Angeles, CA 90057

Matrix Institute
5220 West Washington Boulevard
Suite 101
Los Angeles, CA 90016

MCC Managed Behavioral Care of California
1910 West Sunset Boulevard
Suite 500
Los Angeles, CA 90026

Medical Holistic Center Alcohol/ Drug Detoxification Program (ADDP)
7514 South Vermont Avenue
Los Angeles, CA 90044

Mendes Consultation Services
3660 Wilshire Boulevard
Suite 1114
Los Angeles, CA 90010

Mini Twelve-Step House
303 East 52 Street
Los Angeles, CA 90011

MJB Transitional Recovery
11152 South Main Street
Los Angeles, CA 90061

Narcotic Educational Foundation of America
5055 Sunset Boulevard
Los Angeles, CA 90027

Narcotics Prevention Project Methadone Maintenance Treatment Program
942 South Atlantic Boulevard
Los Angeles, CA 90022

National Council on Alcoholism and Drug Dependency/Los Angeles County (NCADDLA)
600 South New Hampshire Avenue
Los Angeles, CA 90015

Neighborhood Youth Association Youth Services
3877 Grandview Boulevard
Los Angeles, CA 90066

Open Quest Institute/West Hollywood
8702 Santa Monica Boulevard
Los Angeles, CA 90069

People Coordinated Services of Southern California, Inc.
Castle Substance Abuse Program
4771 South Main Street
Los Angeles, CA 90037

Outpatient
3021 South Vermont Avenue
Los Angeles, CA 90007

Residential
1319 South Manhattan Place
Los Angeles, CA 90019

People in Progress Nonresidential Recovery Services
426 South Spring Street
2nd Floor
Los Angeles, CA 90013

Pizarro Treatment Center Outpatient Methadone Maintenance
1525 Pizarro Street
Los Angeles, CA 90026

Plaza Community Center The Esperanza Project
648 South Indiana Street
Los Angeles, CA 90023

Project Heavy/West
11818 Wilshire Boulevard
Suite 209
Los Angeles, CA 90025

Safety Consultant Services
161 West 33 Street
Los Angeles, CA 90007

Salvation Army
Harbor Light Center
809 East 5 Street
Los Angeles, CA 90013

Safe Harbor
721 East 5 Street
Los Angeles, CA 90013

Sea Possible Soledad Enrichment Action Program
161 South Fetterly Avenue
Los Angeles, CA 90022

Self Improvement Alternative Measures (SIAM)
3450 West 43 Street
Suite 217
Los Angeles, CA 90008

Shields for Families Project Genesis Family Day Treatment Program
1721 East 120 Street
Los Angeles, CA 90059

Special Service for Groups Pacific Asian Alcohol Program
1720 West Beverly Boulevard
Los Angeles, CA 90026

Spencer Recovery Centers/Silver Lake
2530 Hyperion Avenue
Los Angeles, CA 90027

Sunrise Community Counseling Center In/Outpatient
1925 West Temple Street
Suite 205
Los Angeles, CA 90026

The Original Men's Twelfth-Step House, Inc.
946 South Menlo Avenue
Los Angeles, CA 90006

Travelers Aid Society of Los Angeles Teen Canteen
6363 Santa Monica Boulevard
Los Angeles, CA 90038

Union Rescue Mission Overcomers Program
226 South Main Street
Los Angeles, CA 90012

Van Ness Recovery House
1919 North Beachwood Drive
Los Angeles, CA 90068

Volunteers of America of Los Angeles Alcohol Services
515 East 6 Street
Los Angeles, CA 90021

Watts Health Foundation
House of Uhuru
8005 South Figueroa Street
Los Angeles, CA 90003

Uhuru Counseling Center
8732 South Western Avenue
Los Angeles, CA 90047

Uhuru Family Center
8520 South Broadway
Los Angeles, CA 90003

Weingart Medical Clinic
515 East 6 Street
Los Angeles, CA 90021

West Los Angeles Treatment Program Clinic I and Clinic II
2321 Pontius Avenue
Los Angeles, CA 90064

Westside Sober Living Centers Promises Residential Treatment Centers
3743 1/2 South Barrington Avenue
Los Angeles, CA 90066

LOS GATOS

South Bay Teen Challenge
16735 Lark Avenue
Los Gatos, CA 95030

West Valley Center Outpatient
375 Knowles Drive
Los Gatos, CA 95030

LOTUS

Rational Recovery Systems
Lotus, CA 95651

LOYALTON

Sierra County Human Services Alcohol and Drug Department
202 Front Street
Loyalton, CA 96118

LYNWOOD

Los Angeles Health Services Lynwood Clinic
11315 South Atlantic Avenue
Lynwood, CA 90262

MADERA

Kings View Corporation Kings View Community Service
968 Emily Way
Suite 101
Madera, CA 93637

Madera Counseling Center Substance Abuse Services
14277 Road 28
Madera, CA 93639

Yosemite Women's Center
420 West Yosemite Street
Madera, CA 93637

MALIBU

Client Appraisal Resources/ Malibu
23525 West Civic Center Way
Suite 106
Malibu, CA 90265

MAMMOTH LAKES

Mono County Alcohol and Drug Program
Sierra Centre Mall
3rd Floor
Mammoth Lakes, CA 93546

MANHATTAN BEACH

MCC Managed Behavioral Care of California
400 South Spulveda Street
Manhattan Beach, CA 90026

MANTECA

Valley Community Counseling Services
110 North Sherman Avenue
Manteca, CA 95336

MARINA DEL RAY

Daniel Freeman Marina Hospital Exodus Recovery Center
4650 Lincoln Boulevard
Marina Del Ray, CA 90292

MARIPOSA

Mariposa Counseling Center Alcohol/Drug Services for Mariposa County
5085 Bullion Street
Mariposa, CA 95338

MARKLEEVILLE

Alpine County Mental Health Department Alcohol and Drug Program
260 Laramie Street
Markleeville, CA 96120

MARTINEZ

Alcohol Intervention/Recovery Services Post Conviction Drinking Driver Program
10 Douglass Drive
Suite 130
Martinez, CA 94553

Born Free
111 Allen Street
Martinez, CA 94533

Contra Costa County Detention Facility Methadone Treatment Program
1000 Ward Street
Martinez, CA 94553

Discovery Program Discovery House
4639 Pacheco Boulevard
Martinez, CA 94553

Options for Recovery La Casa Ujima
904 Mellus Street
Martinez, CA 94553

Veterans Affairs Medical Center Substance Abuse Treatment Program
150 Muir Road
Martinez, CA 94553

MARYSVILLE

Pathways
House I/Recovery Home for Men
605 4 Street
Marysville, CA 95901

House II/Recovery Home for Women
806 5 Street
Marysville, CA 95901

MAYWOOD

Tri City Family Guidance Center
6153 Woodward Avenue
Maywood, CA 90270

MENLO PARK

Moriah House
3406 Alameda de las Pulgas
Menlo Park, CA 94025

Veterans' Affairs Medical Center Drug and Alcohol Recovery Unit
795 Willow Road
Menlo Park, CA 94025

MERCED

Community/Social Model Advocates, Inc.
Hobie House
1301 Yosemite Parkway
Merced, CA 95340

Tranquility House
2410 Edward Avenue
Merced, CA 95340

Lifestyle Management/Drydock Drinking Driver Program
1521 West Main Street
Merced, CA 95340

Merced Alcohol and Drug Services Power Center
1836 K Street
Merced, CA 95340

Merced College Intoxicated Driver Education and Assessment Program
644 West Main Street
Merced, CA 95340

Merced County Alcohol and Drug Abuse Services The Center
658 West Main Street
Merced, CA 95340

MILPITAS

Elmwood Deuce Program
701 South Abel Street
Milpitas, CA 95035

MODESTO

Care Schools
3719-A Tully Road
Modesto, CA 95356

Center for Human Services
Residential Treatment/Choices
2215 Blue Gum Avenue
Modesto, CA 95350

Youth Services Bureau
1700 McHenry Village Way
Suite 11-B
Modesto, CA 95350

Family Service Agency of Stanislaus County
First-Step Program
1100 Kansas Street
Suites I and J
Modesto, CA 95351

True Colors Program
8224 West Grayson Road
Modesto, CA 95351

Memorial Hospital Association Recovery Resources Outpatient Services
1800 Coffee Road
Suite 93
Modesto, CA 95355

Modesto Psychiatric Center Recovery Center
1501 Claus Road
Modesto, CA 95355

New Hope Recovery House
1406 Fordham Avenue
Modesto, CA 95350

Safety Center Stanislaus
1100 Kansas Avenue
Suite F
Modesto, CA 95351

Stanislaus County Dept. of Mental Health
Alcohol Treatment Program
800 Scenic Drive
Building D North
Modesto, CA 95350

Heroin Treatment Services
800 Scenic Drive SW
Building D South
Modesto, CA 95350

Substance Abuse Services
Outpatient Drug Free
1501 F Street
Modesto, CA 95354

MOJAVE

Council on Substance Abuse Awareness Mojave
15734 K Street
Mojave, CA 93501

Desert Counseling Clinic Mojave Branch
16914 1/2 Highway 14
Mojave, CA 93501

MONROVIA

Spencer Recovery Hospital/ Monrovia Crossroads Alcoholism Center
345 West Foothill Boulevard
Monrovia, CA 91016

MONTCLAIR

Aspen Community Counseling Center
9625 Monte Vista Street
Suite 101
Montclair, CA 91763

Inland Health Services (IHS) Montclair
4761 Arrow Highway
Montclair, CA 91763

MONTEREY

Alcohol Programs Management
1193B 10 Street
Monterey, CA 93940

Alcoholics Anonymous Intergroup Monterey Bay Area
1015 Cass Street
Suite 4
Monterey, CA 93940

Community Hospital Recovery Center and Clint Eastwood Youth Program
576 Hartnell Street
Monterey, CA 93940

MONTEREY PARK

Alhambra Safety Services
926 East Garvey Avenue
Suite A
Monterey Park, CA 91754

MONTGOMERY CREEK

Wilderness Recovery Center
19650 Cove Road
Montgomery Creek, CA 96065

MORENO VALLEY

Koalacare of California Ace Program
23846 Sunnymead Boulevard
Suite 5
Moreno Valley, CA 92553

Pathways to Recovery Chemical Dependency Program
23110 Atlantic Circle
Suite E
Moreno Valley, CA 92553

MORGAN HILL

Morgan Hill Community Adult School
1505 East Main Avenue
Morgan Hill, CA 95037

MOUNTAIN VIEW

Community Health Awareness Program
711 Church Street
Mountain View, CA 94041

El Camino Hospital Chemical Dependency Services
2500 Grant Road
Mountain View, CA 94040

Santa Clara County Bureau of Alcohol/Drug North County Center/Mountain View Office
101 Stierlin Road
Mountain View, CA 94043

Total Life Care (TLC) Growth and Leadership Center
1451 Grant Road
Suite 102
Mountain View, CA 94040

MURRIETA

Anderson and Associates Counseling Services
26811 Hobie Circle
Suite 02
Murrieta, CA 92362

NAPA

Alcoholics Anonymous Napa Intergroup
Napa, CA 94558

Family Service of North Bay
1157 Division Street
Napa, CA 94559

Napa County Drinking Driver Program
300 Coombs Street
Suite M16
Napa, CA 94559

Napa County Human Services Alcohol and Drug Program
2344 Old Sonoma Road
Napa, CA 94559

Our Family, Inc.
Napa State Hospital
Evergreen Drive/D Ward
Napa, CA 94559

NATIONAL CITY

Mental Health Systems, Inc. Probationers in Recovery/ South
43 East 12th Street
Suite A
National City, CA 91950

NEEDLES

Mental Health Systems, Inc. Needles Counseling Center
1406 Bailey Avenue
Suite E
Needles, CA 92362

NEVADA CITY

Nevada County Dept. of Mental Health
Lovett Recovery Center
10075 Bost Avenue
Nevada City, CA 95959

Mental Health Services
10433 Willow Valley Road
Nevada City, CA 95959

NEWARK

Second Chance, Inc.
6330 Thornton Avenue
Newark, CA 94560

NEWHALL

Family Center
24880 North Apple Street
Newhall, CA 91321

I-Adarp Newhall
23236 Lyons Avenue
Suite 209
Newhall, CA 91321

National Council on Alcoholism and Drug Dependence of San Fernando Valley
24416 Walnut Street
Newhall, CA 91321

Safety Consultant Services
24506 1/2 Lyons Avenue
Newhall, CA 91321

NEWPORT BEACH

Hoag Memorial Hospital Chemical Dependency Center
301 Newport Boulevard
Newport Beach, CA 92660

Sober Living by the Sea
4260 Unit A 31 Street
Newport Beach, CA 92663

NORCO

California Department of Corrections CA Rehab Center Civil Addict Program
5 and Western Streets
Norco, CA 91760

NORDEN

Eagle Recovery Lodge
58300 Donner Summit Road
Norden, CA 95724

NORTH HOLLYWOOD

CRI Help/North Hollywood
11027 Burbank Boulevard
North Hollywood, CA 91601

Dial Alcohol and Drug Education Center North Hollywood
12034 Vanowen Street
Suite 2
North Hollywood, CA 91605

Gay and Lesbian Community Services Center Addiction Recovery Services/San Fernando Valley
5652 Cahuenga Boulevard
North Hollywood, CA 91601

Studio 12
12406 Magnolia Boulevard
North Hollywood, CA 91607

Vesper House
6301 Cahuenga Boulevard
North Hollywood, CA 91606

NORTHRIDGE

Alternative Recovery Centers Valley Treatment Center/ Balboa House
8941 Balboa Boulevard
Northridge, CA 91325

Drug Consultants California Drug Consultants
9420 Reseda Boulevard
Suite 510
Northridge, CA 91324

Northridge Hospital Medical Center Program for Problem Dependencies
18420 Roscoe Boulevard
Northridge, CA 91328

NORWALK

Los Angeles Centers for Alcohol/ Drug Abuse Recovery House
11400 Norwalk Boulevard
Suite 305
Norwalk, CA 90650

Southeast Council on Alcohol/ Drug Problems Cider House
11400 Norwalk Boulevard
Building 209/313
Norwalk, CA 90650

Western Pacific Norwalk Program Outpatient Detox and Methadone Maintenance
11902 Rosecrans Boulevard
Norwalk, CA 90650

NOVATO

Henry Ohlhoff House/North
5394 Nave Drive
Novato, CA 94949

Threshold for Change
619 Canyon Road
Novato, CA 94947

OAKHURST

Kings View Community Services
49269 Golden Oak Drive
Suite 203
Oakhurst, CA 93644

OAKLAND

Alcoholics Anonymous
East Bay Intergroup Central Office
2910 Telegraph Avenue
Suite 100
Oakland, CA 94609

Allen Temple Haight Ashbury Recovery Center
8500 A Street
Oakland, CA 94621

American Indian Family Healing Center New Dawn Lodge/ White Cloud Lodges
1815 39 Avenue
Suites D and C
Oakland, CA 94601

Asian Community Mental Health Services Recovery For East Bay Asian Youth
310 8 Street
Suite 201
Oakland, CA 94607

Assemblies of God East Bay Rehabilitation
543 8 Street
Oakland, CA 94607

Bi Bett Corporation
East Oakland Recovery Center
7227 East 14 Street
Oakland, CA 94621

Orchid Women's Recovery Center
1342 East 27 Street
Oakland, CA 94606

Catholic Counseling Services
433 Jefferson Street
Oakland, CA 94607

Chemical Dependency
3014 Lake Shore Avenue
Oakland, CA 94610

East Office
6226 Camden Street
Oakland, CA 94605

Chemical Dependency Recovery Hospital Thunder Road
390 40 Street
Oakland, CA 94609

East Bay Community Recovery Project
128 East 14 Street
Oakland, CA 94606

Eastlake YMCA Center Youth Services Program
1612 45 Avenue
Oakland, CA 94601

Family Service of the East Bay
401 Grand Avenue
Suite 200
Oakland, CA 94610

Growth Reorientation Opportunities Unlimited Project Women/Harrison
2941 Harrison Street
Oakland, CA 94611

Highland General Hospital Healthy Infant Program
1411 East 31 Street
Oakland, CA 94602

Horizon Services Chrysalis
3845 Telegraph Avenue
Oakland, CA 94609

Mandana House Community Recovery Center
541 Mandana Boulevard
Oakland, CA 94610

Merritt Peralta Institute Treatment Services
435 Hawthorne Avenue
Oakland, CA 94609

Narcotic Education League
El Chante Alcoholism Recovery Home
425 Vernon Street
Oakland, CA 94610

Centro de Juventud
3209 Galindo Street
Oakland, CA 94601

Drop-in Center/Mujeres Con Esperanza
3315 East 14 Street
Oakland, CA 94601

Minorities Alcohol Treatment Alternatives
1315 Fruitvale Avenue
Oakland, CA 94601

Naval Hospital/Oakland Alcohol Rehab Dept./Level III Residential Treatment
8750 Mountain Boulevard
Building 70-A
Oakland, CA 94627

Oakland Community Counseling Center
2647 East 14 Street
Suite 420
Oakland, CA 94601

Occupational Health Service Drinking Driver Program
340 Pendleton Way
Suite B 129
Oakland, CA 94621

The Solid Foundation
Keller House
353 Athol Avenue
Oakland, CA 94606

Mandela House
3723 Hillview Street
Oakland, CA 94602

West Oakland Health Center
First Step
1531 Jefferson Street
Oakland, CA 94607

Herzog Short Term Recovery Home
6025 Herzog Street
Oakland, CA 94608

Long Term Alcohol Recovery Facility
618 14 Street
Oakland, CA 94612

Methadone Program
614 14 Street
Oakland, CA 94612

Trouble House
3212 San Pablo Avenue
Oakland, CA 94608

14th Street Clinic and Medical Group
1124 East 14 Street
Oakland, CA 94703

OCEANSIDE

McAlister Institute for Treatment and Education (MITE)
North Coastal
514 North Hill Street
Oceanside, CA 92054

North Coastal Detox
4010 Via Serra Street
Oceanside, CA 92056

Mental Health Systems/North Coast Neighborhood Recovery Center
560 Greenbrier Drive
Suite E
Oceanside, CA 92054

OJAI

Hope for Kids
Ojai, CA 93024

OLIVEHURST

House of Hope
1969 17 Street
Olivehurst, CA 95961

ONTARIO

Bilingual Family Counseling Service Center for Recovery
317 West F Street
Ontario, CA 91762

Community Health Projects/ Ontario
324 North Laurel Avenue
Ontario, CA 91762

Counseling Associates
Marin House
1636 North Marin Avenue
Ontario, CA 91764

Orange House/Men's Facility
1003 North Orange Avenue
Ontario, CA 91764

Ephpheta Counseling Center
1003 North Begonia Avenue
Ontario, CA 91762

Reid and Associates Counseling Center
125 West F Street
Suite 108
Ontario, CA 91762

Valley Improvement Programs
210 West B Street
Ontario, CA 91762

ORANGE

City of Orange Police Department Crisis Intervention Unit
1107 North Batavia Street
Orange, CA 92667

Mariposa Women's Center
812 Town and Country Road
Orange, CA 92668

Orange County Community Hospital of Orange Focus on Recovery
401 Tustin Avenue
Orange, CA 92666

Touchstones Social Model Recovery Systems
525 North Parker Street
Orange, CA 92668

ORANGEVALE

Breining Institute
8880 Greenback Lane
Orangevale, CA 95662

Gullhaven
7539 Telegraph Avenue
Orangevale, CA 95662

New Dawn
6043 Roloff Way
Orangevale, CA 95662

Starting Point Behavioral Health Center
8773 Oak Avenue
Orangevale, CA 95662

ORLAND

Glenn County Health Services/ Orland Substance Abuse Department
1187 East South Street
Orland, CA 95963

OXNARD

Alternative Action Programs
2035 South Saviers Road
Suite 5
Oxnard, CA 93033

Community Health Projects
South
D Street Unit
620 South D Street
Oxnard, CA 93030

Ventura
Saviers Road Unit
2055 Saviers Road
Suite 10
Oxnard, CA 93033

Family Counseling Services of Ventura County Alcohol Counseling Center
500 Esplanade Drive
Suite 1200
Oxnard, CA 93030

Miracle House/Phase B
92 South Anacapa Street
Oxnard, CA 93001

Primary Purpose
Detox
603 West 5 Street
Oxnard, CA 93030

Into Action
161 South C Street
Oxnard, CA 93030

Treatment Programs
840 West 5 Street
Oxnard, CA 93030

Rainbow Recovery Centers I
1826 East Channel Island Boulevard
Oxnard, CA 93030

Shamrock House
1334 East Channel Islands Boulevard
Oxnard, CA 93033

Ventura County Alcohol and Drug Programs
Dual Diagnosis Team
1400 Vanguard Road
Oxnard, CA 93033

Oxnard Center
2651 South C Street
Suite 1
Oxnard, CA 93033

Reaching Out to Moms and Kids
2651 South C Street
Room 2
Oxnard, CA 93033

Ventura County Hispanic Commission on Alcohol and Drug Services/Casa Latina
1231 Juneberry Place
Oxnard, CA 93030

PACIFIC GROVE

Beacon House
468 Pine Avenue
Pacific Grove, CA 93950

PACIFICA

Pacific Youth Services Bureau Proyecto Familia
160-A Milagra Drive
Pacifica, CA 94044

Pyramid Alternatives
480 Manor Plaza
Pacifica, CA 94044

PACOIMA

Via Avanta Program
11643 Glenoaks Boulevard
Pacoima, CA 91331

PALM DESERT

Coachella Valley Counseling Palm Desert
73–710 Fred Waring Street
Suite 102
Palm Desert, CA 92260

The Valley Partnership
73–301 Highway 111
Palm Desert, CA 92260

PALM SPRINGS

Alert Program
730 Eugene Road
Palm Springs, CA 92264

Awareness Program
611 South Palm Canyon Road
Suite 204
Palm Springs, CA 92264

Desert Aids Project
750 South Vella Road
Palm Springs, CA 92264

Michael Alan Rosen Foundation Michaels House
430 South Cahuilla Street
Palm Springs, CA 92262

PALMDALE

Antelope Valley Council on Alcoholism and Drug Dependency
1543-F East Palmdale Boulevard
Palmdale, CA 93550

Palmdale Hospital Medical Center Dual Diagnosis
1212 East Avenue S
Palmdale, CA 93550

PALO ALTO

Family Service/Mid Peninsula
375 Cambridge Avenue
Palo Alto, CA 94306

North County Center
270 Grant Avenue
Room 150
Palo Alto, CA 94306

PALOS VERDES PENINSULA

Community Helpline
2161 Via Olivera Street
Palos Verdes Peninsula, CA 90274

PASADENA

Addiction Counseling Treatment (ACT) Family Treatment Program
33 South Catalina Avenue
Suites 103 and 105
Pasadena, CA 91104

Bishop Gooden Home
191 North El Molino Avenue
Pasadena, CA 91101

Casa de las Amigas
160 North El Molino Avenue
Pasadena, CA 91101

City of Pasadena Alcohol and Drug Dependencies Program
1020 North Fair Oaks
Pasadena, CA 91103

Community Health Projects/ Pasadena
1724 East Washington Boulevard
Pasadena, CA 91101

Grandview Foundation
1230 North Marengo Avenue
Pasadena, CA 91103

225 Grandview Street
Pasadena, CA 91104

Haven House
Pasadena, CA 91115

Pasadena Council on Alcoholism and Drug Depedency/Referral Agency
131 North El Molino Avenue
Suite 320
Pasadena, CA 91101

Principles Impact Drug/Alcohol Treatment Center
1680 North Fair Oaks
Pasadena, CA 91103

Saint Luke Medical Center Share Unit
2632 East Washington Boulevard
Pasadena, CA 91107

Starting Over Straight
35 North Craig Avenue
Pasadena, CA 91107

The High Road Program
700 South Arroyo Parkway
Pasadena, CA 91105

Walter Hoving Home
218 South Madison Avenue
Pasadena, CA 91101

PERRIS

National Council on Alcoholism and Drug Dependence of Inland Counties
277 East 4 Street
Perris, CA 92370

PETALUMA

Family Education Centers
629 East D Street
Petaluma, CA 94952

Petaluma People Services Center
1500 A Petaluma Boulevard South
Petaluma, CA 94952

PICO RIVERA

Community Service Organization (CSO) Pico Rivera
6505 Rosemead Boulevard
Suite 301
Pico Rivera, CA 90660

Cornerstone Health Services Outpatient Methadone Clinic
8207 Whittier Boulevard
Pico Rivera, CA 90660

PINOLE

Catholic Counseling Services
2095 San Pablo Street
Pinole, CA 94564

Doctors Hospital of Pinole New Beginnings Program
2151 Appian Way
Pinole, CA 94564

Tri-Cities Discovery Center
2586 Appian Way
Pinole, CA 94564

PIRU

Piru Hotel Sober Living House
550 Temescal Street
Piru, CA 93040

PITTSBURG

Bay Area Addiction Research/ Treatment Pittsburg
45 Civic Avenue
Room 128
Pittsburg, CA 94565

Born Free/Pittsburg
550 School Street
Pittsburg, CA 94565

Catholic Counseling Services
455 West 4 Street
Pittsburg, CA 94565

East County Boys' and Girls' Club
Drug and Alcohol Prevention Program
335 East Leland Road
Pittsburg, CA 94565

El Pueblo Neighborhood Program
85 Treatro Avenue
Suite 283
Pittsburg, CA 94565

East County Community Detox Center, Inc.
500 School Street
Pittsburg, CA 94565

Options for Recovery/Ujima East Intensive Day Program
369 East Leland Road
Pittsburg, CA 94565

PLACERVILLE

El Dorado Council on Alcoholism (EDCA) Lifeskills
2810 Coloma Road
Placerville, CA 95667

New Morning Youth and Family Services
6765 Green Valley Road
Placerville, CA 95667

Progress House Outpatient Program
2914 B Cold Springs Road
Placerville, CA 95667

PLEASANT HILL

Alcohol and Drug Abuse Council of Contra Costa, Inc.
171 Mayhew Way
Suite 210
Pleasant Hill, CA 94523

Bi Bett Corporation First Offender Level One
3478 Buskirk Avenue
Suite 1032
Pleasant Hill, CA 94523

Drake House
808 Grayson Road
Pleasant Hill, CA 94523

PLEASANTON

Federal Correctional Institution Pleasanton
5701 8 Street
Camp Parks
Pleasanton, CA 94568

Valley Care Recovery Center
5720 Stoneridge Mall Road
Suite 300
Pleasanton, CA 94588

POINT REYES STATION

Marin County Community Outreach Project
40 4 Street
Point Reyes Station, CA 94956

POMONA

American Hospital Substance Abuse Services
2180 West Valley Boulevard
Pomona, CA 91768

Community Health Projects/Los Angeles
Garey Clinic
1050 North Garey Avenue
Pomona, CA 91767

Pomona Unit
152 West Artesia Street
Pomona, CA 91768

Counseling Associates
435 West Mission Boulevard
Pomona, CA 91766

Laredo House
787 Laredo Street
Pomona, CA 91768

Park House
720 North Park Avenue
Pomona, CA 91767

End of the Trail Transition House
Pomona, CA 91769

Pomona Community Crisis Center
163 West 21 Street
Pomona, CA 91766

Prototypes Women's Center
845 East Arrow Highway
Pomona, CA 91767

Rehabilitation Alcohol Program (RAP)
2055 North Gary Avenue
Suite 2
Pomona, CA 91767

Serenity House Outpatient Chemical Dependency Program
337 East Foothill Boulevard
Pomona, CA 91767

Whitney Group Alameda House
1530 Alameda Street
Pomona, CA 91768

PORT HUENEME

Steps Program Anacapa by the Sea
224 East Clara Street
Port Hueneme, CA 93044

PORTERVILLE

Paar Center Alcoholism Services
Porterville Halfway House
232 West Belleview Avenue
Porterville, CA 93257

Paar Center Drug Abuse Services
237 West Belleview Avenue
Porterville, CA 93257

The Paar East
114 North Fig Street
Porterville, CA 93257

POTTER VALLEY

North Coast Opportunities Windo
Potter Valley, CA 95469

POWAY

Pomerado Hospital Sunrise Center Chemical Dependency Services
15615 Pomerado Road
Poway, CA 92064

QUINCY

Alcoholics Anonymous/Narcotics Anonymous Adult Children of Alcoholics
Quincy, CA 95971

Plumas County Alcohol and Drug Dept.
Courthouse Annex and Highway 70
Quincy, CA 95971

RAMONA

Group Conscience/Pemarro
1482 Kings Villa Road
Ramona, CA 92065

Mental Health Systems North Rural Recovery Center
323 Hunter Street
Ramona, CA 92065

RANCHO CUCAMONGA

Matrix Institute on Addictions
9375 Archibald Avenue
Suite 204
Rancho Cucamonga, CA 91730

Valley Improvement Programs
8540 Archibald Avenue
Building 18 Suite A
Rancho Cucamonga, CA 91730

RANCHO MIRAGE

Betty Ford Center at Eisenhower
39000 Bob Hope Drive
Rancho Mirage, CA 92270

RED BLUFF

Tehama County Mental Health Alcohol and Drug Abuse Services
1860 Walnut Street
Red Bluff, CA 96080

REDDING

Alcoholics Anonymous Intergroup
1050 State Street
Redding, CA 96001

Care Schools Drinking Driver Program
1445 Market Street
Redding, CA 96001

Empire Recovery Center
1237 California Street
Redding, CA 96001

North Star Transitional Living Center
3715 Capricorn Street
Redding, CA 96001

Redding Specialty Center New Beginnings
2801 Eureka Way
Redding, CA 96001

Shasta County Perinatal Program
1326 Trinity Street
Redding, CA 96001

Shasta County Substance Abuse Program
1525 Pine Street
Suite 1
Redding, CA 96001

Shasta County Youth Services
Outpatient Drug Free
1317 Court Street
Redding, CA 96001

Shasta Options
2530 Larkspur Lane
Redding, CA 96002

Shasta Sierra Work Furlough Program
1727 South Street
Redding, CA 96001

REDLANDS

Jackson/Bibby Awareness Group, Inc.
1200 Arizona Street
Suite B-10
Redlands, CA 92374

Redlands Northside Counseling Services
320 West Union Avenue
Redlands, CA 92374

Redlands Yucaipa Guidance Clinic Redlands Open Center
604 East State Street
Redlands, CA 92374

REDWOOD CITY

Archway/Redwood City Youth and Family Assistance
609 Price Avenue
Suite 205
Redwood City, CA 94063

Daytop Village Adolescent
631 Woodside Road
Redwood City, CA 94061

Professional Treatment Redwood City Treatment Clinic
500 Arguello Street
Redwood City, CA 94063

Redwood Center Chemical Awareness and Treatment Services
100 Edmonds Road
Redwood City, CA 94062

Service League of San Mateo County Hope House
3789 Hoover Street
Redwood City, CA 94063

RESEDA

ASAP Family Outpatient Center Adolescent Substance Abuse Program
18528 Galt Street
Reseda, CA 91335

Fully Alive Center
18554 Sherman Way
Reseda, CA 91335

I-ADARP
18210 Sherman Way
Suite 215
Reseda, CA 91335

Safety Education Center
18309A Sherman Way
Reseda, CA 91335

Western Pacific Reseda Program Outpatient Detox and Methadone Maintenance
18437 Saticoy Street
Reseda, CA 91335

RICHMOND

Alcohol Intervention/Recovery Services Post Conviction Drinking Driver Program
3043 Research Drive
Suite 100
Richmond, CA 94804

Bay Area Addiction Research/ Treatment/Richmond
2910 Cutting Boulevard
Richmond, CA 94804

Born Free/Richmond
100 37 Street
Room 1608
Richmond, CA 94804

Catholic Counseling Services
225 Civic Center Street
Richmond, CA 94804

Intensive Day Treatment
Ujima West
3939 Bissel Street
Richmond, CA 94805

Kaiser Permanente Medical Center Alcohol and Drug Abuse Program
1300 Potrero Avenue
Richmond, CA 94804

Neighborhood House of North Richmond
Hollomon/Fauerso New Way Center
208 23 Street
Richmond, CA 94804

North Star Treatment Center
1744 4 Street
Richmond, CA 94801

Sojourne Community Counseling Center
3029 Mac Donald Avenue
Richmond, CA 94804

RIDGECREST

Alcohol Drug Alert Program
308 South Norma Street
Ridgecrest, CA 93555

Council on Substance Abuse Awareness
314 South Norma Street
Ridgecrest, CA 93555

Desert Counseling Clinic Ridgecrest Unit
814 North Norma Street
Ridgecrest, CA 93555

RIO VISTA

Rio Vista Care
125 Sacramento Street
Rio Vista, CA 94571

RIVERSIDE

Community Behavioral Health Group, Inc.
5995 Brockton Avenue
Suite B
Riverside, CA 92506

Community Settlement Association Drinking Driver Programs
4366 Bermuda Avenue
Riverside, CA 92507

Inland Health Services
1021 West La Cadena Drive
Riverside, CA 92501

March AFB Social Actions Office Drug and Alcohol Abuse Control Program
22 CSG/SLD
Building 466
Riverside, CA 92518

My Family, Inc.
Recovery Center
17270 Roosevelt Avenue
Riverside, CA 92508

Recovery Women's Program
7223 Magnolia Avenue
Riverside, CA 92504

National Council on Alcoholism and Drug Dependence of the Inland Counties
3767 Elizabeth Street
Riverside, CA 92506

Nelson House Foundation
3685 15 Street
Riverside, CA 92501

Riverside County Alcohol Control Program/Riverside
1970 University Avenue
Riverside, CA 92507

Riverside County Drug Abuse Program
Outpatient Drug Free
3925 Orange Street
Suite 26
Riverside, CA 92501

Western Riverside Methadone Program
3929 Orange Street
Riverside, CA 92501

Teen Challenge Christian Life School Men's Facility
5445 Chicago Avenue
Riverside, CA 92507

Whiteside Manor, Inc.
Alcoholic Recovery Home
2743 Orange Street
Riverside, CA 92501

Sammon House
2709 Orange Street
Riverside, CA 92501

Youth Service Center of Riverside
3847 Terracina Drive
Riverside, CA 92506

ROSEMEAD

ABC Traffic Safety Programs
8623A Garvey Avenue
Rosemead, CA 91770

Asian Youth Center
9032 Mission Road
Rosemead, CA 91770

ROSEVILLE

Sierra Council on Alcohol and Drug Dependency Roseville Service Center
1A Sierragate Plaza
Suite 110
Roseville, CA 95678

Sierra Family Services/Roseville
424 Vernon Street
Roseville, CA 95678

RUNNING SPRINGS

Pine Ridge Treatment Center
2727 Highland Drive
Running Springs, CA 92382

SACRAMENTO

Adult Services Center
4875 Broadway
Sacramento, CA 95820

Alcoholics Anonymous Central California Fellowship
7500 14 Avenue
Suite 27
Sacramento, CA 95820

American Indian Substance Abuse Program, Inc. Turquoise Indian Lodge
2727 P Street
Sacramento, CA 95816

Bi-Valley Medical Clinic
2100 Capitol Avenue
Sacramento, CA 95816

Norwood
310 Harris Avenue
Suite A
Sacramento, CA 95838

Bridge to Change
2848 Arden Way
Suite 103
Sacramento, CA 95825

Chemical Dependency Center for Women
1507 21 Street
Suite 100
Sacramento, CA 95814

Family Maintenance Bureau
3970 Research Drive
Sacramento, CA 95838

Family Service Agency of Greater Sacramento Area
8912 Volunteer Lane
Suite 100
Sacramento, CA 95826

Gateway Foundation, Inc. Gateway Recovery House
4049 Miller Way
Sacramento, CA 95817

Maguire House
2130 22 Street
Sacramento, CA 95818

Mexican American Alcoholism Program
7000 Franklin Boulevard
Suite 210
Sacramento, CA 95823

Mi Casa Recovery Home
2515 48 Avenue
Sacramento, CA 95822

National Council on Alcohol and Drug Dependency/Sacramento County Affiliate
650 Howe Avenue
Suite 1055
Sacramento, CA 95825

River City Recovery Center
E Street Unit
2218 E Street
Sacramento, CA 95816

G Street Unit
2217 G Street
Sacramento, CA 95816

Sacramento Urban Indian Health Leo Camp Alcohol Program
801 Broadway
Sacramento, CA 95818

Sacramento Black Alcoholism Center (SBAC)
2425 Alhambra Boulevard
Suite F
Sacramento, CA 95817

Sacramento Recovery House, Inc.
1914 22 Street
Sacramento, CA 95816

Safety Center Drinking Driver Program
3909 Bradshaw Road
Sacramento, CA 95827

Sutter Outpatient Drug and Alcohol Program
7700 Folsom Boulevard
Sacramento, CA 95826

The Effort
Stimulant Residential Detox
Sacramento, CA 95818

Alternative House
Opiate Detoxification Program
Residential
1550 Juliesse Avenue
Sacramento, CA 95815

Counseling Center
1820 J Street
Sacramento, CA 95814

SALINAS

Alcohol Programs Management
837 South Main Street
Salinas, CA 93901

Alcoholics Anonymous
9 West Gabilan Street
Suite 11
Salinas, CA 93901

Community Human Services Methadone Clinic
1101 F North Main Street
Salinas, CA 93906

Door to Hope
Halfway House
165 Clay Street
Salinas, CA 93901

Sober Living
257 Clay Street
Salinas, CA 93901

Gente Del Sol Community Recovery Center
5 Williams Road
Salinas, CA 93905

Monterey County Health Department Patterns Program
140 West Gabilan Street
Salinas, CA 93901

Sun Street Centers II Residential Recovery Program
8 Sun Street
Salinas, CA 93901

Trucha
727 East Market Street
Salinas, CA 93905

Valley Health Associates Medetrac
622–6 East Alisal Street
Salinas, CA 93905

Victory Outreach Homes Rehabilitation Services of Salinas
325 North Main Street
Salinas, CA 93906

SAN ANDREAS

Calaveras County Alcohol/Drug Abuse Program
891 Mountain Ranch Road
Government Center Department 64–66
San Andreas, CA 95249

SAN BERNARDINO

Casa de Ayuda/San Bernardino
7255 Garden Drive
San Bernardino, CA 92404

7274 Garden Drive
San Bernardino, CA 92404

Casa de San Bernardino
735 North D Street
San Bernardino, CA 92401

Center for Community Counseling and Education (AGAPE)
607 East Highland Avenue
San Bernardino, CA 92404

Hase and Associates Systems
353 West 6 Street
San Bernardino, CA 92401

Industry Community Interface Projects (ICI Projects)
265 East Mill Street
Suite 1
San Bernardino, CA 92408

Inland Behavioral Services (IHS)
1963 North E Street
San Bernardino, CA 92405

Methadone Maintenance
1236 North Waterman Avenue
San Bernardino, CA 92404

Loma Linda Center for Health Promotion Clearview Outpatient Services
1898 Business Center Drive
Suite 101
San Bernardino, CA 92408

Mental Health Systems
Pegasus
1905 North E Street
San Bernardino, CA 92405

Pride
108 North E Street
San Bernardino, CA 92401

New House, Inc.
Men's Program
840 North Arrowhead Avenue
San Bernardino, CA 92401

Women with Children Under Five Years and Pregnant Women
856 North Arrowhead Avenue
San Bernardino, CA 92401

Rolling Start/Choices
536 West 11 Street
Suite B
San Bernardino, CA 92410

San Bernardino Cnty Dept. Public Health Alcohol and Drub Abuse Prevention Program
351 Mountain View Avenue
San Bernardino, CA 92415

San Bernardino County Office of Alcohol and Drug Programs Treatment Services
565 North Mount Vernon Avenue
Suite 200
San Bernardino, CA 92411

Veterans' Alcohol Rehabilitation Program (ARP)
Stoddard Street I
1087 North Stoddard Street
San Bernardino, CA 92410

Belleview House
916 Belleview Avenue
San Bernardino, CA 92410

Gibson House
1100 North D Street
San Bernardino, CA 92410

Gibson Recovery Home for Women
1135 North D Street
San Bernardino, CA 92410

Harris House
907 West Rialto Avenue
San Bernardino, CA 92410

Rialto House
921 Rialto Avenue
San Bernardino, CA 92410

11th Street B/House
349 West 11 Street
San Bernardino, CA 92405

SAN DIEGO

Alcoholics Anonymous
7075 Mission Gorge Road
Suite B
San Diego, CA 92120

Behavioral Health Group Women's Outpatient Recovery Center
1400 6 Avenue
Suite 104
San Diego, CA 92101

Caring Ministries of San Diego
Serenity House 1
4163 Illinois Street
San Diego, CA 92104

Serenity House 2
4060 Illinois Street
San Diego, CA 92104

Serenity House 3
4035 Illinois Street
San Diego, CA 92104

Serenity House 4
4048 Illinois Street
San Diego, CA 92104

Serenity House 5
4370 Oregon Street
San Diego, CA 92104

Serenity House 6
3830 Ray Street
San Diego, CA 92104

Catholic Charities Rachels Women's Center
759 8 Avenue
San Diego, CA 92101

Cobar House
1140 Beverly Street
San Diego, CA 92114

Counseling and Recovery Institute
3233 3 Avenue
San Diego, CA 92103

Crash
Golden Hill House
2410 E Street
San Diego, CA 92102

Options for Recovery Central
5605 El Cajon Boulevard
San Diego, CA 92115

Short Term/East
4890 67 Street
San Diego, CA 92105

Short Term/RDF
4161 Marlborough Avenue
San Diego, CA 92105

Southeast Recovery Program
220 North Euclid Avenue
Suite 120
San Diego, CA 92114

Ikunu African American Program
220 North Euclid Avenue
Suite 130
San Diego, CA 92114

Crossroads Foundation
3594 4 Avenue
San Diego, CA 92103

Deaf Community Services of San Diego Center for Empowerment of Deaf Alcoholics in Recovery
3041 University Avenue
San Diego, CA 92104

Episcopal Community Service (ECS)
One Flight Up
4092 University Avenue
San Diego, CA 92105

Clairemont Neighborhood Recovery Center
7601–7603 Convoy Court
San Diego, CA 92111

Mid-City Drug Abuse Services
4094 University Avenue
San Diego, CA 92105

North City Drug Abuse Services
5560 Ruffin Road
Suite 4
San Diego, CA 92123

Project Para
3785 Fairmount Avenue
San Diego, CA 92105

House of Metamorphosis
Parolee Partnership Program
3010 Elm Street
San Diego, CA 92102

Residential
2970 Market Street
San Diego, CA 92102

Indochinese Mutual Assistance Assoc.
2171 Ulric Street
Suite 100
San Diego, CA 92111

Kaiser Permanente Medical Group Chemical Dependency Recovery Program
328 Maple Street
San Diego, CA 92103

MAAC Project Recovery Home Casa de Milagros
1127 38 Street
San Diego, CA 92113

MCC Managed Behavioral Care of California
9040 Friars Road
Suite 320
San Diego, CA 92108

Mental Health Systems Mid-Coast Regional Recovery Center
1633 Garnet Avenue
San Diego, CA 92109

Mesa Vista Hospital Chemical Dependency Program
7850 Vista Hill Avenue
San Diego, CA 92123

Metropolitan Correctional Center Substance Abuse Services
808 Union Street
San Diego, CA 92101

Narcotics Anonymous
Business Office
4689 Felton Street
San Diego, CA 92116

Naval Air Station
Miramar Counseling and Assistance Center (CAAC)
19913 Mitscher Way
Suite 3
San Diego, CA 92145

North Island Counseling and Assistance Center (CAAC)
North Island
Building 708
San Diego, CA 92135

Naval Submarine Base Point Loma Counseling and Assistance Center (CAAC)
140 Sylvester Road
San Diego, CA 92106

Naval Training Center Counseling and Assistance Center (CAAC)
Building 304
San Diego, CA 92133

Ocean Beach Recovery and Referral
5029 West Point Loma Boulevard
San Diego, CA 92107

Pathfinders of San Diego, Inc.
Recovery Home
2980 Cedar Street
San Diego, CA 92102

Service Center
3049 University Avenue
San Diego, CA 92104

San Diego Health Alliance
West Office
3293 Greyling Drive
San Diego, CA 92123

San Diego State University Center on Substance Abuse/ Drinking Driver Program
7875 Convoy Court
Suite 3
San Diego, CA 92111

San Diego Treatment Services Home Avenue Clinic
3940 Home Avenue
San Diego, CA 92105

San Diego Youth and Community Services
Teen Options
3923 Adams Avenue
San Diego, CA 92116

Teen Recovery Center
3937 Adams Avenue
San Diego, CA 92116

Sharp Cabrillo Hospital Alcohol and Drug Treatment Program
3475 Kenyon Street
San Diego, CA 92110

Stepping Stone of San Diego
Long Term Rehab
3767 Central Avenue
San Diego, CA 92105

Nonresidential
3425 5 Avenue
San Diego, CA 92103

Telesis II of California, Inc.
409 Camino de Rio South
Suite 205
San Diego, CA 92108

The Palavra Tree
1212 South 43 Street
Suite D
San Diego, CA 92113

The Way Back
2516 A Street
San Diego, CA 92102

Tradition One
Men's Facility
4104 Delta Street
San Diego, CA 92113

Women's Facility
3895 Newton Avenue
San Diego, CA 92113

Turning Point Home of San Diego
1315 25 Street
San Diego, CA 92102

Twelve-Step House Heartland House
5855 Streamview Drive
San Diego, CA 92105

U.S. Marine Corps Substance Abuse Control Center
Marine Corps Recruit Depot
Building 8
San Diego, CA 92141

Veterans' Affairs Medical Center Alcohol and Drug Treatment Program
3350 La Jolla Village Drive
116A
San Diego, CA 92161

Vietnam Veterans of San Diego The Landing Zone
4141 Pacific Highway
San Diego, CA 92110

Villa View Community Hospital Alert Unit
5550 University Avenue
San Diego, CA 92105

Vista Pacifica
7989 Linda Vista Road
San Diego, CA 92111

Volunteers of America Alcohol Services Center
Amigos Sobrios at Parrick House
741 11 Avenue
San Diego, CA 92101

Sobriety House/Detoxification
1111 Island Avenue
San Diego, CA 92101

Women's Action Committee on Alcohol Harmony Recovery Center
6150 Mission Gorge Road
Suite 116
San Diego, CA 92120

SAN FERNANDO

Lazarus Foundation/A Consulting Ref. Alcoholism and Drug Addiction Intervention
14850 Mayall Street
San Fernando, CA 91345

Maclay House
13770 Sayre Street
San Fernando, CA 91342

Northeast Valley Health Corporation Alcoholism Program/DUI
1024 North Maclay Street
Suite J
San Fernando, CA 91340

SAN FRANCISCO

Acceptance Place
673 San Jose Avenue
San Francisco, CA 94110

Alcoholics Anonymous
Intergroup Fellowship
1540 Market Street
Suite 150
San Francisco, CA 94122

Oficina Hispana Paso 12
383 Valencia Street
San Francisco, CA 94103

Alcoholics Rehabilitation Association First Step Home
1035 Haight Street
San Francisco, CA 94117

Amethyst Counseling
4104 24 Street
Suite 107
San Francisco, CA 94114

Asian American Residential Recovery Services
2024 Hayes Street
San Francisco, CA 94117

Bakers New Place
1375 Grove Street
San Francisco, CA 94117

Bay Area Addiction Research/ Treatment, Inc.
Embarcadero Clinic
75 Townsend Street
San Francisco, CA 94107

Geary Street Clinic
1040 Geary Street
San Francisco, CA 94109

Bayview Hunters Point Foundation
Alice Griffith Clinic
43 Nichols Street
San Francisco, CA 94124

Bayview Hunters Point Center
1625 Carrol Street
2nd Floor
San Francisco, CA 94124

Youth Services Day Treatment
5033 3 Street
San Francisco, CA 94124

Catholic Charities Jelani House
1601 Quesada Avenue
San Francisco, CA 94124

Chinatown Youth Center
1693 Polk Street
San Francisco, CA 94109

Counseling and Assistance Center (CAAC) Level II Outpatient
636 C Avenue
Treasure Island Naval Station
San Francisco, CA 94130

Counseling Services for Drinking Drivers
43 Fell Street
San Francisco, CA 94102

Driver Performance Institutes
330 Townsend Street
Suite 106
San Francisco, CA 94107

Forensic Health Care
110 Gough Street
San Francisco, CA 94102

Fort Help Methadone Program
495 3 Street
San Francisco, CA 94107

Freedom from Alcohol and Drugs
1353 48 Avenue
San Francisco, CA 94122

Friendship House Association of American Indians
80 Julian Avenue
San Francisco, CA 94103

Futures in Recovery
3601 Taraval Street
San Francisco, CA 94116

Golden Gate for Seniors
637 South Van Ness Avenue
San Francisco, CA 94110

Haight Ashbury Free Clinics
Alcohol Treatment Services
1698 Haight Street
San Francisco, CA 94117

Bill Pone Memorial Unit
1779 Haight Street
San Francisco, CA 94117

Black Extended Family
330 Ellis Street
San Francisco, CA 94102

Drug Detoxification Project
529 Clayton Street
San Francisco, CA 94117

Moving Addicted Mother Ahead (MAMA)
1696 Haight Street
San Francisco, CA 94117

Smith House
766 Stanyan Street
San Francisco, CA 94117

Harriet Street Center
444 6 Street
San Francisco, CA 94103

Henry Ohlhoff Outpatient Programs
2418 Clement Street
San Francisco, CA 94121

Horizons Unlimited of San Francisco Juventud Program
440 Potrero Avenue
San Francisco, CA 94110

Iadet First Offender Program
2141-C Mission Street
San Francisco, CA 94110

Intensive Substance Abuse Service
1153 Guerrero Street
San Francisco, CA 94110

Kaiser Permanente Hospital Department of Psychiatry
2350 Geary Boulevard
San Francisco, CA 94115

Liberation House Programs
1724 Steiner Street
San Francisco, CA 94115

Mission Council on Alcohol Abuse for the Spanish Speaking
820 Valencia Street
San Francisco, CA 94110

Mobile Assistance Patrol (MAP)
43 Fell Street
San Francisco, CA 94102

National Council on Alcoholism and Other Drug Addictions/ Bay Area
944 Market Street
3rd Floor
San Francisco, CA 94102

North of Market
Senior Alcohol Program
333 Turk Street
San Francisco, CA 94102

Senior Services/Senior Sobriety Center
291 Eddy Street
San Francisco, CA 94102

Operation Recovery Operation Concern
1853 Market Street
San Francisco, CA 94103

Presidio Counseling Center Alcohol and Drug Abuse Branch
Corner of Mauldin and MacDonald Sts.
Presidio/San Francisco Building 910
San Francisco, CA 94129

Saint Anthony's Foundation Covenant House
818 Steiner Street
San Francisco, CA 94117

Saint Mary's Hospital Chemical Dependency Recovery Center
450 Stanyan Street
San Francisco, CA 94117

Saint Vincent De Paul Society of San Francisco
Arlington Hotel
480 Ellis Street
San Francisco, CA 94102

Ozanam Reception Center
1175 Howard Street
San Francisco, CA 94103

Salvation Army Harborlight Center
Alcohol Unit
1255 Harrison Street
San Francisco, CA 94103

Detox/Primary/Recovery
1275 Harrison Street
San Francisco, CA 94103

Salvation Army Turk Street Corps The Bridgeway Project/ San Francisco
242 Turk Street
Suite 210
San Francisco, CA 94102

San Francisco General Hospital
Stimulant Treatment Outpatient Program
1001 Potrero Avenue
Building 1 Room 203
San Francisco, CA 94110

Outpatient Detox Clinic
Substance Abuse Services/ Methadone Maintenance
1001 Potrero Avenue
Building 90 Ward 93
San Francisco, CA 94110

Tom Smith Substance Abuse Treatment Center
1001 Potrero Avenue
Building 20 3rd Floor
San Francisco, CA 94110

San Francisco Pretrial Diversion Project Substance Abuse Referral Unit (SARU)
885 Bryant Street
2nd Floor
San Francisco, CA 94103

San Francisco Professional Clinic
1 Daniel Burnham Court
Suite 350-C
San Francisco, CA 94109

San Francisco Suicide Prevention Drugline/Relapse Prevention Line
3940 Geary Boulevard
San Francisco, CA 94118

San Francisco Women's Rehab Foundation The Stepping Stone
255 10 Avenue
San Francisco, CA 94118

Swords to Plowshares Veterans Rights Organization
400 Valencia Street
San Francisco, CA 94103

Twelve-Step Programs
4037 Judah Street
San Francisco, CA 94122

Walden House
890 Hayes Street
San Francisco, CA 94117

Adult Residential
815 Buena Vista West
San Francisco, CA 94117

Adult Residential Detox
1840 Van Ness Avenue
San Francisco, CA 94109

Walden Multi-Service Center
1885 Mission Street
San Francisco, CA 94110

Adolescent Program
214 Haight Street
San Francisco, CA 94102

Westside Community Mental Health Center
Westside Methadone Treatment Program
1301 Pierce Street
San Francisco, CA 94115

Westside Tenderloin Detox Program
183 Golden Gate Avenue
San Francisco, CA 94102

Westside Youth Awareness Program
1140 Oak Street
San Francisco, CA 94117

Inner City Outpatient Services
1049 Market Street
Mezzanine Level
San Francisco, CA 94103

Women's Alcoholism Center
Aviva House Recovery Home
1724 Bryant Street
San Francisco, CA 94110

Florette Pomeroy House
2263 Bryant Street
San Francisco, CA 94110

Lee Woodward Counseling Center
2201 Sutter Street
San Francisco, CA 95115

Women's Institute for Mental Health Iris Center
333 Valencia Street
Suite 222
San Francisco, CA 94103

18th Street Services
217 Church Street
San Francisco, CA 94114

SAN GABRIEL

Family Counseling Services
314 East Mission Drive
San Gabriel, CA 91776

San Gabriel Valley Driver Improvement
121 South Del Mar Avenue
Suite D
San Gabriel, CA 91776

SAN JACINTO

Ahmium Education
166 East Main Street
Suite 2
San Jacinto, CA 92586

La Vista Women's Alcoholic Recovery Center
2220 Girard Street
San Jacinto, CA 92581

Riverside County Alcohol Program
950 Ramona Boulevard
Suites 1 and 2
San Jacinto, CA 92583

SAN JOSE

Alcoholics Anonymous Intergroup Santa Clara County Central Office
535 Race Street
Suite 130
San Jose, CA 95126

Alexian Family Psychology and Counseling Associates
3110 Provo Court
Suite A
San Jose, CA 95127

Alum Rock Counseling Center Substance Abuse Program
1935 Clarice Drive
San Jose, CA 95122

ARH Recovery Homes Benny McKeown Center
1281 Fleming Avenue
San Jose, CA 95127

Asian Americans for Community Involvement
232 East Gish Road
Suite 200
San Jose, CA 95112

Comadre's Womens Polydrug Abuse Program
55 East Empire Street
San Jose, CA 95112

Combined Addicts and Professional Services (CAPS)
Residential Unit
398 South 12 Street
San Jose, CA 95112

Women's Relapse Program
135 East Gish Road
Suite 200
San Jose, CA 95112

Downtown Fellowship Alcohol and Drug Recovery Home
561 South Almaden Avenue
San Jose, CA 95110

Economic Social Opportunities/ Steps
1445–1447 Oakland Road
San Jose, CA 95112

Fortunes Inn Alcoholic Recovery Center
52 South 12 Street
San Jose, CA 95112

Four Winds Lodge Primary Alcoholism Treatment Program
919 The Alameda
San Jose, CA 95126

Horizon Services Horizon South
650 South Bascom Avenue
San Jose, CA 95128

Justice Services Outpatient Substance Abuse Services
614 Tully Road
San Jose, CA 95111

Kids Are Special
535 Race Street
Suite 190
San Jose, CA 95126

Mariposa Lodge
9500 Malech Road
San Jose, CA 95151

Muriel Wright Ranch
298 Bernal Road
San Jose, CA 95119

National Council on Alcohol and Drug Dependency Santa Clara County Affiliate
1922 The Alameda
Suite 212
San Jose, CA 95126

O'Connor Hospital Recovery Center
2105 Forest Avenue
San Jose, CA 95128

Pacific Center of San Jose Medical Center/Triad
675 East Santa Clara Street
San Jose, CA 95112

Pate House Recovery Home
35 South 12 Street
San Jose, CA 95112

Pathway
Day Treatment
1088 Broadway Avenue
San Jose, CA 95125

Pathway House
102 South 11 Street
San Jose, CA 95112

Proyecto Primavera Outpatient Alcohol Services
614 Tully Road
San Jose, CA 95111

Salvation Army Adult Rehabilitation Center
760 West Taylor Street
San Jose, CA 95126

San Jose Medical Center Triad Community Services
675 East Santa Clara Street
Suite 1577
San Jose, CA 95112

Santa Clara County Central Center
976 Lenzen Avenue
San Jose, CA 95126

Santa Clara County Bureau of Alcohol/Drug Programs Moorpark Methadone Clinic
2220 Moorpark Avenue
Building H-11
San Jose, CA 95128

Santa Clara County Methadone Treatment Program Hillview Clinic/Stride Program
1675 Burdette Drive
Suites A and B
San Jose, CA 95121

Social Advocates for Youth Alameda School
11 Cleaves Street
San Jose, CA 95126

Twelfth-Step House
619 North 4 Street
San Jose, CA 95112

Victory Outreach Men's Home
1215 Karl Street
San Jose, CA 95122

Vida Nueva Alcohol/Drug Recovery Home
2212 Quimby Road
San Jose, CA 95122

Volunteers of America/Bay Area
Odyssey Recovery Home
390 West Court
San Jose, CA 95116

Sullivan Recovery Home
2345 Mather Drive
San Jose, CA 95116

Willow Home
808 Palm Street
San Jose, CA 95110

SAN JUAN CAPISTRANO

Academy of Defensive Driving
31726 Rancho Viejo Road
Suite 120
San Juan Capistrano, CA 92675

Community Counseling Center Neighborhood Recovery Center
27514 Calle Arroyo Street
San Juan Capistrano, CA 92675

SAN LEANDRO

Family Service of the East Bay
2208 San Leandro Boulevard
San Leandro, CA 94577

Horizon Services
Horizon Community Center
1403 164 Avenue
San Leandro, CA 94578

Horizon House
1430 168 Avenue
San Leandro, CA 94578

Humanistic Alternatives to Addiction Research and Treatment, Inc. (HAART)
15400 Foothill Boulevard
San Leandro, CA 94578

San Leandro Community Counseling
296 Broadmoor Boulevard
San Leandro, CA 94577

SAN LUIS OBISPO

San Luis Obispo County
Drug and Alcohol Services
994 Mill Street
Suite 201
San Luis Obispo, CA 93401

Drug and Alcohol Treatment Services
1102A Laurel Lane
San Luis Obispo, CA 93408

SAN MARCOS

Escondido Youth Encounter (EYE) Options Recovery North
340 Rancheros Drive
Suite 103
San Marcos, CA 92069

Occupational Health Services
1637 Capalina Road
San Marcos, CA 92069

San Diego Health Alliance North Office
1560 Capalina Street
Suite A
San Marcos, CA 92069

SAN MARTIN

Santa Clara Bureau of Alcohol and Drug Programs/South County Clinic
80 West Highland Avenue
San Martin, CA 95046

SAN MATEO

Alcoholics Anonymous
1941 Ofarrell Street
Suite 7
San Mateo, CA 94403

Archway/San Mateo Youth and Family Assistance
2121 South El Camino Real
Suite 301
San Mateo, CA 94403

Interagency Perinatal Substance Abuse Team
3080 La Selva Street
San Mateo, CA 94403

Occupational Health Services San Mateo County Programs
475 Concar Drive
Suite 206
San Mateo, CA 94402

Palm Avenue Detoxification
2251 Palm Avenue
San Mateo, CA 94403

Project Ninety
15 9 Avenue
San Mateo, CA 94401

Solidarity Family Center
1885 South Norfolk Avenue
San Mateo, CA 94403

SAN PABLO

Bridge Over Troubled Waters
2302 Del Monte Avenue
San Pablo, CA 94806

Family and Community Services of Contra Costa
2523 El Portal Drive
Suite 201
San Pablo, CA 94806

San Pablo Discovery Center
2523 El Portal Drive
Suite 102
San Pablo, CA 94806

Tri County Women's Recovery Services The Rectory
1901 Church Lane
San Pablo, CA 94806

SAN PEDRO

Abstinent Living Centers
Men and Women's Programs
305 West 14 Street
San Pedro, CA 90731

West Sepulveda Boulevard Services
1445 West Sepulveda Boulevard
San Pedro, CA 90731

Women's Sober Living
270 West 14 Street
San Pedro, CA 90731

19th Street Services
856 19 Street
San Pedro, CA 90731

Beacon House Association of San Pedro
1003 South Beacon Street
San Pedro, CA 90731

Fred Brown's Recovery Services
14th Street Services
349 West 14 Street
San Pedro, CA 90731

House of Hope Foundation
235 West 9 Street
San Pedro, CA 90731

Joint Efforts, Inc. Outpatient Services
505 South Pacific Avenue
Suite 205
San Pedro, CA 90731

South Bay Recovery, Inc. Serenity House for Women
563 North Leland Street
San Pedro, CA 90732

SAN RAFAEL

Bay Area Institute for Family Therapy
2400 Las Gallinas Street
Suite 260
San Rafael, CA 94903

Center Point, Inc.
Lifestart Perinatal Services
805 D Street
San Rafael, CA 94901

Detoxification
1634 5 Avenue
San Rafael, CA 94901

Intensive Outpatient
86C Belvedere Street
San Rafael, CA 94901

Reilly House
812 D Street
San Rafael, CA 94901

The Manor
603 D Street
San Rafael, CA 94901

Family Service Agency of Marin
1005 A Street
Suite 307
San Rafael, CA 94901

Henry Ohlhoff Outpatient Programs Marin County Office
121 Knight Drive
San Rafael, CA 94901

Kaiser Permanente Medical Group Alcohol and Drug Abuse Program (ADAP)
820 Las Gallinas Avenue
San Rafael, CA 94903

Marin Community Resource
1623 B 5 Avenue
San Rafael, CA 94901

Marin County Drinking Driver Program
118 Alto Street
San Rafael, CA 94901

Marin Grove
34–42 Grove Street
San Rafael, CA 94901

Marin Services for Men
424 Mission Avenue
San Rafael, CA 94901

Marin Treatment Center Outpatient Services
1466 Lincoln Avenue
San Rafael, CA 94901

Personal Support Group
55 Mitchell Road
Suite 22
San Rafael, CA 94945

Ross Hospital Chemical Dependency Services
1150 Sir Francis Drake Boulevard
San Rafael, CA 94904

Sober Classroom of Marin County Office of Education
160 B North San Pedro Road
San Rafael, CA 94903

SAN RAMON

John Muir Medical Center Adolescent Treatment Center
2501 Deerwood Drive
San Ramon, CA 94583

San Ramon Regional Medical Center New Beginnings
6001 Norris Canyon Road
San Ramon, CA 94583

SANTA ANA

Alcoholics Anonymous Central Office Orange County
2712 South Grand Avenue
Santa Ana, CA 92705

Asian American Substance Abuse Treatment Services
412 South Lyon Street
Suite 103
Santa Ana, CA 92701

Breakaway Health Corporation
20271 SW Birch Street
Suite 201
Santa Ana, CA 92707

California Treatment Services Third Street Clinic
717 East 3 Street
Santa Ana, CA 92701

Cooper Fellowship Men's Services
417 North Cooper Street
Santa Ana, CA 92703

Hispanic Alcoholism Services Center of Orange County
1717 North Broadway Street
Santa Ana, CA 92706

Industry Community Interface Projects (ICI Projects)
1638 17th Street
Santa Ana, CA 92701

MCC Managed Behavioral Care of California
1551 North Tustin Avenue
Suite 600
Santa Ana, CA 92701

Narcotics Anonymous Helpline Orange County
217 North Main Street
In the Basement Suite LL-10
Santa Ana, CA 92705

National Traffic Safety Institute
1901 East 4 Street
Suite 311
Santa Ana, CA 92705

Orange County Alcohol Program East Central Alcohol Services
1200 North Main Street
Suite 100-B
Santa Ana, CA 92701

Orange County Health Care Agency Perinatal Treatment Program/Drug Abuse Services
1200 North Main Street
Suite 630
Santa Ana, CA 92706

Santa Ana Drug Abuse Services
1725 West 17 Street
Room 101-B/146-B
Santa Ana, CA 92706

Webster Health Care/La Vida
1517 North La Bonita Street
Santa Ana, CA 92703

Phoenix House Orange County Adult and Adolescent Programs
1207 East Fruit Street
Santa Ana, CA 92701

School Ten Main Street Unit
1772 South Main Street
Santa Ana, CA 92707

Straight Talk Gerry House
1225–1227 West 6 Street
Santa Ana, CA 92703

Teen Challenge of Orange County
418 South Main Street
Santa Ana, CA 92702

The Villa Orange County Women's Alcohol Rehab Center
910 North French Street
Santa Ana, CA 92701

SANTA BARBARA

Casa Serena
1515 Bath Street
Santa Barbara, CA 93101

Community Health Projects
217 Camino Del Remedio Street
Santa Barbara, CA 93110

Drug Abuse Preventive Society of Santa Barbara
24 West Arrellaga Street
Santa Barbara, CA 93101

Gay and Lesbian Resource Center Counseling and Recovery Services
126 East Haley Street
Suite A-17
Santa Barbara, CA 93101

Isla Vista Health Project/Casa Rosa
1850 North Jameson Lane
Santa Barbara, CA 93108

Klein Bottle Counseling Center Youth Programs
401 North Milpas Street
Santa Barbara, CA 93103

Santa Barbara Council on Alcohol and Drug Abuse/Project Recovery
133 East Haley Street
Santa Barbara, CA 93102

Santa Barbara New House I
509 Chapala Street
Santa Barbara, CA 93101

Santa Barbara New House II
227 West Haley Street
Santa Barbara, CA 93101

Santa Barbara New House III
2434 Bath Street
Santa Barbara, CA 93105

Santa Barbara Rescue Mission and Bethel House
535 Eats Yanonali Street
Santa Barbara, CA 93102

Victory Outreach Rehabilitation Home
822 West Islay Street
Santa Barbara, CA 93101

Villa Esperanza Counseling and Education Center
4500 Hollister Avenue
Santa Barbara, CA 93110

Zona Seca
Alcohol/Drug Abuse Counseling Agency
119 North Milpas Street
Santa Barbara, CA 93103

Jail II Program
4434 Calle Real Street
Santa Barbara, CA 93110

SANTA CLARA

Kaiser Permanente Medical Group Alcohol and Drug Treatment Program
1333 Lawrence Expressway
Suite 350
Santa Clara, CA 95051

Pathway Society/Basin
1659 Scott Boulevard
Suite 32
Santa Clara, CA 95050

SANTA CRUZ

Alcoholics Anonymous Santa Cruz County Intergroup
1509 Seabright Avenue
Suite C-1
Santa Cruz, CA 95062

Janus of Santa Cruz
200 7 Avenue
Suite 150
Santa Cruz, CA 95062

Perinatal Program
1314 Ocean Street
Santa Cruz, CA 95060

Santa Cruz Community Counseling Center
Alto Counseling Center/North
271 Water Street
Santa Cruz, CA 95060

Sunflower House
125 Rigg Street
Santa Cruz, CA 95060

Youth Services North County
709 Mission Street
Santa Cruz, CA 95060

Triad Santa Cruz Clinic Outpatient Methadone Maintenance
1000-A Emeline Avenue
Santa Cruz, CA 95060

Women's Crisis Support
1025 Center Street
Santa Cruz, CA 95060

SANTA FE SPRINGS

Los Angeles Centers for Alcohol/ Drug Abuse (LACADA) Santa Fe Springs
11769 Telegraph Road
Santa Fe Springs, CA 90670

SANTA MARIA

Camino Segundo Counseling and Educational Center
2121 South Center Pointe Parkway
Santa Maria, CA 93455

Central Coast Headway
Day Treatment Program and DDP
117 West Bunny Street
Santa Maria, CA 93454

Santa Maria Drug and Alcohol Awareness Program
400 North McClelland Street
Santa Maria, CA 93454

Community Health Projects Santa Maria Unit
115 East Fesler Street
Santa Maria, CA 93454

Good Samaritan Homeless Shelter Transitional Center for Women/Children
830 West Church Street
Santa Maria, CA 93454

Klein Bottle Youth Programs
412 East Tunnell Street
Santa Maria, CA 93454

National Council on Alcoholism North County Office
500 South Broadway
Suite 110
Santa Maria, CA 93454

Recovery Point
406 South Pine Street
Santa Maria, CA 93454

Santa Maria Valley Youth and Family Center/Santa Maria
305 West Church Street
Santa Maria, CA 93454

SANTA MONICA

Alcoholism Council West Area High Gain Project
1424 4 Street
Suite 205
Santa Monica, CA 90401

Clare Foundation
Adult Recovery Home
1871 9 Street
Santa Monica, CA 90405

Court Referral DWI
909 Pico Boulevard
Santa Monica, CA 90405

Santa Monica Recovery/Detox Center
907 Pico Boulevard
Santa Monica, CA 90405

Signs of Recovery Program
1027 Pico Boulevard
Santa Monica, CA 90405

Teen Center
1002 Pico Boulevard
Santa Monica, CA 90405

Youth Recovery Home
844 Pico Boulevard
Santa Monica, CA 90405

Mellen Recovery Center
1341 Centinela Avenue
Suite 101
Santa Monica, CA 90404

Saint John's Hospital and Health Center Chemical Dependency Center
1328 22 Street
Santa Monica, CA 90404

Santa Monica Bay Area Drug Abuse Council/New Start
612 Colorado Avenue
Suite 108
Santa Monica, CA 90401

SANTA PAULA

Santa Clara Valley Alcoholism Services United
Outpatient
951 East Main Street
Santa Paula, CA 93060

Recovery Home/Un Paso Adelante
307 North Ojai Road
Santa Paula, CA 93060

SANTA ROSA

California Human Development Corp Athena House
1539 Humboldt Street
Santa Rosa, CA 95404

Campobello Chemical Dependency Recovery Center
3400 Guerneville Road
Santa Rosa, CA 95401

Council on Addiction Sonoma County
2455 Bennett Valley Road
Suite 116B
Santa Rosa, CA 95404

Drug Abuse Alternatives Center
Redwood Empire Addictions Program (REAP)
2999 Cleveland Avenue
Suite B
Santa Rosa, CA 95403

Outpatient Treatment Turning Point
2800 Cleveland Avenue
Suite 11
Santa Rosa, CA 95403

Perinatal Substance Abuse Program
2800 Cleveland Avenue
Suites 6 and 7
Santa Rosa, CA 95404

R House
Outpatient
4527 Montgomery Drive
Suite A
Santa Rosa, CA 95405

Rincon Program
429 Speers Road
Santa Rosa, CA 95405

Santa Rosa Treatment Program
1901 Cleveland Avenue
Unit B
Santa Rosa, CA 95403

Social Advocates for Youth, Inc.
1303 College Avenue
Santa Rosa, CA 95404

Sonoma County Alcohol Services
Orenda Center
2759 Bennett Valley Road
Santa Rosa, CA 95404

Ruth Place
1018 Ruth Place
Santa Rosa, CA 95401

Unity House
920 West 8 Street
Santa Rosa, CA 95401

Sonoma County Indian Health Project Behavioral Health Department
791 Lombardi Court
Suite 101
Santa Rosa, CA 95407

Unique Place
Alcohol/Drug Recovery Home for Women
713 Tupper Street
Santa Rosa, CA 95404

Bridge House Women's/Children's Residence for Alcohol Recovery
2602 Giffen Avenue
Santa Rosa, CA 95407

Hope House
Women's Transition Residence
924 West 8 Street
Santa Rosa, CA 95401

Villa Lodge
3640 Stony Point Road
Santa Rosa, CA 95407

SANTEE

Santee Connection Santee Substance Abuse Counseling Project
9302 Carlton Hills Boulevard
Santee, CA 92071

SARATOGA

West Valley College Substance Abuse Services
14000 Fruitvale Avenue
Saratoga, CA 95070

SAUGUS

Live Again Recovery Home
38215 North San Francisquito Canyon Road
Saugus, CA 91350

SAUSALITO

Marin City Community Development Corp. Marin City Alcohol/Drug Outpatient Services
109 Drake Avenue
Sausalito, CA 94965

Sausalito Professional Clinic
2401 Marinship Way
Suite 300
Sausalito, CA 94965

SCOTTS VALLEY

Alcohol Abuse Hospitals The Camp
3192 Glen Canyon Road
Scotts Valley, CA 95066

Triad Community Services Outpatient Drug and Alcohol Treatment Program
5321 Scotts Valley Drive
Suite 200
Scotts Valley, CA 95066

SEASIDE

Community Human Services
Alcohol and Drug Center
1001 Elm Avenue
Seaside, CA 93955

Genesis Residential Center
1152 Sonoma Avenue
Seaside, CA 93955

SEBASTOPOL

Azure Acres Chemical Dependency Recovery Center
2264 Green Hill Road
Sebastopol, CA 95472

SEPULVEDA

Veterans' Affairs Medical Center Chemical Dependency Treatment Program
16111 Plummer Street
Ward 41C
Sepulveda, CA 91343

SHAFTER

Catholic Social Services Kaleidoscope/Shafter
558 Central Avenue
Shafter, CA 93263

SIMI VALLEY

ACT Family Treatment Program Chemical Dependency
1919 Williams Street
Suite 220
Simi Valley, CA 93065

Alcoholics Anonymous Simi Valley Central Office
1730 Los Angeles Avenue
Simi Valley, CA 93065

Alternative Action Program
2245 First Street
Suite 209A
Simi Valley, CA 93065

Community Health Projects/Simi Valley
2943 Sycamore Drive
Suite 1
Simi Valley, CA 93065

Namaskara/Twelve Friends House
1186 Appleton Road
Simi Valley, CA 93065

Rainbow Recovery Centers II
3165 Tapo Canyon Road
Simi Valley, CA 93063

Ventura County Alcohol and Drug Programs Simi Valley Center
4322 Eileen Street
Simi Valley, CA 93063

SOLVANG

Wallace and Associates Rehab and Workers Compensation
540 Alisal Road
Suite 7
Solvang, CA 93463

SONORA

Alcohol/Drug and Perinatal Services for Tuolumne County
12801 Cabezut Road
Sonora, CA 95370

Care Schools
14647 Mono Way
Sonora, CA 95370

Maynords Chemical Dependency Recovery Center Meadows for Women
16185 Tuolumne Road
Sonora, CA 95370

SOUTH GATE

ICI Projects
5202 Firestone Place
South Gate, CA 90280

Safety Consultant Services
4120 Tweedy Boulevard
South Gate, CA 90280

SOUTH LAKE TAHOE

Sierra Recovery Center
931 Macinaw Street
South Lake Tahoe, CA 96150

2677 Reaves Street
South Lake Tahoe, CA 96150

972 Tallac Avenue
Suite B
South Lake Tahoe, CA 96150

Tahoe Turning Point Juvenile Residential Treatment Centers
1415 Keller Road
South Lake Tahoe, CA 96151

562 Tehema Street
South Lake Tahoe, CA 96151

Tahoe Youth and Family Services, Inc.
1021 Fremont Avenue
South Lake Tahoe, CA 95705

SOUTH SAN FRANCISCO

Sitike Counseling Center
1211 Mission Road
South San Francisco, CA 94080

STANFORD

Stanford Alcohol and Drug Treatment Center
300 Pasteur Drive
Stanford, CA 94305

STANTON

Stanton Detox
Roque Center
10936 Dale Street
Stanton, CA 90680

Western Pacific Stanton Medical Clinic
10751 Dale Street
Stanton, CA 90680

STOCKTON

Alcohol Recovery Center and Drinking Driver Programs
7273 Murray Drive
Suite 1
Stockton, CA 95210

Alcoholics Anonymous Central Office
1125 North Hunter Street
Stockton, CA 95202

Alliance for Infants and Mothers
548 East Park Street
Stockton, CA 95202

Council for the Spanish Speaking
511 East Magnolia Street
4th Floor
Stockton, CA 95202

Jesus Saves Ministries
438 South Sutter Street
Stockton, CA 95203

Saint Joseph's Health Care Corporation Saint Joseph's Behavioral Health Center
2510 North California Street
Stockton, CA 95204

San Joaquin County
Chemical Dependency Counseling Center
640 North Union Street
Stockton, CA 95205

Prevention Services
640 North San Joaquin Street
2nd Floor Room 7
Stockton, CA 95202

San Joaquin Safety Council DUI First Offender Program
1221 North El Dorado Street
Stockton, CA 95202

Starting Point
701 East Park Street
Stockton, CA 95202

SUISUN CITY

Solano Co Economic Opportunity Council Suisun Recovery Services
205-A Marina Center
Suisun City, CA 94585

SUN VALLEY

People in Progress Sun Valley Community Rehab Center
8140 Sunland Boulevard
Sun Valley, CA 91352

SUSANVILLE

Alcoholics Anonymous
476 Alexander Avenue
Susanville, CA 96130

Lassen County Alcohol and Drug Program
476 Alexander Avenue
Susanville, CA 96130

TAFT

Catholic Social Services Kaleidoscope/Taft
915 North 10th Street
Rooms 12A 14 and 16
Taft, CA 93268

Project Aware for Drug and Alcohol Abuse
501 Kern Street
Taft, CA 93268

TAHOE CITY

Sierra Council on Alcohol and Drug Dependency
Tahoe City Service Center
605 Westlake Boulevard
Suite 5
Tahoe City, CA 96145

Sierra Family Services/Tahoe
3080 North Lake Boulevard
Suite C
Tahoe City, CA 96145

TARZANA

Center for Counseling and Education
6025 Etiwanda Street
Tarzana, CA 91356

Looking Glass Counseling Center
19318 Ventura Boulevard
Suite 206
Tarzana, CA 91356

Tarzana Treatment Center
18646 Oxnard Street
Tarzana, CA 91356

Valley Women's Center and Family Recovery Center
5530 Corbin Avenue
Suite 325
Tarzana, CA 91356

TEHACHAPI

Council on Substance Abuse Awareness Family/Adolescent Counseling
112 East F Street
Tehachapi, CA 93561

TEMECULA

Riverside County Dept. of Mental Health Alcohol Program Mid County/Temecula
41002 County Center Drive
Temecula, CA 92390

Solutions Recovery Services
29373 Rancho California Road
Temecula, CA 92592

THOUSAND OAKS

A Center for Creative Change
3537 Old Conejo Road
Suite 113
Thousand Oaks, CA 91320

TORRANCE

Del Amo Hospital
23700 Camino Del Sol
Torrance, CA 90505

Gratitude Retreat
1729 Cabrillo Avenue
Torrance, CA 90501

High Gain Drinking Driver Program
1332 Post Avenue
Torrance, CA 90501

National Council on Alcoholism and Drug Dependence of the South Bay
1334 Post Avenue
Torrance, CA 90501

Options for Recovery
The Stork Club
1124 West Carson Street
Suite N33
Torrance, CA 90502

South Bay Human Service Center
2370 West Carson Street
Suite 136
Torrance, CA 90501

Southwest Driver Benefits Program
2370 West Carson Street
Suite 150
Torrance, CA 90501

TRACY

Valley Community Counseling Center
1018 E Street
Tracy, CA 95376

TRAVIS AIR FORCE BASE

Travis Air Force Base Social Actions Office
530 Hickam Avenue
60 MSSQ/MDLD
Travis Air Force Base, CA 94535

TRUCKEE

Genesis Outpatient Chemical Dependency Treatment
10121 Pine Avenue
Truckee, CA 96161

Nevada County Council on Alcoholism Truckee Service Center
12070 Donner Pass Road
Truckee, CA 96162

Nevada County Mental Health/ Truckee Substance Abuse Services
10075 Levon Avenue
Suite 102
Truckee, CA 96161

TULARE

Alcohol Center for Teenagers (ACT)
23393 Road 68
Tulare, CA 93274

Kings View Substance Abuse Program Tulare County
559 East Bardsley Avenue
Tulare, CA 93275

National Council on Alcoholism and Drug Dependence/Tulare County
525 East Bardsley Avenue
Tulare, CA 93275

TUOLUMNE

Maynords Chemical Dependency Recovery Center Ranch for Men
19325 Cherokee Road
Tuolumne, CA 95379

Tuolumne Rural Indian Health Program Substance Abuse Services
Mi Wuk Street
Tuolumne Rancheria
Tuolumne, CA 95379

TURLOCK

Emanuel Medical Center Chemical Dependency Center
825 Delbon Avenue
Turlock, CA 95380

Tuum Est, Inc./North Phoenix House
219 South Broadway
Turlock, CA 95380

TUSTIN

Recovery Homes of America Cornerstone
13682 Yorba Street
Tustin, CA 92680

Tustin Hospital Medical Center Chemical Recovery Services
14662 Newport Avenue
Tustin, CA 92680

UKIAH

Ford Street Project
139 Ford Street
Ukiah, CA 95482

Mendocino County Public Health Dept. Division of Alcohol and Other Drug Programs
302 West Henry Street
Ukiah, CA 95482

Mendocino County Youth Project
202 South State Street
Ukiah, CA 95482

Narcotics Anonymous
52 Lorraine Street
Ukiah, CA 95482

R House
Talmage House
1201 Talmage Road
Ukiah, CA 95482

UNION CITY

Occupational Health Services
32980 Alvarado-Niles Road
Suite 812
Union City, CA 94587

UPLAND

Counseling Associates
Arrow House/Women's Facility
1439 West Arrow Highway
Upland, CA 91786

The Recovery Center
934 North Mountain Avenue
Upland, CA 91786

New Beginnings/Upland
323 North Mountain Avenue
Suite B
Upland, CA 91786

Reach Out West End
123 East 9 Street
Suite 102
Upland, CA 91786

Valley Improvement Programs Abuse Services Center
414 East 9 Street
Upland, CA 91786

VACAVILLE

Solano County Economic Opport Council Vacaville Recovery Services
831 Alamo Drive
Suite A
Vacaville, CA 95687

Youth and Family Services Vacaville
1707 California Drive
Vacaville, CA 95688

VALLEJO

First Call for Help
401 Amador Street
Vallejo, CA 94590

First Hospital/Vallejo Dual Diagnosis Adult/Adoles Services
525 Oregon Street
Vallejo, CA 94590

Genesis House
1149 Warren Avenue
Vallejo, CA 94591

Youth and Family Services Project Vallejo/Project Aurora
1200 Marin Street
Vallejo, CA 94590

VAN NUYS

California Women's Commission on Alcohol and Drug Dependencies
14622 Victory Boulevard
Suite 100
Van Nuys, CA 91411

Crossroads School
6305 Woodman Avenue
Van Nuys, CA 91401

Health Care Delivery Services Pride House
7447 Sepulveda Boulevard
Van Nuys, CA 91405

I-ADARP Van Nuys Clinic
14517 Victory Boulevard
Van Nuys, CA 91411

Narcotics Anonymous World Service Office, Inc.
16155 Wyandotte Street
Van Nuys, CA 91406

National Council on Alcoholism and Drug Dependency of San Fernando Valley/Unit 2
14557 Friar Street
Van Nuys, CA 91411

New Directions for Youth
7400 Van Nuys Boulevard
Suite 203
Van Nuys, CA 91405

Northeast Valley Health Corporation Mi Descanso
6819 Sepulveda Boulevard
Suite 102
Van Nuys, CA 91405

San Fernando Valley CMHC Youth Contact Component
14531 Hamlin Street
Suite 101
Van Nuys, CA 91411

The High Road Program
14430 Sherman Way
Van Nuys, CA 91405

Van Nuys Hospital Substance Abuse Services
15220 Van Owen Street
Van Nuys, CA 91401

Western Pacific Panorama Program Outpatient Detox and Methadone Maintenance
9462 Van Nuys Boulevard
Van Nuys, CA 91402

VENICE

Didi Hirsch CMHC Outpatient Drug Abuse Services
318 Lincoln Boulevard
Venice, CA 90291

Phoenix Houses of Los Angeles
503 Ocean Front Walk
Venice, CA 90291

Social Model Recovery Systems Nightmoves
1522 Abbot Kinney Boulevard
Venice, CA 90291

VENTURA

Fresh Start Counseling Center
261 Youmans Street
Ventura, CA 93003

Khepra House
105 West Harrison Avenue
Ventura, CA 93001

Medical Support Services to Substance Abusers
3291 Loma Vista Road
Ventura, CA 93003

Miracle House/Ventura
94 South Anacapa Street
Ventura, CA 93001

Rainbow Recovery Youth Center
192 Reata Avenue
Ventura, CA 93004

Turning Point Foundation DBA Hacienda Help Services
Ventura, CA 93001

Ventura County Drinking Driver Program
4651 Telephone Road
Suite 210
Ventura, CA 93003

Ventura County Alcohol and Drug Programs
Division of Special Programs
955 East Thompson Boulevard
Ventura, CA 93001

Moms' and Kids' Recovery Center
801 Poinsettia Place
Ventura, CA 93001

Ventura Center
739 East Main Street
Ventura, CA 93001

VICTORVILLE

High Desert Child/Adolescent and Family Service Center
16248 Victor Street
Victorville, CA 92392

Saint John of God Health Care Services How House Men's Program
15534 6 Street
Victorville, CA 92392

VISALIA

Alcohol and Drug Services of Tulare County Alternative Services
2015 West Tulare Street
Visalia, CA 93277

Care Schools
420 East Murray Street
Visalia, CA 93291

Kings/Tulare Area Agency on Aging Older Persons Alcohol and Drug Project
1920 West Princeton Street
Visalia, CA 93277

Mothering Heights
504 South Locust Street
Visalia, CA 93277

New Generation
420 North Church Street
Visalia, CA 93277

Tulare County Alcoholism Council, Inc.
New Visions for Women
1425 East Walnut Street
Visalia, CA 93292

Pine Recovery Center
120 West School Street
Visalia, CA 93291

Turning Point of Central California Turning Point Youth Services
119 South Locust Street
Visalia, CA 93291

VISTA

Mental Health Systems Probationers in Recovery
538 West Vista Way
Vista, CA 92084

WALNUT CREEK

Alcoholics Anonymous Contra Costa Service Center
185 Mayhew Way
Walnut Creek, CA 94596

Alpha System
1225 Alpine Road
Suite 208
Walnut Creek, CA 94596

Bi Bett Corporation Gregory Recovery Center
1860 3 Avenue
Walnut Creek, CA 94596

Counterpoint Center/Walnut Creek Hospital Dual Diagnosis
175 La Casa Via
Walnut Creek, CA 94598

Family and Community Services of Contra Costa
1300 Civic Drive
Walnut Creek, CA 94596

Kaiser Permanente Medical Center Alcohol and Drug Abuse Program
1425 South Main Street
Walnut Creek, CA 94596

WATSONVILLE

Fenix Family
Alcoholism Service Center
406 Main Street
Suite 403
Watsonville, CA 95076

Hermanas Recovery Home
321 East Beach Street
Watsonville, CA 95076

Santa Cruz Community Counseling Center
Alto Counseling Center/South
11-D Alexander Street
Watsonville, CA 95076

Si Se Puede
161 Miles Lane
Watsonville, CA 95076

Sunflower Youth House
187-A San Andreas Road
Watsonville, CA 95076

Youth Services/South
107 California Street
Watsonville, CA 95076

Watsonville Community Hospital Alcohol and Drug Treatment Center
298 Green Valley Road
Watsonville, CA 95076

WEAVERVILLE

Trinity County Counseling Center
1 Industrial Park Way
Weaverville, CA 96093

WEST COVINA

Community Health Projects/Los Angeles
Glendora Unit
336 1/2 South Glendora Avenue
West Covina, CA 91790

West Covina Unit
1825 East Thelborn Street
West Covina, CA 91790

Crossroads
405 Lyall Street
West Covina, CA 91790

Industry Community Interface Projects (ICI Projects)
1901 West Pacific Street
West Covina, CA 91793

Los Angeles Centers for Alcohol/Drug Abuse (LACADA) West Covina
1323 West Covina Parkway
West Covina, CA 91790

Rickman Recovery Center/West Covina
1107 South Glendora Avenue
West Covina, CA 91790

Safety Education Center
1500 West Covina Parkway
Suite 105
West Covina, CA 91790

San Gabriel Valley Driver Improvement Program
1502 West Covina Parkway
Suite 207
West Covina, CA 91790

WEST SACRAMENTO

California Freedom House Fellowship
1101 Pierce Street
West Sacramento, CA 95605

Community Clinic/West Sacramento Drug Treatment Program
950 Sacramento Avenue
West Sacramento, CA 95605

Detox Center of America
820 West Acre Road
West Sacramento, CA 95691

East Yolo Information Center for Alcohol and Other Drug Abuse
1109 West Capitol Avenue
West Sacramento, CA 95691

Yolo County Department of Alcohol/Drug Programs/West Sacramento
350 C Street
West Sacramento, CA 95691

WESTCHESTER

National Council on Alcoholism/South Bay High Gain Program
9100 South Sepulveda Boulevard
Suite 105
Westchester, CA 90045

WESTLAKE VILLAGE

Conejo Counseling Center Be Free
3609 Thousand Oaks Boulevard
Suite 110
Westlake Village, CA 91362

WESTMINSTER

Orange County Alcohol Program West Region
14180 Beach Boulevard
Suite 206
Westminster, CA 92683

Orange County Health Care Agency Westminster Drug Abuse Services
14600 Goldenwest Street
Suite 112
Westminster, CA 92683

School Ten
Garden Grove Unit
6156 Garden Grove Boulevard
Westminster, CA 92683

Westminster Avenue Unit
6926 Westminster Avenue
Westminster, CA 92683

WHITTIER

Awakenings Program for Deaf and Hard of Hearing Persons/Residential
9608 Regatta Avenue
Whittier, CA 90604

Catholic Rainbow Outreach
11419 Carmenita Road
Whittier, CA 90605

Community Health Projects/Whittier Methadone Treatment Program
11738 Valley View Avenue
Suite B
Whittier, CA 90604

Fred C. Nelles School Lazarus House Anti-Substance Abuse Program
11850 East Whittier Boulevard
Whittier, CA 90601

Presbyterian Intercommunity Hospital The Chemical Dependency Center
12401 East Washington Boulevard
Whittier, CA 90602

Project Info Family/Youth Program
9401 South Painter Avenue
Whittier, CA 90605

Public Interest
7336 South Painter Avenue
Whittier, CA 90602

Safety Consultant Services
13501 East Whittier Boulevard
Whittier, CA 90605

Southeast Council on Alcohol/Drug Problems Foley House
10511 Mills Avenue
Whittier, CA 90604

Whitney Group Norwalk House
7320 Norwalk Boulevard
Whittier, CA 90606

WILLITS

Lucky Deuce DUI Program
110 South Main Street
Suite E
Willits, CA 95490

WILMINGTON

Behavioral Health Services Wilmington Comm. Prevention/Recovery Center
531 South Marine Avenue
Wilmington, CA 90744

Community Health Projects Wilmington Clinic/OMM
936 North Wilmington Boulevard
Wilmington, CA 90744

La Clinica Del Pueblo
402 A West Anaheim Street
Wilmington, CA 90744

Neighborhood Youth Association Wilmington Drug Program
1323 North Avalon Boulevard
Wilmington, CA 90744

Transcultural Health Development
117 East B Street
Wilmington, CA 90744

WINDSOR

Sonoma County Alcohol Services First Offender Drinking Driver Program
9047 Old Redwood Highway
Windsor, CA 95492

WINTERHAVEN

Fort Yuma Alcohol and Drug Abuse Prevention Program
375 Picacho Road
Winterhaven, CA 92283

WOODLAND

Beamer Street Detoxification and Residential Treatment Center
4 North Cottonwood Street
Woodland, CA 95695

Yolo Alcoholic Recovery Center Cache Creek Lodge
Route 2
Woodland, CA 95695

Yolo County
Department of Alcohol and Drug Programs
201 West Beamer Street
Woodland, CA 95695

Drinking Driver Program
825 East Street
Suite 123
Woodland, CA 95776

WOODLAND HILLS

Matrix Center/Woodland Hills
6300 Variel Avenue
Suite C
Woodland Hills, CA 91367

MCC Managed Behavioral Care of California
6300 Variel Avenue
Suite B
Woodland Hills, CA 91367

YOUNTVILLE

Veterans' Home of California Alcohol Program Service
Yountville Veterans Home
Lincoln Hall Section A
Yountville, CA 94599

YREKA

Siskiyou Alcohol and Drug Abuse Services
804 South Main Street
Yreka, CA 96097

YUBA CITY

Pathways
Drinking Driver Programs
Midvalley Recovery Facilities
Youth and Family
430 Teegarden Avenue
Yuba City, CA 95991

Sutter Yuba Mental Health Services Drug Alternatives Program
1965 Live Oak Boulevard
Yuba City, CA 95991

YUCAIPA

Redlands Yucaipa Guidance Clinic Valley Guidance Center
32371 Yucaipa Boulevard
Yucaipa, CA 92399

COLORADO

ALAMOSA

Professional Counseling Services
11825 Henry Lane
Alamosa, CO 81101

San Luis Valley Comp. CMHC Substance Abuse Services
1015 4 Street
Alamosa, CO 81101

ARVADA

Alcohol Behavior Information, Inc.
7550 Grant Place
Arvada, CO 80002

ASPEN

Applied Health Resources DBA Springs Counseling Center
605 East Main Street
Suite 7
Aspen, CO 81612

AURORA

Careunit Hospital of Colorado
1290 South Potomac Street
Aurora, CO 80012

Fitzsimmons Army Medical Center Alcohol/Drug Abuse Prevention and Control Program Commander Fitzsimmons Army Medical Center
Building 317
Aurora, CO 80045

Professional Psychotherapy Associates Countermeasures
11100 East Mississippi Street
Suite 100
Aurora, CO 80012

Rangeview Counseling Center
14501 East Alameda Avenue
Suite 204
Aurora, CO 80012

BOULDER

Boulder Alcohol Education Center
1525 Spruce Street
Suite 100
Boulder, CO 80302

Boulder Community Hospital Behavioral Health Services/ Recovery Program
311 Mapleton Avenue
Boulder, CO 80302

Boulder County Health Department Substance Abuse Program
3450 Broadway
Boulder, CO 80304

Boulder Mental Health Center Drug Abuse Program
1333 Iris Avenue
Boulder, CO 80302

Personal Growth Services
1722 14 Street
Room 260
Boulder, CO 80302

Serenity Center for Personal Growth, Inc.
1790 30 Street
Sussex One Suite 301
Boulder, CO 80301

Whole Person Health Center DUI/Alcohol Outpatient
2969 Baseline Road
Boulder, CO 80303

BRIGHTON

Educational Center for Addictions
710 South Main Street
Brighton, CO 80601

CANON CITY

Addiction Recovery Programs, Inc.
U.S. Highway 50 West
Canon City, CO 81212

Colorado Department of Corrections Drug and Alcohol Abuse, Inc.
107 North First Street
Canon City, CO 81212

Drug and Alcohol Abuse, Inc.
618 Main Street
Canon City, CO 81215

COLORADO SPRINGS

Cate Alcohol Education Program
4740 Flintridge Drive
Suite 201-B
Colorado Springs, CO 80918

Department of Health and Environment McMaster Center for Alcohol/Drug Treatment
301 South Union Boulevard
Colorado Springs, CO 80910

Emerging Reality
2910 North Academy Boulevard
Suite 103
Colorado Springs, CO 80917

Gordon S. Riegel Chemical Dependency Center
825 East Pikes Peak Avenue
Colorado Springs, CO 80903

Health Challenge Counseling Center, Inc.
1715 Monterey Road
Suite 198
Colorado Springs, CO 80910

Introspect Counseling Services
984 Chapman Drive
Suite 5
Colorado Springs, CO 80916

Pathways Confidential Counseling, Inc.
304 South 8 Street
Suite 101C
Colorado Springs, CO 80905

Pike's Peak Mental Health Center
Alcohol Emergency Services
2741 East Las Vegas Street
Colorado Springs, CO 80906

Positive Change
1425 North Union Street
Suite 201
Colorado Springs, CO 80909

Some Place Else/Genesis
179 Parkside Drive
Suite 200
Colorado Springs, CO 80910

Terrace Gardens
2438 East Fountain Boulevard
Colorado Springs, CO 80910

The Ark, Inc. Outpatient Chem Dependency Treatment
423 South Cascade Street
Colorado Springs, CO 80903

Turning Point Alternatives
2862 South Circle Drive
Springs Office Park Suite 205
Colorado Springs, CO 80906

Uplift Awareness Center Alcohol Abuse Program
432 West Bijou Street
Colorado Springs, CO 80905

COMMERCE CITY

Washington House, Inc. Parkside Lutheran Recovery Center
7373 Birch Street
Commerce City, CO 80022

CRAIG

Gary Gurney Psychotherapist and Associates/Substance Abuse Services
439 Breeze Street
Suite 101
Craig, CO 81625

DENVER

Alcohol Awareness Information Center, Inc.
10248 East Jewell Avenue
Suite 30
Denver, CO 80231

Alcohol Counseling Services of Colorado, Inc.
1300 South Lafayette Street
Denver, CO 80210

Alpar Human Development Services Alcohol Outpatient Treatment
1315 Krameria Street
Suite 103
Denver, CO 80222

Anchor Counseling, Inc.
1009 Grant Street
Suite 202
Denver, CO 80203

Community Alcohol/Drug Rehab and Education Center (CADREC)
3315 Gilpin Street
Denver, CO 80205

Community Resource for Alcohol and Family Treatment (CRAFT)
1060 Bannock Street
Suite 314
Denver, CO 80204

Comprehensive Addiction Treatment Services
2222 East 18 Avenue
Denver, CO 80206

Denver Area Youth Services/ Days Adolescent Substance Abuse Treatment
1240 West Bayaud Avenue
Denver, CO 80223

Denver Health and Hospitals Denver Cares
1155 Cherokee Street
Denver, CO 80204

Denver Treatment Services for Alcohol and Drug Abuse
2231 Federal Boulevard
Denver, CO 80211

Division of Youth Services
4255 South Knox Court
Denver, CO 80236

Driver Development Center
120 West 5 Avenue
Denver, CO 80204

Eagle Lodge, Inc.
1264 Race Street
Denver, CO 80206

Essex Growth Center
280 South Federal Street
Denver, CO 80219

Gateway Treatment Center
1191 South Parker Road
Suite 100
Denver, CO 80231

Mercy Medical Center Behavioral Medicine Department
1650 Fillmore Street
Denver, CO 80206

Metropolitan Counseling Services
1601 South Federal Boulevard
Heritage Plaza Suite 115
Denver, CO 80219

Mile High Club Alcohol Abuse Halfway House
2260 Larimer
Denver, CO 80205

Milestone Counseling Services, Inc.
1210 East Colfax Street
Suite 308
Denver, CO 80218

Multi-Services Clinic, Inc.
3010 West 16 Avenue
Suite 207
Denver, CO 80204

Porter Centre for Behavioral Health
2525 South Downing Street
Porter Memorial Hospital
Denver, CO 80210

Salvation Army Adult Rehab Center Alcohol Abuse Halfway House Care
4751 Broadway
Denver, CO 80216

Sobriety House, Inc. Stepping Stone
107 Acoma Street
Denver, CO 80223

Special Services Clinic, Inc.
301 Knox Court
Denver, CO 80219

Substance Treatment Services (STS)
320 West 8 Avenue
Unit 2
Denver, CO 80204

UCHSC Addiction Research TRT Services (ARTS)
3738 West Princeton Circle
Denver, CO 80236

Veterans' Affairs Medical Center
Substance Abuse Outpatient/ Building A
Substance Abuse Services/7 West
1055 Clermont Street
Denver, CO 80220

Western Clinical Health Services
1038 Bannock Street
Denver, CO 80204

DURANGO

Mercy Medical Center The Durango Centre
3801 North Main Avenue
Durango, CO 81301

SW Colorado Community Corrections Center Hilltop Treatment Services
1045 1/2 East 2 Avenue
Durango, CO 81302

ELIZABETH

Running Creek Counseling Service
313 East Kiowa Street
Elizabeth, CO 80107

ENGLEWOOD

Englewood Counseling Center
3969 South Broadway
Englewood, CO 80110

Federal Correctional Institution Substance Abuse Services
Englewood, CO 80123

ESTES PARK

Harmony Foundation, Inc. Alcohol/Drug Abuse Program
1600 Fish Hatchery Road
Estes Park, CO 80517

EVERGREEN

Bellwood Resources Center, Inc.
7592 South Gartner Road
Evergreen, CO 80439

FLORENCE

Clear View Center
521 West 5 Street
Florence, CO 81226

FORT COLLINS

Center for Life Skills Education
400 West Magnolia Street
Fort Collins, CO 80521

Hope Counseling Center
1644 South College Avenue
Fort Collins, CO 80525

Larimer County Institute for Alcohol Awareness
253 Linden Street
Suite 206
Fort Collins, CO 80524

Mountain Crest Hospital Addictive Disease Unit (ADU)
4601 Corbett Drive
Fort Collins, CO 80525

New Beginnings Fort Collins
1225 Redwood Street
Fort Collins, CO 80524

Seven Lakes Recovery Program
2362 East Prospect Avenue
Fort Collins, CO 80525

GLENWOOD SPRINGS

Colorado West Regional Mental Health Substance Abuse Services
711 Grand Avenue
Glenwood Springs, CO 81601

Valley View Youth Recovery Center
1906 Blake Avenue
Glenwood Springs, CO 81601

GOLDEN

Corrections Evaluation and Treatment Program, Inc.
2433 Ford Street
Golden, CO 80401

GRAND JUNCTION

Adult Adolescent Alcohol Treatment (AAAT)
2600 North 12 Street
Grand Junction, CO 81506

Aru/Grand Junction Alcohol and Drug Abuse Program
436 South 7 Street
Grand Junction, CO 81501

ASET Clinic, Inc. DBA Addiction Services Education/Treatment Clinic
844 Grand Avenue
Grand Junction, CO 81501

In Roads Counseling
1141 North 25 Street
Suite F
Grand Junction, CO 81501

National Council on Alcohol and Drug Abuse Mesa County, Inc.
136 North 7 Street
Grand Junction, CO 81501

Veterans' Affairs Medical Center Substance Abuse Treatment Program (SATP)
2121 North Avenue
Building 6
Grand Junction, CO 81501

GREELEY

Addiction Recovery Center
800 8 Avenue
Suite 200
Greeley, CO 80631

Institute for Alcohol Awareness of Greeley
920 11 Street
Greeley, CO 80631

Island Grove Regional Treatment Center, Inc.
421 North 15 Avenue
Greeley, CO 80631

North Colorado Medical Center Family Recovery Center
1910 15 Street
Greeley, CO 80631

Residential Treatment Center
1776 6 Avenue
Greeley, CO 80631

IDAHO SPRINGS

Clear Creek Counseling
2401 Colorado Boulevard
Idaho Springs, CO 80452

IGNACIO

Southern Ute Alcohol Recovery Center Peaceful Spirit
Ignacio, CO 81137

LA JUNTA

Pathfinders
207 1/2 Colorado Avenue
La Junta, CO 81050

LAFAYETTE

Lafayette Alcohol Education and Therapy
201 East Simpson Street
Suite 201-B
Lafayette, CO 80026

LAKEWOOD

Alternative Behaviors Counseling
1949 Wadsworth Street
Suite 206
Lakewood, CO 80215

Bethesda Behavioral Management Counseling of Colorado
390 Union Boulevard
Suite 120
Lakewood, CO 80228

Cenikor Foundation, Inc. Alcohol and Drug Abuse Program
1533 Glen Ayr Drive
Lakewood, CO 80215

Crossroads Counseling, Ltd.
8000 West 14 Avenue
Suite 1
Lakewood, CO 80215

Empowerment Counseling Services, Inc.
7580 West 16 Avenue
Suite 206
Lakewood, CO 80215

Jefferson County Health Department Substance Abuse Counseling Program
260 South Kipling Street
Lakewood, CO 80226

Serenity Counseling
1880 South Pierce Street
Suite 11
Lakewood, CO 80226

Touchstone Counseling Center
777 South Wadsworth Boulevard
Irongate 2 Suite 205
Lakewood, CO 80226

LAMAR

Se Colorado for Drug Free Communities
2401 South Main Street
Lamar, CO 81052

LAS ANIMAS

Resada Alcohol and Drug Abuse Program
11000 RD GG 5
Las Animas, CO 81054

LEADVILLE

Leadville Counseling Centre
130 West 5 Street
Suite 122
Leadville, CO 80461

LITTLETON

Arapahoe Mental Health Center Aquarius Center/Sami
2100 West Littleton Boulevard
Littleton, CO 80120

Columbine Psychiatric Center Challenging Directions
8565 South Poplar Way
Littleton, CO 80126

Dry Creek Treatment Center
1562 East Mineral Place
Littleton, CO 80122

LONGMONT

Arvada/Longmont Counseling Center Alcoholism Program
514 Kimbark Street
Longmont, CO 80501

Palmer Drug Abuse Program Boulder County, Inc.
1050 Washley Street
Longmont, CO 80501

LOUISVILLE

Patterns for Positive Living
1017 South Boulder Road
Room F
Louisville, CO 80027

MONTE VISTA

Creative Resource Center, Inc.
4979 North County Road 3 West
Corporate Office
Monte Vista, CO 81144

MONTROSE

Midwestern Colorado Mental Health Center The Center for Mental Health/Substance Abuse Services
605 East Miami Road
Montrose, CO 81401

Montrose Memorial Hospital Care Center
945 South 4 Street
Montrose, CO 81401

North Fork Counseling
636 South 2 Street
Montrose, CO 81401

Touch Stone Counseling
118 North Cascade Street
Montrose, CO 81401

MORRISON

Lost and Found, Inc. Substance Abuse Program
9189 South Turkey Creek Road
Morrison, CO 80465

NORTHGLENN

Counseling Dimensions, Inc.
11658 Huron Street
Northglenn, CO 80234

Insights Counseling Services, Inc.
2200 East 104 Avenue
Suite 213
Northglenn, CO 80233

PAGOSA SPRINGS

Rio Blanco Counseling Center
Route 2
Pagosa Springs, CO 81147

The Gate Counseling Center
42 Pagosa Street
2nd Floor
Pagosa Springs, CO 81147

PARKER

First Step Counseling
10521 South Parker Road
Suite F
Parker, CO 80134

Parker Valley Hope
22422 East Main Street
Parker, CO 80134

PUEBLO

ADAPCP Counseling Center Pueblo Depot Activity
SDSTE-PU-ADCO
Pueblo, CO 81001

Awareness Institute, Inc.
635 West Corona Avenue
Suite 210
Pueblo, CO 81004

Colorado Mental Health Institute at Pueblo Psychiatric Substance Abuse Program (PSAP)
1600 West 24 Street
Building 116 WD 79
Pueblo, CO 81003

Parkview Episcopal Medical Center Chemical Dependency Program
58 Club Manor Drive
Pueblo, CO 81008

Pueblo Treatment Services, Inc.
509 East 13 Street
Pueblo, CO 81004

Saint Mary/Corwin Regional Medical Center Recovery Center
1008 Minnequa Avenue
Pueblo, CO 81004

Serenity Systems, Inc.
1624 Bonforte Boulevard
Suite B
Pueblo, CO 81001

RIFLE

White River Counseling
400 South 7 Street
Suite 3200
Rifle, CO 81650

SHERIDAN

Porter Memorial Hospital Adventures in Change
3445 West Mansfield Street
Sheridan, CO 80110

STERLING

Ajax, Inc.
217 North Front Street
Sterling, CO 80751

Centennial Mental Health Sterling Office Substance Abuse Services
211 West Main Street
Sterling, CO 80751

THORNTON

Arapahoe House, Inc.
8801 Lipan Street
Thornton, CO 80221

WESTMINSTER

Choices in Living
5005 West 81 Place
Suite 300
Westminster, CO 80030

WHEAT RIDGE

Adolescent and Family Institute of Colorado, Inc.
10001 West 32 Avenue
Wheat Ridge, CO 80033

Choices in Living Counseling Center, Inc.
6415 West 44 Avenue
Suite 101
Wheat Ridge, CO 80033

Treatment Services, Inc.
3805 Newland Street
Wheat Ridge, CO 80033

West Pines/A Psychiatric Hospital Chemical Dependency Unit
3400 Lutheran Parkway
Wheat Ridge, CO 80033

WOODLAND PARK

Journeys Counseling and Education Center
700 Valley View Drive
Woodland Park, CO 80866

CONNECTICUT

ANSONIA

L N Valley Council on Alcohol/ Drug Abuse Alcoholism and Drug Abuse Program
75 Liberty Street
Ansonia, CT 06401

AVON

Reid Treatment Center
Intensive Treatment
121 West Avon Road
Avon, CT 06001

Non-Hospital Medical Unit
121 West Avon Road
Avon, CT 06001

Partial Hospitalization
121 West Avon Road
Avon, CT 06001

BLOOMFIELD

Blue Ridge Center Administrative Unit
1095 Blue Hills Avenue
Bloomfield, CT 06002

BRANFORD

Branford Counseling Center Outpatient
342 Harbor Street
Branford, CT 06405

BRIDGEPORT

Chemical Abuse Services Agency, Inc. (CASA) Administrative Unit
690 Artic Street
Bridgeport, CT 06604

Community Addiction Services Bridgeport Unit
40 Fairfield Avenue
Bridgeport, CT 06604

Family Services Woodfield, Inc. Employee Assistance Youth Evaluation Program
475 Clinton Avenue
Bridgeport, CT 06605

Guenster Rehabilitation Center, Inc.
Day Treatment Unit
Evening Program
Intensive Residential Unit
276 Union Avenue
Bridgeport, CT 06607

Helping Hand Center
488 Stratford Avenue
Bridgeport, CT 06608

Horizons
Intensive Residential Drug Free Program
Intermediate Residential Drug Free Program
1635 Fairfield Avenue
Bridgeport, CT 06605

Regional Network of Programs, Inc.
Administrative Unit
144 Golden Hill Street
Suite 301
Bridgeport, CT 06604

Center for Human Services
1549 Fairfield Avenue
Bridgeport, CT 06605

Golden Hill Treatment Center/ Detox
Methadone Maintenance
Regional Adolescent Program (RAP)
Regional Counseling Services (RCS) Outpatient/DF
171 Golden Hill Street
Bridgeport, CT 06604

Regional Alcoholism Services
480 Bond Street
Bridgeport, CT 06610

Southern Connecticut Mental Health and Substance Abuse Treatment Center
4920 Main Street
Suite 310
Bridgeport, CT 06606

United Way Regional Youth Adult Substance Abuse Project
75 Washington Avenue
Bridgeport, CT 06604

BRIDGEWATER

Midwestern Connecticut Council on Alcoholism
McDonough House/Intensive Residential
McDonough House/Intermediate Residential
132 Hut Hill Road
Bridgewater, CT 06752

BRISTOL

Bristol Hospital Evening Chemical Dependency Program
440 C North Main Street
Bristol, CT 06010

CANAAN

Parkside Lodge of Connecticut
Route 7
Canaan, CT 06018

Parkside Lodge of Connecticut Alcoholism Treatment Center/ Nonhospital Medical
Route 7
Canaan, CT 06018

COS COB

Education Center of the Alcohol and Drug Abuse Council (TADAC)
521 Post Road
Cos Cob, CT 06807

DANBURY

Danbury Hospital Danbury Treatment Center
24 Hospital Avenue
Danbury, CT 06810

Danbury Youth Services, Inc.
32 Stevens Street
Danbury, CT 06810

Federal Correctional Institution Substance Abuse Services
Route 37
Danbury, CT 06810

Midwestern Connecticut Council on Alcoholism Outpatient Unit
238 White Street
Danbury, CT 06810

DARIEN

Youth Options Darien Unit
120 Brookside Road
Darien, CT 06820

DAYVILLE

United Services, Inc. Alcohol and Drug Abuse Services
1007 North Main Street
Dayville, CT 06241

DERBY

Griffin Hospital Alcohol Chem Dependency Services Evening Treatment Program
130 Division Street
Department of Psychiatry
Derby, CT 06418

ENFIELD

New Directions, Inc. of North Central Connecticut Alcohol and Drug Services
5102 Bigelow Commons
Enfield, CT 06082

FAIRFIELD

Fairfield Community Services
370 Beach Road
Fairfield, CT 06430

FARMINGTON

John Dempsey Hospital Alcohol and Drug Abuse Treatment Center
Farmington Avenue
Farmington, CT 06030

GLASTONBURY

Clayton House
203–205 Williams Street
Glastonbury, CT 06033

GREENWICH

Youth Options Greenwich Unit
75 Mason Street
2nd Floor
Greenwich, CT 06830

GROTON

Connecticut Counseling Associates Outpatient Substance Abuse Treatment
333 Long Hill Road
Groton, CT 06340

HAMDEN

Hamden Youth Services Bureau Substance Abuse Education
490 Newhall Street
Hamden, CT 06517

HARTFORD

Alcohol and Drug Recovery Centers, Inc. Detoxification Center
500 Vine Street
Hartford, CT 06112

Blue Hills Hospital Alcohol and Drug Services
51 Coventry Street
Hartford, CT 06112

Community Addiction Services Hartford Unit
233 Washington Street
Hartford, CT 06106

Community Health Services, Inc. Chemical Dependency Program
520 Albany Avenue
Hartford, CT 06120

Community Renewal Team of Greater Hartford, Inc. Youth Services
555 Windsor Street
Hartford, CT 06120

Connecticut Department of Corrections Addiction Services Unit
90 Brainard Road
Hartford, CT 06114

Hartford Dispensary Clinic
345 Main Street
Hartford, CT 06106

Henderson/Johnson Clinic
Ambulatory Detoxification Program
14 Weston Street
Hartford, CT 06120

Methadone Maintenance Program
12 Weston Street
Hartford, CT 06120

Hispanic Alcoholism Program Institute for the Hispanic Family
80 Jefferson Street
Hartford, CT 06106

Mercy Housing and Shelter Corporation Project Mercy
118 Main Street
Hartford, CT 06106

Methadone to Abstinence Program Outpatient Methadone Maintenance
14 Weston Street
Hartford, CT 06120

Open Hearth Mission
437 Sheldon Street
Hartford, CT 06106

Parents and Students in Training Together
c/o Hartford Schools
5 Cone Street
Hartford, CT 06105

Regional Alcohol and Drug Abuse Resources, Inc.
645 Farmington Avenue
Hartford, CT 06105

Salvation Army Adult Rehab Center Alcohol Abuse Program
333 Homestead Avenue
Hartford, CT 06112

United Way of Connecticut
Infoline/Administration
Infoline/North Central
900 Asylum Avenue
Hartford, CT 06105

Youth Challenge of Greater Hartford
15–17 May Street
Hartford, CT 06105

LEBANON

SE Council on Alcohol and Drug Dependency, Inc. (SCADD) Lebanon Pines
Camp Moween Road
Lebanon, CT 06249

MANCHESTER

Crossroads Prevention Program of New Hope Manor, Inc.
20 Hartford Road
Manchester, CT 06040

New Hope Manor, Inc. Residential
48 Hartford Road
Manchester, CT 06040

MERIDEN

Veterans' Memorial Medical Center Department of Alcohol/ Substance Abuse
1 Kings Place
Meriden, CT 06450

MIDDLEBURY

Parkside of Middlebury
850 Straits Turnpike/Route 63
Crossroads Complex Suite 202
Middlebury, CT 06762

MIDDLETOWN

Dutcher Chem Dependence Treatment Center
Holmes Drive
Middletown, CT 06457

Rushford Center, Inc.
Administrative and Prevention Unit
Intensive Outpatient/Day/Evening
Intensive Residential Unit
Intermediate Residential Unit
MISA/Healthy Living Program
Non-Hospital Medical Unit
1250 Silver Street
Middletown, CT 06457

The Connection, Inc.
Connection House
163–167 Liberty Street
Middletown, CT 06457

Greater Middletown Counseling Center
196 Court Street
Middletown, CT 06457

Women and Children's Center
99 Eastern Drive
Middletown, CT 06457

MILFORD

Milford Mental Health Clinic, Inc.
Administrative Unit
Outpatient
Peer Counseling Program
949 Bridgeport Avenue
Milford, CT 06460

MONROE

Regional Network of Programs, Inc. Monroe Builds Communication
1014 Monroe Turnpike
Masuk High School
Monroe, CT 06468

MOOSUP

Youth Challenge Bible Training Center
111 North Sterling Road
Moosup, CT 06354

NEW BRITAIN

Farrell Treatment Center
Outpatient Unit
Intensive Residential Unit
586 Main Street
New Britain, CT 06051

New Britain General Hospital Dept. of Substance Abuse Services/Hillcrest
100 Grand Street
New Britain, CT 06050

Parkview Counseling Center Evening Component/New Britain
91 Lexington Street
New Britain, CT 06052

Wheeler Clinic, Inc. Lifeline/ Pregnant Women's Program
35 Russell Street
New Britain, CT 06052

NEW CANAAN

Connecticut Communities for Drug Free Youth, Inc.
11 Farm Road
Room 149
New Canaan, CT 06840

New Canaan Cares, Inc. Helpline and Library
Farm Road
New Canaan High School
New Canaan, CT 06840

Youth Options New Canaan Unit
11 Forest Street
New Canaan, CT 06840

NEW HAVEN

Affiliates for Consultation and Therapy
389 Orange Street
New Haven, CT 06511

Alcohol Services Organization of South Central Connecticut, Inc.
871 State Street
New Haven, CT 06511

Alcohol Treatment Unit Substance Abuse Treatment Unit/OP TRT
285 Orchard Street
New Haven, CT 06511

APT Foundation, Inc.
904 Howard Avenue
New Haven, CT 06519

Central Treatment Unit/Cocaine and Pregnant Women
914 1/2 Howard Avenue
New Haven, CT 06519

Legion Avenue Clinic/Methadone
60–62 Legion Avenue
New Haven, CT 06519

Community Addiction Services New Haven Unit
830 Grand Avenue
New Haven, CT 06511

Connecticut AIDS Residence Program (CARP) Outpatient
254 College Street
Suite 205
New Haven, CT 06510

Connecticut Mental Health Center Hispanic Clinic
1 Long Wharf Drive
New Haven, CT 06511

Crossroads, Inc.
54 East Ramsdell Street
New Haven, CT 06515

Evaluation and Brief Treatment Unit
285 Orchard Street
New Haven, CT 06511

Latino Youth Development, Inc.
155 Minor Street
New Haven, CT 06519

Orchard Clinic Methadone
540 Ella T Grasso Boulevard
New Haven, CT 06519

Park Hill Clinic Outpatient Methadone
540 Ella T Grasso Boulevard
New Haven, CT 06519

Psychological Evaluation Services of New Haven
436 Orange Street
New Haven, CT 06511

Research Buprenorphine Study Outpatient Maintenance
60–62 Legion Avenue
New Haven, CT 06519

Satu Outpatient Services
904 Howard Avenue
New Haven, CT 06519

Shirley Frank Foundation
Administrative Unit
Outpatient
659 George Street
New Haven, CT 06511

United Way of Connecticut Infoline/South Central
419 Whalley Avenue
New Haven, CT 06511

Veterans' Affairs Medical Center Substance Abuse Treatment Program
950 Campbell Avenue
Building 36 Room 11
New Haven, CT 06516

Wakeman Hall at the Children's Center
1400 Whitney Avenue
New Haven, CT 06517

NEW LONDON

Care Center
516 Vauxhall Street
Suite 102
New London, CT 06320

Community Addiction Services New London Unit
153 Williams Street
New London, CT 06320

Family Services Association of Southern New London County
11 Granite Street
New London, CT 06320

Hartford Dispensary New London Clinic
931 Bank Street
New London, CT 06320

SE Council on Alcohol and Drug Dependency, Inc. (SCADD)
Altruism House/Male
189 Howard Street
New London, CT 06320

Altruism House/Women
1000 Bank Street
New London, CT 06320

Outpatient Program Detox
47 Coit Street
New London, CT 06320

NEW MILFORD

New Milford Youth Agency
50 East Street
New Milford, CT 06776

NEWTOWN

Alpha House Residential
Mile Hill Road
Newtown, CT 06470

Berkshire Woods Chemical Dependency Treatment Center
Detoxification Unit
Fairfield Hills Hospital
Kent House
Newtown, CT 06470

Residential Drug Free Unit
Litchfield Building
Newtown, CT 06470

Daytop Residential Drug Free Unit
Mile Hill Road
Newtown, CT 06470

NORTH STONINGTON

Stonington Institute
Administrative Unit
Infirmary
Intensive Residential Unit
Partial Hospitalization Program
Swantown Hill Road
North Stonington, CT 06359

NORWALK

Connecticut Counseling Centers, Inc.
Norwalk Methadone Program
Norwalk Outpatient Treatment Program
31 West Avenue
Norwalk, CT 06854

Connecticut Renaissance, Inc.
Administrative Unit
Norwalk Outpatient Unit
83 Wall Street
Norwalk, CT 06850

United Way of Connecticut Infoline/Southwest
83 East Avenue
Norwalk, CT 06851

Vitam Center, Inc.
Administrative Unit
Residential Drug Free Unit
57 West Rocks Road
Norwalk, CT 06852

NORWICH

Eugene T. Boneski Chemical Dependence Treatment Center
Route 12
Gallup Building
Norwich, CT 06360

Hartford Dispensary Norwich Clinic
Norwich Hospital
Lippett Building
Norwich, CT 06360

SE Council on Alcohol and Drug Dependency, Inc. (SCADD) Altruism House/Male
313 Main Street
Norwich, CT 06360

United Way of Connecticut Infoline/East
74 West Main Street
Norwich, CT 06360

OLD SAYBROOK

The Connection, Inc. Valley Shore Counseling Center
263 Main Street
Suite 108
Old Saybrook, CT 06475

PLAINVILLE

Wheeler Clinic, Inc.
Alcohol and Drug Abuse Unit
Night Treatment Program
91 Northwest Drive
Plainville, CT 06062

PORTLAND

Stonehaven
325 Main Street
Portland, CT 06480

Stonehaven for Women
315 Main Street
Portland, CT 06480

PUTNAM

Milestone
Intensive Program
Intermediate Program
391 Pomfret Street
Putnam, CT 06260

SANDY HOOK

Parkside Lodge of Eagle Hill
28 Alberts Hill Road
Sandy Hook, CT 06482

SHARON

Midwestern Connecticut Council on Alcoholism Trinity Glen
149 West Cornwall Road
Sharon, CT 06069

The McCall Foundation, Inc. Sharon Office/Hamlin House
Upper Main Street
Sharon, CT 06069

SOUTH WINDSOR

Connecticut North Treatment Center
15 Morgan Farms Drive
South Windsor, CT 06074

STAFFORD SPRINGS

Stafford Human Services
c/o Warren Memorial Town Hall
Stafford Springs, CT 06076

STAMFORD

Alcohol and Drug Abuse Council
Administrative Unit
Intensive Outpatient
Outpatient Unit
159 Colonial Road
Stamford, CT 06902

Liberation Programs, Inc.
Administrative Unit/Liberation House
119 Main Street
Stamford, CT 06901

Liberation Clinic
125 Main Street
Stamford, CT 06901

Main Street Clinic Treatment Program
115 Main Street
Stamford, CT 06901

McKinney House
8 Woodland Place
Stamford, CT 06902

Saint Luke's Community Services
8 Clinton Avenue
Stamford, CT 06902

Meridian House
929 Newfield Avenue
Stamford, CT 06905

Viewpoint Recovery Program Intermediate Long Term
104–106 Richmond Hill Avenue
Stamford, CT 06902

Youth Options Stamford Unit
141 Franklin Street
Stamford, CT 06901

STRATFORD

Stratford Community Services
2730 Main Street
Stratford, CT 06497

TORRINGTON

Catholic Family Services
132 Grove Street
Torrington, CT 06790

McCall Foundation, Inc.
Administrative Unit
Evening Program
Outpatient Program
58 High Street
Torrington, CT 06790

McCall House/Intensive Residential
McCall House/Intermediate Residential
127 Midgeon Avenue
Torrington, CT 06790

WATERBURY

Central Naugatuck Valley Help, Inc.
Administrative Unit
Non Residential Program
Residential Unit
900 Watertown Avenue
Waterbury, CT 06708

Community Addiction Services Waterbury Unit
232 North Elm Street
Waterbury, CT 06702

Connecticut Counseling Centers, Inc.
Waterbury Methadone Program
Waterbury Outpatient Program
951 Chase Parkway
Waterbury, CT 06708

Connecticut Renaissance, Inc. Residential Treatment Facility
31 Wolcott Street
Waterbury, CT 06702

Family Intervention Center
1875 Thomaston Avenue
Waterbury, CT 06704

Family Service of Greater Waterbury
34 Murray Street
Waterbury, CT 06710

Morris Foundation, Inc.
Administrative Unit
Center for Alcohol and Drug-Free Living
26 North Elm Street
Waterbury, CT 06702

Driving While Intoxicated Therapeutic Shelter
142 Griggs Street
Waterbury, CT 06702

Morris/Kendall House
28 North Elm Street
Waterbury, CT 06702

United Way of Connecticut Infoline/Northwest
232 North Elm Street
Waterbury, CT 06702

WESTPORT

Alcohol and Drug Dependency Council, Inc.
One Kings Highway North
Westport, CT 06880

WILLIMANTIC

New Perceptions Counseling Service
1003 Main Street
Willimantic, CT 06226

Northeast Connecticut Alcohol Council
Outpatient Program
Thomas Murphy Center
1491 West Main Street
Willimantic, CT 06226

Perception House Residential Drug Free
134 Church Street
Willimantic, CT 06226

United Services, Inc. Addiction Recovery Services
132 Mansfield Avenue
Willimantic, CT 06226

WINDHAM

Hartford Dispensary Willimantic Clinic
54–56 Boston Post Road
Windham, CT 06280

WINSTED

The McCall Foundation, Inc. Winsted Office
231 North Main Street
Winsted, CT 06098

DELAWARE

CLAYMONT

Open Door, Inc.
3301 Green Street
Claymont, DE 19703

DELAWARE CITY

Glass House
Governor Bacon Health Center
Delaware City, DE 19706

Northeast Treatment Centers
Long Term Care Programs
Recovery Center of Delaware
Sentac Program
Governor Bacon Health Center
Cottage 5
Delaware City, DE 19706

DOVER

APR Counseling Associates
71 South Shore Drive
Dover, DE 19901

Because We Care II
48A McKee Road
Dover, DE 19901

Delaware Drinking Driver Program Kent County Unit
1661 South Dupont Highway
Dover, DE 19901

Delmarva Rural Ministries Hispanic Outreach Project
26 Wyoming Avenue
Dover, DE 19901

Dover Psychiatry Services Substance Abuse Services
1001 South Bradford Street
Dover, DE 19901

Kent County Counseling Services
1525 Lebanon Road
Route 10
Dover, DE 19901

Kent General Hospital Adult/ Adolescent Psychiatric Treatment Services/Substance Abuse Services
640 South State Street
Dover, DE 19901

New Beginnings/Dover
707 Walker Road
Dover, DE 19901

Serenity Place
327 Martin Street
Dover, DE 19901

YMCA of Delaware/Dover Resource Center
120 North State Street
Wesley College
Dover, DE 19901

DOVER AIR FORCE BASE

Dover Air Force Base Social Actions
13 Street
Building 520 Room 112
Dover Air Force Base, DE 19902

ELLENDALE

Kent/Sussex Detoxification Center
Main Street
Ellendale School House
Ellendale, DE 19941

GEORGETOWN

Corinthian House
219–221 South Race Street
Georgetown, DE 19947

Delaware Drinking Driver Program Sussex County Unit
510 North Dupont Highway
The 113 Professional Building
Georgetown, DE 19947

Houston Hall
431 East Market Street
Georgetown, DE 19947

Tau House
11 West Pine Street
Georgetown, DE 19947

Turnabout Counseling Center
8 North Race Street
Georgetown, DE 19947

HOCKESSIN

Greenwood
Old Lancaster Pike
Hockessin, DE 19707

LEWES

Beebe Hospital Psychiatric Service Substance Abuse Services
424 Savannah Road
Lewes, DE 19958

Dry Dock, Inc. Inebriate Transportation Program
Angola Grange
Lewes, DE 19958

MILFORD

People's Place Counseling Center
219 South Walnut Street
Milford, DE 19963

NEW CASTLE

Meadow Wood Hospital for Adolescents Substance Abuse Treatment Program
575 South Dupont Highway
New Castle, DE 19720

NEWARK

Eugenia Counseling Center
Route 273 Christiana Village
Professional Center Suite 200
Newark, DE 19702

Newark Family Counseling Center
501 Ogletown Road
Hudson State Service Center
Newark, DE 19711

SMYRNA

Greentree Drug and Alcohol Program Delaware Correction Center
Route 1
Smyrna, DE 19977

WILMINGTON

A and D Associates The Promises
5211 West Woodmill Drive
Suite 36
Wilmington, DE 19808

American Council on Alcoholism and Drug Addiction
1225 King Street
Wilmington, DE 19801

Big Brothers/Big Sisters of Delaware Early Intervention Project
102 Middleboro Road
Wilmington, DE 19804

Brandywine Counseling, Inc.
2400 West 4 Street
Wilmington, DE 19805

Catholic Social Services Center for Pastoral Care
8 Old Church Road
Wilmington, DE 19807

Delaware Drinking Driver Program New Castle County Unit
5193 West Woodmill Drive
Woodmill Corporate Center Suite 28
Wilmington, DE 19805

Delaware Safety Council Social Drinkers Program
3836 Kennett Pike
Powder Mill Square
Wilmington, DE 19807

Family and Children's Services of Delaware
2005 Baynard Boulevard
Wilmington, DE 19802

Foundations
2901 Northeast Boulevard
Sojourners Place
Wilmington, DE 19802

Inroads
2110 Duncan Road
Wilmington, DE 19808

Key Program
1301 East 12 Street
Wilmington, DE 19809

Limen House for Men
903 Madison Street
Wilmington, DE 19801

Limen House for Women
624 North Broom Street
Wilmington, DE 19805

MCC Behavioral Care Substance Abuse Services
6 Denny Road
2 Fox Point Center
Wilmington, DE 19809

Net Counseling Center
2055 Limestone Road
Suite 201
Wilmington, DE 19808

New Beginnings/Hiddenbrook of Delaware Family Center
5205 West Woodmill Drive
Woodmill Corporate Center Suite 33
Wilmington, DE 19808

Pace, Inc.
5307 Limestone Road
Wilmington, DE 19808

People's Settlement
408 East 8 Street
Wilmington, DE 19801

Psychotherapeutic Services, Inc. Relapse Prevention/ Continuous Treatment
5149 West Woodmill Drive
Wilmington, DE 19808

Recovery Center of Delaware Kirkwood Detox
3315 Kirkwood Highway
Wilmington, DE 19808

Sodat Counseling and Evaluation Center
625 Orange Street
Wilmington, DE 19801

Union Baptist Community Services, Inc. Anti-Drug Outreach Program
2801 North Pine Street
Wilmington, DE 19802

Veterans' Affairs Medical Center Substance Abuse Treatment Program
1601 Kirkwood Highway
Wilmington, DE 19805

Women's Place Won
1011 North Clayton Street
Wilmington, DE 19805

YMCA of Delaware/Wilmington Resource Center
501 West 11 Street
Wilmington, DE 19801

DISTRICT OF COLUMBIA

WASHINGTON

A. L. Nellum and Associates, Inc.
Adero House
2700 Martin Luther King Avenue, SE
Saint Elizabeth's Hospital, Building N
Washington, DC 20036

Comp. Alcohol and Drug Abuse Treatment Center (CADAC)
2700 Martin Luther King Avenue SE
Saint Elizabeth's Hospital, Building P
Washington, DC 20032

Adams Mill Alcohol Treatment Center
1808 Adams Mill Road NW
Washington, DC 20009

ADASA/CPH/Bureau of Alcoholism Treatment Services Pregnant/Postpartum Women and Infants
1900 Massachusetts Avenue SE
DC General Hospital Karrick Hall
Washington, DC 20003

Against Drugs and Alcohol with Planned Treatment (ADAPT)
33 N Street NE
Washington, DC 20002

Alcohol and Drug Abuse Services Admin. (ADASA)
Substance Abuse Detox Center
1900 Massachusetts Avenue SE
Building 12
Washington, DC 20003

Youth Abstinence Program
2146 24 Place NE
Washington, DC 20005

Andromeda Transcultural Hispanic Mental Health Center
1400 Decatur Street NW
Washington, DC 20011

Bureau of Rehabilitation, Inc.
Community Care Center
3301 16 Street NW
Washington, DC 20010

Shaw Residence 1
1770 Park Road NW
Washington, DC 20010

Shaw Residence 2
1740 Park Road NW
Washington, DC 20010

Shaw Residence 3
1301 Clifton Street NW
Washington, DC 20009

Center for Child Protection and Family Support
714 G Street SE
Washington, DC 20036

Central Intake Division Site A
1300 First Street NE
Washington, DC 20002

Change, Inc. Consulting and Mental Health Services
5000 Nannie Helen Burroughs Avenue NE
Washington, DC 20019

Church Association for Community Services
712 Randolph Street NW
Washington, DC 20011

Coalition for the Homeless Return from Addiction
1234 Massachusetts Avenue NW
Washington, DC 20005

Community Research, Inc. Go-Getters Entrepreneurial Youth Club
1840-B Fenwick Street NE
Washington, DC 20002

Concerned Citizens on Alcohol and Drug Abuse, Inc. (CCADA)
3115 Martin Luther King Avenue SE
Washington, DC 20032

Corporation Against Drug Abuse
1010 Wisconsin Avenue NW
Suite 250
Washington, DC 20007

Criminal Justice Division Client Tracking Information Branch
33 N Street NE
2nd Floor
Washington, DC 20002

DC Department of Corrections Office of Substance Abuse Services
1901 D Street SE
Washington, DC 20003

DC Employee Consultation and Counseling Service
1825 Connecticut Avenue NW
Suite 101
Washington, DC 20009

DC General Hospital
Dept. of Psychiatry Substance Abuse Program
1900 Massachusetts Avenue SE
DC General Hospital Unit 42
Washington, DC 20003

Prenatal Substance Abuse Program
19 Street and Massachusetts Avenue SE
OB-GYN Department
Washington, DC 20003

DC Public Schools Substance Abuse Prevention Education
3rd and G Streets SE
Giddings School
Washington, DC 20003

DC Superior Court Social Services Division Probation/ Parole Resource Center (PPRC)
409 E Street NW
Building B Room 205
Washington, DC 20001

Deafpride, Inc. Project A Second Chance
1350 Potomac Avenue SE
Washington, DC 20003

Foundation for Contemporary Mental Health Next Step
2112 F Street NW
Suite 404
Washington, DC 20037

Georgetown University Alcohol and Drug Abuse Program
3800 Reservoir Road NW
Washington, DC 20007

Greater Southeast Community Hospital Alcohol and Drug Treatment Unit
1310 Southern Avenue SE
Washington, DC 20032

Holy Comforter/Saint Cyprian Community Action Group/ Carriage House
901 Pennsylvania Avenue SE
Washington, DC 20003

Karrick Hall Residential Treatment Unit
1900 Massachusetts Avenue SE
Building 17
Washington, DC 20003

Kenilworth Parkside
Resident Management Treatment Program
4500 Quarles Street NE
Washington, DC 20019

Substance Abuse Prevention Program
4513 Quarles Street NE
Washington, DC 20019

Koba Associates Diagnostic Unit
1300 First Street NE
Suite 214
Washington, DC 20003

Kolmac Clinic
1411 K Street NW
Suite 703
Washington, DC 20005

Latin American Youth Center Drug Treatment Program
3045 15 Street NW
Washington, DC 20009

Links Foundation Project Lead/ High Expectations
1200 Massachusetts Avenue NW
Washington, DC 20005

Living Stage Theatre Company
1901 14 Street NW
Washington, DC 20009

Mary E. Herring Aftercare Facility
700 Monroe Street NE
Washington, DC 20017

Moving Addicts Toward Self Sufficiency
33 N Street NE
Washington, DC 20002

Naval District Washington Counseling and Assistance Center
2701 South Capitol Street SE
Building 72 Anacostia
Washington, DC 20374

Parklands Community Center, Inc.
3320 Stanton Road SE
Lower Level
Washington, DC 20020

Partners in Drug Abuse Rehabilitation and Counseling (PIDARC)
2112 F Street NW
Suite 101
Washington, DC 20037

Pettsons, Inc. Ward 8 Substance Abuse Prevention Initiative
2804 Martin Luther King Jr. Avenue SE
Washington, DC 20032

Project Care
415 Edgewood Street NE
Suite B-3
Washington, DC 20017

Rap, Inc.
3451 Holmead Place NW
Washington, DC 20010

Sasha Bruce Youthwork, Inc.
1022 Maryland Avenue NE
Washington, DC 20002

Second Genesis Residential Therapeutic Community DC Clinic
1320 Harvard Street NW
Washington, DC 20009

Service Helping Addicts Come Klean (SHACK)
123 K Street SE
Washington, DC 20003

Shaw Abstinence Program
1300 First Street NE
Washington, DC 20002

So Others May Eat, Inc. (SOME) Addiction Treatment Project
71 O Street NW
Washington, DC 20001

The Model Treatment Program Tyrone Patterson
1300 First Street NE
Washington, DC 20002

Therapeutic Outcomes Through Prescribed Strategies
25 K Street NE
3rd Floor
Washington, DC 20002

Treatment and Rehabilitation for Addicts in Need (Train II)
1905 E Street SE
Building 14
Washington, DC 20003

United Planning Organization Substance Abuse Program
810 Potomac Avenue SE
Washington, DC 20003

Upper Cardozo Abstinence Program
3020 14 Street NW
Washington, DC 20010

Urban Resources, Inc.
2909 Georgia Avenue NW
Washington, DC 20001

Washington Area Council on Alcoholism and Drug Abuse Inc./Comp Counseling Center
2813 12 Street NE
Washington, DC 20017

Washington Assessment/Therapy Services
3801 Connecticut Avenue NW
Suite 203
Washington, DC 20008

Washington Hospital Center Chemical Dependency Detoxification Unit 2K
110 Irving Street NW
Washington, DC 20010

Whitman Walker Clinic, Inc. Alcoholism and Substance Abuse Services
1407 S Street NW
Washington, DC 20009

Women's Services Clinic
1900 Massachusetts Avenue SE
Building 13
Washington, DC 20003

FEDERATED MICRONESIA

KOSRAE

Department of Health Services Substance Abuse Services
Kosrae State Hospital
Kosrae, FM 96944

Kosrae Community Action Program
Kosrae, FM 96944

PONAPE

Department of Health Services Community Mental Health Center
Ponape, FM 96941

Micronesia Bound, Inc. Outward Bound Aramas KAPW
Ponape, FM 96941

TRUK

Department of Health Services Substance Abuse Services
Chuuk State Hospital
Truk, FM 96942

YAP

Department of Youth and Civic Affairs Yap Youth Substance Abuse Prevention
Yap, FM 96943

Yap Memorial Hospital Department of Health Services
Yap, FM 96943

FLORIDA

ALTAMONTE SPRINGS

Addictions Compulsions Treatment Center (ACT, Inc.) Altamonte Springs Outpatient
801 State Road 436
Suite 1083
Altamonte Springs, FL 32714

Christian Family Institute
375 Douglas Avenue
Suite 1007
Altamonte Springs, FL 32714

Cornerstone Institute, Inc.
400 Maitland Avenue
Altamonte Springs, FL 32701

Grove Counseling Center, Inc. TCEP Altamonte Springs
927 Hickory Street
Altamonte Springs, FL 32701

Lifeworks Center, Inc./Altamonte Springs
385 Whooping Loop
Suite 1303
Altamonte Springs, FL 32701

New Leaf Center, Inc.
600 South North Lake Boulevard
Suite 200
Altamonte Springs, FL 32701

Quest Counseling Centre Young Recovery
711 Ballard Street
Altamonte Springs, FL 32701

Renaissance Counseling Center, Inc.
370 Whooping Loop
Suite 1154
Altamonte Springs, FL 32701

Seminole Community Mental Health Center Seminole Children's Resource Center
417 Whooping Loop
Suite 1721
Altamonte Springs, FL 32701

APALACHICOLA

Apalachee Center for Human Services Apalachicola
159 12 Street
Apalachicola, FL 32320

APOPKA

Center for Drug Free Living, Inc. South Apopka
445 West 13 Street
Apopka, FL 32703

Temporary Living Center/ Apopka
1717 Piedmont/Wekiva Roads
Apopka, FL 32703

AVON PARK

Tri-County Addictions Rehab Services, Inc.
Avon Park Outpatient
2801 U.S. 27 South
Avon Park, FL 33825

Florida Center
100 West College Drive
Avon Park, FL 33825

The Retreat
Trout Lake Camp/Adolescent Services
Route 6
Avon Park, FL 33825

BARTOW

Program to Aid Drug Abusers, Inc. (PAD) TASC
180 North Central Avenue
Bartow, FL 33830

Tri-County Addictions Rehab Services, Inc.
Bartow Adolescent Center
Detoxification Unit
2725 Highway 60 East
Bartow, FL 33830

BAY PINES

Veterans' Affairs Medical Center Substance Abuse Treatment Program
10000 Bay Pines Boulevard
Bay Pines, FL 33504

BELLE GLADE

Comp. Alcoholism Rehab Programs, Inc. (CARP)
Adolescent Residential Treatment
Glades Area Programs/Outpatient
125 NW 2 Street
Belle Glade, FL 33430

Glades Area Programs
141 NW 2 Street
Belle Glade, FL 33430

Drug Abuse Foundation of Palm Beach County/Belle Glade
38840 State Road 80
Belle Glade, FL 33480

West County Mental Health
Alpha I/Lake Shore Middle School
1102 West Avenue A
Belle Glade, FL 33430

Alpha II/Belle Glade Elementary School
7 and Canal Streets NW
Belle Glade, FL 33430

Glades Crisis Stabilization Unit
808 NW Avenue D
Belle Glade, FL 33430

Outpatient Substance Abuse Services
1024 NW Avenue D
Belle Glade, FL 33430

Panda
572 SW 2 Street
Belle Glade, FL 33430

BELLEVIEW

New Beginnings Belleview Unit
10242 South Highway 441
Suite B-1
Belleview, FL 32620

BLOUNTSTOWN

Chemical Addictions Recovery Effort Calhoun County Outpatient Office
425 East Central Avenue
Room G-8
Blountstown, FL 32424

BOCA RATON

Alternatives in Treatment, Inc.
7601 North Federal Highway
Suite 260-B
Boca Raton, FL 33487

Center for Family Services Boca Raton Unit
500 NE Spanish River Boulevard
Suite 14
Boca Raton, FL 33401

Drug Abuse Foundation of Palm Beach County
Boca Raton I
1251 NW 8 Street
Boca Raton, FL 33432

Boca Raton II
5775 Jog Road
Boca Raton, FL 33496

Interphase Recovery Program, Inc.
23120 Sandlefoot Plaza Drive
Boca Raton, FL 33428

Lifeskills of Boca Raton
7301A West Palmetto Park Road
Suite 300B
Boca Raton, FL 33433

National Recovery Institute
Outpatient Ambulatory
Boca Raton/Residential Level I
1000 NW 15 Street
Boca Raton, FL 33486

Renaissance Institute of Palm Beach
7300 North Federal Highway
Suite 201
Boca Raton, FL 33487

BOYNTON BEACH

Atlantic Counseling
200 Knuth Road
Suite 238
Boynton Beach, FL 33436

Drug Abuse Foundation of Palm Beach County/Boynton Beach
101 East Congress Avenue
Boynton Beach, FL 33432

BRADENTON

Coastal Recovery Centers Clinic IV
410 Cortez Road West
Suite 410
Bradenton, FL 34207

Manatee Community College
Institute of Alcohol and Drug Education
26 Street At 60 Avenue
Bradenton, FL 34206

Adolescent Recovery Center
1819 5 Street West
Bradenton, FL 34205

Alpha/Daughtrey
515 63 Avenue East
Bradenton, FL 34203

Alpha/Samoset
1720 33 Avenue East
Bradenton, FL 34208

Employee Assistance Program (EAP)
239–301 Boulevard East
Bradenton, FL 34208

Glen Oaks Hospital and Crisis Center/Detox
2020 26 Avenue East
Bradenton, FL 34208

Outpatient/Youth and Adult
375 6 Avenue West
Bradenton, FL 34209

Manatee Palms Adolescent Specialty Hospital Residential Treatment Center
1324 37 Avenue East
Bradenton, FL 34208

Operation PAR, Inc. Narc Addiction Treatment Center/ Bradenton
3915 8 Avenue West
Bradenton, FL 34205

BRANDON

Drug Abuse Comp. Coord Office (DACCO) Brandon Alternative
521 Wilbur Street
Brandon, FL 33511

BRISTOL

Apalachee Center for Human Services Bristol Outpatient
Bristol, FL 32321

BRONSON

Mental Health Services of North Central Florida DBA Community Counseling Center/ Bronson
CR 32
Bronson, FL 32621

BROOKSVILLE

Brooksville Drug Treatment Center
16415 Springhill Drive
Brooksville, FL 32609

Hernando County DUI Programs
1036 Varsity Drive
Brooksville, FL 34601

Professional Therapy Centers, Inc. Substance Abuse Services
11331 Ponce de Leon Boulevard
Brooksville, FL 34601

BUNNELL

ACT Corporation TASC Program
507 North State Street
Bunnell, FL 32010

Leon F. Stewart Treatment Center, Inc. Palm Coast Outpatient Services
302 1/2 Moody Boulevard
Bunnell, FL 32110

BUSHNELL

New Beginnings
317 North Highway 301
Bushnell, FL 33513

CANTONMENT

Community Drug and Alcohol Commission Beta Project
1000 West Kingsfield Road
Cantonment, FL 32533

CAPE CORAL

Outreach, Inc.
1105 6 Court SE
Cape Coral, FL 33990

CASSELBERRY

Addictions Compulsions Treatment Center (ACT, Inc.) Casselberry Outpatient
4300 South Highway 17-92
Casselberry, FL 32707

Sunrise Counseling Center, Inc. Unit 1
274 Wilshire Boulevard
Suite 221
Casselberry, FL 32707

CECIL FIELD NAVAL AIR STATION

Navy Counseling and Assistance Center Level II Treatment Program/Outpatient
Naval Air Station Cecil Field
Building 24
Cecil Field Naval Air Station, FL 322

CHATTAHOOCHEE

Florida State Hospital Addiction Services
Chattahoochee, FL 32324

CHIEFLAND

Henry and Rilla White Foundation Florida School for Youth Achievement
Levy County Road 218
Chiefland, FL 32626

CHIPLEY

Chemical Addictions Recovery Effort Holmes/Washington County Outpatient Office
995 Highway 77 South
Chipley, FL 32428

CLEARWATER

Community Counseling/ Diagnostic Services
4625 East Bay Drive
Suite 107
Clearwater, FL 34624

Fairwinds Treatment Center Residential
1569 South Fort Harrison Street
Clearwater, FL 34616

Family Resources, Inc. Lincoln Avenue Prevention and Early Intervention
626 Lakeview Road
Clearwater, FL 34616

Family Resources, Inc. Residential Services/North
1622 Turner Street
Clearwater, FL 34616

Family Services Centers
Clearwater
2960 Roosevelt Boulevard
Clearwater, FL 34620

North Clearwater
29605 U.S. 19 North
Suite 110
Clearwater, FL 34621

Operation PAR, Inc.
DUI Central Diagnostic Unit
Fact Team
4900 Creekside Drive
Suite 4908-B
Clearwater, FL 34620

Largo Community Correctional Center
5201 Ulmerton Road
Clearwater, FL 34620

Offender Treatment
TASC in Jail Program
TASC/Juvenile/Juvenile Diversion
4400 140 Avenue North
Clearwater, FL 34622

Prevention Programs
Primary Prevention Program
19329 U.S. 19 North #415
Arbor Shoreline Office Park
Clearwater, FL 34624

The Gardens
2960 Tanglewood Drive
Units G and H
Clearwater, FL 34622

Professional Comp. Addictive Services, Inc.
Criminal Justice Liaison
Detoxification Unit
Opportunity House
Residential
6150 150 Avenue North
Clearwater, FL 34620

Vantage Point Counseling and Consulting, Inc.
2753 State Road 580
Suite 212
Clearwater, FL 34621

COCOA

Central Florida Substance Abuse Treatment Centers, Inc. Cocoa Treatment Center
1048-A Dixon Boulevard
Cocoa, FL 32922

Crisis Services of Brevard, Inc.
865 North Cocoa Boulevard
Cocoa, FL 32922

COOPER CITY

High Point
5960 SW 106 Avenue
Cooper City, FL 33328

CRAWFORDVILLE

Apalachee Center for Human Services Wakulla Outpatient Office
1 Harper Street
Crawfordville, FL 32327

CROSS CITY

Mental Health Services of North Central Florida Community Counseling Center/Cross City
Airport Prison Road
Cross City, FL 32628

DADE CITY

Human Development Center of Pasco County East Pasco Outpatient
608 West Howard Avenue
Dade City, FL 33525

Pride, Inc. Dade City Outpatient
104 North 5 Street
Dade City, FL 33525

Youth and Family Alternatives, Inc. Focus East/Dade City Drug Prevention Services
2023 U.S. Highway 301 South
Dade City, FL 33525

DAVIE

Seafield Center South, Inc.
5151 SW 61 Avenue
Davie, FL 33314

DAYTONA BEACH

ACT Corporation
Criminal Justice Services
948 Orange Avenue
Daytona Beach, FL 32114

Reality House
1341 Indian Lake Road
Daytona Beach, FL 32114

TASC Center
440 1/2 Beach Street
Daytona Beach, FL 32014

Anti-Recidivist Effort, Inc. (ARE)
Administrative Unit
308 South Martin Luther King Boulevard
Daytona Beach, FL 32114

Assessment and Evaluation Center
Outpatient
841 Bethune Village
Daytona Beach, FL 32114

Hope House
854 Pine Haven Street
Daytona Beach, FL 32114

Atlantic Treatment Center, Inc. DBA Atlantic Shores Hospital/ Breaking Away
841 Jimmy Ann Drive
Daytona Beach, FL 32117

Daytona Methadone Treatment Center
737 Volusia Avenue
Daytona Beach, FL 32114

I Care
1122 2 Avenue
Daytona Beach, FL 32114

Leon F. Stewart Treatment Center, Inc.
Detox Unit
1200 Red John Road
Daytona Beach, FL 32114

Outpatient Unit 2
129 Michigan Avenue
Daytona Beach, FL 32114

Outpatient Unit 3
543 Orange Avenue
Daytona Beach, FL 32114

Residential Treatment
3875 Tiger Bay Road
Daytona Beach, FL 32124

Women's Halfway House
417 Daytona Street
Daytona Beach, FL 32114

Miles and Associates/Daytona Beach Alcohol/Drug Intervention/Prevention Services
955 Orange Avenue
Suite M
Daytona Beach, FL 32114

Milestones, Inc. Center for Substance Abuse Intervention
1501 Ridgewood Avenue
Daytona Beach, FL 32117

Pioneer Healthcare, Inc.
433 Silver Beach Avenue
Suite 103
Daytona Beach, FL 32118

Salvation Army/Serenity House
560 Ballough Road
Daytona Beach, FL 32114

Serenity House of Volusia, Inc. Serenity House
547 High Street
Daytona Beach, FL 32114

Volusia/Flagler Safety Council, Inc.
65 Coral Sea Avenue
Daytona Beach, FL 32114

DE FUNIAK SPRINGS

Cope Alcohol and Drug Program
South 331 and Coy Burgess Road
De Funiak Springs, FL 32433

DE LAND

ACT Corporation De Land Outpatient Treatment
803 Woodland Boulevard
De Land, FL 32720

Community Outreach Services, Inc. De Land Residential/ Outpatient Unit 1
245 South Amelia Street
De Land, FL 32724

Lifeline Counseling Associates, Inc.
125 West Plymouth Avenue
Suite C
De Land, FL 32720

Miles and Associates/De Land Alcohol/Drug Intervention/ Prevention Services
620 East New York Avenue
Suite A
De Land, FL 32720

The House Next Door
121 West Pennsylvania Avenue
De Land, FL 32720

West Volusia Memorial Hospital Psychiatric Services/Substance Abuse Services
701 West Plymouth Avenue
De Land, FL 32721

DELRAY BEACH

American Biodyne, Inc. Delray Beach Unit
1300 NW 17 Avenue
Suite 140
Delray Beach, FL 33435

Drug Abuse Foundation of Palm Beach County
Linton Blvd. Unit
400 South Swinton Avenue
Delray Beach, FL 33444

Oak Square
2702 North Federal Highway
Delray Beach, FL 33444

RTC
400-B South Swinton Avenue
Delray Beach, FL 33444

14th Avenue Unit
301 SW 14 Avenue
Delray Beach, FL 33444

Pathways to Recovery, Inc. Extended Care Facility for Adult Men
13132 Barwick Road
Delray Beach, FL 33445

South County Mental Health Center Substance Abuse Treatment Program Unit 1
16155 South Military Trail
Delray Beach, FL 33484

The Beachcomber
4493 North Ocean Boulevard
Delray Beach, FL 33483

Wayside House
378 NE 6 Avenue
Delray Beach, FL 33483

DUNEDIN

Mac Jacobs Ma Cap
1340 Bayshore Boulevard
Dunedin, FL 34697

Professional Comp. Addiction Services Wellness Center/ Dunedin
537 Douglas Avenue
Suites 17-A and 18-A
Dunedin, FL 33528

EGLIN AIR FORCE BASE

Federal Prison Camp Drug Education Class
Eglin Air Force Base, FL 32542

EUSTIS

Lake/Sumter CMHC Crossroads II
115 Citrus Avenue
Eustis, FL 32726

FORT LAUDERDALE

Alternative Substance Abuse Systems, Inc.
613 SE First Avenue
Fort Lauderdale, FL 33301

Broward Addiction Recovery Center (BARC)
1000 SW 2 Street
Fort Lauderdale, FL 33312

ATACC
Drug Court Treatment
601 South Andrews Street
Fort Lauderdale, FL 33301

West/Lauderdale Lakes
4487 North State Road 7
Fort Lauderdale, FL 33319

Alcohol/Drug Abuse Education and Prevention
600 SE 3 Avenue
4th Floor
Fort Lauderdale, FL 33301

DUI Facility
5400 NW 9 Avenue
Fort Lauderdale, FL 33309

Coral Ridge Psychiatric Hospital Alcohol and Addiction Treatment and Rehab Program
4545 North Federal Highway
Fort Lauderdale, FL 33308

Court Alcohol and Substance Abuse Program
624 SW First Avenue
Fort Lauderdale, FL 33301

Covenant House
Addictions Management Project
733 Breakers Avenue
Fort Lauderdale, FL 33304

Non-Residential
311 NE 3 Street
Fort Lauderdale, FL 33304

CPC Fort Lauderdale Hospital Counterpoint
1601 East Las Olas Boulevard
Fort Lauderdale, FL 33301

Family Institute/Fort Lauderdale
1144 SE 3 Avenue
Fort Lauderdale, FL 33316

Fort Lauderdale Counseling Services
1215 SE 2 Avenue
Fort Lauderdale, FL 33316

House of Hope, Inc. Stepping Stones Residential/First Street
908 SW First Street
Fort Lauderdale, FL 33312

MCC Behavioral Care, Inc. Fort Lauderdale Substance Abuse Services
3313 West Commercial Boulevard
Suite 112
Fort Lauderdale, FL 33309

North Broward Hospital District Children's Diagnostic and Treatment Center
303 SE 17 Street
Fort Lauderdale, FL 33316

Nova University Counseling Program
4801 South University Drive
Suite 258
Fort Lauderdale, FL 33328

Recovery Resources
7200 Griffin Road
Suite 6
Fort Lauderdale, FL 33314

Retreat/Sunrise Adult Chemical Dependency Program
555 SW 148 Avenue
Fort Lauderdale, FL 33325

Spectrum Programs, Inc.
Broward Outpatient
2455 East Sunrise Boulevard
Suite 416
Fort Lauderdale, FL 33304

Residential Level 2
2301 Wilton Drive
Fort Lauderdale, FL 33305

The Seed, Inc.
1313 South Andrews Avenue
Fort Lauderdale, FL 33316

Urban League of Broward County
11 NW 36 Avenue
Fort Lauderdale, FL 33311

FORT MYERS

Addiction Recovery Center
3949 Evans Avenue
Suite 203
Fort Myers, FL 33901

Age Link of Lee County, Inc.
6309 Corporate Court
Suite E
Fort Myers, FL 33919

Charter Glade Hospital Chemical Dependency Unit
3550 Colonial Boulevard
Fort Myers, FL 33906

Lee County Sheriff's Office Lee County Jail Substance Abuse Services
2115 Dr. Martin Luther King, Jr. Blvd
Fort Myers, FL 33901

Lee Mental Health Center, Inc.
Alpha II
4524 Tice Street
Fort Myers, FL 33905

Alpha Program
1858 Suncoast Lane
Fort Myers, FL 33917

Drug Abuse Unit
2789 Ortiz Avenue SE
Fort Myers, FL 33905

Family Life Center
4424 Michigan Avenue
Apartment 507
Fort Myers, FL 33916

Pride, Inc.
3049 Cleveland Avenue
Suite 101
Fort Myers, FL 33901

Southwest Florida Addiction Services, Inc.
Detoxification
2562 Dixie Parkway
Fort Myers, FL 33901

Residential and Outpatient
2101 McGregor Boulevard
Fort Myers, FL 33901

Residential Level 2
2450 Prince Street
Fort Myers, FL 33901

FORT PIERCE

Alpha Health Services
1025 Orange Avenue
Fort Pierce, FL 34950

Drug Abuse Treatment Association, Inc. (DATA) Norman C. Hayslip Treatment Center
4590 Selvitz Road
Fort Pierce, FL 34981

New Horizons of the Treasure Coast, Inc.
Children's Center
8247 South U.S. 1
Suite 100
Fort Pierce, FL 34952

Detoxification Unit
800 Avenue H
Fort Pierce, FL 33950

New Life
610 North 7 Street
Fort Pierce, FL 34950

Saint Lucie County Outpatient Branch
901 North 7 Street
Fort Pierce, FL 34950

TASC Program
417 Avenue H
Fort Pierce, FL 34950

Recovery Associates, Inc.
8000 South U.S. 1
Suite 202
Fort Pierce, FL 34952

FORT WALTON BEACH

Bridgeway Center Alcohol/Drug Program
205 Shell Avenue SE
Fort Walton Beach, FL 32548

FOUNTAIN

Chemical Addictions Recovery Effort Starting Over Straight (SOS)
Van Doren Lane
Fountain, FL 32438

GAINESVILLE

Campus Alcohol and Drug Resource Center Bacchus Prevention Program
University of Florida
P202 Peabody Hall
Gainesville, FL 32611

Charter Counseling Center/ Gainesville
611 NW 60 Street
Suite C
Gainesville, FL 32607

Corner Drug Store, Inc.
Myra Terwilliger Alpha Program
301 NW 62 Street
Gainesville, FL 32607

Outpatient Services
Prevention Project
1300 NW 6 Street
Gainesville, FL 32601

Crest Services, Inc.
1717 NE 9 Street
Suite 104
Gainesville, FL 32609

Mental Health Services of North Central Florida
Acute Care Services/Detox Unit
Administration/Treatment
4300 SW 13 Street
Gainesville, FL 32608

DBA Addictions and Family Health
DBA Bridgehouse
4400 SW 13 Street
Gainesville, FL 32608

DBA Alachua Associates
3601 SW 2 Avenue
Suite V
Gainesville, FL 32607

Metamorphosis Alachua County Drug Abuse Program
306 NE 7 Street
Gainesville, FL 32601

North Central Florida Safety Council DUI Program
3710 NW 51 Street
Suite A
Gainesville, FL 32606

Veterans' Affairs Medical Center Alcohol/Drug Dependence Treatment Program
1601 SW Archer Road
116A ADTP
Gainesville, FL 32602

GREEN COVE SPRINGS

River Region Human Services, Inc. Clay High School Prevention Program
2025 Highway 16 West
Green Cove Springs, FL 32043

GULF BREEZE

Twelve Oaks Alcohol and Drug Recovery Center Detox
2068 Healthcare Avenue
Route 1
Gulf Breeze, FL 32566

HIALEAH

Counseling and Evaluation Center/Hialeah
1490 West 49 Place
Suite 390
Hialeah, FL 33012

Dade Family Counseling, Inc. Hialeah Unit
60 East 3 Street
Hialeah, FL 33010

Economic Opportunity Family Health Center
Day Night Program
490 East Hialeah Drive
Hialeah, FL 33010

Reaves House/Women's Residential
2985 NW 54 Street
Hialeah, FL 33010

Hialeah Hospital Crossroads
651 East 25 Street
Hialeah, FL 33013

Northwest Dade Center, Inc.
4175 West 20 Avenue
Hialeah, FL 33012

HOLLYWOOD

Broward Addiction Recovery Center (BARC) South
4035 North 29 Avenue
Hollywood, FL 33020

Broward Methadone Maintenance Rehab and Research Facility
1101 South 21 Avenue
Hollywood, FL 33020

Lock Towns CMHC Sub. Arts Project/Dade County Dual Diagnosis
1000 SW 84 Avenue
Hollywood, FL 33025

Spectrum Programs, Inc. Hollywood Outpatient
2219 Hollywood Boulevard
Suite 102
Hollywood, FL 33020

The Starting Place, Inc. Residential and Outpatient
2057 Coolidge Street
Hollywood, FL 33020

HOMESTEAD

Associates for Psychological Services Homestead Alcohol Abuse Program
225 NE 8 Street
Suite 3
Homestead, FL 33030

Metro Dade Office of Rehab Services Jack Orr Ranch
31601 SW 197 Avenue
Homestead, FL 33030

IMMOKALEE

David Lawrence Center The Pines
425 North First Street
Immokalee, FL 33934

INDIANTOWN

Martin Drug Treatment Center
1175 SW Allapattah Road
Indiantown, FL 34956

INVERNESS

Marion/Citrus Mental Health Center, Inc. Residential Program
701 White Boulevard
Inverness, FL 32650

New Beginnings Inverness Unit
943 South Highway 41
Suite B
Inverness, FL 32650

JACKSONVILLE

Baptist Medical Center Substance Abuse Services
800 Prudential Drive
Jacksonville, FL 32207

Boys' and Girls' Clubs of Northeastern Florida
Substance Abuse Prevention/Newton Road
8711 Newton Road
Jacksonville, FL 32211

Substance Abuse Prevention/Romana Blvd.
6750 Ramona Boulevard
Jacksonville, FL 32204

Substance Abuse Prevention/10th Street
313 East 10 Street
Jacksonville, FL 32206

Charter by the Sea in Jacksonville Alcohol and Drug Outpatient
2636 Oak Street
Jacksonville, FL 32204

Endpoint
6484 Fort Caroline Road
Jacksonville, FL 32211

Gateway Community Services, Inc.
Administrative Unit and Outpatient
Adolescent Unit/Outpatient
Adult Intensive Residential Program
555 Stockton Street
Jacksonville, FL 32204

Adolescent/Residential
2671 Huffman Boulevard
Jacksonville, FL 32216

Detoxification Unit
1245 Jesse Street
Jacksonville, FL 32206

Outpatient/Edgewood
1105 West Edgewood Avenue
Jacksonville, FL 32208

Outpatient/Lexington Avenue
4814 Lexington Avenue
Jacksonville, FL 32210

Outpatient/University
2422 West University Boulevard
Jacksonville, FL 32217

Greenfield Center
1820 Barrs Street
Suite 640
Jacksonville, FL 32204

Help Center
743 West Ashley Street
Jacksonville, FL 32202

Jacksonville Metro Treatment Center
5830 North Main Street
Jacksonville, FL 32208

Mental Health Resource Center
Administrative Unit
New Directions
11820 Beach Boulevard
Jacksonville, FL 32216

Outpatient
6316 San Juan Avenue
Suite 36
Jacksonville, FL 32210

Methodist Regional Hospital Systems, Inc. Methodist Pathway Center
580 West 8 Street
Plaza 1 Suite 510
Jacksonville, FL 32209

Milestones in Recovery, Inc.
3333 Hendricks Avenue
Jacksonville, FL 32207

NE Florida Safety Council, Inc.
1725 Art Museum Drive
Jacksonville, FL 32207

Next Step
6428 Beach Boulevard
Jacksonville, FL 32216

Oak Center Substance Abuse Services
8889 Corporate Square Boulevard
Jacksonville, FL 32216

River Region Human Services, Inc.
Alpha Program
1137 Cleveland Street
John E. Ford Elementary School
Jacksonville, FL 32209

Criminal Justice Substance Abuse Program
451 Catherine Street/CCD
Jacksonville, FL 32202

Female Section
Male Section
4727 Lannie Road
Jacksonville, FL 32218

Jacksonville Recovery Center/ Unit 1
577 College Street
Jacksonville, FL 32204

Jacksonville Recovery Center/ Unit 2
5045 Soutel Drive
Suite 25
Jacksonville, FL 32208

Oak Hill School-Based Intervention Program
6910 South Daughtry Boulevard
Jacksonville, FL 32210

Pretrial Detention Center
500 East Adams Street
Jacksonville, FL 32202

Residential Treatment Center
2981 Parental Home Road
Jacksonville, FL 32216

Salvation Army
900 West Adams Street
Jacksonville, FL 32202

San Pablo School-Based Intervention/Kid Power
801 North 18 Avenue
Jacksonville, FL 32250

Southside/Non-Residential
3728 Philips Highway
Suite 220
Jacksonville, FL 32207

Teen Alliance Center
3550 Brentwood Avenue
Jacksonville, FL 32206

Tier 4/Dinsmore
13200 Old Kemp Road
Jacksonville, FL 32219

Tier 4/DOC
2830 Park Street
Jacksonville, FL 32073

Windy Hill School-Based Intervention Program
3831 Forest Boulevard
Jacksonville, FL 32216

Riverside Tradition House
2911 Riverside Avenue
Jacksonville, FL 32205

Saint John's River Hospital Counterpoint Center Adult Unit
6300 Beach Boulevard
Jacksonville, FL 32216

Suicide Prevention Center, Inc.
325 East Duval Street
Jacksonville, FL 32202

University of North Florida Center for Alcohol and Drug Studies
4567 South Saint Johns Bluff Road
Jacksonville, FL 32224

JASPER

North Florida Mental Health Center, Inc. Hamilton Counseling Center
Highway 41 South
Jasper, FL 32052

JUPITER

Center for Family Services Jupiter Unit
825 U.S. Highway 1
Suite 260
Jupiter, FL 33477

Comp. Alcoholism Rehab Programs, Inc. (CARP) Outpatient Treatment
1070 East Indiantown Road
Suite 204
Jupiter, FL 33477

KEY WEST

Care Center for Mental Health Outpatient
1205 4 Street
Key West, FL 33040

Helpline, Inc.
3314 Northside Drive
Suite 18A
Key West, FL 33040

Lower Florida Keys Health System, Inc.
1200 Kennedy Drive
Key West, FL 33040

KISSIMMEE

Addictions Compulsions Treatment Center (ACT, Inc.) Kissimmee Outpatient
800 Office Plaza Boulevard
Suite 401
Kissimmee, FL 34744

Center for Drug Free Living, Inc.
Alpha/Central Avenue
1502 Central Avenue
Kissimmee, FL 34741

Osceola Counseling Center
1200 Central Avenue
Suite 212
Kissimmee, FL 34741

Colonial Counseling Associates Outpatient/Kissimmee
3501 West Vine Street
Suite 290
Kissimmee, FL 34741

Mental Health Services of Osceola County
Adult Outpatient Substance Abuse Services
917 Emmett Street
Kissimmee, FL 34741

Waterfront Center/Psychology
220 East Monument Avenue
Suite C
Kissimmee, FL 34741

Parkside Lodge of Florida
Kissimmee, FL 34742

LAKE BUTLER

Bradford/Putnam/Union Guidance Clinic Union Guidance Clinic
105 North Lake Avenue
Lake Butler, FL 32054

LAKE CITY

North Florida Mental Health Center, Inc.
Columbia Counseling Center
Counseling Services of Suwannee Valley
3900 South First Street
Lake City, FL 32055

Gateway Alcoholism/Psychiatric Treatment
Route 10
Lake City, FL 32055

Veterans' Affairs Medical Center Substance Abuse Services
Lake City, FL 32055

LAKE MARY

Pineview Academy
Lake Mary Boulevard
Lake Mary, FL 32795

LAKE WALES

Lake Wales Area Drug Awareness Council
200 East Orange Avenue
Lake Wales, FL 33853

LAKE WORTH

Comp. Alcoholism Rehab Programs, Inc. (CARP)
Domiciliary
1532 South Federal Highway
Lake Worth, FL 33460

Lake Worth Treatment Center
3153 Canada Court
Lake Worth, FL 33462

Outpatient Treatment
5700 Lake Worth Road
Lake Worth, FL 33463

Family Alternatives/Counseling and Treatment Services, Inc.
3230 Lake Worth Road
Lake Worth, FL 33461

Growing Together, Inc.
1013 Lucerne Avenue
Lake Worth, FL 33460

Loretta Farren
138 J. F. Kennedy Circle
Lake Worth, FL 33462

LAKELAND

Central Florida Human Services Centers Tom Mims Outpatient Office
1104 North Martin Luther King Blvd
Lakeland, FL 33805

Drug Prevention Resource Center, Inc.
1835 North Crystal Lake Drive
Lakeland, FL 33801

Lakeland Hills Treatment Center
3506 Lakeland Hill Boulevard
Lakeland, FL 33805

Lakeland Regional Medical Center Chemical Dependency Unit
1324 Lakeland Hills Boulevard
Lakeland, FL 33802

Palmview Hospital Adult Services
2510 Nort Florida Avenue
Lakeland, FL 33805

Program to Aid Drug Abusers, Inc. (PAD)
Administrative Unit
Eaton Park Outpatient
2920 Franklin Street
Lakeland, FL 33803

Tri-County Addictions Rehab Services, Inc. Lakeland Outpatient
1831 North Crystal Lake Drive
Lakeland, FL 33801

LANTANA

Crisis Line Information and Referral Services, Inc.
415 Gator Drive
Lantana, FL 33465

LARGO

Bay Harbor Residential Treatment Center Dual Diagnosis Tract
12895 Seminole Boulevard
Largo, FL 34648

Counseling Center for New Direction Ken Donaldson and Jim Sullivan
10823 Seminole Boulevard
Suite 3B
Largo, FL 34648

Employee Assistance Services
10225 Ulmerton Road
Largo, FL 34641

Medfield Hospital Dual Diagnosis
12891 Seminole Boulevard
Largo, FL 34648

Operation PAR, Inc.
Adolescent Residential Center
Day Treatment
PAR Detoxification Center
PAR Therapeutic Community
13800 66 Street North
Largo, FL 34641

Professional Comp. Addiction Services, Inc.
Prevention
Wellness Center/Largo
131 First Street NW
Largo, FL 34640

Recovery Bridge at Sun Coast Hospital
2025 Indian Rocks Road
Largo, FL 34644

LECANTO

Marion/Citrus Mental Health Center, Inc. Citrus Alcoholism Program
3238 South Lecanto Highway
Lecanto, FL 34461

Tri County Rehab Center
1645 West Gulf to Lakes Highway
Lecanto, FL 32661

LEESBURG

Lake/Sumter CMHC and Hospital Leesburg Outpatient
800 North Lee Street
Leesburg, FL 34748

LIVE OAK

North Florida Mental Health Center, Inc. Suwannee River Counseling Center Alcohol Program
Nobles Ferry Road
Live Oak, FL 32060

LONGWOOD

Human Service and Resources and Associates, Inc.
530 South Country Road 427
Suite 100
Longwood, FL 32750

Rational Therapy Center, Inc.
465 West Warren Avenue
Longwood, FL 32750

LOWELL

Marion Correctional Institute
Tier III/Doing Time Getting Straight
Lowell, FL 32663

Tier II/Prison Health Services
Lowell, FL 32663

MACCLENNY

Gateway Community Services, Inc. Outpatient/Baker
U.S. Highway 90 West
Agricultural Building
MacClenny, FL 32063

MADISON

Apalachee Center for Human Services Madison Outpatient
225 Sumatra Road
Madison, FL 32340

MAITLAND

Orlando Health Care Group Pru Care
2301 Lucien Way
Suite 145
Maitland, FL 32751

MARATHON

Comprehensive Psychiatric Center/Keys
11399 Overseas Highway
Marathon, FL 33050

Guidance Clinic of Middle Keys
3000 41 Street/Ocean
Marathon, FL 33050

MARIANNA

Chemical Addictions Recovery Effort Jackson County Outpatient Office
4150 Hollis Drive
Marianna, FL 32446

Federal Correctional Institution Psychology Services/Drug Abuse Program
3625 Fed. Correctional Institution Road
Marianna, FL 32446

MAYO

North Florida Mental Health Center, Inc. Suwannee River Counseling/Outpatient
Highway 27
Mayo, FL 32066

MELBOURNE

Brevard Outpatient Alternative Treatment (BOAT)
1127 South Patrick Drive
Suite 24
Melbourne, FL 32937

Circles of Care, Inc. Melbourne Detox/Residential
400 East Sheridan Road
Melbourne, FL 32901

Cope of Brevard County, Inc.
1948 Pineapple Avenue
Melbourne, FL 32935

Grove Counseling Center, Inc. TASC Melbourne
21-B West Fee Avenue
Melbourne, FL 32901

Harbor City Counseling Center
668 West Eau Gallie Boulevard
Melbourne, FL 32935

Heritage Family Treatment Center Heritage Health
2000 Commerce Drive
Melbourne, FL 32904

The Weiss Foundation Outpatient
541 East New Haven Avenue
Melbourne, FL 32901

MERRITT ISLAND

Sunrise Substance Abuse Program of Wuesthoff Hospital
2400 North Courtnay Parkway
Merritt Island, FL 32953

MIAMI

Aspira of Florida, Inc.
3650 North Miami Avenue
Miami, FL 33137

Associates for Psychological Services Miami Beach Substance Abuse Services
2301 Collins Avenue
Suite M-123
Miami, FL 33139

Bayview Centers, Inc.
Division of Outpatient Services
12550 Biscayne Boulevard
Miami, FL 33150

The Shores Center
9325 Park Drive
Suite F
Miami, FL 33138

Better Way of Miami, Inc.
800 NW 28 Street
Miami, FL 33127

Cambridge Foundation
782 NW Le Jeune Road
Suite 537
Miami, FL 33126

Community Crusade Against Drugs
11380 NW 27 Avenue
Room 1389
Miami, FL 33168

Community Health of South Dade, Inc. CMHC Substance Abuse Services
10300 SW 216 Street
Miami, FL 33190

Comprehensive Psychiatric Center/North
838 NW 183 Street
Suite 203
Miami, FL 33169

Comprehensive Psychiatric Center/South
9735 East Fern Street
Miami, FL 33157

Concept House, Inc.
162 NE 49 Street
Miami, FL 33137

Counseling and Evaluation Center/Kendall
8740 North Kendall Drive
Suite 212
Miami, FL 33176

Dade Family Counseling, Inc. Miami Unit
6850 SW 24 Street
Suite 500
Miami, FL 33155

Deering Hospital Dual Diagnosed
9333 SW 152 Street
Miami, FL 33157

Family Counseling Services
Alternatives/Transitions
West Dade
8900 SW 107 Avenue
Suite 200
Miami, FL 33176

111 NW 183 Street
Suite 408
Miami, FL 33169

75 SW 8 Street
Suite 301
Miami, FL 33130

Family Health Center Alpha Program
1775 NW 60 Street
Miami, FL 33142

Health Crisis Network
1351 NW 20 Street
Miami, FL 33142

Here's Help, Inc. Day Care/ Outpatient South
12645 South Dixie Highway
Miami, FL 33156

Humana Hospital/Biscayne Somerset Treatment Services
20900 Biscayne Boulevard
Suite 4 SE
Miami, FL 33180

Informed Families, Inc.
9200 South Dadeland Boulevard
Suite 509
Miami, FL 33156

Jackson Memorial Hospital/ Highland Park Pavilion/ Maternal Addiction Program
1660 NW 7 Court
Miami, FL 33136

Jewish Family Service of Greater Miami
27 Avenue Unit
1790 SW 27 Avenue
Miami, FL 33145

Kedem Counseling Center, Inc. Outpatient Substance Abuse Treatment
5730 Bird Road
Miami, FL 33155

Lifeline of Miami, Inc.
9380 Sunset Drive
Suite B-240
Miami, FL 33173

Lock Towns CMHC Outpatient Substance Abuse Services/Main
18475 NW 2 Avenue
Miami, FL 33169

Mailman Center for Child Development Gilbert's Angels
3038 NW 48 Terrace
Miami, FL 33142

Metro Dade Dept. of Human Resources Office of Rehabilitative Services/Admin.
111 NW First Street
Suite 2150
Miami, FL 33128

Metro Dade Office of Rehab Services
Bay House Residential Treatment Program
600 NE 27 Street
Miami, FL 33137

Central Intake/Detox/Observation Division
2500 NW 22 Avenue
Miami, FL 33142

Central Methadone Clinic
Central Receiving and Treatment
2500 NW 62 Street
Miami, FL 33147

CSARAP/Stockade Treatment Program
6950 NW 41 Street
Miami, FL 33166

Diversion and Treatment Program/South
10300 SW 216 Street
Building 75
Miami, FL 33170

Diversion and Treatment Program/Model Cities
8500 NW 27 Avenue
Miami, FL 33147

Employee Assistance Program
140 West Flagler Street
Suite 1001
Miami, FL 33130

Juvenile Overlay Youth Intervention
Seymour Gelber Adolescent Treatment Center
11025 SW 84 Street
Building 12
Miami, FL 33173

Juvenile TASC
2700 NW 36 Street
Room 2-2
Miami, FL 33142

Life Enrichment Program
11306 SW 214 Street
Miami, FL 33170

Miami Beach Treatment Center
2902 NW 2 Avenue
Miami, FL 33137

New Directions
3140 NW 76 Street
Miami, FL 33147

New Opportunity House
777 NW 30 Street
Miami, FL 33127

Rehab and Aftercare Center/North
3190 NW 116 Street
Miami, FL 33167

Rehab and Aftercare Center/South
11011 SW 104 Street
Building 75
Miami, FL 33176

T/G/K Correctional Facility A/C Program/Men
T/G/K Correctional Facility A/C Program/Women
7000 NW 41 Street
Miami, FL 33166

TASC Assessment and Referral Services
1500 NW 12 Avenue
Suite 715
Miami, FL 33136

TASC Court Evaluation Services
1515 NW 7 Street
Room 213
Miami, FL 33125

Miami Counseling Services
13831 SW 59 Street
Suite 101
Miami, FL 33183

Miccosukee Tribe of Florida Health Department
U.S. Route 41
Mile Marker 70
Miami, FL 33109

Mount Sinai Medical Center Addiction Treatment Program
4300 Alton Road
Miami, FL 33140

New Horizons CMHC
Diagnosed Homeless
1475 NE 36 Street
Suite K3
Miami, FL 33142

3rd Avenue Unit
1600 NW 3 Avenue
Miami, FL 33142

36th Street Substance Abuse Unit
1469 NW 36 Street
Miami, FL 33142

Regis House Prevention Services
2010 NW 7 Street
Miami, FL 33125

Saint Luke's Addiction Recovery
Outpatient Services
3271 NW 7 Street
Miami, FL 33125

Residential Services
7707 NW 2 Avenue
Miami, FL 33150

Saint Luke's Center DARE
9401 Biscayne Boulevard
Miami, FL 33138

Somerset Laurel Group, Inc. Outpatient Treatment
67 NE 168 Street
Miami, FL 33162

South Miami Hospital Addiction Treatment Program
7400 SW 62 Avenue
Miami, FL 33143

Spectrum Programs, Inc.
Dade Outpatient
11049 NE 6 Avenue
Miami, FL 33161

Dade Residential
140 NW 59 Street
Miami, FL 33127

Nontreatment Services
11055 NE 6 Avenue
Miami, FL 33161

Switchboard of Miami New Incentives
75 SW 8 Street
Suite 401
Miami, FL 33130

The Belafonte Tacolcy Center, Inc. Prevention/Education/ Outpatient
6161 NW 9 Avenue
Miami, FL 33127

The Village South, Inc.
400 NE 31 Street
Miami, FL 33137

The Village South, Inc. Addiction Treatment Center
4900 NE 2 Avenue
Miami, FL 33137

Total Rehab Services
7171 SW 24 Street
Suite 503
Miami, FL 33155

Treatment Company
725 NE 125 Street
Suite 100
Miami, FL 33161

Up Front Drug Information
5701 Biscayne Boulevard
Suite 9-PH
Miami, FL 33137

Veterans' Affairs Medical Center Substance Abuse Clinic (SAC) Outpatient
1201 NW 16 Street
Miami, FL 33125

MIAMI BEACH

Douglas Gardens CMHC Substance Abuse Services
701 Lincoln Road
2nd Floor
Miami Beach, FL 33139

MIDDLEBURG

River Region Human Services, Inc.
Middleburg High School Prevention Program
3750 County Road 220
Middleburg, FL 32068

Wilkinson High School Prevention Program
5025 State Route 218
Middleburg, FL 32068

MILTON

Avalon Center, Inc. Substance Abuse Department
1101 Old Bagdad Highway
Milton, FL 32572

MONTICELLO

Apalachee Center for Human Services Monticello
950 West Mahan Drive
Monticello, FL 32344

MULBERRY

Program to Aid Drug Abusers, Inc. (PAD) Bradley Oaks Juvenile Res. Treatment Center
6980 State Road 37 South
Mulberry, FL 33860

NAPLES

A Kind Ear
5495 16 Place SW
Naples, FL 33999

Alternatives Chemical Dependency Consultant Services, Inc.
3071 Terrace Avenue
Naples, FL 33942

David Lawrence Center
Chemical Dependency Services
6075 Golden Gate Parkway
Naples, FL 33999

Court Related Substance Abuse Services
2806 South Horseshoe Drive
Naples, FL 33942

EAP of Southwestern Florida
2400 North Tamiami Trail
Naples, FL 33940

Banyan Pavilion
6075 Golden Gate Parkway
Naples, FL 33999

Family Works
2335 Tamiami Trail North
Suite 309
Naples, FL 33940

Naples Research and Counseling Center Willough at Naples
9001 Tamiami Trail East
Naples, FL 33962

Project Help, Inc. Hotline and Referral
2900 14 Street North
Suite 40
Naples, FL 33940

NARANJA

Metatherapy Institute, Inc.
27200 Old Dixie Highway
Naranja, FL 33032

NEW PORT RICHEY

Human Development Center of Pasco County Outpatient and Receiving Center
8251 Arevee Drive
New Port Richey, FL 34653

Pride, Inc. Track Programs
8018 State Road 54
New Port Richey, FL 34653

Youth and Family Alternatives, Inc.
Drug Prevention Services
7524 Plathe Road
New Port Richey, FL 34653

Runaway Alternatives Program
11451 Wildcat Lane
New Port Richey, FL 34654

NEW SMYRNA BEACH

ACT Corporation Outpatient
143 Canal Street
New Smyrna Beach, FL 32069

Leon F. Stewart Treatment Center, Inc. New Smyrna Beach Outpatient Services
515 Canal Street
New Smyrna Beach, FL 32169

Turning Point
237 North Causeway
New Smyrna Beach, FL 32169

NORTH MIAMI

Transitions Recovery Program
13499 Biscayne Boulevard
Suite M-1
North Miami, FL 33181

NORTH PALM BEACH

Counseling and Consulting Professionals, Inc.
321 Northlake Boulevard
Suite 114
North Palm Beach, FL 33408

Synergy
860 U.S. 1
Suite 104
North Palm Beach, FL 33408

OCALA

Charter Springs Hospital Addictive Disease Unit
3130 SW 27 Avenue
Ocala, FL 32674

Comp. Addiction Treatment Services (CATS)
2105 SW College Road
Ocala, FL 32674

Fountain Counseling Center
2353 SE 17 Street
Ocala, FL 32671

Keeton Corrections, Inc. Ocala Community Service Center
3820 NE 41 Street
Ocala, FL 34479

Mad Dads of Greater Ocala Connection
1510 NW 4 Street
Ocala, FL 32675

Marion County Jail Sheriff's Department Relapse Prevention
700 NW 30 Avenue
Ocala, FL 34475

Marion/Citrus Mental Health Center, Inc.
Children's Program
324 SE 24 Street
Ocala, FL 34471

Day Treatment
108 NW Pine Avenue
Ocala, FL 34470

Detox Unit
Ocala, FL 34474

Halfway House
243 NW 4 Terrace
Ocala, FL 34474

Outpatient Program
Ocala, FL 34470

Transitional Living Center
Ocala, FL 34470

Quad County Treatment Center
913 East Silver Springs Boulevard
Ocala, FL 32670

OCOEE

Addictions Compulsions Treatment Center (ACT Inc) Ocoee Unit
2 South Cumberland Street
Ocoee, FL 34761

OKEECHOBEE

New Horizons of the Treasure Coast, Inc.
Alpha Program
610 SW 2 Avenue
Okeechobee, FL 34974

Okeechobee County Outpatient
605 East North Park Street
Okeechobee, FL 34972

OPA LOCKA

Here's Help, Inc. Residential
15100 NW 27 Avenue
Opa Locka, FL 33054

Lock Towns, CMHC
Opa Locka Substance Abuse Outpatient
15055 NW 27 Avenue
Opa Locka, FL 33054

Project Impact/Daybreak
16555 NW 25 Avenue
Opa Locka, FL 33054

Metro Date Office of Rehab Services North County Treatment Center
490 Opa Locka Boulevard
Opa Locka, FL 33054

ORANGE PARK

Clay County Community Services, Inc.
1532 Kingsley Avenue
Suite 107
Orange Park, FL 32073

Crossroads Community Services, Inc.
2301 Park Avenue
Suite 405
Orange Park, FL 32073

River Region Human Services, Inc.
Lakeside High School Prevention Program
2750 Moody Road
Orange Park, FL 32073

Orange Park High School
Prevention Program
2300 Kingsley Avenue
Orange Park, FL 32073

ORLANDO

Addictions Compulsions Treatment Center (ACT Inc)
Command Plaza
750 South Orange Blossom Trail
Orlando, FL 32805

Orlando Outpatient Unit 2
4300 South Semoran Boulevard
Suite 207
Orlando, FL 32822

Orlando Outpatient Unit 3
5761 South Orange Blossom Trail
Suite 2
Orlando, FL 32810

Orlando Outpatient Unit 4
1510 East Colonial Drive
Suite 101
Orlando, FL 32803

Orlando Outpatient Unit 5
8617 East Colonial Drive
Suite 1200
Orlando, FL 32817

Outpatient Unit 1
2700 South Orange Avenue
Suite 230
Orlando, FL 32806

Center for Drug Free Living, Inc.
Alpha I/McCoy Center
8433 Daetwyler Drive
Orlando, FL 32827

Alpha II/Silver Star Center
1600 Silver Star Road
Orlando, FL 32804

Alpha IV/Chickasaw Center
6900 Autumnvale Drive
Orlando, FL 32822

Department of Corrections Units 2–5
Methadone Clinic
New Horizons Middle School/High School
Outpatient/Assessment/Evaluation
Prevention Institute
100 West Columbia Street
Orlando, FL 32806

Griffin Park Enrichment Center
744 Dunbar Street
Apartment 2
Orlando, FL 32805

Lake Mann Enrichment Center
3600 Eccleston Street
Building 18, Apartment 1
Orlando, FL 32805

Murchison Terrace Enrichment Center
3332 Wilts Circle North
Building 57, Apartment 1
Orlando, FL 32805

Phoenix South/Men's Residential
8301 East Colonial Drive
Orlando, FL 32817

The Center at Hillcrest
1200 East Hillcrest Drive
Suite 301
Orlando, FL 32803

Treatment Alternatives to Street Crime
632 South Hughey Avenue
Orlando, FL 32801

Women's Residential Program
1780 North Mercy Drive
Orlando, FL 32808

Central FL Substance Abuse Treatment Centers, Inc. Outpatient Methadone Maintenance
1800 West Colonial Drive
Orlando, FL 32804

Central Florida Safety Council DUI Counterattack Programs
427 North Primrose Drive
Orlando, FL 32803

Christian Prison Ministries The Bridge
2100 Brengle Avenue
Orlando, FL 32808

Colonial Counseling Associates
9446 East Colonial Drive
Orlando, FL 32817

Colonial Counseling Center/West Colonial
6905B West Colonial Drive
Orlando, FL 32818

Outpatient/Hillcrest
1400 Hillcrest Street
Orlando, FL 32803

Florida Hospital Center for Psychiatry
Addictions Treatment Unit
East Orlando Breaking Free Outpatient Services
7727 Lake Underhill Drive
Orlando, FL 32822

Florida Psychiatric Associates Orlando Outpatient
7300 Sandlake Commons Boulevard
Suite 112
Orlando, FL 32819

Glenbeigh Hospital of Orlando
7450 Sandlake Commons Boulevard
Orlando, FL 32819

Lakeside Alternatives, Inc. Delta Program
434 West Kennedy Boulevard
Orlando, FL 32810

Metropolitan Clinic of Counseling (MCC) Orlando Substance Abuse Services
600 East Colonial Drive
Suite 200
Orlando, FL 32803

Naval Training Center Counseling and Assistance Center
Building 2095
Orlando, FL 32813

Orlando General Hospital Addiction Unit/Admin. Unit
7727 Lake Underhill Drive
Orlando, FL 32822

Orlando Methadone Treatment Center
601 South Semoran Boulevard
Orlando, FL 32807

Project III of Central Florida, Inc.
Adult Detox Center
Clarcona Point
5258 Clarcona-Ocoee Road
Orlando, FL 32808

Evans Community Center
3101 North Pine Hills Road
Orlando, FL 32808

Family Assessment/Detox/Adolescent Treatment Unit
Therapeutic Community Enrichment Program
712 West Gore Street
Orlando, FL 32805

Freedom Haus
5302 Setel Drive
Orlando, FL 32810

Lucerne House
6415 West Michigan Street
Orlando, FL 32806

Men's Long-Term Treatment
2600 South Nashville Street
Orlando, FL 32805

Orange Group
7500 Silver Star Road
Orlando, FL 32808

Orange Halfway House
5275 South Orange Blossom Trail
Orlando, FL 32839

Prevention Services
1400 West Colonial Drive
Orlando, FL 32804

Snowbabies, Inc.
2515 East Pine Street
Orlando, FL 32803

Substance Abuse Family Education (SAFE)
2400 Silver Star Road
Orlando, FL 32804

Sunrise Counseling Center, Inc.
2500 Discovery Drive
Orlando, FL 32826

University of Central Florida Campus Drug and Alcohol Awareness Center Health Resource Center
Room 107
Orlando, FL 32816

Victory Over Drugs, Inc.
725 South Goldwyn Street
Orlando, FL 32805

ORMOND BEACH

Ted Curry M Ed Cap
115 East Granada Boulevard
Ormond Beach, FL 32176

OSPREY

Life Is For Everyone, Inc. (LIFE)
Drug Rehab and Counseling Program
Prevention and Education Center
803 South Tamiami Trail
Osprey, FL 34229

OVIEDO

Human Service and Resources and Associates, Inc.
830 Eyre Drive
Suite 5
Oviedo, FL 32765

PALATKA

Bradford/Putnam/Union Guidance Clinic Putnam Guidance Clinic
3001 Kennedy Street
Palatka, FL 32178

Putnam County Alcohol and Drug Council Putnam House
Route 6
Highway 19 North
Palatka, FL 32177

PALM BAY

Broken Glass, Inc.
4175 Steele Street
Palm Bay, FL 32905

PALM BEACH GARDENS

Alcoholism Counseling and Treatment (ACT)
4362 North Lake Boulevard
Suite 209
Palm Beach Gardens, FL 33400

Counseling/Psychotherapy Associates Drug and Alcohol Program
600 Sandtree Drive
Palm Beach Gardens, FL 33403

PALM HARBOR

Currents Counseling Services
38517 U.S. 19 North
Connell Square
Palm Harbor, FL 34684

Darlene Ruth Worley
1022 Nebraska Avenue
Palm Harbor, FL 34682

PANAMA CITY

Chemical Addictions Recovery Effort
A Woman's Addiction Recovery Effort (AWARE)
3407 North East Avenue
Panama City, FL 32405

Bay County Outpatient Office
Prevention Services
School Prevention
4000 East 3 Street
Suite 200
Panama City, FL 32404

DUI Program
420 West Beach Drive
Panama City, FL 32401

Primary Care/Detox Unit
Reliance House/Halfway
619 North Cove Boulevard
Panama City, FL 32401

Federal Prison Camp Substance Abuse Services
Tyndall Air Force Base
Panama City, FL 32403

PEMBROKE PINES

Memorial Hospital Share Program
801 SW Douglas Road
Pembroke Pines, FL 33025

PENSACOLA

Community Drug and Alcohol Commission
Alpha I Project
Escambia Drug Education Program
803 North Palafox Street
Pensacola, FL 32501

Counseling and Assistance Center Naval Air Station
Building 654
Pensacola, FL 32508

Federal Prison Camp Substance Abuse Services
Saufley Field
Pensacola, FL 32509

Lakeview Center, Inc.
Adolescent Day Treatment/Residential
Adult Residential/Methadone/Primary Care
Help Line
Outpatient Counseling
Pathway
1221 West Lakeview Avenue
Building D
Pensacola, FL 32501

Day Treatment
3300 North Pace Boulevard
Town and Country Plaza, Suite H
Pensacola, FL 32501

In Jail Substance Abuse Program
1190 West Leonard Street
Pensacola, FL 32501

West Florida Regional Medical Center The Pavilion
2191 Johnson Avenue
Pensacola, FL 32514

PERRY

Apalachee Center for Human Services
301 Industrial Park Drive
Perry, FL 32347

PINELAND

Cloisters at Pine Island
13771 Waterfront Drive
Pineland, FL 33945

PINELLAS PARK

Bay Area Treatment Center (BATC)
6328 Park Boulevard North
Suite 4
Pinellas Park, FL 34665

Future Steps at Metropolitan General Hospital
7950 66 Street North
Pinellas Park, FL 34665

Pinellas Emergency Mental Health Services
Acute Care Unit/North
Crisis Stabilization Unit
11254 58 Street North
Pinellas Park, FL 34666

Professional Comp. Addiction Services, Inc. Wellness Center/ Pinellas Park
5931 Park Boulevard
Pinellas Park, FL 34665

PLANT CITY

Drug Abuse Comp. Coord. Office (DACCO) Plant City Outpatient/Counseling Unit
4288 U.S. Highway 92 West
Suite 2
Plant City, FL 33567

PLANTATION

American Biodyne, Inc. Addictions/Intensive Outpatient
8211 West Broward Boulevard
Suite 330
Plantation, FL 33324

POMPANO BEACH

Bridges of America/Turning Point
400 SW 2 Street
Pompano Beach, FL 33060

Broward Addiction Recovery Center (BARC) North/ Residential
803 NW 2 Avenue
Pompano Beach, FL 33060

Pompano Treatment Center, Inc. Methadone Maintenance
380 SW 12 Avenue
Pompano Beach, FL 33069

Spectrum Programs, Inc. Spectrum Adolescent Services Unit
450 East Atlantic Boulevard
Pompano Beach, FL 33060

PORT CHARLOTTE

Life Transitions, Inc.
319 A Elmira Boulevard
Port Charlotte, FL 33952

PORT SAINT JOE

Chemical Addictions Recovery Effort Gulf County Outpatient Office
302 3 Street
Port Saint Joe, FL 32456

PUNTA GORDA

Charlotte County Community Mental Health Services, Inc. Substance Abuse Services
1700 Education Avenue
Punta Gorda, FL 33950

Charlotte County Probation Office
254 West Marion Avenue
Punta Gorda, FL 33950

Coastal Recovery Centers Kelly Hall Residential Treatment Center
2208 Castilla Avenue
Punta Gorda, FL 33950

QUINCY

Apalachee Center for Human Services Area 1/Quincy Office
363 Crawford Street
Quincy, FL 32351

RIVERVIEW

Operation PAR, Inc. Hillsborough Correctional Institute
1135 Balm Road
Riverview, FL 33569

ROCKLEDGE

Circles of Care, Inc. Outpatient Services
1770 Cedar Street
Rockledge, FL 32955

Counterattack/Brevard
1239 South Florida Avenue
Rockledge, FL 32955

Family Counseling Center of Brevard, Inc.
220 Coral Sands Drive
Rockledge, FL 32955

Grove Counseling Center, Inc.
TASC Rockledge
208 Hardee Lane
Rockledge, FL 32955

SAINT AUGUSTINE

Drug Education and Prevention Center
99 Orange Street
Saint Augustine, FL 32084

Teen Alliance Center
545 West King Street
Saint Augustine, FL 32084

Mental Health Resource Center, Inc. Saint John's County Community Mental Health Services
179 Marine Street
Saint Augustine, FL 32084

SAINT PETERSBURG

Behavioral Sciences Center Structured Outpatient Chemical Dependency Treatment Program
5100 First Avenue North
Saint Petersburg, FL 33710

Boley, Inc. Owl's Nest
1147 16 Street North
Saint Petersburg, FL 33705

Family Resources, Inc. Helpline
5235 16 Street North
Saint Petersburg, FL 33733

Family Services Centers/St. Petersburg
1208 66 Street North
Saint Petersburg, FL 33710

Focus One, Inc.
Saint Petersburg, FL 33733

MCC Behavioral Care, Inc.
888 Executive Center Drive West
Saint Petersburg, FL 33702

Operation PAR, Inc.
Administrative Unit
Community/Elder Education Program
Outpatient Counseling Program
10901-C Roosevelt Boulevard
Suite 1000
Saint Petersburg, FL 33716

Alpha Program
Blanton Elementary School
6400 54 Avenue North
Saint Petersburg, FL 33709

Beta Program
2100 4 Street South
Saint Petersburg, FL 33701

Children of Substance Abusers (COSA)
I Team
2000 4 Street South
Saint Petersburg, FL 33705

Multi-Cultural Resource Center
1267 22 Lane South
Suite 332
Saint Petersburg, FL 33712

Narcotics Addiction Treatment Center
1900 9 Street South
Saint Petersburg, FL 33705

Sports Camp
2267 22 Lane South
Saint Petersburg, FL 33712

Pinellas County Urban League Peer Power
333 31st Street North
Saint Petersburg, FL 33713

Professional Comp. Addiction Services, Inc. Wellness Center/ Saint Petersburg
25-A 9 Street
Saint Petersburg, FL 33705

Saint Anthony's Hospital Behavioral Medicine
1200 7 Avenue North
5th Floor SE
Saint Petersburg, FL 33705

Sharing Counseling and Consulting Service
432 Pasadena Avenue South
Saint Petersburg, FL 33707

Stepping Stone Inc. of Tampa
4615 Gulf Boulevard
Dolphin Village Suite 213
Saint Petersburg, FL 33706

Straight/Tampa Bay
3001 Gandy Boulevard
Saint Petersburg, FL 33702

Suncoast Center for Community Mental Health, Inc.
4024 Central Avenue
Saint Petersburg, FL 33733

Tampa Bay Community Correctional Center Partnership
10596 Gandy Boulevard
Saint Petersburg, FL 33733

Youth and Family Connection Residential Services/South
3821 5 Avenue North
Saint Petersburg, FL 33733

SANFORD

Grove Counseling Center, Inc. Halfway House
591 Lake Minnie Drive
Sanford, FL 32771

Grove Counseling Center, Inc. TCEP Midway
2851 Midway Avenue
Sanford, FL 32771

Seminole Community Mental Health Center Crossroads Alcohol/Poly Drug Abuse Treatment Center
300 South Bay Avenue
Sanford, FL 32771

SARASOTA

A Woman's Place of Sarasota, Inc.
1275 2 Street
Sarasota, FL 34236

Anabasis, Inc.
1084 South Briggs Avenue
Sarasota, FL 34237

Coastal Recovery Centers
Clinic I
1750 17 Street
Building C
Sarasota, FL 34234

Day Treatment/Alpha Program
3801 Bee Ridge Road
Suite 9
Sarasota, FL 34233

TASC
2080 Ringling Boulevard
Suite 201B
Sarasota, FL 34237

Colene West, MS. CAP Substance Abuse Services
1608 Oak Street
Sarasota, FL 34236

Doctors' Hospital of Sarasota, Ltd. Genesis Center
2750 Bahia Vista Street
Sarasota, FL 34239

First Step of Sarasota, Inc.
Residential Center
4613 North Washington Boulevard
Sarasota, FL 34234

18th Street Branch
1726 18 Street
Sarasota, FL 34234

Sarasota Memorial New Dawn Center for Alcohol and Drug Treatment
1700 South Tamiami Trail
Sarasota, FL 34239

Sarasota Palms Hospital Dual Diagnosis Unit
1650 South Osprey Avenue
Sarasota, FL 34239

SHALIMAR

Chemical Health Awareness/ Information Networking Systems, Inc. (CHAINS)
60 2 Street
Suite 302
Shalimar, FL 32579

SHARPES

Brevard Correctional Institution Substance Abuse Treatment Program
340 Camp Road
Sharpes, FL 32959

The Weiss Foundation Intervention
870 Camp Road
Sharpes, FL 32959

SPRING HILL

Bi-County Center for Psychotherapy and Counseling
5327 Commercial Way
Park Place Suite A-102
Spring Hill, FL 34606

New Beginnings Springhill Unit
5331 Commercial Way
Suite 108
Spring Hill, FL 34606

Professional Therapy Centers, Inc.
7537 Forest Oaks Boulevard
Spring Hill, FL 34606

Youth and Family Alternatives, Inc. Shady Hills Elementary Alpha Program
18000 Shady Hills Road
Spring Hill, FL 34610

STARKE

Bradford/Putnam/Union Guidance Clinic Bradford Guidance Clinic
945 Grand Street
Starke, FL 32091

STUART

Adventures in Recovery
2215 South Kanner Highway
Stuart, FL 34994

Alcohol and Drug Abuse Program Prevention Programs
451 Riverside Drive
Stuart, FL 34994

Martin County Substance Abuse Treatment Assistance Program
400 East Osceola Street
Suite 2
Stuart, FL 34994

New Horizons of the Treasure Coast, Inc. Martin County Outpatient Branch
2440 SE Federal Highway
Regency Plaza Suite 100
Stuart, FL 34994

Prevention Resource Center, Inc.
120 West 6 Street
Stuart, FL 34994

Substance Treatment Outpatient Program (STOP)
1111 South Federal Highway
Suite 106
Stuart, FL 34994

TALLAHASSEE

A Life Recovery Center
1212 South Monroe Street
Tallahassee, FL 32301

Apalachee Center for Human Services
Chemical Dependency Outpatient
625 East Tennessee Drive
Tallahassee, FL 32308

Chemical Dependency Outpatient/Pride
438 West Brevard Street
Lincoln Center
Tallahassee, FL 32304

Detox
2634 Capitol Circle NE
Tallahassee, FL 32308

Disc Village, Inc. Tallahassee/ Leon County Human Services
3333 West Pensacola Street
Suite 140
Tallahassee, FL 32304

Federal Correctional Institution Drug Abuse Program
501 Capital Circle NE
Tallahassee, FL 32311

Florida A and M University Prevention Center
William H. Gray Building
Section A
Tallahassee, FL 32307

Florida Informed Parents for Drug Free Youth
241 John Knox Road
Suite 200
Tallahassee, FL 32303

Salvita, Inc.
Administrative Unit
Outpatient Services
Salvita Lodge
419 East Georgia Street
Tallahassee, FL 32301

Telephone Counseling and Referral Service, Inc.
Tallahassee, FL 32316

Turn About, Inc.
2051 A Tech Place
Tallahassee, FL 32308

TAMPA

Agency for Community Treatment Services, Inc. (ACTS)
Administrative Offices
North Hillsborough Outpatient
4211 East Busch Boulevard
Tampa, FL 33617

Adolescent Receiving Facility
Youth Outpatient Services
8620 North Dixon Avenue
Tampa, FL 33604

Adult Residential Treatment Unit
Detox Program
6806 North Nebraska Avenue
Tampa, FL 33604

Extended Care Domiciliary
Halfway House
4403 West Buffalo Avenue
Tampa, FL 33614

TASC/DAT
3806 West Dr. Martin Luther King Jr. Boulevard
Tampa, FL 33614

W. T. Edwards Group Home
3810 West Dr. Martin Luther King Jr. Boulevard
Tampa, FL 33614

West Hillsborough Outpatient
1815 West Sligh Avenue
Tampa, FL 33604

AMI Town and Country Hospital Alcohol and Drug Recovery Center
6001 Webb Road
Tampa, FL 33615

C. E. Mendez Foundation Choices and Challenges
601 South Magnolia Avenue
Tampa, FL 33606

Cambridge Foundation
2203 North Lois Avenue
Suite 1100
Tampa, FL 33607

Careunit of South Florida/Tampa
12220 Bruce B. Downs Boulevard
Tampa, FL 33612

Centre for Women Project Recovery
305 South Hyde Park Avenue
Tampa, FL 33606

Drug Abuse Comp. Coord. Office (DACCO)
4422 East Columbus Drive
Tampa, FL 33605

Alternative Program
8325 North Packwood
Tampa, FL 33604

Chemotreatment Center/ Methadone Maintenance/Detox
Tampa Outpatient Counseling Unit
2511 Swann Avenue
Tampa, FL 33609

MacFarlane Alternative School
1721 North MacDill Avenue
Tampa, FL 33607

Male and Female Residential
4424 East Columbus Drive
Tampa, FL 33605

Residential Treatment Facility II
3636 North 50 Street
Tampa, FL 33619

Substance Abusing Mothers/ Infants (SAMI)
301–309 South Caesar Street
Tampa, FL 33602

DUI Counterattack Hillsborough, Inc.
4711 North Hubert Avenue
Tampa, FL 33614

Florida Counseling, Inc.
5401 West Kennedy Boulevard
Suite 480
Tampa, FL 33609

Florida Mental Health Institute Substance Abuse Program for the Elderly
13301 Bruce B. Downs Boulevard
Tampa, FL 33612

Healthcare Connection of Tampa, Inc.
8019 North Himes Avenue
Suite 202
Tampa, FL 33614

Hillsborough County Crisis Center, Inc.
2214 East Henry Avenue
Tampa, FL 33610

James A. Haley Veterans' Hospital Alcohol and Drug Abuse Treatment Program
13000 North 30 Street
Tampa, FL 33612

Michael G. Vennekotter, CAP
11814 North 56 Street
Suite A
Tampa, FL 33617

Neuropsychiatric Institute Substance Abuse Services
2203 North Lois Avenue
Suite 1100
Tampa, FL 33607

Northside Centers, Inc.
13301 North Bruce B. Downs Boulevard
Tampa, FL 33612

Plaza Therapy Associates
4950 West Kennedy Boulevard
Suite 207
Tampa, FL 33609

Tampa Bay Outpatient Center
240 Plant Avenue
Suite B-202
Tampa, FL 33606

Tampa Crossroads
202 West Columbus Drive
Tampa, FL 33602

Tampa Metro Treatment Center
5202-C East Busch Boulevard
Tampa, FL 33617

Tampa/Hillsborough Action Plan, Inc.
Nebraska Avenue Outpatient Services
Nebraska Avenue Prevention Services
1495 North Nebraska Avenue
Tampa, FL 33602

Turning Point of Tampa
6301 Memorial Highway
Suite 201
Tampa, FL 33615

Univ. of South Florida Tampa General Healthcare Univ. Psych. Center Alcohol and Drug Abuse Recovery
3515 East Fletcher Avenue
Tampa, FL 33613

TARPON SPRINGS

Agency for Community Treatment Services, Inc. (ACTS) Pinellas Domiciliary
3575 Old Keystone Road
Tarpon Springs, FL 34689

TAVARES

Lake/Sumter CMHC Treatment Alternatives to Street Crime
544 Duncan Drive
Tavares, FL 32784

New Beginnings
323 West Alfred Street
Tavares, FL 32778

Teleios Ministries, Inc.
551 West Main Street
Tavares, FL 32778

TAVERNIER

Guidance Clinic of the Upper Keys Outpatient
92140 Overseas Highway
Suite 5
Tavernier, FL 33070

THONOTOSASSA

Agency for Community Treatment Services, Inc.
Adolescent Intensive Residential Treatment Unit
Adolescent Group Home
11309 Tom Folsom Road
Thonotosassa, FL 33592

TITUSVILLE

Circles of Care, Inc. North County Clinic
6700 South U.S. Highway 1
Titusville, FL 32780

Grove Counseling Center, Inc. TASC Titusville
524 South Hopkins Avenue
Titusville, FL 32708

How House, Inc.
1116 Main Street
Titusville, FL 32796

TRENTON

Mental Health Services of North Central Florida DBA Community Counseling Center/Trenton
115 NW First Avenue
Trenton, FL 32693

VENICE

Coastal Recovery Centers South County Clinic III
119 Corporation Way
Venice, FL 34292

First Step of Sarasota, Inc. Venice Office
2201 Tamiami Trail South
Suite 9
Venice, FL 34293

VERO BEACH

Center for Counseling and Addiction Recovery
1031 18 Street
Suite K
Vero Beach, FL 32961

New Horizons of the Treasure Coast, Inc.
Alcohope Residential Substance Abuse
5925 37 Street
Vero Beach, FL 32966

Indian River County Outpatient Branch
2300 3 Court
Suite C
Vero Beach, FL 32960

WEBSTER

Lake/Sumter CMHC Alpha Program
300 South Market Boulevard
Webster Elementary School
Webster, FL 33597

WEST PALM BEACH

American Biodyne, Inc. West Palm Beach Center
2090 Palm Beach Lake Boulevard
Suite 808
West Palm Beach, FL 33409

Beverly C. O'Neill Outpatient Treatment
1551 Forum Place
Plaza 1551 Suite 400-D
West Palm Beach, FL 33401

Center for Alcohol and Drug Studies
321 North Lake Boulevard
Suite 214A
West Palm Beach, FL 33408

Center for Family Services West Palm Beach Unit
2218 South Dixie Highway
West Palm Beach, FL 33401

Comp. Alcoholism Rehab Programs, Inc. (CARP)
Adolescent Outpatient Treatment
Medical Admissions Program
Outpatient Treatment
5400 East Avenue
West Palm Beach, FL 33402

Chris House
3217 Broadway
West Palm Beach, FL 33404

Peter Fairclough Residence
421 Iris Street
West Palm Beach, FL 33402

Drug Abuse Treatment Association, Inc. (DATA)
Outpatient
1720 East Tiffany Drive
Suite 102
West Palm Beach, FL 33407

Walter D. Kelly Treatment Center
1041 45 Street
West Palm Beach, FL 33407

Glenbeigh Hospital of Palm Beach, Inc.
4700 Congress Avenue
West Palm Beach, FL 33407

Gratitude House
317 North Lakeside Court
West Palm Beach, FL 33407

Hanley Hazelden Center at Saint Mary's
5200 East Avenue
West Palm Beach, FL 33407

Lee Ballard, RN. CD. CAP
1408 North Killian Drive
Suite 208
West Palm Beach, FL 33403

Marlys A. Maury RN. CAP MSW
600 Sandtree Drive
Suite 106B
West Palm Beach, FL 33410

Nina de Gerome MSW/F. Edward McCabe Substance Abuse Services
333 Southern Boulevard
Suite 204
West Palm Beach, FL 33405

Palm Beach Treatment Center
1771 South Congress Avenue
Congress Plaza Unit 7
West Palm Beach, FL 33406

Palm Beach Wellness Center
10111 Forest Hill Boulevard
West Palm Beach, FL 33414

Pride, Inc.
DUI Program
2715 Australian Avenue
Suite 101
West Palm Beach, FL 33407

Track Programs
2711 Exchange Court
West Palm Beach, FL 33409

Professional Educational Consultants, Inc.
4623 Forest Hill Boulevard
Suite 110
West Palm Beach, FL 33415

Schocoff Center
2601 North Flagler Drive
Suite 104
West Palm Beach, FL 33407

Wellington Regional Medical Center Cornerstone Program
10101 Forest Hill Boulevard
West Palm Beach, FL 33414

WILLISTON

Corner Drug Store, Inc. Williston Middle School Beta Program
1345 NE 3 Avenue
Williston, FL 32696

WINTER GARDEN

Center for Drug Free Living, Inc. Alpha III/Dillard Center
310 North Dillard Street
Winter Garden, FL 34787

WINTER HAVEN

Parkside Lodge of Winter Haven
175 5 Street
Suite 300
Winter Haven, FL 33883

Pride of Polk County, Inc.
65 3 Street NW
Suite 200
Winter Haven, FL 33881

Program to Aid Drug Abusers, Inc. (PAD) Winter Haven Outpatient
147 Avenue A NW
Winter Haven, FL 33881

Tri-County Addictions Rehab Services
Employee Assistance Program
Intensive Outpatient Program
Outpatient/Adolescent
Outpatient/Adult
37 3 Street SW
Winter Haven, FL 33880

WINTER PARK

Addictions Compulsions Treatment Center (ACT Inc) Winter Park Outpatient
2354 B Winter Woods Boulevard
Winter Park, FL 32792

Co-Dependency Counseling Center, Inc.
1850 Lee Road
Suite 132
Winter Park, FL 32789

Interact Counseling Associates Winter Park Unit
2211 Lee Road
Suite 210
Winter Park, FL 32789

Parkside Lodge of Winter Park Outpatient
1400 South Orlando Avenue
Suite 100
Winter Park, FL 32789

Project III of Central Florida, Inc.
Employee Assistance Program
Outpatient Services
1408 Gay Road
Winter Park, FL 32789

WINTER SPRINGS

Grove Counseling Center, Inc.
Adolescent
Adult Outpatient
580 Old Sanford Oviedo Road
Winter Springs, FL 32708

WOODVILLE

Disc Village, Inc.
Adolescent Treatment Program
Woodville, FL 32362

Natural Bridge Treatment Center
Raft
Natural Bridge Road
Woodville, FL 32362

YULEE

River Region Human Services, Inc. Yulee Alpha Program
Highway A1A
Yulee, FL 32097

GEORGIA

AMERICUS

Middle Flint MH/MR Substance Abuse Program
Substance Abuse Detoxification Unit
Substance Abuse Services Outpatient/Residential
425 North Lee Street
Americus, GA 31709

ATHENS

Athens Regional Medical Center Commencement Center
1199 Prince Avenue
Athens, GA 30613

Northeast Georgia Alcohol and Drug Abuse Prevention and Treatment
250 North Avenue
Athens, GA 30601

ATLANTA

Atlanta West Treatment Center
3201 Atlanta Industrial Parkway
Building 100 Suite 101
Atlanta, GA 30331

Dekalb Addiction Clinic
1260 Briarcliff Road NE
Atlanta, GA 30306

Dekalb County Board of Health Kirkwood Substance Abuse Clinic
30 Warren Street SE
Dekalb/Atlanta Human Services Center
Atlanta, GA 30317

Fulton County Drug and Alcohol Treatment Center
265 Boulevard Street NE
Atlanta, GA 30312

Grady Memorial Hospital Drug Dependence Unit
60 Coca Cola Place SE
Atlanta, GA 30303

Natl. Parents Resource Institute for Drug Education, Inc. (PRIDE)
Hurt Plaza
Hurt Building Suite 450
Atlanta, GA 30303

New Start Community Residential Treatment Center
138 Douglas Street SE
Atlanta, GA 30317

Northside Comprehensive CMHC Alcohol and Drug Abuse Treatment Program
975 Johnson Ferry Road NE
Suite 220
Atlanta, GA 30342

Odyssey Family Counseling Center
3578 South Fulton Avenue
Atlanta, GA 30354

Psychiatric Institute of Atlanta
811 Juniper Street NE
Atlanta, GA 30308

Safe Recovery System of Atlanta, Inc.
2300 Peachford Road
Suite 2000
Atlanta, GA 30338

Southside Healthcare Substance Abuse Unit Methadone/ Alcoholism/Cocaine/Drug Free
1660 Lakewood Avenue SW
Atlanta, GA 30315

United States Penitentiary Substance Abuse Services
Atlanta, GA 30315

AUGUSTA

CMHC of East Central Georgia Alcohol and Drug Services
3421 Old Savannah Road
Augusta, GA 30906

University Hospital Behavioral Health Center Chemical Dependency Program
1350 Walton Way
Augusta, GA 30902

Veterans' Affairs Medical Center Substance Abuse Treatment Program Uptown Division
One Freedom Way
Augusta, GA 30904

BRUNSWICK

Coastal Area Alcohol and Drug Abuse Program
1609 Newcastle Street
Winchester Center
Brunswick, GA 31520

BUFORD

Gwinnett Hospital System
Substance Abuse Services
Gwinnett Treatment Center
55 Morningside Drive
Buford, GA 30518

CARROLLTON

Chatt/Flint Area Substance Abuse Services Carroll County Substance Abuse Clinic
122 Lee Street
Carrollton, GA 30117

CEDARTOWN

Residential Treatment Unit
180 Wateroak Drive
Cedartown, GA 30125

CLAYTON

Woodridge Hospital Substance Abuse Services
Germany Road
Clayton, GA 30525

COCHRAN

Middle Georgia Adolescent Residential Center
408 Peacock Street
Cochran, GA 31014

COLLEGE PARK

Anchor Hospital Substance Abuse Services
5454 Yorktowne Drive
College Park, GA 30349

COLUMBUS

Alchemy Therapeutic Comm. Columbus TC
8134 Blythe Street
Columbus, GA 31909

Alcohol and Drug Services
1334 2 Avenue
Columbus, GA 31901

Saint Francis Hospital The Pathway
2122 Manchester Expressway
Columbus, GA 31904

Turning Point Alcohol and Drug Residential Facility
919 Lawyers Lane
Columbus, GA 31906

Turning Point Detoxification Unit
919 Lawyers Lane
Columbus, GA 31906

DALTON

Georgia Highlands Treatment Services
900 Shugart Road
Dalton, GA 30720

Hamilton Medical Center Westcott Center
Burleyson Drive
Dalton, GA 30720

DECATUR

Biobehavioral Associates
625 Dekalb Industrial Way
Decatur, GA 30033

Carp
2145 Candler Road
Decatur, GA 30032

Dekalb Substance Abuse Services Clifton Springs Subst Abuse Services
3110 Clifton Springs Road
Suite A
Decatur, GA 30034

Fox Recovery Center Alcohol and Drug Abuse Program
3100 Clifton Springs Road
Decatur, GA 30034

Georgia Regional Hospital at Atlanta Alcohol and Drug Unit
3073 Panthersville Road
Decatur, GA 30034

Veterans' Affairs Medical Center Drug Dependence Treatment Program
1670 Clairmont Road
Decatur, GA 30033

DOUGLAS

Satilla Community Mental Health Substance Abuse Clinic
1005 Shirley Avenue
Douglas, GA 31533

DUBLIN

Middle Georgia Alcohol and Drug Clinic
600 North Jefferson Street
Dublin, GA 31021

Twin Oaks Recovery Center
2121A Belevue Street
Dublin, GA 31021

Veterans' Affairs Medical Center Substance Abuse Treatment Program
1826 Veterans Boulevard
Dublin, GA 31021

FORT GORDON

Fort Gordon Community Counseling Center Prevention and Control Program
32502 Brainard Avenue
Fort Gordon, GA 30905

GAINESVILLE

North Georgia MH/MR Center Alcohol and Drug Program
472 South Enota Street
Gainesville, GA 30501

GRIFFIN

McIntosh MH/MR Substance Abuse Services

Adolescent Substance Abuse Day Treatment
431 West Poplar Street
Griffin, GA 30223

Sub-Acute Care
125 South 13 Street
Griffin, GA 30223

Katharos
1459 Williamson Road
Griffin, GA 30223

Substance Abuse Outpatient Services
141 West Solomon Street
Griffin, GA 30223

HAPEVILLE

Decatur Hospital Substance Abuse Program
450 North Candler Street
Hapeville, GA 30030

JEFFERSON

The Potter's House Christian Rehabilitation Center
655 Potters House Road
Route 2
Jefferson, GA 30549

JONESBORO

Clayton Mental Health Center Alcohol and Drug Program
853 Battle Creek Road
Jonesboro, GA 30236

KENNESAW

Cobb/Douglas CMHC First Step Recovery Center
260 Hawkins Store Road
Kennesaw, GA 30144

LA FAYETTE

Vista Community Programs Substance Abuse Services
702 East Villanow Street
La Fayette, GA 30728

LAWRENCEVILLE

Gwinnett/Rockdale/Newton Alcohol and Drug Abuse Program
175 Gwinnett Drive
Lawrenceville, GA 30245

LEESBURG

The Anchorage, Inc.
Route 2
Leesburg, GA 31763

MACON

Charter Lake Hospital Addictive Disease Unit
3500 Riverside Drive
Macon, GA 31209

Crisis Stabilization Program
3575 Fulton Mill Road
Macon, GA 31206

Macon/Bibb County Substance Abuse
Methadone Program
657 Hemlock Street
Macon, GA 31201

Outpatient Program
624 New Street
Macon, GA 31201

MARIETTA

Kennestone Hospital Mental Health Unit
737 Church Street
Marietta, GA 30060

Straight, Inc.
2221 Austell Road
Marietta, GA 30060

MILLEDGEVILLE

Central State Hospital Regional Psychiatric Division
Swint Avenue
Milledgeville, GA 31062

Oconee Alcohol and Drug Program
900 Barrows Ferry Road
Milledgeville, GA 31061

Pete Wheeler Domiciliary Georgia War Veterans' Home
Milledgeville, GA 31062

NEWNAN

Chat/Flint Area Substance Abuse Services Coweta Substance Abuse Center
107 Jefferson Street
Newnan, GA 30263

RIVERDALE

Clayton Mental Health Center Drug/Alcohol Program for Children and Adolescents
6315 Don Hastings Road
Flint River Center
Riverdale, GA 30274

ROME

Northwest Georgia Regional Hospital Admission and Evaluation Unit
1305 Redmond Street
Building 103
Rome, GA 30165

Star House, Inc. Halfway House
212 1/2 North 5 Avenue
Rome, GA 30161

SAVANNAH

Chatham Clinic for Addictions
607 Abercorn Street
Savannah, GA 31401

SMYRNA

Ridgeview Institute Adult Addictions Medicine
3995 South Cobb Drive
Smyrna, GA 30080

STATESBORO

Pineland Community Mental Health Center Alcohol/Drug Programs
407 South Zetterower Avenue
Statesboro, GA 30458

Willingway Hospital Substance Abuse Services
311 Jones Mill Road
Statesboro, GA 30458

THOMASVILLE

Southwestern State Hospital Evaluation Unit/Substance Abuse Services
Thomasville, GA 31799

TIFTON

MH/MR Department of Human Resources Midstep Intensive Residential Care Facility
283 Love Avenue
Tifton, GA 31794

VALDOSTA

District Substance Abuse Services
206 South Patterson Street
Human Resource Building 3rd Floor
Valdosta, GA 31601

Greenleaf Center, Inc. Substance Abuse Treatment Program
2209 Pineview Drive
Valdosta, GA 31602

WARNER ROBINS

Peachbelt Community Mental Health Center Substance Abuse Services
202 North Davis Drive
Warner Robins, GA 31099

WAYCROSS

Satilla Area Substance Abuse Program
1305 Pendergast Street
Waycross, GA 31501

WINDER

Project Adam Community Assistance Center, Inc.
Lanthier Street
Winder, GA 30680

GUAM

TAMUNING

Dept. of Mental Health and Substance Abuse Drug and Alcohol Program
788 Chalan San Antonio Street
Tamuning, GU 96911

HAWAII

EWA BEACH

Kahi Mohala Chemical Dependency Services
91-2301 Fort Weaver Road
Ewa Beach, HI 96706

HALEIWA

Central Oahu Youth Services Assoc., Inc. Haleiwa Emergency Shelter
66-528 Haleiwa Road
Haleiwa, HI 96712

HILO

Awareness House, Inc.
190 Keawe Street
Suite 25
Hilo, HI 96720

Big Island Substance Abuse Council (BISAC)
Adolescent Residential
Adult Residential
1190 Waianuenue Avenue
Hilo, HI 96721

Castle Medical Center Outpatient Alcohol and Addictions Program Hilo
305 Wailuku Drive
Room 4
Hilo, HI 96720

HONOLULU

Alu Like, Inc.
1427 Dillingham Boulevard
Suite 205-B
Honolulu, HI 96817

American Lung Association of Hawaii Public Health Education Program
245 North Kukui Street
Suite 100
Honolulu, HI 96817

Catholic Services to Families Na Ohana Pulama
200 North Vineyard Boulevard
Suite 200
Honolulu, HI 96817

Coalition for a Drug Free Hawaii
1218 Waimanu Street
Honolulu, HI 96814

Department of Education Youth Traffic Safety Project Driver Educ.
2530 10th Avenue
Room A-12
Honolulu, HI 96816

Department of Public Safety Project Bridge
677 Ala Moana Boulevard
Honolulu, HI 96813

Dept. of Health School Health Services Branch Peer Education Program
741-A Sunset Avenue
Room 106
Honolulu, HI 96816

Drug Addiction Services of Hawaii, Inc. (DASH)
Hotline
Drug Free
Methadone Maintenance
1031 Auahi Street
Honolulu, HI 96814

Hawaii Alcoholism Foundation Sand Island Treatment Center
Outpatient
Residential
12–40 Sand Island Access Road
Honolulu, HI 96819

Hawaii Conf. of Seventh Day Adventists Community Crusade Against Drugs
2728 Pali Highway
Honolulu, HI 96817

Hawaii Youth at Risk
770 Kapiolani Boulevard
Suite 701
Honolulu, HI 96813

Honolulu Police Department Drug Awareness/Drug Abuse Resistance Education
801 South Beretania Street
Honolulu, HI 96813

Kalihi Palama Health Clinic Health Care for Homeless Project
766 North King Street
Honolulu, HI 96817

Kapiolani Medical Center for Women/Children Teen Intervention Program
1319 Punahou Street
Honolulu, HI 96826

Mothers Against Drunk Driving
1108 Fort Street Mall
Room 18
Honolulu, HI 96813

Native Hawaiian Drug Free Schools and Community Program
1850 Makuakane Street
Building B
Honolulu, HI 96817

Naval Counseling and Assistance Center Alcohol and Drug Safety Action Program
Comnavbase Pearl Harbor
Honolulu, HI 96860

PACT Parents and Children Together Community Training Program
1475 Linapuni Street
Suite 117A
Honolulu, HI 96819

Queen's Medical Center Psychiatric Day Hospital
1301 Punchbowl Street
Honolulu, HI 96813

Saint Francis Medical Center
Women's Addiction Treatment Center of Hawaii
Outpatient/Outreach/Residential
2230 Liliha Street
Honolulu, HI 96817

Salvation Army Addiction Treatment
Alcohol Treatment Program
Detox Unit
Eureka House
Outpatient Services Program
3624 Waokanaka Street
Honolulu, HI 96817

Salvation Army Treatment Facilities For Children and Youth/Women's Way
2950 Manoa Road
Cottage D
Honolulu, HI 96822

Salvation Army Treatment Facilities for Children and Youth/Male O Ka Ohana
2950 Manoa Road
Cottage A
Honolulu, HI 96822

State Department of Education Office of Instructional Services/General Education Branch
1390 Miller Street
Honolulu, HI 96813

Straub Clinic/Hospital Corp. Health Dept. Resource Employee Assistance Program
888 South King Street
Honolulu, HI 96813

Susannah Wesley Community Center Alternative Skills/ Communication for Youth
1117 Kaili Street
Honolulu, HI 96819

University of Hawaii at Manoa Center for Student Development/Substance Abuse Education/Prevention Program
2440 Campus Road
Honolulu, HI 96822

Veterans' Affairs Substance Abuse Treatment Program
300 Ala Moana Boulevard
Suite 1126
Honolulu, HI 96813

Western Regional Center for Drug Free Schools and Communities
1164 Bishop Street
Suite 1409
Honolulu, HI 96813

Winners Camp Foundation
1016 Kapahulu Avenue, 2nd Floor
Kilohana Square
Honolulu, HI 96816

YMCA Kaimuki-Waialae Palolo Youth Program
4835 Kilauea Avenue
Honolulu, HI 96816

YMCA Outreach Services
1335 Kalihi Street
Honolulu, HI 96819

KAHUKU

Bobby Benson Center
Kahuku, HI 96731

Castle Medical Center/Bobby Benson Center Adolescent Outpatient Program/Big Island
Kahuku, HI 96731

KAHULUI

Aloha House Outpatient Services
220 Lalo Place
Suite 2-A
Kahului, HI 96732

KAILUA

Alcoholic Rehab Services of Hawaii, Inc. DBA Hina Mauka
Public Safety
Teen Care Program
43 Oneawa Street
Suite 204
Kailua, HI 96734

Hawaii Counseling and Education Center Inc. Chemical Dependency Outpatient Treatment
970 North Kalaheo Avenue
Suite C-214
Kailua, HI 96734

Poailani, Inc. Dual Diagnosis Program
1396 Onioni Street
Kailua, HI 96734

KANEOHE

Castle Medical Center Outpatient Alcohol and Addictions Program/Windward
46-001 Kamehameha Highway
Kaneohe, HI 96744

KAPAA

Kauai Non-Profit Resource Center Community Youth Activity Program/(CYAP)
4565 Mamanr Street
Kapaa, HI 96746

Serenity House
Adolescent Residential Program
Adult Residential Program
4800-B Kawaihau Road
Kapaa, HI 96746

KEALAKEKUA

Castle Medical Center Outpatient Alcohol and Addictions Program/Kona
Mamalahoa Highway
Honalo Business Center Room 6
Kealakekua, HI 96750

LIHUE

Kauai Outreach Program
4444 Rice Street
Lihue, HI 96766

Young Women's Christian Association Adolescent Outpatient Drug/Alcohol Program
3094 Elua Street
Lihue, HI 96766

MAKAHA

New Horizons/Hawaii Substance Abuse Treatment
84–183 Makau Street
Suite 100
Makaha, HI 96792

MAKAWAO

Aloha House Adult Residential Treatment
4395 Ike Drive
Maunaolu Campus
Makawao, HI 96768

NANAKULI

Hale Ola Hoopakolea, Inc./Oahu
89–137 Nanakuli Avenue
Nanakuli, HI 96792

WAHIAWA

Central Oahu Youth Services Assoc., Inc.
Coysa Storefront Alcohol Abuse Program
Wilderness/Ocean Experience
801 Center Street
Wahiawa, HI 96786

WAIANAE

Hawaii Addiction Center
84–998 Farrington Highway
Waianae, HI 96792

Ohana Hale
86–631 Puuhulu Road
Waianae, HI 96792

Waianae Coast Community Mental Health Center Substance Abuse Program
85–670 Farrington Highway
Waianae, HI 96792

WAILUKU

Castle Medical Center Outpatient Alcohol and Addictions Program/Maui
270 Hookahi Street
Suite 302
Wailuku, HI 96793

Maui Kokua Services
Wailuku, HI 96793

Maui Youth and Family Services Adolescent Outpatient Treatment
16 South Market Street
Suite K
Wailuku, HI 96793

Students Staying Straight
1325 Lower Main Street
Suite 107
Wailuku, HI 96793

WAIPAHU

Castle Medical Center Outpatient Alcohol and Addictions Program
94–239 Waipahu Depot Road
Room 216
Waipahu, HI 96797

Teen Challenge Oahu
94–560 Kamehameha Highway
Waipahu, HI 96797

IDAHO

ASHTON

Ashton Memorial Chemical Dependency Center
801 Main Street
Ashton, ID 83420

BLACKFOOT

Road to Recovery, Inc.
583 West Sexton Street
Blackfoot, ID 83221

BOISE

Alcoholism Intervention Services
8436 Fairview Avenue
Fairview Plaza C
Boise, ID 83704

Care Institute Starting Point
4696 Overland Street
Suite 460
Boise, ID 83705

Careunit Outpatient Services
410 South Orchard Street
Suite 132
Boise, ID 83705

CPC Intermountain Hospital of Boise
303 North Allumbaugh Street
Boise, ID 83704

First Step for Women First Step for Men
1818 West State Street
Boise, ID 83702

Gemhaven Psychological
2300 West Boise Avenue
Boise, ID 83702

Nelson Institute
1010 North Orchard Street
Suite 1
Boise, ID 83706

New Life Counseling Centers
7247 Potomac Drive
Boise, ID 83704

Northview Hospital DBA Northview
8050 Northview Street
Boise, ID 83704

Northwest Passages and Counseling Center
131 North Allumbaugh Street
Boise, ID 83704

The Aerie, Inc.
9600 Brookside Lane
Boise, ID 83703

Veterans' Affairs Medical Center Substance Abuse Treatment Programs
500 West Fort Street
Boise, ID 83702

COEUR D'ALENE

Idaho Youth Ranch Anchor House
1609 Government Way
Coeur D'Alene, ID 83814

Port of Hope North
218 North 23 Street
Coeur D'Alene, ID 83814

GOODING

Walker Center
1120A Montana Street
Gooding, ID 83330

IDAHO FALLS

Alcohol Rehabilitation Association Phoenix Center
163 East Elva Street
Idaho Falls, ID 83401

Aspen Crest Counseling Center
1970 East 17 Street
Suite 206
Idaho Falls, ID 83404

LEWISTON

Saint Joseph's Hospital Substance Abuse Services
415 6 Street
Lewiston, ID 83501

MOSCOW

Lakeside Recovery Center
316 South Washington Street
Moscow, ID 83843

NAMPA

Care Institute Starting Point
508 East Florida Street
Nampa, ID 83686

Mercy Medical Center Careunits
1512 12 Avenue
Nampa, ID 83686

OROFINO

State Hospital North Alcoholism Treatment Unit
State Hospital North Drive
Orofino, ID 83544

PLUMMER

Coeur D'Alene Tribe Benewah Medical Center Family Healing Center
1115 B Street
Plummer, ID 83851

POCATELLO

Aspen Crest Hospital and Counseling Centers Life Works Program
797 Hospital Way
Pocatello, ID 83201

Pocatello Regional Medical Center Dayspring
777 Hospital Way
Pocatello, ID 83201

TWIN FALLS

Port of Hope Centers Alcoholic Recovery Center, Inc.
425 2 Avenue North
Twin Falls, ID 83301

ILLINOIS

ADDISON

Serenity House, Inc.
891 South Route 53
Addison, IL 60101

ALTON

Saint Clare's Hospital Chemical Dependency Treatment Center
915 East 5 Street
Alton, IL 62002

ANNA

Fellowship House
800 North Main Street
Anna, IL 62906

Union County Counseling
204 South Street
Anna, IL 62906

ARLINGTON HEIGHTS

Mercy Counseling at Arlington Heights
115 South Wilke Road
Suite 101
Arlington Heights, IL 60005

AURORA

Association for Individual Development
229A West Galena Boulevard
Aurora, IL 60506

Attitude/Behavior Modification Systems, Inc.
31 West Downer Place
Aurora, IL 60506

Breaking Free, Inc. Outpatient Services
250 West Downer Place
Aurora, IL 60506

Community Counseling Center of the Fox Valley, Inc.
Opportunity House
469 North Lake Street
Aurora, IL 60506

Outpatient Services
400 Mercy Lane
Aurora, IL 60506

Project Safe
479 North Lake Street
Aurora, IL 60505

Reese Clinical and Consulting Services
205 North Lake Street
Suite 103
Aurora, IL 60506

BEARDSTOWN

Cass County Mental Health Center Alcoholism Treatment Program
121 East 2 Street
Beardstown, IL 62618

BELLEVILLE

Belleville Mental Health Outpatient Center Comprehensive Prevention and Treatment Program
200 North Illinois Street
Belleville, IL 62220

Gateway Foundation, Inc. Belleville Unit
101 West Main Street
Belleville, IL 62220

Saint Claire County Regional Superintendant of Schools Intouch PSA 16
500 Wilshire Drive
Educational Cooperative Building
Belleville, IL 62223

Saint Elizabeth's Hospital Chemical Dependence Program
211 South 3 Street
Belleville, IL 62222

BENSENVILLE

Bensenville Home Society Lifelink
331 South York Road
Bensenville, IL 60106

BLOOMINGDALE

Bloomingdale Township Committee on Youth
123 North Rosedale Street
Suite 100
Bloomingdale, IL 60108

BLOOMINGTON

Chestnut Health Systems, Inc.
Lighthouse/Bloomington Youth
702 West Chestnut Street
Bloomington, IL 61701

Lighthouse/Adult
1003 Martin Luther King Jr. Drive
Bloomington, IL 61701

Project Oz
502 South Morris Street
Bloomington, IL 61701

BLUE ISLAND

Guildhaus Halfway House
2413 South Canal Street
Blue Island, IL 60406

BOLINGBROOK

Indizi International Institute for Addictions Counseling and Prevention/Education
3 H Wildwood Lane
Bolingbrook, IL 60440

Life Works Chemical Dependency Treatment Centers
420 Medical Center Drive
Suite 230
Bolingbrook, IL 60440

Will County Health Department Addiction Services
241 Canterbury Lane
Bolingbrook, IL 60440

BOURBONNAIS

Brooks and Johnson Associates Treatment Services
19 Heritage Plaza
Suite 208
Bourbonnais, IL 60914

BUFFALO GROVE

Omni Youth Services Substance Abuse Treatment Program
1111 Lake Cook Road
Buffalo Grove, IL 60089

CALUMET CITY

Success Center The City of Calumet City
145 167 Street
Calumet City, IL 60409

CANTON

Community Mental Health Center of Fulton and McDonough Counties
229 Martin Avenue
Canton, IL 61520

Proctor Chemical Dependency Center at Graham
210 West Walnut Street
Canton, IL 61520

CARBONDALE

Jackson County Mental Health Center Alcohol/Drug Abuse Prevention/Treatment
604 East College Street
Arlington Building
Carbondale, IL 62901

CARLINVILLE

Macoupin County Mental Health Center Alcoholism Outpatient Center
100 North Side Square
Carlinville, IL 62626

CAROL STREAM

Outreach Community Center Comprehensive Prevention
345 South President Street
Carol Stream, IL 60188

CARROLLTON

Tri-County Counseling Center Outpost
302 5 Street
Carrollton, IL 62016

CARY

Advantage Group, Inc.
8807 Cary/Algonquin Road
Cary, IL 60013

CASEYVILLE

Gateway Foundation, Inc.
Caseyville Facility
600 West Lincoln Street
Caseyville, IL 62232

CENTRALIA

Community Resource Center
101 South Locust Street
Centralia, IL 62801

CHAMPAIGN

Carle Clinic Association New Choice Adult Outpatient/ Alcohol/Drug Recovery
809 West Church Street
Carle Pavilion
Champaign, IL 61820

LWS Place Alcohol/Drug Education and Outpatient Counseling
403 South State Street
Champaign, IL 61820

Prairie Center for Substance Abuse Hill Street Unit
122 West Hill Street
Champaign, IL 61820

CHARLESTON

Central East Alcohol and Drug Council
Family Outpatient Clinic
Hour House Residential and Detox
635 Division Street
Charleston, IL 61920

Women's Project
1501 1/2 18 Street
Charleston, IL 61920

CHESTER

Chester Memorial Hospital The Newark Center
1900 State Street
Chester, IL 62233

CHICAGO

Alexian Brothers/Bonaventure House
825 West Wellington Street
Chicago, IL 60657

Alternatives, Inc.
1126 West Granville Avenue
2nd Floor
Chicago, IL 60660

Archdiocese of Chicago Office of Catholic Education/Substance Abuse Prevention
155 East Superior Street
Chicago, IL 60611

Association House of Chicago
2650 West Hirsch Street
Chicago, IL 60622

Brass Foundation, Inc.
Brass Essence House
1223 West Marquette Road
Chicago, IL 60636

Brass II
8000 South Racine Avenue
Chicago, IL 60620

Brass Tapestry Youth Services
950 East 61 Street
Lower Level
Chicago, IL 60637

Brass I/Alcohol and Drug Abuse and Drug Free Families with a Future
514 East 50 Place
Chicago, IL 60615

Brotherhood Against Slavery to Addiction (BASTA) Drug Abuse Program
3054–56 West Cermak Road
Chicago, IL 60623

Cathedral Shelter of Chicago
Higgins Halfway House
207 South Ashland Boulevard
Chicago, IL 60607

Higgins Recovery Home Phase II
1668 West Ogden Avenue
Chicago, IL 60607

Center for Addictive Problems
609 North Wells Street
Chicago, IL 60610

Center for Rehabilitation and Training Prevention Center
2001 North Clybourn Street
Chicago, IL 60614

Chicago Clergy Association
Haymarket House
750 West Montrose Street
Chicago, IL 60613

Haymarket House Unit I
108 North Sangamon Street
Chicago, IL 60607

Haymarket House Unit II
120 North Sangamon Street
Chicago, IL 60607

Chicago Public Schools Phillips Room 127 Intouch
244 East Pershing Road
Chicago, IL 60653

Chicagoland Chamber of Commerce Drug Free Work Place Program
200 North La Salle Street
6th Floor
Chicago, IL 60601

City of Chicago
Bethel New Life
367 North Karlov Street
Chicago, IL 60624

East Garfield Park
10 South Kedzie Avenue
Room 200C
Chicago, IL 60612

Healthy Moms and Healthy Kids
6241 South Halsted Street
Chicago, IL 60621

Westside Association Community Action
3600 West Ogden Avenue
Chicago, IL 60623

Westside Futures YMCA
1001 West Roosevelt Road
Chicago, IL 60608

Comprand, Inc. Youth and Women's Service
6857 South Halsted Street
Chicago, IL 60621

Cook County Hospital Integrated Care Substance Abuse Treatment
1835 West Harrison Street
B Building 2nd Floor
Chicago, IL 60612

Center for Rehab/Training Persons Disabled Centers Addiction Recovery of the Deaf
108 North Sangamon Street
Chicago, IL 60607

El Rincon Community Clinic
1874 North Milwaukee Avenue
Chicago, IL 60647

Englewood Comm. Health Organization (ECHO)
Halfway House
1223 West 87 Street
Chicago, IL 60620

Outpatient/Intensive Outpatient/Prevention
945 West 69 Street
Chicago, IL 60621

Recovery Home Program
1503–05 West 68 Street
Chicago, IL 60636

Family Guidance Center, Inc. Adult Outpatient
737 North La Salle Street
3rd Floor
Chicago, IL 60610

Fulfilling Our Responsibility Unto Mankind (FORUM) Substance Abuse Prevention Services
7510 South Saginaw Street
Chicago, IL 60649

Garfield Counseling Center Full Service/Women's Project/Adolescent
4132 West Madison Street
Chicago, IL 60624

Gateway Foundation, Inc.
Chicago Outpatient
2855 North Sheffield Avenue
Chicago, IL 60657

Chicago Outpatient/South
2615 West 63 Street
Chicago, IL 60629

Cook County Jail/SATC
2700 South California Avenue
Chicago, IL 60608

Kedzie Facility
1706 North Kedzie Avenue
Chicago, IL 60647

Health Care Alternative Systems, Inc.
2755 West Armitage Avenue
Chicago, IL 60647

Outpatient
1736 West 47 Street
Chicago, IL 60609

Residential
1949 North Humboldt Avenue
Chicago, IL 60647

Human Resources Development Institute
Billie Holiday Center
131 East 111 Street
Chicago, IL 60628

Inner City Youth Leader Institute
417 South Dearborn Street
Suite 300
Chicago, IL 60605

Residential Services West
2207 West 18 Street
Chicago, IL 60608

Southeast Residential Services
8731 South Exchange Avenue
Chicago, IL 60615

Women's Treatment Services at Roseland
11352 South State Street
Chicago, IL 60628

Illinois Masonic Medical Center Center for Addiction Medicine/Cleanstart
919 West Wellington Avenue
Chicago, IL 60657

Interventions
Central Intake
1234 South Michigan Avenue
Suite 100
Chicago, IL 60605

Crossroads
3738 West 103 Street
Chicago, IL 60655

Lincoln Park Programs
2043 North Sheffield Street
Chicago, IL 60614

Northside Clinic
2723 North Clark Street
1st and 2nd Floors
Chicago, IL 60614

Southwood
5701 South Wood Street
Chicago, IL 60636

King Drive Counseling and Referral Services
6252 South Martin Luther King Drive
Chicago, IL 60637

Lakeside Community Committee Drug Prevention
4414 South Cottage Grove Avenue
Suite 110
Chicago, IL 60653

Lutheran Social Services of Illinois
Addiction in Edgewater
1758 West Devon Street
Chicago, IL 60660

Addiction in Portage/Cragin
5825 West Belmont Street
Chicago, IL 60634

Addiction Program Walsh Residence
5517 North Kenmore Avenue
Chicago, IL 60640

Alcohol/Drug Dependency Program/South
3220 West 115 Street
Chicago, IL 60655

The Residence for Men
1640 West Morse Avenue
Chicago, IL 60626

The Residence South
7843 South Essex Avenue
Chicago, IL 60649

The Women's Residence
1710 West Lunt Avenue
Chicago, IL 60626

Mercy Medical at Presidential Towers
614 West Monroe Tower 3
Chicago, IL 60606

Mercy Medical on Pulaski
5635 South Pulaski Road
Chicago, IL 60629

Mexican Community Committee of South Chicago/Substance Abuse Prevention
2939 East 91 Street
Chicago, IL 60617

Mount Sinai Hospital/Medical Center Careunit Substance Abuse Services
California and 15 Streets
Chicago, IL 60608

Near North Health Services Winfield Moody Health Center
1276 North Clybourn Street
Chicago, IL 60610

New City YMCA
755 West North Avenue
Chicago, IL 60605

NIA Comp. Center for Developmental Disabilities Developmentally Disabled Drinker
151–153 West 75 Street
Chicago, IL 60620

Northern District of Illinois U.S. Prob. Office Drug Aftercare Service
219 South Dearborn Street
Room 1100
Chicago, IL 60604

Northwest Youth Outreach Logan Square/ARI/OSAP
6417 West Irving Park Road
Chicago, IL 60634

Northwestern Memorial Hospital Chemical Dependence Program/Admin.
448 East Ontario Street
8th Floor
Chicago, IL 60611

Pilsen Little Village CMHC
1858 West Cermak Road
Chicago, IL 60608

Polish Welfare Association Starting Point
3834 North Cicero Avenue
Chicago, IL 60641

Prevention Partnership
5936 West Lake Street
Chicago, IL 60644

Rush/Presbyterian St. Luke's Medical Center Rush Addiction Management and Prevention
1720 West Polk Street
Marshall Field IV Center
Chicago, IL 60612

Safer Foundation Substance Abuse Prevention Services
571 West Jackson Boulevard
Chicago, IL 60661

Saint Augustine's Center for American Indians, Inc.
4512 North Sheridan Road
Chicago, IL 60640

Salvation Army Harbor Light Center
1515 West Monroe Street
Chicago, IL 60607

South East Alcohol and Drug Abuse Center
9101 South Exchange Avenue
Chicago, IL 60617

South Shore Youth Program
7257 South Jeffery Boulevard
Chicago, IL 60649

Southwest YMCA Adolescent Outpatient Treatment Program
3801 West 127 Street
Chicago, IL 60658

Substance Abuse and Alcoholism Treatment Center, Inc.
701 West Roosevelt Road
Chicago, IL 60607

Substance Abuse Services, Inc.
Outpatient Unit
2101 South Indiana Avenue
Chicago, IL 60616

Project Success and Infant Mortality
2126 South Prairie Avenue
Chicago, IL 60616

Tarnowski Counseling and Clinical Services
5642 West Diversey Street
Room 107
Chicago, IL 60639

Veterans' Affairs Medical Center
Alcohol Dependence Treatment Program
Drug Dependence Treatment Program
820 South Damen Avenue
Chicago, IL 60680

West Side Holistic Family Center Ujima House Outpatient Unit
839 North Central Avenue
Chicago, IL 60651

Western Clinical Health Services, Inc. (WCHS)
63 East Adams Street
Chicago, IL 60603

Woodlawn Organization
Entry House
Woodlawn Residential Rehab Program
1447 East 65 Street
Chicago, IL 60637

Youth Outreach Services, Inc.
Northwest Youth Outreach/Austin
5912 West Division Street
Chicago, IL 60651

Northwest Youth Outreach/Irving Park
6417 West Irving Park Road
Chicago, IL 60634

Northwest Youth Outreach/Kedzie/OP/IOP
3841 North Kedzie Avenue
Chicago, IL 60618

Youth Service Project, Inc.
3942 West North Avenue
Chicago, IL 60647

CHICAGO (EVERGREEN PARK)

Illinois Biodyne Evergreen Park Center
9415 South Western Avenue
Suite 100
Chicago (Evergreen Park), IL 60620

Interaction Institute Alderian Family Counseling Service
2400 West 95 Street
Suite 401
Chicago (Evergreen Park), IL 60642

CLINTON

Dewitt County Human Resource Center Substance Abuse Treatment Program
1150 Route 54 West
Clinton, IL 61727

DANVILLE

Crosspoint Human Services
309 North Logan Avenue
Danville, IL 61832

Housing Authority of City of Danville Substance Abuse Prevention Services
1607 Clyman Lane
Danville, IL 61832

Vermilion County Health Department Drug Free Families with a Future
Tilton Road
Rural Route 1
Danville, IL 61832

Veterans' Affairs Medical Center Alcohol/Drug Dependence Treatment Program
1900 East Main Street
Danville, IL 61832

DE KALB

Ben Gordon Community Mental Health Center Substance Abuse Services Program
12 Health Services Drive
De Kalb, IL 60115

DECATUR

Decatur Mental Health Center Critical Populations
403 Longview Place
Decatur, IL 62521

Decatur Mental Health Center, Inc. Geoffrey M. Geoghegan Recovery Center
2300 North Edward Street
Building 5
Decatur, IL 62526

Saint Mary's Treatment Center Alcohol/Drug Addiction Treatment Center
1800 East Lakeshore Drive
Decatur, IL 62521

DES PLAINES

Partners in Psychiatry Step One Recovery Center
1695 Elk Boulevard
Des Plaines, IL 60016

DIXON

Sinnissippi Centers, Inc. Center for Addictions
325 Illinois Route 2
Dixon, IL 61021

DOWNERS GROVE

Linda Savage Alcohol Counseling
5329 Main Street
Downers Grove, IL 60515

Township of Downers Grove Comprehensive Prevention
4340 Prince Street
Downers Grove, IL 60515

DU QUOIN

Perry County Counseling Center, Inc.
RR 1
Du Quoin, IL 62832

DUNDEE

Professional Consultations
825 Village Quarter Road
Suite A-2
Dundee, IL 60118

Renz Addiction Counseling Center Outpatient Substance Abuse Services
514 Market Loop Street
Suite 109
Dundee, IL 60118

EAST SAINT LOUIS

Comp. Mental Health Center of Saint Clair County, Inc.
Alcoholism and Substance Abuse Programs
913 Martin Luther King Drive
East Saint Louis, IL 62201

Serenity Women's Recovery Home
1308 Cleveland Street
East Saint Louis, IL 62205

Gateway East Health Services, Inc. Youth Leadership Program
327 Missouri Avenue
First Illinois Bank Building Room 420
East Saint Louis, IL 62201

EDWARDSVILLE

Saint Elizabeth Medical Center The Edgewood Program
1121 University Drive
Edwardsville, IL 62025

EFFINGHAM

Heartland Human Services Guidance and Counseling Center
1108 South Willow Street
Effingham, IL 62401

ELDORADO

Egyptian Public and Mental Health Dept. Alcohol Outpatient
Rural Route 3
Eldorado, IL 62930

ELGIN

Latino Treatment Center
54 Fountain Square Plaza
Elgin, IL 60120

Lutheran Social Services of Illinois
Addiction Center
The Residence West for Men
675 Varsity Drive
Elgin, IL 60120

Prevention Education and Resource Consultants
76 Fountain Square Plaza
Elgin, IL 60120

Renz Addiction Counseling Center
Outpatient Substance Abuse Services
76 Fountain Square Plaza
Elgin, IL 60120

Passage Program
80 Fountain Square Plaza
Elgin, IL 60120

ELK GROVE VILLAGE

Kenneth W. Young Center In Touch
1001 Rohlwing Road
Elk Grove Village, IL 60007

ELMHURST

Life Education Center Foundation
180 West Park Avenue
Suite 160
Elmhurst, IL 60126

EVANSTON

Saint Francis Hospital Outpatient Addiction Treatment/Education Services (OATES)
355 Ridge Avenue
Evanston, IL 60202

FOREST PARK

Suburban Clinical Services PC
8300 West Roosevelt Road
Forest Park, IL 60130

FOX LAKE

Western Lake County Alcohol and Drug Dependency Treatment Program
17 West Grand Avenue
Fox Lake, IL 60020

FRANKLIN PARK

Leyden Family Service/Mental Health Center Alcoholism Services
10001 West Grand Avenue
Franklin Park, IL 60131

Youth Outreach Services, Inc. Northwest Youth Outreach/Leyden
10013–15 West Grand Avenue
Franklin Park, IL 60131

FREEPORT

Freeport Drug Education and Prevention Project/Youth Services
1133 West Stephenson Street
Freeport, IL 61032

Jane Addams Family Services Intouch
1133 West Stephenson Street
Freeport, IL 61032

Martin Luther King Jr. Community Services Substance Abuse Prevention Services
511 South Liberty Street
Freeport, IL 61032

Sojourn House, Inc.
565 North Turner Avenue
Freeport, IL 61032

GENEVA

Attitude Behavior Modification Systems, Inc.
324 West State Street
Geneva, IL 60134

GLENDALE HEIGHTS

Leyden Family Service/Mental Health Center Share Program
2040 Glen Ellyn Road
Glendale Heights, IL 60139

GOLCONDA

Family Counseling Center, Inc.
Market and Washington Streets
Golconda, IL 62938

GRANITE CITY

Alcoholic Rehab Community Home Arch House
1313 21st Street
Granite City, IL 62040

Piasa Health Care
11 Nameoki Village Shopping Center
Granite City, IL 62040

GRAYSLAKE

Lake County Health Dept. Mental Health Division Prevention Services/In Touch Project
19361 West Washington Street
College of Lake County
Grayslake, IL 60030

GREAT LAKES

Counseling and Assistance Center (CAAC) Naval Training Center
Building 42/Topside
Great Lakes, IL 60088

GREENVILLE

Bond County Health Department Alcohol/Drug Services and Adolescent Treatment Service
503 South Prairie Street
Greenville, IL 62246

HANOVER PARK

Renz Addiction Counseling Center
7431 Astor Street
Hanover Park, IL 60103

HARVEY

Foundation I Center for Human Development Methadone Treatment Unit
15400 South Page Avenue
Harvey, IL 60426

Mercy Counseling at Markham
16601 South Kedzie Street
Suite 106
Harvey, IL 60426

HAZEL CREST

South Suburban Council on Alcoholism and Substance Abuse
1909 Cheker Square
Section D
Hazel Crest, IL 60429

HERRIN

Associated Psychotherapists
120 West Walnut Street
Herrin, IL 62948

HILLSBORO

Montgomery County Prevention and Treatment Program
Route 185
Hillsboro, IL 62049

HINES

Veterans' Affairs Edward Hines Jr. Hospital Drug Dependency Treatment Center
Department 116C
Hines, IL 60141

HINSDALE

Interventions Du Page Adolescent Program
11 South 250 Route 83
Hinsdale, IL 60521

New Day Center of Hinsdale Hospital
120 North Oak Street
Hinsdale, IL 60521

HOFFMAN ESTATES

Spectrum Youth and Family Services Agency of the Township of Schaumburg
25 Illinois Boulevard
Hoffman Estates, IL 60194

HOPEDALE

Hopedale Hall Alcoholism Rehab Program
Railroad and Tremont Streets
Hopedale Medical Complex
Hopedale, IL 61747

IRVING

Continuing Recovery Center
Central at Vine
Irving, IL 62051

JACKSONVILLE

Jacksonville Community Center for the Deaf
907 West Superior Street
Jacksonville, IL 62650

The Wells Center
Outpatient Program
Residential Program
1300 Lincoln Avenue
Jacksonville, IL 62650

JERSEYVILLE

Recovery Center Outpatient Program
301 South Jefferson Street
Jerseyville, IL 62052

Tri-County Counseling Center Drug Abuse Program
104 North State Street
Jerseyville, IL 62052

JOLIET

Healy and Associates
121 Springfield Avenue
Joliet, IL 60455

Lifeworks Chemical Dependency Center Ottawa Substance Abuse Services
214 North Ottawa Street
Joliet, IL 60431

Saint Joseph Medical Center Substance Abuse Program
333 North Madison Street
Joliet, IL 60435

Silver Cross Hospital Chemical Dependency Unit
1200 Maple Road
Joliet, IL 60432

Stepping Stones, Inc.
Men's Halfway House
Outpatient Services
Residential Short Term
Women's Extended Care
1621 Theodore Street
Joliet, IL 60435

Will County Health Department Addiction Services
501 Ella Avenue
Joliet, IL 60433

JUSTICE

Mercy Medical In Justice
81 Street and Kean Avenue
Justice, IL 60458

KANKAKEE

Aunt Martha's Youth Service Center, Inc.
187 South Schuyler Street
Suite 420
Kankakee, IL 60901

Duane Dean Recovery Clinic Threshold MM/Hope Polydrug
700 East Court Street
Kankakee, IL 60901

Kankakee County Regional Office of Education/In Touch
189 East Court Street
Suite 400
Kankakee, IL 60901

New Hope Counseling Center
150 North Schulyer Avenue
Suite 1002
Kankakee, IL 60901

Parkside Lodge South Resolve Center
401 North Wall Street
Kankakee, IL 60901

KEWANEE

Housing Authority of Henry County Comprehensive Prevention
Fairview Apartments
Administration Building
Kewanee, IL 61443

LA SALLE

La Salle County Council for Alcohol and Drug Abuse
535 3rd Street
La Salle, IL 61301

LAKE VILLA

Gateway Foundation, Inc. Gateway Youth Care Foundation
Lake Villa Facility
25480 West Cedarcrest Lane
Lake Villa, IL 60046

Lake County Health Dept. Mental Health Division Outpatient Substance Abuse NW Satellite
121 East Grand Avenue
Lake Villa, IL 60046

LIBERTYVILLE

Alliance Institute for the Treatment of Chemical Dependency
501 West Peterson Road
Libertyville, IL 60048

Condell Medical Center Living Free/Outpatient Addiction Recovery Program
345 North Milwaukee Avenue
Libertyville, IL 60048

Lake County Health Dept. Mental Health Division Women's Residential Services
1125 North Milwaukee Road
Libertyville, IL 60048

LINCOLN

Abraham Lincoln Memorial Hospital Outpatient Substance Abuse Treatment
315 8 Street
Lincoln, IL 62656

LOMBARD

Du Page Counseling and Referral Services, Inc.
1156 South Main Street
Lombard, IL 60148

Patricia Ely and Associates Alcoholism Treatment Alternatives
450 22 Street
Suite 170
Lombard, IL 60148

MACOMB

CMHC of Fulton/McDonough Counties Substance Abuse Services
301 East Jefferson Street
Macomb, IL 61455

MANTENO

Parkside Lodge South/Resolve Center
411 Division Street
Manteno, IL 60950

Substance Abuse Services, Inc. Branden House
800 Bramble Street
Manteno, IL 60950

MARION

Franklin/Williamson Human Services, Inc. Substance Abuse Services
1305–07 West Main Street
Marion, IL 62959

United States Penitentiary and Camp Substance Abuse Services/Psychology Services
R R 5
Marion, IL 62959

MARYVILLE

Chestnut Health Systems, Inc. Lighthouse/Maryville Youth
21487 Vadalabene Road
Maryville, IL 62062

MATTOON

Central East Alcohol and Drug Council
Adolescent Outpatient Services
513 North 13 Street
Mattoon, IL 61938

Outpatient Services
416 North 19 Street
Mattoon, IL 61938

Choice
Mattoon, IL 61938

MAYWOOD

Cook County Sheriff's Youth Services In Touch
1401 Maybrook Drive
Maywood, IL 60153

Health Improvement Program Hip House
308 South 5 Avenue
Maywood, IL 60153

The Way Back Inn, Inc.
Halfway House I
104 Oak Street
Maywood, IL 60153

Halfway House II
201 South 2 Avenue
Maywood, IL 60153

Youth Outreach Services, Inc. Northwest Youth Outreach/Proviso
1701 South First Avenue
Maywood, IL 60153

MCHENRY

Family Service and CMHC For McHenry County
5320 West Elm Street
McHenry, IL 60050

MELROSE PARK

Gottlieb Memorial Hospital Lifestyle Institute
675 West North Avenue
Melrose Park, IL 60160

Westlake Community Hospital Substance Abuse Center
1225 Lake Street
Melrose Park, IL 60160

MENDOTA

Mendota Community Hospital DUI/Outpatient Services
1315 Memorial Drive
Mendota, IL 61342

MOLINE

Parkside Recovery Center at Trinity Medical Center
501 10 Avenue
Moline, IL 61265

MONTICELLO

Piatt County Mental Health Center Substance Abuse Services
125 West Lafayette Street
Monticello, IL 61856

MOUNT CARMEL

Wabash Country Health Department Counseling Services Division
130 West 7 Street
Mount Carmel, IL 62863

MOUNT PROSPECT

Bryant and Associates
1060 West NW Highway
Suite 108
Mount Prospect, IL 60056

MOUNT STERLING

Brown County Mental Health Center Alcoholism Services
111 West Washington Street
Mount Sterling, IL 62353

MOUNT VERNON

Jefferson County Comp. Services, Inc. Vantage Point
Route 37 North
Mount Vernon, IL 62864

MUNDELEIN

Parkside Recovery Program at Mundelein
24647 North Highway 21
Mundelein, IL 60060

NAPERVILLE

FVSC Care Clinics, Inc.
600 South Washington Street
Naperville, IL 60540

Naperville Task Force for Drug Free Youth
Naperville, IL 60567

NASHVILLE

Washington County Outpatient Alcoholism Program
Holzhauer Drive
Nashville, IL 62263

NORTH CHICAGO

Veterans' Affairs Medical Center Substance Abuse Program
Building 11
North Chicago, IL 60064

OAK FOREST

Bremen Youth Services
15350 Oak Park Avenue
Oak Forest, IL 60452

OAK LAWN

Associates in Alcohol and Drug Counseling
8938 South Ridgeland Avenue
Suite 100
Oak Lawn, IL 60453

Phoenix Counseling and Education Foundation
8938 South Ridgeland Avenue
Suite 100
Oak Lawn, IL 60453

OAK PARK

Grateful Hand Foundation, Inc. Grateful House
412 South Wesley Avenue
Oak Park, IL 60302

Illinois Biodyne Oak Park Center
1146 Westgate Avenue
Suite 205
Oak Park, IL 60301

OAKBROOK TERRACE

Alexander Zubenko and Associates
17 West 620 14 Street
Suite 202
Oakbrook Terrace, IL 60181

OLNEY

Southeastern Illinois Counseling Centers, Inc. Alcohol Outpatient Services
4 Micah Drive
Olney, IL 62450

OLYMPIA FIELDS

Olympia Fields Addictions Counseling and Family Recovery Program
2400 West Lincoln Highway
Olympia Fields, IL 60461

OTTAWA

DUI Assessments and Services
417 West Madison Street
Suite 206 A–B
Ottawa, IL 61350

James R. Gage and Associates
417 West Madison Street
Ottawa, IL 61350

La Salle County Council for Alcohol and Drug Abuse
776 Centennial Drive
Ottawa, IL 61350

PALATINE

Lutheran Social Services of Illinois Add in the Northwest Suburbs
4811 Emerson Avenue
Suite 112
Palatine, IL 60067

Relapse Prevention Center
1613 Colonial Parkway
Palatine, IL 60067

The Bridge Youth and Family Services Comprehensive Prevention
721 South Quintin Road
Palatine, IL 60067

PARK FOREST

Aunt Martha's Youth Enhancement Substance Abuse Program
224 Blackhawk Street
Park Forest, IL 60466

PEKIN

Tazewood Mental Health Center, Inc. Substance Abuse Program/ Treatment Unit
1423 Valle Vista Boulevard
Pekin, IL 61554

PEORIA

Human Service Center
Counseling Family Home
1318 North University Street
Peoria, IL 61606

New Leaf Retreat
3500 West New Leaf Lane
Peoria, IL 61615

White Oaks Center
3400 New Leaf Lane
Peoria, IL 61615

White Oaks Knolls
2101 West Willow Knolls Drive
Peoria, IL 61615

Work Release
218 NE Jefferson Avenue
Peoria, IL 61603

Central Illinois Center for Treatment of Addictions/ Outpatient
130 North Sheridan Road
Peoria, IL 61605

Peoria City/County Health Department Families With a Future
2116 North Sheridan Road
Peoria, IL 61604

Ripper and Associates, Ltd. Junction City Shopping Center
107 Town Hall Building
Peoria, IL 61614

T. W. Mathews and Associates
7501 North University Street
Suite 215
Peoria, IL 61614

PITTSFIELD

Counseling Center of Pike County
121 South Madison Street
Pittsfield, IL 62363

PONTIAC

Institute for Human Resources
310 Torrance Avenue
Pontiac, IL 61764

PRINCETON

Quad County Counseling Center
530 Park Avenue East
Princeton, IL 61356

QUINCY

Adams/Pike Educational Service Region Intouch/Adams County Office Substance Abuse Prevention
237 North 6 Street
Quincy, IL 62301

Family Therapy Associates
200 North 8 Street
Suite 111
Quincy, IL 62301

Great River Recovery Resource
428 South 36 Street
Quincy, IL 62301

Saint Mary Hospital Behavioral Health Services
1415 Vermont Street
Quincy, IL 62301

RED BUD

Human Service Center of Southern Metro/East Substance Abuse Services/Red Bud
Route 1
Red Bud, IL 62278

RIVER GROVE

Northwest Youth Outreach River Grove Unit
2725 Thatcher Avenue
River Grove, IL 60171

ROCK FALLS

Rockford Memorial Hospital Addiction Treatment and Education Program
1503 First Avenue
Suite A
Rock Falls, IL 61071

ROCK ISLAND

Alcohol Information Services, Inc.
3727 Blackhawk Road
Suite 103
Rock Island, IL 61201

Center for Alcohol and Drug Services Freedom House Clinic I
4230 11th Street
Rock Island, IL 61201

Martin Luther King Jr. Community Center Substance Abuse Prevention Services
630 Martin Luther King Drive
Rock Island, IL 61201

Paul A. Hauck, PhD., Ltd. Substance Abuse Services
1800 3 Avenue
Suite 302
Rock Island, IL 61201

Robert Young Center for Community Mental Health Riverside
2701 17 Street
Rock Island, IL 61201

ROCKFORD

Addiction Treatment and Education Program (ATEP)
5758 Elaine Drive
Rockford, IL 61108

Al Tech, Inc.
3415 North Main Street
Rockford, IL 61103

Alpine Park Center
5411 East State Street
Suite 204
Rockford, IL 61108

City of Rockford Comprehensive Prevention
425 East State Street
Rockford, IL 61104

Evergreen Recovery Center
1055 East State Street
Rockford, IL 61104

Family Addiction Instruction Recovery Treatment Center
5301 East State Street
Suite 101
Rockford, IL 61108

Illinois Biodyne Rockford Center
4216 Maray Drive
Suite B2
Rockford, IL 61107

Personal Health Abuse Services and Education (PHASE, Inc.)
319 South Church Street
Rockford, IL 61101

Rockford School District 205 Substance Abuse Programming
201 South Madison Street
Rockford, IL 61104

Rosecrance Center, Inc.
1505 North Alpine Road
Rockford, IL 61107

Winnebago Community Correctional Center
315 South Court Street
Rockford, IL 61102

Winnebago County Dept. of Public Health Comprehensive Prevention Program
401 Division Street
Rockford, IL 61104

ROLLING MEADOWS

Rolling Meadows Counseling Services
1545 Hicks Road
Rolling Meadows, IL 60008

ROUND LAKE

Lake County Health Dept Mental Health Division Avon Township Center
423 East Washington Street
Round Lake, IL 60073

Northern Illinois Council on Alcoholism and Substance Abuse
31979 North Fish Lake Road
Round Lake, IL 60073

RUSHVILLE

Schuyler Counseling and Health Services
127 South Liberty Street
Rushville, IL 62681

SAINT CHARLES

Interventions Valley View
34 W 826 Villa Maria Road
Saint Charles, IL 60174

Renz Addiction Counseling Center Riverside Center
1001 East Main Street
Suite E
Saint Charles, IL 60174

Tri-City Family Services Student Assistance Program
15 North First Avenue
Saint Charles, IL 60174

SCOTT AIR FORCE BASE

Scott Air Force Base Medical Center Alcoholism Rehabilitation Center
Scott/SGHAA
Scott Air Force Base, IL 62225

SHELBYVILLE

Central East Alcohol and Drug Council
155 South Morgan Street
Shelbyville, IL 62565

SHERIDAN

Gateway Foundation, Inc. Sheridan Correctional Facility Substance Abuse Treatment Center
Lasalle County and Highway 3
Sheridan, IL 60551

SKOKIE

Alon Treatment Center
9150 North Crawford Avenue
Skokie, IL 60076

Illinois Biodyne Skokie Center
5215 Old Orchard Road
Suite 390
Skokie, IL 60077

Northern Illinois Counseling Services, Inc. North Suburban Counseling Center
9933 North Lawler Street
Suite 425
Skokie, IL 60077

Rush/North Shore Medical Center Rush Chemical Dependency Program
9600 Gross Point Road
Skokie, IL 60076

SPRING VALLEY

Spring Valley Outpatient Services
213 East Saint Paul Street
Spring Valley, IL 61362

SPRINGFIELD

Alcohol and Alcohol Abuse Associates
1901 Peoria Road
Springfield, IL 62702

Dr. Marion Smith McGeath Health Services Center
3124 Stonehill Drive
Springfield, IL 62704

Gateway Foundation, Inc. Springfield Facility
2200 Lake Victoria Drive
Springfield, IL 62703

Illinois Alcoholism and Drug Dependence Association (IADDA)
500 West Monroe Street
2nd Floor
Springfield, IL 62704

Midwest Psychological Systems DUI and Substance Abuse Services
975 Durkin Drive
Clock Tower Village
Springfield, IL 62704

Personal Consultants
1945 South Spring Street
Springfield, IL 62704

Prevention Resource Center, Inc.
822 South College Street
Springfield, IL 62704

Saint John's Hospital Libertas Program
800 East Carpenter Street
Springfield, IL 62769

Sangamon Menard Triangle Center
120 North 11 Street
Springfield, IL 62703

Springfield Housing Authority Substance Abuse Prevention Services
200 North 11 Street
Springfield, IL 62703

Stillmeadow Counseling Center
833 South 4 Street
Springfield, IL 62703

STREATOR

La Salle County Council for Alcohol and Drug Abuse
104 6 Street
Streator, IL 61364

Saint Mary's Hospital Chemical Dependency Program
111 East Spring Street
Streator, IL 61364

SUMMIT

Des Plaines Valley Community Center Family Outpatient Addiction
7355 West Archer Avenue
Summit, IL 60501

SYCAMORE

Attitude/Behavioral Modification Systems, Inc.
134 West State Street
Sycamore, IL 60178

TREMONT

Tazewell County Health Department Substance Abuse Prevention Services
21306 Illinois Route 9
Tremont, IL 61568

URBANA

Prairie Center for Substance Abuse Killarney Street Unit
718 Killarney Drive
Urbana, IL 61801

University of Illinois Urbana/Champaign Center for Prevention Research and Development
1002 West Nevada Street
Urbana, IL 61801

VILLA PARK

Life Awareness Center, Inc. Adult Outpatient Treatment
53 East Saint Charles Road
Suite 7
Villa Park, IL 60181

WATERLOO

Human Support Services of Monroe County Substance Abuse Alternatives
988 North Market Street
Waterloo, IL 62298

WATSEKA

Iroquois Mental Health Center Outpatient Alcoholism Program
908 East Cherry Street
Watseka, IL 60970

WAUKEGAN

Lake County Community Action Project Project Family Tree
106 South Sheridan Road
Waukegan, IL 60085

Lake County Health Dept. Mental Health Division
Alcoholism Treatment Center
Substance Abuse Program
2400 Belvidere Street
Waukegan, IL 60085

Youth Services Program and MISA Program
3012 Grand Avenue
Waukegan, IL 60085

Northern Illinois Council on Alcoholism and Substance Abuse
1113 Greenwood Avenue
Waukegan, IL 60087

Bridge House
3016 Grand Avenue
Waukegan, IL 60085

WEST CHICAGO

Du Page County Health Department Women's Program
245 West Roosevelt Road
Suite 122
West Chicago, IL 60185

WESTCHESTER

Provisio Family Services Substance Abuse Services
9855 Roosevelt Road
Westchester, IL 60154

WHEATON

Attitude Behavior Modification Systems, Inc.
571 West Liberty Drive
Wheaton, IL 60187

Du Page County Dept. of Human Resources
First Offender Program
Multiple Offender
421 North County Farm Road
Wheaton, IL 60187

Du Page County Health Department West Public Health Center
111 North County Farm Road
Wheaton, IL 60187

Minirth/Meier Day Hospital
2100 Manchester Road
Suite 1410
Wheaton, IL 60187

Pape and Associates
618 South West Street
Wheaton, IL 60187

Wheaton Youth Outreach Comprehensive Prevention
122 West Liberty Drive
Wheaton, IL 60187

WINNETKA

New Trier High School Student Assistance Program
385 Winnetka Avenue
Winnetka, IL 60093

WOOD RIVER

Wood River Township Hospital Flex Care Program
101 East Edwardsville Road
Wood River, IL 62095

WOODRIDGE

Parkside Lodge of Du Page, Inc.
Parkside Residential Outpatient Center
Parkside Youth Center
2221 64 Street
Woodridge, IL 60517

WOODSTOCK

Horizons The Center for Counseling Services
400 Russell Court
Suite D
Woodstock, IL 60098

McHenry County Youth Service Bureau Outpatient Substance Abuse Treatment
101 South Jefferson Street
Woodstock, IL 60098

Memorial Hospital for McHenry County Alcoholism/Substance Abuse Unit
527 West South Street
Woodstock, IL 60098

INDIANA

ANDERSON

Anderson Center of Saint John's Chemical Dependency Treatment Service
2210 Jackson Street
Anderson, IN 46014

Center for Mental Health, Inc. Addiction Services
2020 Brown Street
Anderson, IN 46015

Crestview Center
2201 Hillcrest Drive
Anderson, IN 46012

House of Hope of Madison County, Inc.
902 High Street
Anderson, IN 46012

Madison County Detention Center Substance Abuse Services
217 East 15 Street
Anderson, IN 46016

Women's Alternatives, Inc.
Anderson, IN 46015

ANGOLA

Cameron Memorial Community Hospital Substance Abuse Services
416 East Maumee Avenue
Angola, IN 46703

Metropolitan School Dist. of Steuben County
400 South Martha Street
Angola, IN 46703

ATTICA

Attica City Court Substance Abuse Program
c/o Wabash Valley Outpatient Services
101 Suzie Lane
Attica, IN 47918

Fountain County Court Substance Abuse Program
c/o Wabash Valley Hospital Outpatient Services
101 Suzie Lane
Attica, IN 47918

Warren County Circuit Court Substance Abuse Program
Courthouse
Attica, IN 47918

BEDFORD

Lawrence County Substance Abuse Program
1502 I Street
Stone City Bank Building Room 209
Bedford, IN 47421

BLOOMINGTON

Amethyst House Halfway House/ Recovering Alcoholics
215 North Rogers Street
Bloomington, IN 47404

Argo Counseling, Inc.
118 East 6 Street
Bloomington, IN 47408

Bloomington Meadows Hospital
3800 North Prow Road
Bloomington, IN 47404

Bloomington Parks and Recreation Dept./Substance Abuse Services
349 South Walnut Street
Bloomington, IN 47401

Indiana Univ. Dept. of Applied Health Prevention Services
840 State Road 46 Bypass
Room 110
Bloomington, IN 47405

Monroe County Court Probation Dept. Substance Abuse Division
301 North College Avenue
Bloomington, IN 47404

South Central CMHC Substance Abuse Services
645 South Rogers Street
Bloomington, IN 47402

BOONVILLE

Warrick County Superior Court Drug/Alcohol Program (DAP)
Warrick County Courthouse
Boonville, IN 47601

BROWNSBURG

Brownsburg Community School Corp. Substance Abuse Services
225 South School Street
Brownsburg, IN 46112

Brownsburg Guidance and Counseling Center
23 Boulevard Motif
Brownsburg, IN 46112

CARMEL

Affinity Retreat and Counseling Center
12773 North Meridian Street
Carmel, IN 46032

CLAYTON

Mill Creek Community School Corporation
Clayton, IN 46118

CLINTON

Vermillion County Alcohol/Drug Service
825 South Main Street
Suite 207
Clinton, IN 47842

COLUMBUS

Bartholomew Superior Court 2 Substance Abuse Services
725 3 Street
Columbus, IN 47201

Brumbaugh and Associates
703 1/2 3 Street
Columbus, IN 47201

Family Services of Bartholomew County
331 Franklin Street
Columbus, IN 47201

Girls' Club/Boys' Club
400 North Cherry Street
Columbus, IN 47201

Quinco Consulting Center Bartholomew County Services
2075 Lincoln Park Drive
Columbus, IN 47201

CORYDON

Harrison County Court Substance Abuse Services
117 1/2 East Chestnut Street
Corydon, IN 47112

CRAWFORDSVILLE

Montgomery County Court Alcohol and Drug Services
Courthouse
Crawfordsville, IN 47933

CROWN POINT

Anglican Social Services of Northern Indiana
1119 North Main Street
Crown Point, IN 46307

DANVILLE

Center for Rational Living
1600 East Main Street
Suite 203
Danville, IN 46122

Cummins Mental Health Center, Inc. Addictions Program
6655 East U.S. 36
Danville, IN 46122

Hendricks County Superior Court Probation Department
52 West Main Street
Danville, IN 46122

DELPHI

Carroll County Superior Court Substance Abuse Services
Courthouse
Delphi, IN 46923

DYER

Saint Margaret Mercy Healthcare Centers
South Campus U.S. Highway 30
Dyer, IN 46311

EAST CHICAGO

Tri-City Community Mental Health Center, Inc. Substance Abuse Services
3901 Indianapolis Boulevard
East Chicago, IN 46312

ELKHART

Center for Problem Resolution
211 South 5 Street
Elkhart, IN 46516

City of Elkhart Community Health Org. Interventions Counseling Education
131 West Tyler Street
Suite 7
Elkhart, IN 46516

Elkhart County Court Alcohol and Drug Abuse Program
315 South 2 Street
County Court Building
Elkhart, IN 46516

Elkhart General Hospital Addictions Services
600 East Boulevard
Elkhart, IN 46515

Life House, Inc.
803 West Wolf Avenue
Elkhart, IN 46516

Oaklawn ASAP and Addiction Outpatient
2600 Oakland Avenue
Elkhart, IN 46518

Psychological and Family Consultants
926 East Jackson Boulevard
Elkhart, IN 46516

Renewal Center
208 South 4 Street
Elkhart, IN 46516

EVANSVILLE

Deaconess Hospital Recovery Center
600 Mary Street
Evansville, IN 47747

Evansville Black Coalition, Inc.
625 Bellemeade Avenue
Evansville, IN 47713

Evansville State Hospital Addiction Service Unit
3400 Lincoln Avenue
Evansville, IN 47714

Evansville/Vanderburgh School Corp. Culver School Substance Abuse Services
1 SE 9 Street
Evansville, IN 47708

Saint Mary Medical Center Substance Abuse Services
3700 Washington Avenue
Evansville, IN 47750

SW Indiana Mental Health Center Inc. Vanderburgh County Substance Abuse Services
415 Mulberry Street
Evansville, IN 47713

United Behavioral Clinics, Inc. Chemical Dependency Services
981 Kenmore Drive
Evansville, IN 47715

Vanderburgh County Superior Court Drug/Alcohol Deferral Services, Inc. (DADS)
111 NW 4 Street
Landmark Building Suite 200
Evansville, IN 47708

Welborn Memorial Baptist Hospital Parkside Substance Abuse Services
401 SE 6 Street
Evansville, IN 47713

Women's Alcoholic Halfway House, Inc.
Evansville, IN 47728

Youth Service Bureau, Inc.
1018 Lincoln Avenue
Evansville, IN 47714

FORT BENJAMIN HARRISON

Help Center Alcohol and Drug Abuse Prevention/Control Program
Building 32
Fort Benjamin Harrison, IN 46216

FORT WAYNE

Allen County Alcohol Countermeasures Program
c/o Misdemeanor and Traffic Court
Allen County Courthouse
Fort Wayne, IN 46802

Charter Beacon Hospital Substance Abuse Services
1720 Beacon Street
Fort Wayne, IN 46805

Community Addiction Program
1402 East State Boulevard
Fort Wayne, IN 46805

Family and Children's Services, Inc.
2712 South Calhoun Street
Fort Wayne, IN 46807

Fort Wayne Women's Bureau, Inc.
303 East Washington Boulevard
Fort Wayne, IN 46802

Hope House
1115 Garden Street
Fort Wayne, IN 46802

Life Skills Counseling and Learning Center
3326 South Calhoun Street
Fort Wayne, IN 46807

Lutheran Hospital of Indiana, Inc. Careunit
7930 West Jefferson Boulevard
Fort Wayne, IN 46804

Metropolitan School District Southwest Allen County/ Substance Abuse Services
4510 Homestead Road
Fort Wayne, IN 46804

Park Center, Inc. Alcoholism and Drug Department
909 East State Boulevard
Fort Wayne, IN 46805

Veterans' Affairs Medical Center Alcohol Outpatient Program
2121 Lake Avenue
Fort Wayne, IN 46805

Washington House, Inc.
2720 Culbertson Street
Fort Wayne, IN 46802

FRANKFORT

Clinton County Court Alcohol Substance Program
Courthouse
3rd Floor
Frankfort, IN 46041

FRANKLIN

Johnson County Superior Court II Countermeasures Program
5 East Jefferson Street
Johnson County Courthouse
Franklin, IN 46131

Tara Treatment Center
RR 5
Franklin, IN 46131

GARY

Actualized Counseling and Stress Management Services, Inc.
6992 Broadway
Gary, IN 46410

City of Gary Gary Commission on Status of Women
475 Broadway
Suite 508
Gary, IN 46402

Corinthian Christian Center Senior Day Treatment Project
667 Van Buren Street
Gary, IN 46402

Gary City Court Alcohol and Drug Treatment Service
1301 Broadway
Gary, IN 46402

Gary Community Mental Health Center, Inc. Substance Abuse Services
1100 West 6 Street
Gary, IN 46402

Holliday Health Care Professional Corporation
8410 Maple Avenue
Gary, IN 46403

Lake County Juvenile Court Substance Abuse Services
400 Broadway
Gary, IN 46402

Merrillville Community School Corp. Substance Abuse Services
6701 Delaware Street
Gary, IN 46410

Saint Mary Medical Center Alcoholism Treatment Center
540 Tyler Street
Gary, IN 46402

Serenity House of Gary, Inc.
Gary, IN 46408

Southlake Center for Mental Health, Inc. Substance Abuse Programs
8555 Taft Street
Gary, IN 46410

GRANGER

Charter Medical/St. Joseph County, Inc. Charter Hospital of South Bend
6407 North Gumwood Drive
Granger, IN 46530

GREENCASTLE

Putnam County Court Substance Abuse Program
Courthouse
Room 13
Greencastle, IN 46135

GREENFIELD

Hancock County Court Substance Abuse Program
Masonic Building
Suite 6
Greenfield, IN 46140

GREENSBURG

Decatur County Court Substance Abuse Services
Greensburg, IN 47240

GREENWOOD

Valle Vista Hospital Counter Point Center
898 East Main Street
Greenwood, IN 46013

GROVERTOWN

Interventions Grovertown Youth Program
Rural Route 1
Grovertown, IN 46531

HARTFORD CITY

Blackford County Hospital Substance Abuse Services
503 East Van Cleve Street
Hartford City, IN 47348

HOBART

Charter Medical/Lake County, Inc. Charter Hospital of Northwest Indiana
101 West 61 Avenue and SR 51
Hobart, IN 46342

HUNTINGBURG

Saint Joseph Hospital Chemical Dependency Services
1900 Medical Arts Drive
Huntingburg, IN 47542

HUNTINGTON

Huntington Memorial Hospital Huntington Center
1215 Etna Avenue
Huntington, IN 46750

INDIANAPOLIS

Actions for Achieving Control of Truancy Program (AACT)
901 North Carrollton Avenue
Indianapolis, IN 46202

Adult and Child Mental Health Center Substance Abuse Services
8320 Madison Avenue
Indianapolis, IN 46227

Alliance Health Systems
3850 Priority Way South Drive
Suite 215
Indianapolis, IN 46240

Alpha Resources
539 Turtle Creek South Drive
Suite 18
Indianapolis, IN 46227

Alternative Indianapolis Counseling Center
6214 North Carrollton Street
Suite A
Indianapolis, IN 46220

Auntie Mame's Child Development Center
3120 North Emerson Avenue
Indianapolis, IN 46218

Big Sisters of Central Indiana
615 North Alabama Street
Suite 336
Indianapolis, IN 46204

Broad Ripple Counseling Center
6208 North College Street
Indianapolis, IN 46220

Central State Hospital Addiction Unit
3000 West Washington Street
Indianapolis, IN 46222

Charter Counseling Centers Patriots Place Office Park
5663 Caito Drive
Suite 125
Indianapolis, IN 46226

Community Action Against Poverty of Greater Indianapolis, Inc.
2445 North Meridian Street
Indianapolis, IN 46208

Community Centers of Indianapolis, Inc.
615 North Alabama Street
Suite 312
Indianapolis, IN 46204

Fallcreek Counseling Service
2511 East 46 Street
Building VI
Indianapolis, IN 46205

Family Service Assoc. of Indianapolis Outpatient Chemical Dependency Program
615 North Alabama Street
Room 212
Indianapolis, IN 46204

First Step House
1425 South Mickley Avenue
Indianapolis, IN 46241

Flynn Christian Fellowship House Life Effectiveness Training
4040 West 10 Street
Indianapolis, IN 46222

Gallahue Mental Health Center Alcohol/Drug Services
6934 Hillsdale Court
Indianapolis, IN 46250

Girls, Inc. of Indianapolis
3959 Central Avenue
Indianapolis, IN 46205

Indiana Board of Health Bureau of Family Health Services
1330 West Michigan Street
Suite 232
Indianapolis, IN 46206

Indiana Community AIDS Action Network
1532 North Alabama Street
Indianapolis, IN 46202

Indiana Criminal Justice Institute Substance Abuse Services
302 West Washington Street
Room E209
Indianapolis, IN 46204

Indiana Juvenile Justice Task Force Substance Abuse Services
3050 North Meridian Street
Indianapolis, IN 46208

Indiana University Research and Sponsored Programs
1001 West 10 Street
Indianapolis, IN 46202

Indianapolis Campaign for Healthy Babies
324 East New York Street
Indianapolis, IN 46204

Indianapolis Police Athletic League Substance Abuse Services
50 North Alabama Street
Indianapolis, IN 46204

Indianapolis Public School Substance Abuse Services
120 East Walnut Street
Room 602J
Indianapolis, IN 46204

Indianapolis Urban League
850 North Meridian Street
Indianapolis, IN 46204

Marion County Health Department Substance Abuse Prevention Services
3838 North Rural Street
Indianapolis, IN 46205

Martin Center College Counseling Center
3553 North College Avenue
Indianapolis, IN 46205

Methodist Hospital Substance Abuse Services
1701 West Senate Street
Indianapolis, IN 46206

Metrohealth Chemical Dependency Program
4565 Century Plaza Road
Indianapolis, IN 46254

Midtown Mental Health Center Drug Free/Alcohol Treatment Program
1001 West 10 Street
Indianapolis, IN 46202

Midwest Medical Center Substance Abuse Services
3232 North Meridian Street
Indianapolis, IN 46208

Municipal Court of Marion County Probation Dept. Alcohol/Drug Services
City County Building
Room T-401
Indianapolis, IN 46204

Professional Counseling Centers of Indiana Westside Guidance Center
602 North High School Road
Suite 100 B
Indianapolis, IN 46214

Riverside Community Corrections Corp.
1415 North Pennsylvania Street
Indianapolis, IN 46202

Saint Vincent Hospital and Health Care Center, Inc.
2001 West 86 Street
Indianapolis, IN 46260

Salvation Army Harbor Light Center
927 North Pennsylvania Street
Indianapolis, IN 46204

Sterling Healthcare Corp. of Indiana
1404 South State Street
Indianapolis, IN 46203

Tri-County Center, Inc. Substance Abuse Outpatient Services
8945 North Meridian Street
Indianapolis, IN 46260

Veterans' Affairs Medical Center Chemical Dependency Treatment Section
2601 Cold Spring Road
Unit 116J
Indianapolis, IN 46222

Volunteers of America, Inc.
422 North Capitol Avenue
Indianapolis, IN 46204

JASPER

Dubois County Superior Court Alcohol and Drug Services Program
Dubois County Courthouse
2nd Floor
Jasper, IN 47546

Southern Hills Counseling Center Substance Abuse Program
480 Eversman Drive
Jasper, IN 47546

JEFFERSONVILLE

Clark County Alcohol and Drug Services
501 East Court Avenue
Room 218
Jeffersonville, IN 47130

Greater Clark County Schools Substance Abuse Services
1982 Blue Teal Lane
Jeffersonville, IN 47130

Jefferson Hospital Alcohol and Drug Treatment Unit
2700 River City Park Road
Jeffersonville, IN 47130

Lifespring Mental Health Services Substance Abuse Treatment Program
207 West 13 Street
Jeffersonville, IN 47130

KENDALLVILLE

Northeastern Center Substance Abuse Services
Kendallville, IN 46755

KOKOMO

Howard Community Hospital Mental Health Center
3500 South LaFountain Street
Apollo Professional Building
Kokomo, IN 46902

Howard County Court III Alcohol and Drug Services Program
Howard County Courthouse
Room 10
Kokomo, IN 46901

Kokomo Center Township Cons. School Corp. Washington Street Substance Abuse Services
2200 North Washington Street
Kokomo, IN 46902

Saint Joseph Hospital Trinity House
1907 West Sycamore Street
Kokomo, IN 46901

YWCA Family Intervention Center Substance Abuse Intervention/Prevention
406 East Sycamore Street
Kokomo, IN 46901

LA PORTE

La Porte Superior Court 3 Substance Abuse Prevention Services
Government Complex
Courthouse Square
La Porte, IN 46350

LAFAYETTE

Brines Consulting Service
2075 1/2 Main Street
Lafayette, IN 47904

Charter Hospital of Lafayette Addiction Services
Lafayette, IN 47903

Family Services, Inc.
731 Main Street
Lafayette, IN 47901

Home With Hope, Inc. Transitional Halfway House
1001 Ferry Street
Lafayette, IN 47901

New Directions Treatment Center
360 North 775 East
Lafayette, IN 47905

Tippecanoe County Court II
Tippecanoe County Courthouse
Lafayette, IN 47901

Wabash Valley Hospital, Inc. Riverside
2900 North River Road
Lafayette, IN 47906

LAWRENCEBURG

Community Mental Health Center Recovery Services
285 Bielby Road
Lawrenceburg, IN 47025

LEBANON

Boone County Superior Court III Substance Abuse Services
104 Courthouse Square
Lebanon, IN 46052

LIBERTY

Community Care in Union County
Liberty, IN 47353

LOGANSPORT

Cass County Circuit/Superior Courts Alcohol and Drug Court Program
200 Court Park
Cass County Government Building
Logansport, IN 46947

Four County Comp. Mental Health Center, Inc. Substance Abuse Program
1015 Michigan Avenue
Logansport, IN 46947

Logansport State Hospital Longcliff Center for Chemical Dependency
State Road 25 South
Logansport, IN 46947

LOOGOOTEE

Loogootee Community Schools Substance Abuse Services
201 Brooks Avenue
Loogootee, IN 47553

MADISON

Madison State Hospital Addiction Unit
Highway 7
Madison, IN 47250

MARION

Grant County Community Corrections Substance Abuse Services
401 South Adams Street
Marion, IN 46953

Grant/Blackford Mental Health, Inc.
505 Wabash Avenue
Marion, IN 46952

MARTINSVILLE

Morgan County Court Probation Dept. Substance Abuse Division
Morgan County Courthouse
Room 101
Martinsville, IN 46151

MERRILLVILLE

Lake County Superior Court/ County Division Prevention Services
509 West 84 Drive
Merrillville, IN 46410

MICHIGAN CITY

Kingswood Hospital Substance Abuse Services
3714 Franklin Street
Michigan City, 46360

La Porte County Youth Service Bureau
415 East 4 Street
Harborside Community Building
Michigan City, IN 46360

Memorial Hospital of Michigan City, Inc.
5 and Pine Streets
Michigan City, IN 46360

Swanson Center Satellite Office
501 Marquette Mall
Michigan City, IN 46360

MISHAWAKA

Family and Children Center Addiction Services The Children's Campus
1411 Lincolnway West
Mishawaka, IN 46544

MONROE

Adams Central Schools Substance Abuse Services
222 West Washington Street
Monroe, IN 46772

MONTICELLO

White County Court Substance Abuse Services
Courthouse
Monticello, IN 47960

MOUNT SAINT FRANCIS

Our Place Drug and Alcohol Education Services, Inc.
Mount Saint Francis, IN 47146

MUNCIE

Aquarius House, Inc.
413 South Liberty Street
Muncie, IN 47305

Associates in Mental Health Substance Abuse Services
420 West Washington Street
Muncie, IN 47305

Ball Memorial Hospital Middletown Center for Chemical Dependency
2401 University Avenue
Muncie, IN 47303

Comprehensive Mental Health Services of East Central Indiana/Chemical Dependency Program
240 North Tillotson Avenue
Muncie, IN 47304

Delaware County Court Substance Abuse Countermeasures Program
100 West Main Street
Room 307
Muncie, IN 47305

Family Services of Delaware County, Inc.
806 West Jackson Street
Muncie, IN 47305

MUNSTER

School Town of Munster Substance Abuse Services
8616 Columbia Avenue
Munster, IN 46321

NASHVILLE

Brown County Circuit Court Drug and Alcohol Program
Nashville, IN 47448

NEW ALBANY

Floyd County Alcohol and Drug Services
City County Building
Room 425
New Albany, IN 47150

Hedden House
801 Vincennes Street
New Albany, IN 47150

NEW CASTLE

Aries, Inc.
1000 North 16 Street
c/o Community Mental Health Services
New Castle, IN 47362

NOBLESVILLE

Hamilton County Court Division of Probation Services/ Substance Abuse Program
Hamilton County Courthouse
Noblesville, IN 46060

Prevail, Inc.
212 South 9 Street
Noblesville, IN 46060

NORTH VERNON

Jennings County Court Alcohol and Drug Abuse Program
Courthouse
North Vernon, IN 47282

PERU

Miami Circuit/Superior Courts Miami County Alcohol/Drug Court Program
Courthouse
Peru, IN 46970

PORTAGE

Center of Attention, Inc.
2588 Portage Mall
Portage, IN 46368

RENSSELAER

Jasper County Superior Court II Substance Abuse Program
105 1/2 West Kellner Street
Rensselaer, IN 47978

Newton County Superior Court Substance Abuse Program
Courthouse
Rensselaer, IN 47978

Ryan and Ryan Consulting and Educational Development
125 South McKinley Street
Medical Arts Building
Rensselaer, IN 47978

RICHMOND

Dunn Mental Health Center, Inc. Drug and Alcohol Program Outpatient Services
831 Dillon Drive
Richmond, IN 47374

Reid Memorial Hospital Aurora Program
1401 Chester Boulevard
Richmond, IN 47374

Richmond State Hospital Eastern Indiana Center for Chemical Dependency
498 NW 18 Street
Richmond, IN 47374

Wayne Superior Court III Substance Abuse Program
Wayne County Courthouse
2nd Floor
Richmond, IN 47374

ROCKVILLE

Parke County Circuit Court Alcohol/Drug Services Court Program
Court House
Lower Level West Entrance Room 18
Rockville, IN 47872

SCOTTSBURG

Scott County Superior Court Substance Abuse Services
1 East McClain Street
Scott County Courthouse
Scottsburg, IN 47170

SEYMOUR

Jackson Superior Court Alcohol and Drug Abuse Program
4 and Chestnut Streets
Post Office Building
Seymour, IN 47274

SHELBYVILLE

Shelby County Superior Court I
Shelby County Court II
Substance Abuse Services
Shelby County Courthouse
Shelbyville, IN 46176

SHOALS

Shoals Community School Corporation Substance Abuse Services
R R 2
Shoals, IN 47581

SOUTH BEND

Madison Center, Inc. Aid, Prevention and Treatment
403 East Madison Street
South Bend, IN 46617

Memorial Hospital Pathways Center
615 North Michigan Street
South Bend, IN 46601

Saint Joseph Probate Court Juvenile Division Substance Abuse Services
1921 Northside Boulevard
South Bend, IN 46615

Saint Joseph Superior Court Court Administered Alcohol Program
101 South Main Street
Courthouse
South Bend, IN 46601

Talbot Farm, Inc.
130 South Taylor Street
South Bend, IN 46601

YWCA of Saint Joseph County Chemical Dependency Services
802 Lafayette Boulevard
South Bend, IN 46601

TERRE HAUTE

Charter Hospital of Terre Haute Adult Services
1400 Crossing Boulevard
Terre Haute, IN 47802

Hamilton Center, Inc. Addiction Services
620 8 Avenue
Terre Haute, IN 47804

Recovery Associates, Inc. Fellowship House
2940 Jefferson Street
Terre Haute, IN 47802

Terre Haute Regional Hospital Lamb Center
3901 South 7 Street
Terre Haute, IN 47802

United States Penitentiary Drug Abuse Program/Choices
Highway 63 South
Terre Haute, IN 47808

Vigo County Court Alcohol and Drug Services Program
34 Ohio Street
Terre Haute, IN 47807

Vigo County Juvenile Court Substance Abuse Services
1919 North J. D. Hunt Road
Terre Haute, IN 47805

Vigo County Task Force on Alcohol and Drug Abuse Inc./ High Risk Youth Council
2931 Ohio Boulevard
Suite 100
Terre Haute, IN 47803

VALPARAISO

Boys' and Girls' Clubs of Porter County
354 Jefferson Street
Valparaiso, IN 46384

Moraine House, Inc.
353 West Lincolnway
Valparaiso, IN 46383

Porter County Court 4 Alcohol and Drug Offender Service
157 South Franklin Street
Valparaiso, IN 46383

Porter Memorial Hospital Substance Abuse Program
814 Laporte Avenue
Valparaiso, IN 46383

Porter/Starke CMHC Substance Abuse Services
701 Wall Street
Valparaiso, IN 46383

Youth Service Bureau of Porter County, Inc.
253 West Lincolnway
Valparaiso, IN 46383

VINCENNES

Comprehensive CMHC Vincennes Substance Abuse Program
515 Bayou Street
Vincennes, IN 47591

Knox County YMCA DBA Knox County Youth
c/o Vincennes Univ. Community Services
Vincennes, IN 47591

Knox Superior Court II Alcohol and Drug Abuse Program
620 Busseron Street
Vincennes, IN 47591

Vincennes Univ. Counseling Service Alcoholism Program
1002 North First Street
Vincennes, IN 47591

WABASH

Wabash County Hospital
Addiction Care Center
710 North East Street
Wabash, IN 46992

Addiction Care Center Court Program
670 North East Street
Wabash, IN 46992

WARSAW

Kosciusko Community Hospital Med Park Center
2101 Dubois Street
Warsaw, IN 46580

Otis R. Bowen Center for Human Services, Inc. Kosciusko Cnty Office/Substance Abuse Services
850 North Harrison Street
Warsaw, IN 46580

WINAMAC

Pulaski County Court Pulaski County Alcohol/Drug Court Program
Courthouse
Winamac, IN 46996

ZIONSVILLE

Indiana Federation of Communities for Drug Free Youth, Inc.
39 Boone Village
Zionsville, IN 46077

IOWA

ALLISON

Butler County Alcoholism and Drug Abuse Service Center, Inc.
403 North Main Street
Allison, IA 50602

AMES

Center for Addictions Recovery, Inc.
511 Duff Avenue
Ames, IA 50010

Iowa State University
Employee Assistance Program
Substance Abuse Program
Student Services Building 3rd Floor
Ames, IA 50011

Youth and Shelter Services, Inc.
232 1/2 Main Street
Ames, IA 50010

ANAMOSA

Men's Reformatory Substance Abuse Program
North High Street
Anamosa, IA 52205

ATLANTIC

Alcohol and Drug Assistance Agency, Inc.
320 Walnut Street
Atlantic, IA 50022

BETTENDORF

New Life Outpatient Center, Inc.
2435 Kimberly Road
Bettendorf, IA 52722

BOONE

Boone County Prevention and Community Services
1015 Union Street
Room 386
Boone, IA 50036

BURLINGTON

Alcohol and Drug Dependency Services of Southeast Iowa
1340 Mount Pleasant Street
Lincoln Center
Burlington, IA 52601

Burlington Medical Center Riverview Rehabilitation Center
602 North 3 Street
Burlington, IA 52601

Young House Family Services
105 Valley Street
Burlington, IA 52601

CARROLL

Area 12 Alcohol and Drug Treatment Unit
322 West 3 Street
Carroll, IA 51401

CEDAR FALLS

Area Education Agency 7 Substance Abuse Prevention Education Program
3712 Cedar Heights Drive
Cedar Falls, IA 50613

CEDAR RAPIDS

Area Substance Abuse Council, Inc.
3601 16 Avenue SW
Cedar Rapids, IA 52404

Foundation II, Inc. Crisis Center
1540 2 Avenue SE
Cedar Rapids, IA 52403

Grant Wood Area Education Agency
4401 6 Street SW
Cedar Rapids, IA 52404

Hillcrest Family Services Addictions Program
1727 First Avenue SE
Cedar Rapids, IA 52402

Mercy Medical Center Sedlacek Treatment Center
701 10 Street SE
Cedar Rapids, IA 52403

CLARINDA

Clarinda Correctional Facility The Other Way Substance Abuse Treatment Program
1800 North 16 Street
Clarinda Treatment Complex
Clarinda, IA 51632

CLEAR LAKE

Northern Trails Area Education Agency 2 Health Education Services
Mason City Airport Grounds
Clear Lake, IA 50428

CLINTON

New Directions, Inc. Center for Alcohol and Other Chemical Dependency
217 6 Avenue South
Clinton, IA 52732

Samaritan Health Systems The Bridge
1410 North 4 Street
Clinton, IA 52732

COUNCIL BLUFFS

Family Service Chemical Dependency Program
2 Northcrest Drive
Council Bluffs, IA 51503

Loess Hills Area Education Agency 13
East Highway 92
Halverson Center for Education
Council Bluffs, IA 51501

Mercy Hospital Chemical Dependency Services
800 Mercy Drive
Council Bluffs, IA 51503

CRESTON

Action Now Chemical Dependency Treatment Services
270 North Pine Street
Crossroads Mental Health Center
Creston, IA 50801

Green Valley Area Education Agency 14 Project Save
1405 North Lincoln Street
Creston, IA 50801

DAVENPORT

Center for Alcohol and Drug Services
1523 South Fairmount Street
Davenport, IA 52802

Heartland Place MARC Residential Center
1351 West Central Park Avenue
Suite 4300
Davenport, IA 52804

Mercy Hospital Substance Abuse Program West Central Park at Marquette
Davenport, IA 52805

Wittenmyer Youth Center
2800 Eastern Avenue
Davenport, IA 52803

DECORAH

Helping Services for Northeast Iowa, Inc. Substance Abuse Prevention
421 West Main Street
Decorah, IA 52101

Lutheran Hospital Counseling Services
901 Montgomery Street
Decorah, IA 52101

Northeast Iowa Mental Health Center Alcohol and Related Problems Service Center
305 Montgomery Street
Decorah, IA 52101

DENISON

Midwest Iowa Alcohol and Drug Abuse Center
1233 Broadway
Denison, IA 51442

DES MOINES

Bernie Lorenz Recovery House, Inc.
4014 Kingman Boulevard
Des Moines, IA 50311

Broadlawns Medical Center Dept. of Chemical Dependency Services
1915 Hickman Road
Des Moines, IA 50314

Children and Families of Iowa Our Primary Purpose
1111 University Street
Des Moines, IA 50314

Harold Hughes Centers Corporate Office
600 East 14 Street
Des Moines, IA 50316

Iowa Department of Education Substance Education Program
Grimes Office Building
Des Moines, IA 50319

Iowa Methodist Medical Center Powell Chemical Dependency Center
1313 High Street
Des Moines, IA 50309

Mercy Hospital First Step Mercy Recovery Center
2330 NW 106 Street
Des Moines, IA 50322

National Council on Alcoholism and Other Drug Dependencies (NCA)
218 6 Avenue
706 Fleming Building
Des Moines, IA 50309

United Behavioral Systems
6900 University Street
Suite F
Des Moines, IA 50311

DUBUQUE

Mercy Turning Point Treatment Center
200 Mercy Drive Suite 308
Professional Arts Plaza
Dubuque, IA 52001

Substance Abuse Services Center, Inc.
Nesler Centre Suite 270
Town Clock Plaza
Dubuque, IA 52001

DYERSVILLE

Mercy Health Center St. Mary's Adolescent Substance Abuse Unit
1111 3 Street SW
Dyersville, IA 52040

ELDORA

Addiction Management Systems, Inc. West Edgington Avenue
State Training School
Eldora, IA 50627

Pines Family Recovery Center at Eldora Regional Medical Center
2413 Edgington Avenue
Eldora, IA 50627

ELKADER

Substance Abuse Services for Clayton County, Inc.
431 High Street
Elkader, IA 52043

ESTHERVILLE

Holy Family Hospital Marian Family Recovery Center
826 North 8 Street
Estherville, IA 51334

FORT DODGE

North Central Alcoholism Research Foundation, Inc. (NCARF)
726 South 17 Street
Fort Dodge, IA 50501

FORT MADISON

Iowa State Penitentiary Substance Abuse Program
31 Avenue G
Fort Madison, IA 52627

GRUNDY CENTER

Chemical Dependency Service Center, Inc.
704 1/2 H Avenue
Grundy Center, IA 50638

HAMPTON

Franklin County Alcoholism Service Center
504 2 Avenue SE
Hampton, IA 50441

HARPERS FERRY

Luster Heights Camp
Rural Route 1
Harpers Ferry, IA 52146

HULL

Alcoholism and Drug Abuse Center
1126 Main Street
Hull, IA 51239

INDEPENDENCE

Buchanan County Substance Abuse Agency
209 2 Avenue NE
Independence, IA 50644

INDIANOLA

Prevention Concepts
515 North Jefferson Street
Suite D
Indianola, IA 50125

Warren County Substance Abuse Agency
217 West Salem Street
Courthouse Annex
Indianola, IA 50125

IOWA CITY

Mid Eastern Council on Chemical Abuse (MECCA)
430 Southgate Avenue
Iowa City, IA 52240

United Action for Youth
410 Iowa Avenue
Iowa City, IA 52240

Veterans' Affairs Department Psychiatry Service/ Detoxification
Highway 6 West
116A
Iowa City, IA 52246

IOWA FALLS

Hardin County Alcohol and Drug Services, Inc. We Care
613 1/2 Washington Avenue
Iowa Falls, IA 50126

KNOXVILLE

Marion County Substance Abuse Agency
114 East Robinson Street
Knoxville, IA 50138

Veterans' Affairs Medical Center Addictions Treatment Unit
1515 West Pleasant Street
Knoxville, IA 50138

MANNING

Manning General Hospital Manning Family Recovery Center
410 Main Street
Manning, IA 51455

MARSHALLTOWN

Substance Abuse Treatment Unit of Central Iowa
19 West State Street
Marshalltown, IA 50158

The Listening Post, Inc.
30 West Main Street
Room 106
Marshalltown, IA 50158

MASON CITY

Prairie Ridge
320 North Eisenhower Avenue
Mason City, IA 50401

MITCHELLVILLE

Iowa Correctional Institution for Women/The Recovery Program
300 Elm Street SW
Mitchellville, IA 50169

MOUNT AYR

Ringgold County Hospital Ringgold County Recovery Center
211 Shellway Drive
Mount Ayr, IA 50854

MOUNT PLEASANT

Mental Health Institute Iowa Residential Treatment Center
1200 East Washington Street
Mount Pleasant, IA 52641

Mount Pleasant Correctional Facility Therapeutic Community Program
1200 East Washington Street
Mount Pleasant, IA 52641

MUSCATINE

Muscatine General Hospital New Horizons Outpatient Substance Abuse Program
1518 Mulberry Avenue
Muscatine, IA 52761

NEWTON

Capstone Center, Inc. Substance Abuse Division
306 North 3rd Avenue East
Newton, IA 50208

Correctional Release Center Substance Abuse Treatment Program
321 East 3 Street
Newton, IA 50208

OAKDALE

Univ. of Iowa Hospitals and Clinics Chemical Dependency Center
Oakdale Campus
Oakdale Hall
Oakdale, IA 52319

OTTUMWA

Ottumwa Regional Health Center Family Recovery Center
312 East Alta Vista Avenue
Ottumwa, IA 52501

Southern Iowa Economic Development Assoc. Drug and Alcohol Services
226 West Main Street
Ottumwa, IA 52501

Southern Prairie Area Education Agency 15
Industrial Airport
900 Terminal Avenues Bldgs 40 and 41
Ottumwa, IA 52501

ROCKWELL CITY

Harvest Acres, Inc.
2511 Sigourney Avenue
Rockwell City, IA 50579

Recovery Over Criminality Substance Abuse Program
c/o North Central Correctional Facility
Rockwell City, IA 50579

SERGEANT BLUFF

Native American Alcohol Treatment Program, Inc.
Larpenteur Avenue
Building 544
Sergeant Bluff, IA 51054

SIOUX CITY

Marian Health Center Chemical Dependency Services
2101 Court Street
Sioux City, IA 51104

Saint Luke's Gordon Recovery Centers
2720 Stone Park Boulevard
Sioux City, IA 51104

SPIRIT LAKE

Northwest Iowa Alcoholism and Drug Treatment Unit, Inc.
Dickinson County Memorial Hospital
Spirit Lake, IA 51360

STORM LAKE

Vista Addiction and Recovery Center North Campus of Buena Vista County Hospital
1305 West Milwaukee Street
Storm Lake, IA 50588

TAMA

Mesquakie Alcohol/Drug Abuse Center
3137 F Avenue
Tama, IA 52339

WASHINGTON

Bob Gray Outreach Center Washington County
219 1/2 West Main Street
Washington, IA 52353

WATERLOO

Gordon Recovery Center at Allen Memorial Hospital
1825 Logan Avenue
Waterloo, IA 50703

Northeast Council on Substance Abuse (NECSA)
2222 Falls Avenue
Waterloo, IA 50701

WAUKON

Allamakee County Substance Abuse Prevention Program
Allamakee County Courthouse
Waukon, IA 52172

KANSAS

ABILENE

Dickinson County Council on Alcohol and Other Drugs, Inc.
409 NW 3 Street
Upper E Mezzanine
Abilene, KS 67410

Kansas Mothers and Children Reintegration Treatment, Inc.
409 NW 3 Street
Abilene, KS 67410

ATCHISON

Valley Hope Alcoholism Treatment Center Atchison Valley Hope
1816 North 2 Street
Atchison, KS 66002

AUGUSTA

Valley Hope at Augusta Medical Complex
2101 Dearborn Street
Augusta, KS 67010

BAXTER SPRINGS

MacAuley and Associates
2011 Fairview Avenue
Baxter Springs, KS 66713

COFFEYVILLE

Alcohol/Drug Safety Action Program of Southeast Kansas
808 Willow Street
400 West Building
Coffeyville, KS 67337

COLBY

Citizens Medical Center Substance Abuse Services
100 East College Drive
Colby, KS 67701

Fred Waters Associates Alcohol and Drug Abuse Services
175 South Range Street
Colby, KS 67701

Northwest Kansas Regional Prevention Center at Colby
485 North Court Street
Colby, KS 67701

Thomas County Council on Alcohol/Drug Abuse, Inc.
775 East College Drive
Colby, KS 67701

COLUMBUS

Elm Acres Youth Home for Girls
501 Central Avenue
Columbus, KS 66725

Family Life Center Alcohol and Drug Abuse Program
201 West Walnut Street
Columbus, KS 66725

CONCORDIA

Kerrs Counseling
515 Washington Street
Concordia, KS 66901

DERBY

Community Health Association
111 North Baltimore Street
Derby, KS 67037

DODGE CITY

New Chance, Inc. Community Alcohol/Drug Abuse Treatment Center
201 East Wyatt Earp Boulevard
Dodge City, KS 67801

Saint Josephs Chemical Dependency Program Outpatient
2008 First Avenue
Dodge City, KS 67801

EL DORADO

Parallax Program, Inc. Adapt
El Dorado Correctional Facility
El Dorado, KS 67042

South Central Mental Health, Inc. Counseling Center
2365 West Central Street
El Dorado, KS 67042

ELLSWORTH

Mirror, Inc.
Ellsworth Correction Facility/Adapt
1607 State Street
Ellsworth, KS 67439

EMPORIA

Corner House, Inc.
418 Market Street
Emporia, KS 66801

Counseling and Psychological Services
1024 West 12 Avenue
Emporia, KS 66801

Mental Health Center of East Central Kansas Alcohol and Drug Services
1000 Lincoln Street
Emporia, KS 66801

Newman Memorial County Hospital Recovery Road
1037 Elm Street
Emporia, KS 66801

The Farm, Inc.
528 Commercial Street
Emporia, KS 66801

FORT SCOTT

Bourbon County Alcohol and Drug Counseling Service
11 East First Street
Fort Scott, KS 66701

GARDEN CITY

Saint Joseph's Chemical Dependency Program Outpatient
302 Fleming Street
Suite 3
Garden City, KS 67846

Southwest Kansas Regional Prevention Center
801 Campus Drive
Garden City, KS 67846

Western Kansas Foundation for Alcohol and Chemical Dependency Inc./Crossroads House
811 North Main Street
Garden City, KS 67846

GIRARD

Addiction Treatment Center of Southeast Kansas
810 West Cedar Street
Girard, KS 66743

SE Kansas Regional Prevention Center at the Southeast Kansas Education Services Center
Girard, KS 66743

GODDARD

Judge James V. Riddel Boys Ranch
25331 West 39 Street South
Rural Route 1
Goddard, KS 67052

GOODLAND

Northwest Kansas Medical Center Substance Abuse Unit
First and Sherman Streets
Goodland, KS 67735

GREAT BEND

Center for Counseling and Consultation Substance Abuse Outpatient Treatment Program
5815 Broadway
Great Bend, KS 67530

Central Kansas Psychological Services Eldean Kohrs
925 Patton Street
Great Bend, KS 67530

Saint Joseph's Chemical Dependency Program/ Outpatient
2124 Washington Street
Great Bend, KS 67550

GREENSBURG

Iroquois Center for Human Development
103 South Grove Street
Greensburg, KS 67054

HAYS

High Plains Mental Health Center Alcohol and Drug Abuse Services
208 East 7 Street
Hays, KS 67601

Key Alcohol and Drug Abuse Services
1008 East 17 Street
Hays, KS 67601

Northwest Kansas Regional Prevention Center at Hays
2209 Canterbury Road
Suite C
Hays, KS 67601

Saint Joseph's Chemical Dependency Program
2604 B General Hays Road
Hays, KS 67601

Smoky Hill Foundation for Chemical Dependency, Inc.
2209 Canterbury Road
Suite C
Hays, KS 67601

HIAWATHA

Kanza Mental Health/Guidance Center, Inc. Alcohol and Drug Abuse Services
909 South 2 Street
Hiawatha, KS 66434

HOISINGTON

Clara Barton Hospital Family Recovery Center
250 West 9 Street
Hoisington, KS 67544

HUCHINSON

Mirror, Inc.
Outpatient Office
400 South 2 Street
Suite C
Hutchinson, KS 67501

Adapt
Hutchinson Correctional Facility
500 Reformatory
Hutchinson, KS 67501

Saint Joseph's Chemical Dependency Program/ Outpatient
109 West 5 Street
Suite D
Hutchinson, KS 67502

INDEPENDENCE

Four County Mental Health Center Alcohol and Drug Program
3701 West Main Street
Independence, KS 67301

JUNCTION CITY

Geary Community Hospital Substance Abuse Services
1102 Saint Mary's Road
Junction City, KS 66441

KANSAS CITY

Addiction Stress Center
1330 North 78 Street
Kansas City, KS 66112

Associated Youth Services, Inc.
3111 Strong Avenue
Kansas City, KS 66106

Catholic Social Services
2220 Central Avenue
Kansas City, KS 66101

Heart of America Family Services, Inc.
5424 State Avenue
Kansas City, KS 66102

Kansas City Kansas Spanish Speaking Office
1333 South 27 Street
El Centro Suite 220
Kansas City, KS 66106

Kansas City Kansas Drug and Alcohol Information School/ ASAP
707 Minnesota Avenue
Security National Bank Building/ M-6
Kansas City, KS 66101

Kansas Multicultural A/D Treatment Center
2940 North 17 Street
Kansas City, KS 66104

Kansas University Medical Center Kansas City Metro Methadone Program
39 and Rainbow Boulevard
Room 11G
Kansas City, KS 66103

Kaw Valley Center of Wyandotte House
4300 Brenner Drive
Kansas City, KS 66104

Project Turnaround
815 Ann Avenue
Kansas City, KS 66101

Salvation Army Shield of Service Detox and Reintegration Center
1200 North 7 Street
Kansas City, KS 66101

Wyandotte County Regional Prevention Center at Kansas City Kansas Community College
7250 State Avenue
Kansas City, KS 66112

Wyandotte Mental Health Center, Inc. Alcohol and Drug Abuse Services
Eaton Street at 36 Avenue
Kansas City, KS 66102

LARNED

Larned Correctional Mental Health Facility Substance Abuse Treatment Program
Mental Health Consortium
Route 3
Larned, KS 67550

Saint Joseph's Chemical Dependency Program/ Inpatient
923 Carroll Street
Larned, KS 67550

Sunrise, Inc. Reintegration Program
523 North Main Street
Larned, KS 67550

LAWRENCE

Kansas Women's Substance Abuse Services, Inc. First Step House
345 Florida Street
Lawrence, KS 66044

LEAVENWORTH

Northeast Kansas Mental Health and Guidance Center Recovery Services of Northeast Kansas
818 North 7 Street
Leavenworth, KS 66048

United States Penitentiary Drug Abuse Program
Leavenworth, KS 66048

LIBERAL

Family Alcohol and Drug Services, Inc.
316 West 7 Street
Liberal, KS 67905

MANHATTAN

Greg Potter, Ph.D. Alcohol and Drug Abuse Services
714 Poyntz Street
Suite A
Manhattan, KS 66502

Kansas State University Alcohol and Other Drug Education Services
Lafene Student Health Center
Suite 214
Manhattan, KS 66506

Larry M. Peak Ph.D, Alcohol/ Drug Services at Manhattan Medical Center
1133 College Avenue
Building B Upper Level
Manhattan, KS 66502

Northeast Kansas Regional Prevention Center at Pawnee Mental Health Services, Inc.
2001 Claflin Street
Manhattan, KS 66502

Pawnee Mental Health Center Substance Abuse Services
2001 Claflin Street
Manhattan, KS 66502

MAYETTA

Prairie Band Potawatomi Tribe Wellness Center
Route 2
Mayetta, KS 66509

MCPHERSON

McPherson Area Council for Alcohol and Drug Service, Inc.
1123 South Main Street
McPherson, KS 67460

McPherson County Schools Learning to Live
514 North Main Street
McPherson, KS 67460

NEWTON

Mirror, Inc. South Central Kansas Regional Prevention Center
130 East 5 Street
Newton, KS 67114

Prairie View Mental Health Center Chemical Dependency Treatment
1901 East First Street
Newton, KS 67114

United Methodist Youthville Chemical Dependency Treatment Program
900 West Broadway
Newton, KS 67114

NORTON

Norton Treatment Program
RR 1
Norton, KS 67654

Valley Hope Association Valley Hope Alcoholism Treatment Center
709 West Holme Street
Norton, KS 67654

OLATHE

Cypress Recovery, Inc.
230 South Kansas Street
Olathe, KS 66061

Drug Abuse Education Center, Inc.
807 Clairborne Road
Olathe, KS 66061

Johnson County Adolescent Center for Treatment (ACT)
301 North Monroe Street
Olathe, KS 66061

OTTAWA

Franklin County Mental Health Clinic, Inc. Substance Abuse Program
204 East 15 Street
Ottawa, KS 66067

OVERLAND PARK

Charles Stebbins Counseling Services
5750 West 95 Street
Suite 208
Overland Park, KS 66207

Heart of America Family Services, Inc. Johnson County Office
10500 Barkley Street
Suite 210
Overland Park, KS 66212

Johnson/Leavenworth and Miami Regional Prevention Center
6701 West 64 Street
Suite 222
Overland Park, KS 66202

Overland Park Alcohol Diversion Program DUI Probation/Parole
8826 Santa Fe Street
Beal Building Suite 306
Overland Park, KS 66212

Safe Home, Inc. Counseling Program
Overland Park, KS 66204

PAOLA

Bourbon/Linn/Miami Day Reporting Program
211 North Silver Street
Paola, KS 66071

Miami County Mental Health Center Certified Substance Abuse Program
401 North East Street
Paola, KS 66071

PITTSBURG

Crawford County Mental Health Center
Alcohol and Drug Program
30th and Michigan Streets
Pittsburg, KS 66762

Renewal House
606 East Atchinson Street
Pittsburg, KS 66762

Elm Acres Youth Home, Inc.
1002 East Madison Street
Pittsburg, KS 66762

RESERVE

Sac and Fox of Missouri Substance Abuse Program
Route 1
Reserve, KS 66434

SALINA

North Central Kansas Regional Prevention Center
1805 South Ohio Street
Salina, KS 67401

Smoky Hill Central Kansas Education Service Center Alcohol/Drug Prevention Programs
1648 West Magnolia Street
Salina, KS 67401

SEDAN

Sedan City Hospital Recovery Way Inc./Youth Chemical Dependency
300 North Street
Sedan, KS 67361

SHAWNEE MISSION

Johnson County Substance Abuse Services, Inc.
6221 Richard Drive
Shawnee Mission, KS 66216

SHAWNEE MISSION (MISSION)

Alcohol and Drug Services, Inc.
6005 Martway Street
Suite 100
Shawnee Mission (Mission), KS 66202

MCC Behavioral Care, Inc.
5700 Broadway Drive
Foxridge Towers Suite 920
Shawnee Mission (Mission), KS 66202

Valley Hope Counseling and Referral Center of Greater Kansas City
5410 West 58 Terrace
Shawnee Mission (Mission), KS 66205

SYRACUSE

Syracuse Chemical Addiction Treatment of Kansas, Inc. (SCAT)
Donohue Building
NE Entrance 3rd Floor
Syracuse, KS 67878

TOPEKA

DCCCA Center Youth Services
2914 SW Plass Court
Topeka, KS 66611

National Council on Alcoholism Topeka Division
603 SW Topeka Boulevard
Casson Building
Topeka, KS 66603

Saint Francis Hospital and Medical Center Chemical Dependency Treatment Services
1700 SW 7 Street
3rd Floor
Topeka, KS 66606

Shawnee Community Mental Health Center Services for Alcohol Related Problems
330 SW Oakley Street
Topeka, KS 66606

Shawnee Regional Prevention Center
603 Topeka Boulevard
Casson Building 2nd Floor
Topeka, KS 66603

Sunflower Alcohol Safety Action Project, Inc.
112 SE 7 Street
Suite E
Topeka, KS 66603

Topeka Correctional Facility/ Central Alcohol Drug Abuse Primary Treatment
8 Street and Rice Road
Dorm A
Topeka, KS 66607

Topeka Halfway House East
807 Western Street
Topeka, KS 66606

Veterans' Affairs Medical Center Alcohol/Drug Treatment Unit
2200 Gage Boulevard
Building 2
Topeka, KS 66622

Women's Recovery Center of DCCCA, Inc.
1328 SW Western Street
Topeka, KS 66604

WELLINGTON

Health Network Outpatient Counseling Services
924 South Washington Street
Wellington, KS 67152

WICHITA

Addictive Behavior Consultants
2400 North Woodlawn Street
Suite 210
Wichita, KS 67220

Adolescent/Adult/Family Recovery Program
3540 West Douglas Street
Suites 4 and 6
Wichita, KS 67203

Alcoholism Family Counseling Center
714 South Hillside Street
Wichita, KS 67211

Center for Human Development
915 South Glendale Street
Wichita, KS 67218

Community Health Association
1411 North Saint Paul Street
Wichita, KS 67203

Karen E. Hamilton Counseling Services
5920 East Central Street
Suite 204
Wichita, KS 67208

MCC Behavioral Care, Inc.
216 West Murdock Street
Wichita, KS 67203

Mid-American All-Indian Center Indian Alcoholism Treatment Services
313 North Seneca Street
Suite 109
Wichita, KS 67203

Neva Prosser Tulloch, Ltd.
4805 West Central Street
Wichita, KS 67212

Northeast Drug/Alcohol Referral and Tracking Station, Inc.
1809 North Broadway
Suite C
Wichita, KS 67214

Outpatient Paradigm
1333 North Broadway
Suite D
Wichita, KS 67214

Parallax Program, Inc.
Alcohol/Drug Abuse Treatment
Parallax Reintegration
Department of Corrections Module (DOC)
320 North Market Street
Wichita, KS 67202

Phoenix 7 Counseling Center
9323 East Harry Street
Suite 121C
Wichita, KS 67207

Professional Review Network, Inc.
313 North Seneca Street
Suite 100
Wichita, KS 67203

Recovery Services Council, Inc.
1712 West Douglas Street
Wichita, KS 67202

Relapse and Relapse Prevention Counseling
1333 North Broadway
Suite D
Wichita, KS 67214

Rogers Counseling Service
455 Windsor Street
Wichita, KS 67218

Saint Francis Regional Medical Center Chemical Dependency Services
929 North Saint Francis Street
Wichita, KS 67214

Salvation Army Booth Family Service Center Outpatient Drug/Alcohol Treatment
2050 West 11th Street
Wichita, KS 67203

Sedgwick County Addiction Treatment Services
940 North Waco Street
Wichita, KS 67203

The Lighthouse of Wichita, Inc.
204 South Osage Street
Wichita, KS 67213

Valley Hope Alcoholism Outpatient Counseling and Referral Center
901 West Douglas Street
Wichita, KS 67213

Veterans' Affairs Medical Center Substance Abuse Treatment Unit
5500 East Kellogg Street
Wichita, KS 67208

Victor Montemayor Alcohol and Drug Abuse Services
3081 South Rutan Court
Wichita, KS 67210

Wichita/Sedgwick County Regional Prevention Center The Drug/Alcohol Abuse Prevention Center
1421 East 2nd Street
Wichita, KS 67214

Women's Recovery Center of Central Kansas
309 North Market Street
Wichita, KS 67202

KENTUCKY

ALBANY

Adanta Behavioral Health Services Albany Clinic
507 Cross Street
Albany, KY 42602

DUI Defendant Referral Systems, Inc.
Clinton County Courthouse
Albany, KY 42602

ASHLAND

Federal Correctional Institution Substance Abuse Services
Ashland, KY 41101

Pathways, Inc.
Adkins House
2801 Winchester Avenue
Ashland, KY 41101

Boyd County Outpatient Unit
Withdrawal Unit
201 22 Street
Ashland, KY 41101

BARBOURVILLE

Cumberland River Comprehensive Care Center
317 Cumberland Avenue
Barbourville, KY 40906

BARDSTOWN

Communicare Bardstown Communicare Clinic
331 South 3 Street
Bardstown, KY 40004

BARDWELL

Western Kentucky MH/MR Board Carlisle County MH/MR Services
Highway 62
Bardwell, KY 42023

BARLOW

Western Kentucky MH/MR Board Ballard County MH/MR Services
Highway 60
Barlow, KY 42024

BEATTYVILLE

Kentucky River Community Care, Inc.
Beattyville By Pass
Beattyville, KY 41311

BEDFORD

Seven Counties Services Trimble County Center
Church Street
Bedford, KY 40006

BENHAM

Cumberland River Comp. Care Center Tri-Cities Center
Main Street
Benham, KY 40807

BENTON

Western Kentucky MH/MR Board Benton/Marshall County MH/MR Services
1304 Main Street
Benton, KY 42025

BOWLING GREEN

Lifeskills, Inc.
Bowling Green Center
Park Place
822 Woodway
Bowling Green, KY 42102

Medical Center at Bowling Green Reservoir Hill Care Centre
800 Park Street
Bowling Green, KY 42101

BROOKSVILLE

Comprehend, Inc. Bracken County Community Care Center Outpatient Alcohol/ Drug Services
Grandview Subdivision
Brooksville, KY 41004

BROWNSVILLE

Lifeskills, Inc. Edmonson County Office
1120 South Main Street
Brownsville, KY 42210

BURKESVILLE

Adanta Behavioral Health Services Burkesville Clinic
390 Keen Street
Burkesville, KY 42717

CALHOUN

Green River Comprehensive Care
535 West First Street
Calhoun, KY 42327

CAMPBELLSVILLE

Adanta Behavioral Health Services Campbellsville Clinic
3020 Lebanon Road
Campbellsville, KY 42718

CAMPTON

Kentucky River Community Care, Inc. Wolfe County Health Department
605 Highway 15 South
Suites 1 and 2
Campton, KY 41301

CARLISLE

Bluegrass West Comprehensive Care Center Nicholas County Comprehensive Care
226 Locust Street
Room 4
Carlisle, KY 40311

CARROLLTON

DUI Defendant Referral Systems, Inc.
1302 Highland Avenue
Carroll County Courthouse
Carrollton, KY 41008

CLINTON

Western Kentucky MH/MR Board Clinton/Hickman County MH/ MR Services
South Washington Street
Clinton/Hickman Counties Hospital
Clinton, KY 42031

COLUMBIA

Adanta Behavioral Health Services Columbia Clinic
808 C Jamestown Street
Columbia KY 42728

CORBIN

Baptist Regional Medical Center Adult Chemical Dependency Unit
1 Trillium Way
Corbin, KY 40701

Cumberland River
Comprehensive Care Center
American Greetings Road
Corbin, KY 40701

Independence House
3110 Cumberland Falls Highway
Corbin, KY 40701

DUI Defendant Referral Systems, Inc.
Falls Road
Holiday Inn
Corbin, KY 40701

COVINGTON

DUI Defendant Referral Systems, Inc.
413 Scott Street
Covington, KY 41011

Family Alcohol and Drug Counseling Center
722 Scott Street
Covington, KY 41012

Saint Elizabeth Medical Center Chemical Dependency Units
401 East 20 Street
Covington, KY 41014

Transitions, Inc.
Eighth Street Halfway House
113 East 8 Street
Covington, KY 41011

Women's Residential Addiction Program (WRAP)
1629 Madison Avenue
Covington, KY 41011

CYNTHIANA

Bluegrass West Comprehensive Care Center Harrison County Comprehensive Care
122 East Pleasant Street
Cynthiana, KY 41031

DANVILLE

Bluegrass South Comprehensive Care Danville Comprehensive Care Center
650 High Street
Danville, KY 40422

Collins/Kubale/Miles and Associates
219 South 4 Street
Danville, KY 40422

EDDYVILLE

Western Kentucky Drug and Alcohol Intervention Services, Inc.
1216 Fairview Avenue
Eddyville, KY 42038

EDMONTON

Lifeskills, Inc. Metcalfe County Office
402 Tompkinsville Road
Edmonton, KY 42129

ELIZABETHTOWN

Communicare
Elizabethtown Substance Abuse Services
Communicare Recovery Center
1311 North Dixie Avenue
Elizabethtown, KY 42701

EMINENCE

Seven Counties Services Henry County Mental Health Center
222 South Main Street
Eminence, KY 40019

FALMOUTH

Saint Luke Hospital Alcohol Drug Treatment Center
512 South Maple Avenue
Falmouth, KY 41040

FLEMINGSBURG

Comprehend, Inc. Fleming County CMHC Outpatient Alcohol and Drug Offices
610 Elizaville Road
Flemingsburg, KY 41041

FLORENCE

Comp. Care Centers of Northern Kentucky Boone County Comprehensive Care Center
8172 Mall Road
Suite 239
Florence, KY 41042

FORT CAMPBELL

Community Counseling Services
23 Street and Indiana Avenue
AFZB-PA-D Building 2537
Fort Campbell, KY 42223

FORT KNOX

Fort Knox Alcohol and Drug Counseling Center
Brule Street
Building 6602
Fort Knox, KY 40120

FRANKFORT

Bluegrass Education and Treatment of Alcoholism Program
943 Wash Road
Route 7
Frankfort, KY 40601

Bluegrass West Comprehensive Care Center
Frankfort Office
191 Doctors Drive
Frankfort, KY 40601

Halfway House
Wash Road
Route 7
Frankfort, KY 40601

Counseling Center, Inc.
807 Holmes Street
Frankfort, KY 40601

FRANKLIN

Lifeskills, Inc. Simpson County Office
112 South High Street
Franklin, KY 42134

FRENCHBURG

Pathways, Inc. Menifee County Outpatient Unit
Route 36
Frenchburg, KY 40322

FULTON

Western Kentucky MH/MR Board Fulton County Mental Health Service
350 Browder Street
Fulton, KY 42041

GEORGETOWN

Bluegrass West Comprehensive Care Center Scott County Clinic
1226 Paris Pike
Georgetown, KY 40324

Counseling Center, Inc.
100 Court Street
City Hall Building
Georgetown, KY 40324

GLASGOW

Lifeskills, Inc. Barren County Office
901 Columbia Avenue
Glasgow, KY 42142

GRAYSON

Pathways, Inc. Carter County Outpatient Unit
515 West Main Street
Grayson, KY 41143

GREENSBURG

Adanta Behavioral Health Services Greensburg Clinic
429 Hodgensville Road
Greensburg, KY 42743

GREENUP

Pathways, Inc. Greenup County Outpatient Unit
1018 Walnut Street
Greenup, KY 41144

GREENVILLE

DUI Defendant Referral Systems, Inc.
200 South Martin Street
Muhlenberg County Jail
Greenville, KY 42345

Pennyroyal Mental Health Services Muhlenberg County MH/MR Center
506 Hopkinsville Street
Greenville, KY 42345

HARDINSBURG

Communicare Hardinsburg Communicare Clinic
Route 1
Mattingly Professional Building
Hardinsburg, KY 40143

HARLAN

Cumberland River Comprehensive Care Center
Mounted Route 1
Harlan, KY 40831

HARRODSBURG

Bluegrass South Comprehensive Care Harrodsburg Comprehensive Care Center
352 Mr. Kwik Shopping Center
Harrodsburg, KY 40330

HARTFORD

Green River Comprehensive Care Center
130 East Washington Street
Hartford, KY 42347

HAZARD

Kentucky River Community Care, Inc. Perry County
200 Medical Center Drive
Suite 1B
Hazard, KY 41701

HEBRON

Tri-State Drug Rehabilitation and Counseling Program Inc./Kids Helping Kids
2134 Petersburg Road
Hebron, KY 41048

HENDERSON

Community Methodist Hospital Chemical Dependency Unit
1305 North Elm Street
Henderson, KY 42420

Gerald K. Gaddis Pastorial Counseling Service
110 3 Street
Suite 290
Henderson, KY 42420

Green River Comprehensive Care Elm Street Office
1287 North Elm Street
Henderson, KY 42420

Regional Addiction Resources (RAR)
6347 Highway 60 East
Henderson, KY 42420

HINDMAN

Kentucky River Community Care, Inc.
Courthouse Square
Hindman, KY 41822

HOPKINSVILLE

Pennyroyal Center
735 North Drive
Hopkinsville, KY 42240

Volta Program Substance Abuse Treatment Center
Russellville Road Highway 68
Johnson Building
Hopkinsville, KY 42240

HYDEN

Kentucky River Community Care, Inc. Hyden Unit
Hurts Creek Shopping Center
Hyden, KY 41749

INEZ

Mountain Comprehensive Care Center Martin County Clinic
Rural Route 3
Inez, KY 41224

IRVINE

Bluegrass South Comprehensive Care Irvine Comprehensive Care Center
Star Route
Irvine, KY 40336

JACKSON

Kentucky River Community Care, Inc.
1112 1/2 Main Street
Jackson, KY 41339

JAMESTOWN

Adanta Behavioral Health Services Jamestown Clinic
Highway 127 South
Jamestown, KY 42629

DUI Defendant Referral Services, Inc.
Russell County Courthouse
Jamestown, KY 42629

LA GRANGE

Seven Counties Services Oldham County Center
1919 South Highway 53
La Grange, KY 40031

LANCASTER

Bluegrass South Comprehensive Care Lancaster Comprehensive Care Center
67 Public Square
Lancaster, KY 40444

LAWRENCEBURG

Bluegrass West Comprehensive Care Center Lawrenceburg Comprehensive Care Center
1060 Glensboro Road
Lawrenceburg, KY 40342

LEBANON

Communicare
Lebanon Communicare Clinic
Route 4/Springfield Road
Lebanon, KY 40033

Leitchfield Communicare Clinic
300 South Clinton Street
Health Department Annex
Leitchfield, KY 42754

LEWISPORT

Green River Comprehensive Care Center
Lewisport Shopping Center
Lewisport, KY 42351

LEXINGTON

Alcohol Related Offenders Program
880 Sparta Court
Suite 219
Lexington, KY 40504

Blue Grass East Comprehensive Care
Teen Primary Outpatient Program
Drug and Alcohol Program
200 West 2 Street
Lexington, KY 40507

Detoxification Center
146 East 3 Street
Lexington, KY 40508

Methadone Program
201 Mechanic Street
Lexington, KY 40508

Bluegrass Mental Health Board, Inc. Forensic Services
177 North Upper Street
Lexington, KY 40507

Charles I Schwartz Chemical Dependency Treatment Center
420 South Broadway
Lexington, KY 40508

Chrysalis House, Inc.
251 East Maxwell Street
Lexington, KY 40508

DUI Defendant Referral Systems
431 South Broadway
Suite 331
Lexington, KY 40508

Federal Correctional Institution Atwood Hall
3301 Leestown Road
Lexington, KY 40511

Growth Resources
1517 Nicholasville Road
Suite 404
Lexington, KY 40503

National Council on Alcoholism (NCA) Prevention Research Institute
629 North Broadway
Lexington, KY 40587

Patti Hard Substance Abuse Services
1517 Nicholasville Road
Lexington, KY 40503

Saint Joseph Hospital Center for Chemical Independence
One Saint Joseph Drive
Lexington, KY 40504

Shepherds House, Inc.
154 Bonnie Brae Drive
Lexington, KY 40508

Veterans' Affairs Medical Center Substance Abuse Treatment Program
Leestown Road
Lexington, KY 40511

LIBERTY

Adanta Behavioral Health Services Liberty Clinic
Route 1
Liberty Square
Liberty, KY 42539

LONDON

Cumberland River Comprehensive Care Center
U.S. 25 North
London, KY 40741

Cumberland River Comp Care Center
Crossroad
London, KY 40741

DUI Defendant Referral Systems, Inc.
Marymount Hospital
London, KY 40741

LOUISA

Pathways, Inc. Lawrence County Outpatient Unit
Route 4, Box 44B
Louisa, KY 41230

LOUISVILLE

Baptist Hospital East Chemical Dependency Program Center for Behavioral Health
4000 Kresge Way
Louisville, KY 40207

Council on Prevention and Education/Substances, Inc. (COPES)
1228 East Breckinridge Street
Louisville, KY 40204

Counseling Center, Inc.
1711 Bardstown Road
Louisville, KY 40205

Dismas Charities Drug and Alcohol Treatment Program
1501 Lytle Street
Louisville, KY 40203

Frager Associates
3906 Dupont Square South
Louisville, KY 40207

John P. Sohan Counseling Services
1169 Eastern Parkway
Medical Arts Building Suite 3358
Louisville, KY 40217

Kentucky DUI Institute
1030 West Market Street
Louisville, KY 40202

Kentucky Substance Abuse Programs, Inc.
1030 West Market Street
Louisville, KY 40202

Louisville/Jefferson County Health Dept. Methadone Maintenance Program
1448 South 15 Street
Louisville, KY 40210

Patrick Whelan
1238 East Broadway
Louisville, KY 40204

Seven Counties Services/ Jefferson Alcohol and Drug Abuse Center (JADAC)
600 South Preston Street
Louisville, KY 40202

Talbot House
520 West Saint Catherine Street
Louisville, KY 40203

Ten Broeck Hospital Substance Abuse Services
8521 La Grange Road
Louisville, KY 40242

U.S. Corrections Corporation Lifeline Recovery Center
214 South 8 Street
Louisville, KY 40202

Ultimate Care
10400 Linn Station Road
Atrium Center Suite 226
Louisville, KY 40223

Volunteers of America Kentucky, Inc. Third Step Program
1436 South Shelby Street
Louisville, KY 40217

Wellness Institute
332 West Broadway
Suite 1707
Louisville, KY 40202

Women's Prison Aftercare
1436 South Shelby Street
Louisville, KY 40217

MADISONVILLE

DUI Defendant Referral System, Inc.
254 East Center Street
Room 3
Madisonville, KY 42431

Madisonville Regional Medical Center Alcohol Recovery Center (ARC)
Hospital Drive
Madisonville, KY 42431

Pennyroyal Mental Health Services Madisonville Mental Health Clinic
1303 West Noel Street
Madisonville, KY 42431

MANCHESTER

Cumberland River Comprehensive Care Center
277 White Street
Manchester, KY 40962

MAYFIELD

Western Kentucky Drug and Alcohol Intervention Services, Inc.
820 Paris Road
Mayfield, KY 42066

Western Kentucky MH/MR Board Mayfield/Graves County MH/ MR Services
217 North 7 Street
Mayfield, KY 42066

MAYSVILLE

Comprehend, Inc. Mason County CMHC
611 Forest Avenue
Maysville, KY 41056

MCKEE

Cumberland River Comprehensive Care Center
Main Street
McKee, KY 40447

MCROBERTS

McRoberts Treatment Center
15 Hollow Street
McRoberts, KY 41835

MIDDLESBORO

Cumberland River Comprehensive Care Center
324 North 19 Street
Middlesboro, KY 40965

MONTICELLO

Adanta Behavioral Health Services Wayne County Clinic
Highway 1275
Monticello, KY 42633

DUI Defendant Referral Systems, Inc.
Wayne County Courthouse
Monticello, KY 42633

MOREHEAD

Pathways, Inc. Rowan County Outpatient Unit
321 East Main Street
Morehead, KY 40351

MORGANFIELD

Green River Comprehensive Care Center
233 North Townsend Street
Morganfield, KY 42437

MORGANTOWN

Lifeskills, Inc. Butler County Office
120 West Ohio Street
Morgantown, KY 42261

MOUNT OLIVET

Comprehend, Inc. Robertson County Community Care Center for Outpatient Alcohol/ Drug Services
Highway 62
Mount Olivet, KY 41064

MOUNT STERLING

Pathways, Inc.
Hillcrest Hall
2479 Grassy Lick Road
Mount Sterling, KY 40353

Montgomery County Outpatient Unit
300 Foxglove Drive
Mount Sterling, KY 40353

MOUNT VERNON

Cumberland River Comprehensive Care Center
Old Highway 25
Mount Vernon, KY 40456

MUNFORDVILLE

Lifeskills, Inc. Hart County Office
118 West 3 Street
Munfordville, KY 42765

MURRAY

Western Kentucky MH/MR Board Murray/Calloway County MH/ MR Services
903 Sycamore Street
Murray, KY 42071

NEWPORT

Comp. Care Centers of Northern Kentucky Campbell County
10th and Monmouth Streets
Newport, KY 41071

Transitions, Inc.
Droege House
925 5 Avenue
Newport, KY 41074

Outreach Services
607 York Street
Newport, KY 41071

York Street House
601 York Street
Newport, KY 41071

NICHOLASVILLE

Alcohol Related Offenders Program
926 South Main Street
Faith Baptist Church
Nicholasville, KY 40356

Bluegrass Regional MH/MR Board, Inc. Comprehensive Care Center
102 Lake Street
Nicholasville, KY 40356

OWENSBORO

Green River Comprehensive Care
Ewing Road Center
314 Ewing Road
Owensboro, KY 42301

Morehead Center
1001 Frederica Street
Owensboro, KY 42301

Saradon Center
1100 Walnut Street
Cigar Factory Complex
Owensboro, KY 42301

OWENTON

Comp. Care Centers of Northern Kentucky Owen County
114 West Brown Street
Owenton, KY 40359

OWINGSVILLE

Pathways, Inc. Bath County Outpatient Unit
Route 3
Owingsville, KY 40360

PADUCAH

Behavioral Medicine, Inc.
102 South 31 Street
Paducah, KY 42001

Transition Center
3225 Coleman Road
Paducah, KY 42001

Western Kentucky Drug and Alcohol Intervention Services, Inc.
6 Street
Irvin Cobb Hotel
Paducah, KY 42001

Western Kentucky MH/MR Board
Joseph L. Friedman Substance Abuse Center
1405 South 3 Street
Paducah, KY 42003

Paducah/McCracken County MH/MR Services
1530 Lone Oak Road
Paducah, KY 42001

PAINTSVILLE

Mountain Comprehensive Care Center Johnson County Clinic
1024 Broadway
Paintsville, KY 41240

PARIS

Bluegrass West Comprehensive Care Center Bourbon County
269 East Main Street
Paris, KY 40361

Counseling Center, Inc.
2017 South Main Street
Bourbon Medical Center
Paris, KY 40361

PEWEE VALLEY

Women's Alcohol and Drug Program Kentucky Correctional Institute for Women
Pewee Valley, KY 40031

PIKEVILLE

DUI Defendant Referral Systems, Inc.
319 2 Street
Hopkins Building Room 209
Pikeville, KY 41501

Mountain Comprehensive Care Center Pike County Outpatient Clinic
804 Cline Street
Pikeville, KY 41501

PINEVILLE

Cumberland River Comprehensive Care Center
110 Kentucky Avenue
Pineville, KY 40977

DUI Defendant Referral Systems, Inc.
Bell County Courthouse
Pineville, KY 40977

PRESTONSBURG

Mountain Comprehensive Care Center
Layne House
18 South Front Avenue
Prestonsburg, KY 41653

Outpatient
18 South Front Avenue
Prestonsburg, KY 41653

PRINCETON

Pennyroyal Mental Health Services Caldwell County Mental Health Center
115 McGoodwin Street
Princeton, KY 42445

PROVIDENCE

DUI Defendant Referral Systems, Inc.
230 North Willow Street
City Council Chambers
Providence, KY 42450

Green River Comprehensive Care Center
200 Bradley Street
Providence, KY 42450

RADCLIFF

Communicare Clinic Radcliffe Office
1072 South Dixie Street
Radcliff, KY 40160

RICHMOND

Bluegrass South Comprehensive Care Richmond Comprehensive Care Center
415 Gibson Lane
Richmond, KY 40475

Counseling Center, Inc.
321 North 2 Street
Parks and Recreation Building
Richmond, KY 40475

Professional Consultation Associates
409 Gibson Lane
Richmond, KY 40475

RUSSELLVILLE

Lifeskills, Inc. Logan County Office
237 East 6 Street
Russellville, KY 42276

SALYERSVILLE

Mountain Comprehensive Care Center Magoffin County Clinic
145 Allen Drive
Highway 114
Salyersville, KY 41465

SANDY HOOK

Pathways, Inc. Elliott County Outpatient Unit
Main Street
Sandy Hook, KY 41171

SCOTTSVILLE

Lifeskills, Inc. Scottsville Counseling Center
512 Bowling Green Road
Scottsville, KY 42164

SHELBYVILLE

Seven Counties Services Shelby County Center
25 Village Plaza
Shelbyville, KY 40065

SHEPHERDSVILLE

Seven Counties Services Bullitt County Center
301 Buckman Street
Shepherdsville, KY 40165

SMITHLAND

Western Kentucky MH/MR Board Smithland/Livingston County MH/MR Services
Highway 60
McKinney Building
Smithland, KY 42081

SOMERSET

Adanta Behavioral Health Services Somerset Clinic/ Hardin Lane
101 Hardin Lane
Somerset, KY 42501

DUI Defendant Referral Systems, Inc.
203 West Mount Vernon Street
Suite 108
Somerset, KY 42501

STANFORD

Bluegrass South Comprehensive Care Stanford Comprehensive Care Center
110 Somerset Street
Stanford, KY 40484

STANTON

Bluegrass East Comprehensive Care Center Stanton Unit
354 West College Street
Stanton, KY 40380

TAYLORSVILLE

Seven Counties Services Spencer County Service Center
Reasor Avenue
Taylorsville, KY 40071

TOMPKINSVILLE

Lifeskills, Inc. Monroe County Office
200 East 4 Street
Tompkinsville, KY 42167

VANCEBURG

Comprehend, Inc. Lewis County CMHC Outpatient Alcohol and Drug Office
410 2 Street
Vanceburg, KY 41179

VERSAILLES

Bluegrass West Comp. Care Woodford County Comp. Care Center for Alcohol Abuse Treatment Program
195B Frankfort Street
Medical Terrace
Versailles, KY 40383

Counseling Center
Main Street
Community Education Building
Versailles, KY 40383

WEST LIBERTY

Pathways, Inc. Morgan County Outpatient Unit
280 Prestonsburg Street
Morgan County Office Building
West Liberty, KY 41472

WHITESBURG

Kentucky River Community Care, Inc.
117 Hayes Street
Kraft Building
Whitesburg, KY 41858

WHITLEY CITY

Adanta Behavioral Health Services Whitley City Clinic
South Fork Centera
Highway 27
Whitley City, KY 42653

WILLIAMSBURG

Cumberland River Comprehensive Care Center
Cemetary Road
Williamsburg, KY 40769

WILLIAMSTOWN

DUI Defendant Referral Systems, Inc.
200 Paris Street
Williamstown, KY 41097

WINCHESTER

Bluegrass East Comprehensive Care Center Winchester Unit
26 North Highland Street
Winchester, KY 40391

Counseling Center, Inc.
121 East Lexington Avenue
Winchester, KY 40391

LOUISIANA

ALEXANDRIA

Crossroads Regional Hospital Substance Abuse Services
110 John Eskew Drive
Alexandria, LA 71315

Louisiana Black Alcoholism Council, Inc. Community Intervention Program
2403 Harris Street
Alexandria, LA 71307

Rapides Regional Medical Center Outpatient Chemical Dependency Services
211 4 Street
Alexandria, LA 71301

Veterans' Affairs Medical Center
Chemical Dependency Clinic 116E
Alexandria, LA 71301

BATON ROUGE

Alcohol and Drug Abuse Council of Greater Baton Rouge
1801 Florida Boulevard
Baton Rouge, LA 70802

Alcohol and Drug Abuse Prevention Program I Care Program
1584 North 43 Street
Baton Rouge, LA 70802

Baton Rouge Area Alcohol and Drug Center, Inc.
1819 Florida Boulevard
Baton Rouge, LA 70802

Baton Rouge Substance Abuse Clinic
4615 Government Street
Building A
Baton Rouge, LA 70806

Chemical Dependency Unit of Baton Rouge General Medical Center
3601 North Boulevard
Baton Rouge, LA 70806

Comprehensive Health and Social Services Save A Child
2013 Central Road
Baton Rouge, LA 70807

CPC Meadowwood Hospital Center for Addictive Disorders
9032 Perkins Road
Baton Rouge, LA 70810

Drug Free Schools and Communities Student Services
626 North 4 Street
Baton Rouge, LA 70804

Foundation House for Boys
7324 Alberta Street
Baton Rouge, LA 70808

Foundation House for Girls
455 Lovers Lane
Baton Rouge, LA 70802

O'Brien House
1231 Laurel Street
Baton Rouge, LA 70802

Salvation Army
7361 Airline Highway
Baton Rouge, LA 70805

Serenity House, Inc.
3370 Victoria Drive
Baton Rouge, LA 70805

Shiloh Baptist Church Project Lifeline
185 Eddie Robinson Street
Baton Rouge, LA 70802

Talbot Outpatient Center, Inc.
5414 Brittany Drive
Suite C
Baton Rouge, LA 70808

U.S. Postal Service Employee Assistance Program
750 Florida Boulevard
Room 252
Baton Rouge, LA 70821

BAYOU VISTA

Fairview Treatment Center
915 Southeast Boulevard
2nd Floor
Bayou Vista, LA 70380

Saint Mary Alcohol and Drug Abuse Clinic
915 Southeast Boulevard
Suite B
Bayou Vista, LA 70380

BELLE CHASSE

Plaquemines Alcohol and Drug Abuse Clinic
205 Main Street
Belle Chasse, LA 70037

BOGALUSA

Unity Halfway House of Bogalusa, Inc.
725 East 6 Street
Bogalusa, LA 70427

Washington Parish Alcohol and Drug Abuse Clinic
619 Willis Avenue
Bogalusa, LA 70427

BOSSIER CITY

Sharing Through Examples of Personal Sobriety (STEPS)
201 Monroe Street
Bossier City, LA 71111

CHALMETTE

Saint Bernard Alcohol and Drug Abuse Clinic
2712 Palmisano Boulevard
Building A
Chalmette, LA 70043

COLUMBIA

Columbia Alcohol and Drug Abuse Clinic
Main Street
Courthouse 2nd Floor
Columbia, LA 71418

COVINGTON

Northlake Alcohol and Drug Abuse Clinic
69076 Highway 190 Service Road
Covington, LA 70433

Pride of Saint Tammany, Inc.
832 Boston Street
Unit 4
Covington, LA 70434

CROWLEY

Crowley and Ville Platte Alcohol and Drug Abuse Clinic
703 East 8 Street
Crowley, LA 70526

DONALDSONVILLE

Alcohol and Drug Abuse Council of Ascension
412 Charles Street
Donaldsonville, LA 70346

ELTON

Coushatta Tribe Substance Abuse Prevention Program
Elton, LA 70532

FRANKLINTON

Seven Acres Substance Abuse
Highway 10 at 7 Mile Road
Route 2
Franklinton, LA 70438

Seven Acres Substance Abuse Center
23046 Yacc Road
Franklinton, LA 70438

GONZALES

Powerhouse Services, Inc.
715 Worthey Road
Gonzales, LA 70737

GREENWELL SPRINGS

Greenwell Springs Hospital Young Adult Male Substance Abuse Treatment Services
Greenwell Springs Road
Greenwell Springs, LA 70739

HARVEY

Resources for Human Development, Inc. Family House/Louisiana
1125-B Inca Court
Harvey, LA 70058

West Bank Alcohol and Drug Abuse Clinic
2245 Manhattan Boulevard
Suite 201
Harvey, LA 70058

HOUMA

Alcohol and Drug Abuse Council for South Louisiana (ADAC)
813 Belanger Street
Houma, LA 70360

Bayou Oaks Hospital Life Center
934 East Main Street
Houma, LA 70360

Terrebonne Alcohol and Drug Abuse Clinic
521 Legion Avenue
Houma, LA 70364

INDEPENDENCE

Odyssey House Family Center
900 Highway 51 South
Independence, LA 70443

JACKSON

H. J. Blue Walters Substance Abuse/Pre-Release Treatment Center
Highway 10
Jackson, LA 70748

JENNINGS

Jefferson Davis Chemical Health, Inc.
219 West Nezpique Street
Jennings, LA 70546

LA PLACE

River Parishes Alcohol and Drug Abuse Clinic
421 West Airline Highway
Suite L
La Place, LA 70068

LAFAYETTE

Acadian Oaks Infinity Recovery
310 Youngsville Highway
Lafayette, LA 70508

Alcohol Traffic Action Campaign
527 Evangeline Drive
Lafayette, LA 70501

Cypress Hospital Alcohol and Drug Program
302 Dulles Drive
Lafayette, LA 70506

Gate House Foundation
206 South Magnolia Street
Lafayette, LA 70501

Lafayette Alcohol and Drug Abuse Clinic
400 Saint Julien Street
Suite 1
Lafayette, LA 70506

Saint Francis Foundation
1610 West University Street
Lafayette, LA 70506

Vermilion Hospital for Psychiatric and Addictive Medicine
2520 North University Avenue
Lafayette, LA 70507

LAKE CHARLES

Calcasieu Community Detox Center
1725 Opelousas Street
Lake Charles, LA 70601

Calcasieu Parish School Board Drug Free Programs
1724 Kirkman Street
Lake Charles, LA 70601

Charter Health Systems Addictive Disease Units
4250 5 Avenue South
Lake Charles, LA 70605

Family and Youth Counseling Agency
127 South Ryan Street
Lake Charles, LA 70601

Joseph R. Briscoe Alcohol and Drug Abuse Center
4012 Avenue H
Lake Charles, LA 70601

Lake Charles Substance Abuse Clinic, Inc.
711 North Prater Street
Lake Charles, LA 70601

Lake Charles Memorial Hospital Recovery Center
1701 Oak Park Boulevard
Lake Charles, LA 70601

McNeese State Univ. Counseling Center Bacchus
Ryan Street
Student Life Office
Lake Charles, LA 70609

Saint Patrick Hospital Chemical Dependency Treatment Center
524 South Ryan Street
Lake Charles, LA 70601

MAMOU

SMC New Horizons
120 Country Club Lane
Mamou, LA 70554

MANDEVILLE

Alcohol and Drug Treatment Unit
Highway 190
Mandeville, LA 70448

Bowling Green Hospital of Saint Tammany
701 Florida Avenue
Mandeville, LA 70448

Fontainebleau Treatment Center
Highway 190 West
Mandeville, LA 70470

MARKSVILLE

Hamilton Halfway House
101 South Main Street
Marksville, LA 71351

Tunica/Biloxi Indians of Louisiana Substance Abuse Prevention Program
Marksville, LA 71351

Washington Street Hope Center
106 South Washington Street
Marksville, LA 71351

METAIRIE

Committee on Alcoholism and Drug Abuse for Greater New Orleans (CADA)
476 Metairie Road
Metairie, LA 70005

Drug Education Associates, Inc.
4739 Utica Street
Suite 207
Metairie, LA 70006

Jefferson Substance Abuse Clinic
401 Veterans Boulevard
Suite 102
Metairie, LA 70005

New Freedom, Inc.
3400 Division Street
Metairie, LA 70002

MONROE

Bright House
201 Bright Street
Monroe, LA 71201

Monroe Alcohol and Drug Abuse Clinic
3208 Concordia Street
Monroe, LA 71201

New Way Center
507 Swayze Street
Monroe, LA 71201

Southern Oaks Addiction Recovery Center
4781 South Grand Street
Monroe, LA 71202

Starting Point Detox Center
4781 South Grand Street
Monroe, LA 71202

NATCHITOCHES

Natchitoches Alcohol and Drug Abuse Clinic
116 Highway 1 South
Natchitoches, LA 71457

NEW IBERIA

New Iberia Alcohol and Drug Abuse Clinic
611 West Admiral Doyle Drive
New Iberia, LA 70560

NEW ORLEANS

Brantley Baptist Center
201 Magazine Street
New Orleans, LA 70130

Bridge House, Inc.
1160 Camp Street
New Orleans, LA 70130

Central City Multi Media Center
2020 Jackson Avenue
New Orleans, LA 70113

Coliseum Medical Center Counterpoint Center of New Orleans
3601 Coliseum Street
New Orleans, LA 70115

Desire Narcotic Rehab Center, Inc.
Chemotherapy
3307 Desire Parkway
New Orleans, LA 70126

Outpatient Center
4116 Old Gentilly Road
New Orleans, LA 70126

Division of Addictive Disorders
1542 Tulane Avenue
New Orleans, LA 70112

Dr. O. E. Carter Memorial Rehabilitation Center
5500 North Johnson Street
New Orleans, LA 70117

DRD New Orleans Medical Clinic
530 South Galvez Street
New Orleans, LA 70119

Employee Assistance Program of Louisiana, Inc.
3600 Prytania Street
Suite 72
New Orleans, LA 70115

Family Service of Greater New Orleans Community Care
2515 Canal Street
Suite 201
New Orleans, LA 70119

Foundation House/New Orleans
3942 Laurel Street
New Orleans, LA 70115

Grace House of New Orleans, Inc.
3418 Coliseum Street
New Orleans, LA 70115

Metropolitan Treatment Center, Inc.
3604 Tulane Avenue
New Orleans, LA 70119

New Orleans Education Intervention Center
1322 Aline Street
New Orleans, LA 70115

New Orleans General Hospital Second Chance
625 Jackson Avenue
New Orleans, LA 70130

New Orleans Public Schools Drug Free Schools/Community ACT Program
1815 Saint Claude Avenue
McDonogh 16 Room 6
New Orleans, LA 70116

New Orleans Substance Abuse Clinic
2025 Canal Street
Suite 300
New Orleans, LA 70112

Odyssey House Louisiana, Inc.
1125 North Tonti Street
New Orleans, LA 70119

Orleans Criminal District Court Substance Abuse Services
2700 Tulane Street
Suite 200
New Orleans, LA 70119

River Oaks Center Addictive Disorders Program
1525 River Oaks Road West
New Orleans, LA 70123

The Velocity Foundation, Inc.
1001 Howard Avenue
Suite 3001
New Orleans, LA 70113

Traffic Court Probation Office
727 South Broad Street
Room 210
New Orleans, LA 70119

U.S. Postal Service Employee Assistance Program
701 Loyola Avenue
New Orleans, LA 70113

Veterans' Affairs Medical Center
Alcohol Dependence Treatment Program
Substance Abuse Treatment Program
1601 Perdido Street
Unit 116A
New Orleans, LA 70146

Volunteers of America
Community Correctional Center
1002 Napoleon Avenue
New Orleans, LA 70115

Rehabilitation Services
1523 Constance Street
New Orleans, LA 70130

NEW ROADS

Bonne Sante Chemical Health and Wellness Center
Hospital Road
New Roads, LA 70760

OAKDALE

Federal Correctional Institution Psychology Services/Drug Education Program
Oakdale, LA 71463

OPELOUSAS

New Beginnings of Opelousas
1692 Linwood Loop
Opelousas, LA 70570

Opelousas Alcohol/Drug Abuse Clinic
532 North Court Street
Opelousas, LA 70570

PINEVILLE

Alexandria/Pineville Alcohol and Drug Abuse Clinic
Monroe Highway and Rainbow Drive
Pineville, LA 71361

Cenla Chemical Dependency Council
Bridge House/Phase II
401 Rainbow Drive
Pineville, LA 71361

Gateway Adolescent Unit
Pineville, LA 71360

Rainbow House Detox
Rainbow Drive
Pineville, LA 71361

Red River Treatment Center Central Louisiana State Hospital
Unit 6-D
Pineville, LA 71360

RUSTON

Ruston Alcohol and Drug Abuse Clinic
206 Reynolds Drive
Suite B-3
Ruston, LA 71270

SCHRIEVER

Assisi Bridge House
600 Bull Run Road
Schriever, LA 70395

SCOTT

Opportunities, Inc.
808 Pitt Road
Scott, LA 70583

SHREVEPORT

Caddo/Bossier Center Serenity House/The Cottage
6220 Greenwood Road
Shreveport, LA 71119

Charter Forest Hospital Substance Abuse Services
9320 Linwood Avenue
Shreveport, LA 71106

Council on Alcohol/Drug Abuse of NW Louisiana
Chemical Dependency Recovery Clinic
527 Crockett Street
Shreveport, LA 71101

The Adolescent Center
445 Jordan Street
Shreveport, LA 71101

Council on Alcoholism and Drug Abuse of Northwest Louisiana
820 Jordan Street
Suite 420
Shreveport, LA 71101

Court Alcohol Abuse Program, Inc.
Shreveport, LA 71134

CPC Brentwood Hospital Chemical Dependency Unit
1800 Irving Place
Shreveport, LA 71101

Doctors Hospital Addictive Disease Unit
1130 Louisiana Avenue
Shreveport, LA 71101

Family Counseling and Children's Services CODAC Program
864 Olive Street
Shreveport, LA 71104

First Step Services, Inc.
2000 Creswell Street
Shreveport, LA 71104

Highland Hills Hospital Substance Abuse Services
453 Jordan Street
Shreveport, LA 71101

Mothers Against Drugs of Louisiana, Inc. (MAD)
138 East Columbia Street
Shreveport, LA 71104

Northwest Regional Alcohol and Drug Abuse Clinic
6244 Greenwood Road
Shreveport, LA 71119

Oakwood Home for Women, Inc.
1700 Highland Avenue
Shreveport, LA 71101

Pines Treatment Center
7240 Greenwood Road
Shreveport, LA 71119

Teen Challenge of North Louisiana, Inc.
2631 Barret Street
Shreveport, LA 71134

Veterans' Affairs Medical Center Alcohol Dependence Treatment Program
510 East Stoner Avenue
Unit 10W12
Shreveport, LA 71101

Volunteers of America Madre Program
240 Jordan Street
Shreveport, LA 71101

Willis/Knighton Medical Center Addiction Recovery Center
2701 Portland Avenue
Shreveport, LA 71103

SLIDELL

Slidell Alcohol and Drug Abuse Clinic
2335 Carey Street
Slidell, LA 70458

TALLULAH

Delta Community Action Association Delta Recovery Center
404 East Craig Street
Tallulah, LA 71282

THIBODAUX

Bayou Council on Alcoholism
402 Saint Philip Street
Suite B
Thibodaux, LA 70301

South Louisiana Rehabilitation Center Power House
614 Jackson Street
Thibodaux, LA 70301

Thibodaux Alcohol and Drug Abuse Clinic
303 Hickory Street
Thibodaux, LA 70301

VACHERIE

River Region Hospital Chemical Dependency Treatment Program
22131 Louisiana 20 West
Vacherie, LA 70090

WINNFIELD

Council on Alcoholism and Drug Abuse Winnfield Clinic
905 West Court Street
Winnfield, LA 71483

WINNSBORO

Northeast Louisiana Substance Abuse, Inc.
210 Main Street
Winnsboro, LA 71295

MAINE

ASHLAND

Aroostook Mental Health Center Aroostook Valley Health Center
Walker Street
Ashland, ME 04732

AUBURN

Catholic Charities Maine Saint Francis House
88 3 Street
Auburn, ME 04210

William Hayden LSAC
4 Court Street
Suite 1
Auburn, ME 04210

AUGUSTA

KVRHA/Hearthside
Belgrade Road
Route 27 Sidney
Augusta, ME 04330

Veterans' Affairs Medical Center Chemical Dependence Recovery Program
116A2
Augusta, ME 04330

BANGOR

Albert G. Dietrich
61 Main Street
Room 28
Bangor, ME 04401

Alternative Counseling Services
263 State Street
Suite 6
Bangor, ME 04401

Community Health and Counseling Services Substance Abuse Services/Bangor
42 Cedar Street
Bangor, ME 04401

Maine Counseling Associates
Bangor, ME 04402

Northeast Care Foundation
257 Harlow Street
Suite 201
Bangor, ME 04401

Project Atrium, Inc. Janus House
265 Hammond Street
Bangor, ME 04401

Wellspring, Inc.

Men's Program
98 Cumberland Street
Bangor, ME 04401

Outpatient Services
439 French Street
Professional Office Building
Bangor, ME 04401

Project Rebound
377 Main Avenue
Bangor, ME 04401

Women's Program
319 State Street
Bangor, ME 04401

BAR HARBOR

Jeffery L. Cake, LSW
5 Wayman Lane
Bar Harbor, ME 04609

BELFAST

Waldo County General Hospital Substance Abuse Services
175 High Street
Belfast Center
Belfast, ME 04915

BREWER

JNF Counseling Association
21 North Main Street
Brewer, ME 04412

BRUNSWICK

Counseling and Assistance Center Naval Air Station
Building 12
Brunswick, ME 04011

CALAIS

Calais Regional Hospital Substance Abuse Treatment Facility
50 Franklin Street
Calais, ME 04619

Community Health and Counseling Services Substance Abuse Services/Calais
185 North Street
Calais, ME 04619

CAMDEN

The Community School Substance Abuse Program
79 Washington Street
Camden, ME 04843

CAPE ELIZABETH

Cliff W. Leavis
28 Valley Road
Cape Elizabeth, ME 04107

CARIBOU

Aroostook Mental Health Center Outpatient Substance Abuse Services
Downtown Mall
Saint Peter Building
Caribou, ME 04736

DANFORTH

Aroostook Mental Health Center East Grand Rural Health Center
Houlton Road
Danforth, ME 04424

DOVER-FOXCROFT

Community Health and Counseling Services Substance Abuse Services/Dover
14 Summer Street
Dover-Foxcroft, ME 04426

Mayo Regional Hospital Substance Abuse Services
75 West Main Street
Dover-Foxcroft, ME 04426

EASTPORT

Regional Medical Center at Lubec Eastport Health Care
30 Boynton Street
Eastport, ME 04631

ELLSWORTH

Community Health and Counseling Services Substance Abuse Services/Ellsworth
415 Water Street
Ellsworth, ME 04605

FORT KENT

Aroostook Mental Health Center Outpatient Substance Abuse Services
5 East Main Street
Fort Kent, ME 04743

HALLOWELL

Robert P. Bachand
108 Water Street
Suite 3
Hallowell, ME 04347

HOULTON

Aroostook Mental Health Center Outpatient Substance Abuse Services
2 Kendall Street
Houlton, ME 04730

LEWISTON

Catholic Charities Maine Fellowship House
95 Blake Street
Lewiston, ME 04240

Harbor Light Associates
145 Lisbon Street
Suite 501
Lewiston, ME 04241

Lise Pelletier LSAC
Lewiston, ME 04241

New Beginnings, Inc.
436 Main Street
Lewiston, ME 04240

LIMESTONE

Aroostook Mental Health Center Residential Treatment Facility
Fort Fairfield Road
Limestone, ME 04750

LINCOLN

Community Health and Counseling Services Substance Abuse Services/Lincoln
Transalpine Road
Lincoln, ME 04457

Penobscot Valley Hospital Counseling Alternatives/ Substance Abuse Services
Transalpine Road
Lincoln, ME 04457

LUBEC

Regional Medical Center at Lubec Substance Abuse Services
Lubec, ME 04652

MACHIAS

Community Health and Counseling Services Substance Abuse Services/Machias
Rural Route 1
Machias, ME 04465

MADAWASKA

Aroostook Mental Health Center Outpatient Substance Abuse Services
42 School Street
Madawaska, ME 04756

MILLINOCKET

Millinocket Regional Hospital Outpatient Substance Abuse Treatment Service
200 Somerset Street
Millinocket, ME 04462

NORWAY

Stephens Memorial Hospital Gateway
80 Main Street
Norway, ME 04268

OLD ORCHARD BEACH

Milestone Extended Care
88 Union Avenue
Old Orchard Beach, ME 04064

PATTEN

Aroostook Mental Health Center Katahdin Valley Rural Health Center
Patten, ME 04765

PORTLAND

Casco Bay Substance Abuse Res. Center
205 Ocean Avenue
Portland, ME 04103

McKenney Counseling Service
142 High Street
Suite 306
Portland, ME 04101

Serenity House
30 Mellen Street
Portland, ME 04101

Smith House, Inc.
91–93 State Street
Portland, ME 04101

PRESQUE ISLE

Aroostook Mental Health Center Outpatient Substance Abuse Services
1 Edgemont Drive
Presque Isle, ME 04769

RUMFORD

Rumford Community Hospital Substance Abuse Services
420 Franklin Street
Rumford, ME 04276

SKOWHEGAN

Youth and Family Services, Inc. Substance Abuse Program/ Skowhegan
65 Russell Street
Skowhegan, ME 04976

SOUTHWEST HARBOR

Acadia Family Center
Clark Point Road
Southwest Harbor, ME 04679

VAN BUREN

Aroostook Mental Health Center Outpatient Alcoholism Services
Madawaska Road
Van Buren, ME 04785

WALDOBORO

Alternate Choices Counseling Services
1530 Atlantic Highway
Waldoboro, ME 04572

WESTBROOK

Westbrook Community Hospital Chemical Dependency Treatment Services
40 Park Road
Westbrook, ME 04092

WINDSOR

Dale McGee
Rural Route 1
Windsor, ME 04363

YORK

York Hospital Chemical Dependency Department
15 Hospital Drive
York, ME 03909

MARYLAND

ABERDEEN

Father Martins Ashley Outpatient Program
10 Howard Street
Aberdeen, MD 21001

ANNAPOLIS

Alcohol and Drug Programs Management, Inc.
132 Holiday Court
Suite 200
Annapolis, MD 21401

Anne Arundel County Health Department
Drug and Alcohol Abuse Prevention Program
Open Door/Parole and Glen Burnie
33 Parole Plaza
Suite 203
Annapolis, MD 21401

Open Door/Annapolis
62 Cathedral Street
Annapolis, MD 21401

Open Door/Detention Center
131 Jennifer Road
Annapolis, MD 21401

Family and Children's Society
934 West Street
Annapolis, MD 21401

Greenspring Mental Health Services, Inc. Substance Abuse Services/Annapolis
2525 Riva Road
Annapolis, MD 21401

New Beginnings at Annapolis
135 Old Solomons Island Road
Annapolis, MD 21401

Positive Alternative of Psychological Services, Inc.
111 Annapolis Street
Annapolis, MD 21401

Renewal and Recovery Center of Annapolis
2525 Riva Road
Suite 107
Annapolis, MD 21401

Samaritan House
2610 Greenbrier Lane
Annapolis, MD 21401

Substance Abuse Community Intervention Services
33 Parole Plaza
Suite 203
Annapolis, MD 21401

BALTIMORE

A Woman's Active Recovery Enterprise (AWARE)
7215 York Road
Suite 200
Baltimore, MD 21212

Action Counseling Associates
611 Park Avenue
Suite 2
Baltimore, MD 21201

Adapt Cares/Primary
3101 Towanda Avenue
Baltimore, MD 21215

Addict Referral and Counseling Center, Inc. (ARCC)
21 West 25 Street
Baltimore, MD 21218

Addictions Counseling Service
17 Warren Road
Suite 8A
Baltimore, MD 21208

Aftercare for Continued Recovery
909 Druid Park Lake Drive
Baltimore, MD 21217

Alcohol/Drug/Educational (ADE) Counseling Program
1799 Merritt Boulevard
Baltimore, MD 21222

Alcoholics Anonymous Baltimore Area Intergroup Office
5438 York Road
Suite 202
Baltimore, MD 21212

Alternatives
7215 York Road
Suite 209
Baltimore, MD 21212

American Council on Alcoholism, Inc. Health Education Center
5024 Campbell Boulevard
Suite H
Baltimore, MD 21236

Baltimore Adolescent Treatment Program
4940 Eastern Avenue
B-3-S
Baltimore, MD 21224

Baltimore City Acupuncture Program for Substance Abuse Treatment/Primary
2518 North Charles Street
Baltimore, MD 21218

Baltimore City Health Department
Daybreak Rehabilitation Program
2490 Giles Road
Baltimore, MD 21225

New Visions Counseling Center/Youth/Adults
4 South Frederick Street
3rd Floor
Baltimore, MD 21202

Substance Abuse Bureau
303 East Fayette Street
6th Floor
Baltimore, MD 21202

Baltimore City Partnership for Drug Free Neighborhoods
10 South Street
Suite 404
Baltimore, MD 21202

Baltimore County Office of Substance Abuse
Comprehensive Treatment Program
Prevention Program
401 Washington Avenue
Suite 300
Baltimore, MD 21204

Baltimore Rescue Mission, Inc.
4 North Central Avenue
Baltimore, MD 21202

Bon Secours Hospital Chemical Dependency Program
2000 West Baltimore Street
Baltimore, MD 21223

Bridge House
1516 Madison Avenue
Baltimore, MD 21217

Bright Hope House, Inc.
1611 Baker Street
Baltimore, MD 21217

Broadway East Community Association
1873 North Gay Street
Baltimore, MD 21213

Changing Point Health Services, Inc. Harford Road Unit
4808 Harford Road
Baltimore, MD 21214

Chesapeake Counseling
825 Eastern Boulevard
Baltimore, MD 21221

Church Hospital Corporation Chemical Dependency Unit
100 North Broadway
Baltimore, MD 21231

Clearview Mental Health Services, Inc. Substance Abuse Services
200 East Joppa Road
Suite L101
Baltimore, MD 21204

Community Counseling and Resource Center Intensive Outpatient Cocaine Treatment
208 Washington Avenue
Baltimore, MD 21204

Comprehensive Psycho/Social Services
1401 Reisterstown Road
Suite L1
Baltimore, MD 21208

Crossroads Centers
2 West Madison Street
Baltimore, MD 21201

Damascus House
4203 Ritchie Highway
Baltimore, MD 21225

East Baltimore Drug Abuse Center Treatment Unit
707 Constitution Street
Baltimore, MD 21202

Echo House Multi Service Center Seekers After a New Direction (SAND)
1705 West Fayette Street
Baltimore, MD 21223

Epoch Counseling Center
Dundalk
1107 North Point Boulevard
East Point Office Park Suite 205
Baltimore, MD 21224

Counseling Center/East
621 Stemmers Run Road
Baltimore, MD 21221

Counseling Center/West
22 Bloomsbury Avenue
Baltimore, MD 21228

Family Service Foundation, Inc. Substance Abuse Program
4806 Seton Drive
Suite 204
Baltimore, MD 21215

Fayette House
1844–50 West Baltimore Street
Baltimore, MD 21223

First Step Youth Services Center
8303 Liberty Road
Baltimore, MD 21244

Francis Scott Key Medical Center
Alcohol Treatment Services Outpatient Program
ARC House/Intensive Outpatient/Residential
Center for Addiction/Pregnant Intensive Day Treatment Phase I
4940 Eastern Avenue
D-5-Center
Baltimore, MD 21224

Behavioral Pharmacy Research Unit (BPR)
Southeast Baltimore Drug Treatment Program
5510 Nathan Shock Drive
Baltimore, MD 21224

Franklin Square at White Marsh
8114 Sandpiper Circle
Suite 111
Baltimore, MD 21236

Friendship House
1435 South Hanover Street
Baltimore, MD 21230

Glass Substance Abuse Program, Inc.
821 North Eutaw Street
Suite 101
Baltimore, MD 21201

Glenwood Life Drug Abuse Treatment Program
516 Glenwood Avenue
Baltimore, MD 21212

Greater Baltimore Medical Center Chemical Dependency Inpatient and Outpatient Services
6701 North Charles Street
Unit 51
Baltimore, MD 21204

Greenspring Mental Health Services, Inc. Substance Abuse Services/Towson
7801 York Road
Suite 348
Baltimore, MD 21204

Harbel Substance Abuse Services
5807 Harford Road
Baltimore, MD 21214

Harbel Youth Services
5807 Harford Road
Baltimore, MD 21214

Harbor Mental Health Substance Abuse Services
6310 Harford Road
Baltimore, MD 21214

Help and Recovery Today, Inc. (HART, Inc.)
8200 Harford Road
Suite 200
Baltimore, MD 21234

Innovative Counseling and Rehab
605 Baltimore Avenue
2nd Floor
Baltimore, MD 21204

Institutes for Behavior Resources, Inc. (IBR) Mobile Health Services/Primary
333 Cassell Drive
Triad Tech Center Suite 2400
Baltimore, MD 21224

Jewish Big Brother and Big Sister League of Baltimore
5750 Park Heights Avenue
Baltimore, MD 21215

Johns Hopkins Hospital Comprehensive Women's Center
Intensive Outpatient
Outpatient
Program for Alcohol and Other Drug Dependency
Outpatient Program
Stop Program
911 North Broadway
Baltimore, MD 21205

Jones Falls Community Corporation The Counseling Center
914 West 36 Street
Baltimore, MD 21211

Judith P. Ritchey Youth Services The Youth Services Center
1707 Taylor Avenue
Baltimore, MD 21234

Key Center for Human Services, Inc. Dundalk Unit
6905 Dunmanway Street
First Floor
Baltimore, MD 21222

Liberty Medical Center Substance Abuse Program Overcome
3101 Towanda Avenue
Baltimore, MD 21215

Liberty Medical Center Next Passage Drug Free Substance Abuse Counseling Services
730 Ashburton Street
Baltimore, MD 21216

Man Alive Research, Inc.
2100 North Charles Street
Baltimore, MD 21218

Mayors Coordinating Council on Criminal Justice
10 South Street
Suite 400
Baltimore, MD 21202

Mercy Hospital Operation Recovery
333 Saint Paul Place
Ground Floor Suite B
Baltimore, MD 21202

Morgan State University Alcohol and Drugs Regional Resource Center
Cold Spring Lane and Hillen Road
CGW Building Room 202
Baltimore, MD 21239

Mountain Manor Treatment Center
Outpatient/Residential
3800 Frederick Avenue
Baltimore, MD 21229

Outpatient
1107 North Point Road
Suite 224
Baltimore, MD 21224

New Beginnings at Hidden Brook
1035 North Calvert Street
Baltimore, MD 21202

New Hope Treatment Center Drug Abuse Program
2401 West Baltimore Street
Baltimore, MD 21223

New Outlook
821 North Eutaw Street
Suite 201
Baltimore, MD 21201

Nilsson House
5665 Purdue Avenue
Baltimore, MD 21239

North Baltimore Center, Inc. Substance Abuse Services
2225 North Charles Street
Baltimore, MD 21218

Northwest Baltimore Youth Services, Inc.
3319 West Belvedere Avenue
Baltimore, MD 21215

Park Circle Counseling Center
3000 Druid Park Drive
Baltimore, MD 21215

Performance Dimensions, Inc.
200 East Joppa Road
Baltimore, MD 21286

Quarterway Outpatient Clinic
730 Ashburton Street
Baltimore, MD 21216

S and S Counseling Service
820 Eastern Boulevard
Baltimore, MD 21221

Safe House
7 West Randall Street
Baltimore, MD 21230

Sheppard and Enoch Pratt Hospital Outpatient Recovery Programs
6501 North Charles Street
Baltimore, MD 21204

Sinai Hospital Drug Dependency Program/Primary
Greenspring and Belvedere Avenue
Baltimore, MD 21215

Sinai Hospital Dept. of Psychiatry Substance Abuse Program
2401 West Belvedere Avenue
Baltimore, MD 21215

State of Maryland Dept. of Education/Division of Compensatory Education and Support Services
200 West Baltimore Street
Baltimore, MD 21201

The Resource Group Counseling and Education Center, Inc.
7801 York Road
Suite 215
Baltimore, MD 21204

Total Health Care, Inc. Substance Abuse Services
1609 Druid Hill Avenue
Baltimore, MD 21217

Towson State University Counseling Center
Glen Esk Building
Baltimore, MD 21204

Treatment Alternatives to Street Crime (TASC)
201 West Chesapeake Avenue
Baltimore, MD 21204

Treatment Resources for Youth (TRY)
2517 North Charles Street
Baltimore, MD 21218

Tuerk House Alcohol and Drug Program
730 Ashburton Street
Baltimore, MD 21216

Universal Counseling Services, Inc.
101 West Read Street
Suite 222
Baltimore, MD 21201

University of Maryland
Drug Free and Aftercare Clinic/Federal
Methadone Treatment Program
630 West Fayette Street
First Floor
Baltimore, MD 21201

Alcoholism and Drug Outpatient Clinic
405 West Redwood Street
Baltimore, MD 21201

Valley House
28 South Broadway
Baltimore, MD 21231

Veterans' Affairs Medical Center Substance Abuse Treatment Unit
10 North Green Street
Baltimore, MD 21201

Walter P. Carter MH/MR Center Alcohol and Drug Abuse Program (ADAP)
630 West Fayette Street
Baltimore, MD 21201

Weisman/Kaplan Houses
2521–2523 Maryland Avenue
Baltimore, MD 21218

Whitfield Associates
21 West Road
Suite 150
Baltimore, MD 21204

William Donald Schaefer House
907 Druid Lake Drive
Baltimore, MD 21217

X Cell
Adult Residential Program I
Intensive Halfway Program
Spring Grove State Hospital
Garrett Building
Baltimore, MD 21228

BEL AIR

Harford County Drug and Alcohol Impact Program Prevention
31 West Courtland Street
Bel Air, MD 21014

Harford County Drug Abuse Program Drug Abuse Clinic
715 Shamrock Road
Bel Air-Lee Professional Center
Bel Air, MD 21014

Harford County Health Department Alcoholism Services
5 North Main Street
Bel Air, MD 21014

Help and Recovery Today, Inc. (HART Inc)
112 West Pennsylvania Avenue
Bel Air, MD 21014

Mann House, Inc.
14 Williams Street
Bel Air, MD 21014

New Beginnings at Hidden Brook Residential
522 Thomas Run Road
Bel Air, MD 21015

Recovery with Dignity
2107 Laurel Bush Road
Suite 201
Bel Air, MD 21014

TRW Associates
728 Bel Air Road
Suite 137
Bel Air, MD 21014

BETHESDA

Counseling Institute of Suburban Maryland
4401 East West Highway
Suite 306
Bethesda, MD 20814

Suburban Hospital Addiction Treatment Center
8600 Old Georgetown Road
Wing 2B
Bethesda, MD 20814

BOWIE

McDonald Cassidy Roth and Associates PA/Life Line
14300 Gallant Fox Lane
Suite 112
Bowie, MD 20715

CALIFORNIA

Sierra House
Saint Andrew's Church Road
California, MD 20619

Walden Counseling Center
Saint Andrew's Church Road
California, MD 20619

CAMBRIDGE

Dorchester County Health Department
Addictions Program
751 Woods Road
Cambridge, MD 21613

Addictions Program/Cocaine
443 Race Street
Cambridge, MD 21613

CAMP SPRINGS

Metropolitan Addiction Recovery Strategies (MARS)
5801 Allentown Road
Suite 200
Camp Springs, MD 20746

CAPITOL HEIGHTS

Drug/Alcohol Rehab and Education Systems (DARE Systems)
1 Chamber Avenue
Capitol Heights, MD 20743

CENTREVILLE

Queen Anne's County Health Department Alcohol and Drug Services
206 North Commerce Street
Centreville, MD 21617

CHESAPEAKE BEACH

Calvert County Substance Abuse Program
3849 Harbor Road
Captains Quarters Building
Chesapeake Beach, MD 20732

CHESTERTOWN

A. F. Whitsitt Center Quarterway
Sheeler Road
Chestertown, MD 21620

Kent County Health Department Prevention Program
114A Lynchburg Street
Chestertown, MD 21620

Publick House
114 A South Lynchburg Street
Chestertown Business Park
Chestertown, MD 21620

CHEVERLY

Prince George's County Health Dept. Addictions/Central Region
3003 Hospital Drive
Cheverly, MD 20785

CHURCHVILLE

Addiction Recovery and Related Therapies
3111 Churchville Road
Churchville, MD 21028

CLINTON

Counseling Services Alternatives, Inc.
7900 Old Branch Avenue
Suite 202
Clinton, MD 20735

Prince George's County Health Dept. Addictions/Southern Region
9314 Piscataway Road
Clinton, MD 20735

COCKEYSVILLE

Community Counseling and Resource Center Alcohol and Drug Treatment
10400 Ridgland Road
Cockeysville, MD 21030

Greenside Psychological Associates, Inc.
9727 Greenside Drive
Suite 202
Cockeysville, MD 21030

New Beginnings Family Center
111 Warren Road
Suite 8–9A
Cockeysville, MD 21030

COLLEGE PARK

Changing Point Health Services, Inc. Outpatient
10013 Rhode Island Avenue
College Park, MD 20740

Ethos Foundation
8400 Baltimore Avenue
Suite 106
College Park, MD 20740

Prince George's County Health Dept. Addictions/Northern Region
4810 Greenbelt Road
College Park, MD 20740

Univ. of Maryland Department of Education Regional Alcohol/ Drug Prevention Training/ Resource Center
College Park, MD 20742

University Alcohol and Substance Abuse Program
4700 Berwyn House Road
Univ. Professional Center Suite 201
College Park, MD 20740

COLUMBIA

Columbia Addiction Center
10774 Hickory Ridge Road
Hawthorne Industrial Park
Columbia, MD 21044

Columbia Jewish Congregation Substance Abuse Prevention Services
5885 Robert Oliver Place
Columbia, MD 21045

Greenspring Mental Health Services, Inc. Substance Abuse Services/Columbia
5565 Sterreh Place
Suite 312
Columbia, MD 21044

Oakview Treatment Center Outpatient
10420 Little Patuxent Parkway
Suite 440
Columbia, MD 21044

Orchard Hill Treatment Center
8950 Route 108
Suite 206
Columbia, MD 21045

CROFTON

DWI Assessment and Counseling
1520 Birdwood Court
Crofton, MD 21114

CROWNSVILLE

Hope House
Quarterway
Crownsville, MD 21032

Second Genesis, Inc. Crownsville Unit
107 Circle Drive
Phillips Building
Crownsville, MD 21032

CUMBERLAND

Allegany County Addictions Program
Alcohol and Drug Outpatient
Willowbrook Road
Cumberland MD 21502

Joseph S. Massie Unit
Country Club Road
Thomas B. Finan Center Cottage Four
Cumberland, MD 21502

Lois E. Jackson Unit
Country Club Road
Thomas B. Finan Center Cottage Three
Cumberland, MD 21502

Sacred Heart Hospital Alcoholism Services
915 Bishop Walsh Drive
Cumberland, MD 21502

DENTON

Caroline County Health Department Caroline Counseling Center
104 Franklin Street
Denton, MD 21629

DUNKIRK

J. Russell Horton Associates
202 Dunkirk Professional Building
Dunkirk, MD 20754

EAST NEW MARKET

New Beginnings at Warwick Manor
3680 Warwick Road
East New Market, MD 21631

EASTON

Talbot County Addictions Program
100 South Hanson Street
Easton, MD 21601

Talbot County Health Department Prevention Program
100 South Hanson Street
Easton, MD 21601

ELDERSBURG

Adapt Counseling, Inc.
1425 Liberty Road
Suite 202
Eldersburg, MD 21784

Metwork Health Service, Inc.
2120-A Liberty Road
Eldersburg, MD 21784

ELKTON

Cecil Community College Project Card
105 Railroad Avenue
Elkton, MD 21921

Cecil County Health Department Alcohol and Drug Center
401 Bow Street
Elkton, MD 21921

Haven House, Inc.
Outpatient Unit
Residential Program
253 South Bridge Street
Elkton, MD 21921

ELLICOTT CITY

Changing Point Health Services, Inc. Outpatient
4109 College Avenue
Ellicott City, MD 21043

Changing Point Health Services, Inc. Residential
4100 College Avenue
Ellicott City, MD 21043

Howard County Addictions Services Center
3545 Ellicott Mills Drive
Unit C
Ellicott City, MD 21043

Oakview Treatment Center Residential
3100 North Ride Road
Ellicott City, MD 21043

Psychological Health Associates PA
3691 Park Avenue
Ellicott City, MD 21043

Taylor Manor Hospital Dual Diagnosis Program
4100 College Avenue
Ellicott City, MD 21041

EMMITSBURG

Mountain Manor Treatment Center Emmitsburg Rehabilitation/Outpatient
Route 15
Emmitsburg, MD 21727

FORESTVILLE

Comprehensive Alcohol/Drug Counseling Service, Inc.
2810 Walters Lane
Room 10
Forestville, MD 20747

FORT GEORGE G. MEADE

Fort George G. Meade Comm. Counseling Center Alcohol/ Drug Office
4 1/2 Street
Building 2456
Fort George G. Meade, MD 20755

FORT HOWARD

Veterans' Affairs Medical Center Substance Abuse Treatment Program
Addiction Medicine Section
Unit 111M
Fort Howard, MD 21052

FREDERICK

Allied Counseling Group Drug and Alcohol Treatment
14 West Patrick Street
Frederick, MD 21701

Frederick Counseling Center
405 West 7 Street
Frederick, MD 21701

Frederick County Substance Abuse Services Alcohol/Drug Abuse/Project 103/Prevention Service
350 Montevue Lane
Frederick, MD 21702

Gale House, Inc.
Gale House
608 East Patrick Street
Frederick, MD 21701

Olson House
336 North Market Street
Frederick, MD 21701

Guidelines Counseling Program, Inc.
309 West Patrick Street
Frederick, MD 21701

Karma Academy for Girls KHI Services, Inc.
13 West 3rd Street
Frederick, MD 21701

Mountain Manor Treatment Center Outpatient Services
335 West Patrick Street
Suite 2D
Frederick, MD 21701

FROSTBURG

Frostburg State University
Substance Abuse Facts and Education Program (SAFE)
Library Room 509
Frostburg, MD 21532

University Counseling Center
Pullin Hall
Room 109
Frostburg, MD 21532

GAITHERSBURG

Circle Treatment Center
424 North Frederick Avenue
Suite A
Gaithersburg, MD 20877

Ethos Foundation
1 Bank Street
Gaithersburg, MD 20878

Guide Program Montgomery County, Inc. Adolescent Treatment Program
1 West Deer Park Drive
Room 101
Gaithersburg, MD 20877

GAMBRILLS

New Beginnings at Meadows
730 Maryland Route 3
Gambrills, MD 21054

GERMANTOWN

Alcohol/Drug Education Counseling Center
20120 Timber Oak Lane
Germantown, MD 20874

GLEN BURNIE

Anne Arundel County Health Department Open Door/ Langley
120 North Langley Road
Suite 203
Glen Burnie, MD 21060

Creative Counseling Center, Inc. Outpatient Addictions Services
30 Greenway Street NW
Suite 1
Glen Burnie, MD 21061

North Arundel Hospital Chemical Dependency Unit
301 Hospital Drive
5th Floor
Glen Burnie, MD 21061

GRANTSVILLE

Meadow Mountain Drug Treatment Program
Route 2
Grantsville, MD 21536

GREENBELT

Addiction Assessment and Rehab Service
7713 Belle Point Drive
Greenbelt, MD 20770

HAGERSTOWN

Addiction Specialist Associates
138 East Antietam Street
Suite 201
Hagerstown, MD 21740

Jail Substance Abuse Program (JSAP) Aftercare
13126 Pennsylvania Avenue
Hagerstown, MD 21742

W House, Inc.
37 East Antietam Street
Hagerstown, MD 21740

Washington Board of Education Substance Abuse Prevention Services
Commonwealth Avenue
Hagerstown, MD 21741

Washington County Health Department
Comprehensive Alcohol Program
Division of Addiction Outpatient
1302 Pennsylvania Avenue
Hagerstown, MD 21742

Evening Substance Abuse Program (ESAP)
603 Oak Hill Avenue
Hagerstown, MD 21740

Jail Substance Abuse Program
500 Western Maryland Parkway
Hagerstown, MD 21740

Washington County Hospital Association Alcohol and Drug Treatment Services
322 East Antietam Street
Suite 306
Hagerstown, MD 21740

Wells House Residential Facility
324 North Locust Street
Hagerstown, MD 21740

HAVRE DE GRACE

Ashley, Inc. Quarterway Unit
800 Tydings Lane
Havre de Grace, MD 21078

HOLLYWOOD

Changing Point Health Systems, Inc.
Route 235
Hollywood, MD 20636

HYATTSVILLE

Guide Psychological Services, Inc. Guide Federal Projects/ Maryland
5126 Baltimore Avenue
Hyattsville, MD 20781

Prince George's County Health Dept. DWI Program
5000 Rhode Island Avenue
Community Corrections Center
Hyattsville, MD 20781

JESSUP

Clifton T. Perkins Hospital Center Alcohol and Drug Abuse Services
8450 Dorsey Run Road
Jessup, MD 20794

Howard County Detention Center Classific Treatment and Referral Program
7301 Waterloo Road
Jessup, MD 20794

KENSINGTON

Vaughn J. Howland Intervention Center
3704 Perry Avenue
Kensington, MD 20895

LA PLATA

Charles County Community College (CCCC)
Smart Center
Mitchell Road
La Plata, MD 20646

Alcohol Programs
612 East Charles Street
La Plata, MD 20646

Horizon Center
404 East Charles Street
La Plata, MD 20646

The Jude House, Inc. Long Term Residential Treatment Services
La Plata, MD 20646

LANDOVER HILLS

Commission for Families Clinical Addictions Services
6200 Annapolis Road
Capital Plaza Mall Suite 410
Landover Hills, MD 20784

LANHAM

Prince George's County Hotline and Suicide Prevention
9300 Annapolis Road
Suite 100
Lanham, MD 20706

LAUREL

Counseling Services, Inc.
150 Washington Boulevard
Suite 200
Laurel, MD 20707

Flynn/Lang Counseling Center
13 C Street
Suite H
Laurel, MD 20707

Greater Luarel Beltsville Hospital Renaissance Treatment Program/Outpatient
7100 Contee Road
Laurel, MD 20707

Holistic Counseling and Therapies, Inc. (HCTI) Addiction Program
369 Main Street
Laurel, MD 20707

Key Center for Human Services, Inc.
300 Thomas Drive
Suite 4
Laurel, MD 20707

Reality, Inc.
Aftercare
Quarterway House
419 Main Street
Laurel, MD 20707

Continuing Care Facility for Women
309 Laurel Avenue
Laurel, MD 20707

Halfway House/Male
Outpatient Treatment/Regular
429 Main Street
Laurel, MD 20707

Traditional Rehabilitation Residence
200 Laurel Avenue
Laurel, MD 20707

LEONARDTOWN

Marcey Halfway House
Leonardtown, MD 20650

Office of Community Services Alcohol/Drug Abuse Prevention
Route 245 Governmental Center
Leonardtown, MD 20650

Relapse Prevention Education Center and Addiction Services
Route 5
Ragan Building North End
Leonardtown, MD 20650

LINTHICUM HEIGHTS

Maryland Alliance for Drug Free Youth
404 Shipley Road
Linthicum Heights, MD 21090

LUSBY

Calvert County Substance Abuse Program South Maryland Community Center
Lusby, MD 20657

MOUNT RAINIER

C. A. Mayo and Associates, Inc.
3403 Perry Street
Mount Rainier, MD 20712

OAKLAND

Garrett County Health Department Addictions Service
221 South 3 Street
Oakland, MD 21550

OLNEY

Montgomery General Hospital Addiction Treatment Center
18101 Prince Philip Drive
Olney, MD 20832

OWINGS MILLS

Epoch Counseling Center/Central
9199 Reisterstown Road
Suite 215C
Owings Mills, MD 21117

OXON HILL

Institute of Life and Health Alcohol/Drug Assessment and Therapy Program
6188 Oxon Hill Road
Suite 801
Oxon Hill, MD 20745

Williams Center Addictions Treatment Services
7100 Oxon Hill Road
Oxon Hill, MD 20745

PASADENA

Chrysalis House
8148 Jumpers Hole Road
Pasadena, MD 21122

New Life Addiction Counseling Services
2528 Mountain Road
Pasadena, MD 21122

PATUXENT RIVER NAVAL AIR TEST

Naval Air Station Counseling and Assistance Center
Building 438
Patuxent River Naval Air Test, MD 206

POCOMOKE CITY

Save the Youth, Inc.
2 Street
Pocomoke City, MD 21851

PRINCE FREDERICK

Calvert County Government Substance Abuse Program
Prince Frederick, MD 20678

Calvert County Health Department New Leaf Counseling Center
Route 4 and Stokely Road
Prince Frederick, MD 20678

DWI Services, Inc. Calvert County Treatment Facility
315 Stafford Road
Prince Frederick, MD 20678

RIVERDALE

Directorate of Addictions Prevention Program
6201 Riverdale Road
Riverdale Prof Building Suite 102
Riverdale, MD 20737

Prince George's County Health Dept. Addictions/Northern Region
6201 Riverdale Road
3rd Floor
Riverdale, MD 20737

ROCK HALL

Counseling Resources, Inc.
21997 Kelly Park Road
Rock Hall, MD 21661

ROCKVILLE

Another Path
14901 Broschart Road
Rockville, MD 20850

Avery Road Treatment Center
Detoxification Program
Intermediate Care Facility
14703 Avery Road
Rockville, MD 20853

Dept. of Correction and Rehabilitation Methadone Detoxification Program
1307 Seven Locks Road
Rockville, MD 20854

Ethos Foundation
10701 Old Georgetown Road
Rockville, MD 20852

Family Therapy Institute of Washington DC
5850 Hubbard Drive
Rockville, MD 20852

Jail Addictions Services
1307 Seven Locks Road
Rockville, MD 20850

Karma Academy for Boys
175 Watts Branch Parkway
Rockville, MD 20850

Metro Alcohol and Drug Abuse Services, Inc.
15719 Crabbs Branch Way
Rockville, MD 20855

Montgomery County Dept. Addiction/Victim/Mental Health Services
Child and Adolescent Program
The Other Way Day Treatment Program
401 Hungerford Drive
6th Floor
Rockville, MD 20850

Outpatient Addiction Services
751 Twinbrook Parkway
Rockville, MD 20851

Lawrence Court Halfway House
1 Lawrence Court
Rockville, MD 20850

Montgomery County DWI Treatment Services Program Adult Addiction Programs
751 Twinbrook Parkway
Suite B-26
Rockville, MD 20851

Montgomery County Government Substance Abuse Prevention
401 Fleet Street
Rockville, MD 20850

OACES Corporation
330A Hungerford Drive
Rockville, MD 20850

Second Genesis, Inc. Montgomery County
14701 Avery Road
Rockville, MD 20853

The Recovery Connection
15020 Shady Grove Road
Suite 500
Rockville, MD 20850

White Flint Recovery, Inc.
1335 Rockville Pike
Suite 106
Rockville, MD 20852

SABILLASVILLE

Catoctin Summit Adolescent Program
5980 Cullen Drive
Sabillasville, MD 21780

SALISBURY

Alcohol and Drug Prevention Resource Center
1101 Camden Avenue
Salisbury State University
Salisbury, MD 21801

New Beginnings Salisbury
1202 Old Ocean City Road
Salisbury, MD 21801

Second Wind, Inc.
309 Newton Street
Salisbury, MD 21801

White Flint Recovery, Inc. of the Eastern Shore
1918B Northwood Drive
Salisbury, MD 21801

Wicomico Addictions Center
300 West Carroll Street
Salisbury, MD 21801

Wicomico County Drug Prevention Office
1508 Riverside Drive
Salisbury, MD 21801

Willis W. Hudson Center Residential Alcohol/Drug Treatment Center
Hudson Place
Salisbury, MD 21802

SEVERNA PARK

Stress and Health Management Center Inc.
540 Ritchie Highway
Suite 101
Severna Park, MD 21146

SILVER SPRING

Counseling Plus, Inc.
11141 Georgia Avenue
Suite A24
Silver Spring, MD 20902

D. A. Wynne and Associates Inc. The Silver Spring Program
1709 Elton Road
Silver Spring, MD 23903

Kolmac Clinic
1003 Spring Street
Silver Spring, MD 20910

National Hispanic Council on Aging Substance Abuse Prevention Services
1913 Alabaster Drive
Silver Spring, MD 20904

Thomas Comp. Counseling Services, Inc.
800 Pershing Drive
Suite 105A
Silver Spring, MD 20910

SNOW HILL

Worcester County Health Department Alcohol and Other Drug Services
6040 Public Landing Road
Snow Hill, MD 21863

SUITLAND

Saint Luke Institute
2420 Brooks Drive
Suitland, MD 20746

TAKOMA PARK

Washington Adventist Hospital
Unit 2100
7600 Carroll Avenue
Takoma Park, MD 20912

TIMONIUM

Awakenings Counseling Program
2 West Aylesbury Road
Timonium, MD 21093

UPPER MARLBORO

Second Genesis, Inc.
Upper Marlboro Unit
4620 Mellwood Road
Upper Marlboro, MD 20772

The Awakenings
13400 Dille Drive
Upper Marlboro, MD 20772

WALDORF

Changing Point Health Services
Outpatient
Adolescent Program
Residential
7900 Billingsley Road
Waldorf, MD 20601

Charles County Health Department Adolescent Substance Abuse Program
21 Industrial Park Drive
Waldorf, MD 20604

Waldorf Counseling Services Charles Professional Center
Suite 511
Waldorf, MD 20601

WESTMINSTER

Carroll County Health Department Outpatient Addictions Treatment Services
412 Malcolm Drive
Suite 304
Westminster, MD 21157

Junction, Inc. Drug and Alcohol Abuse Treatment Program
98 North Court Street
Westminster, MD 21157

Mountain Manor Treatment Center Carroll Plaza
Suite 2
Westminster, MD 21158

Reentry Mental Health Services Addiction Services
40 South Church Street
Suite 105
Westminster, MD 21157

Shoemaker Center
540 Washington Road
Westminster, MD 21158

Westminster Rescue Mission
685 Lucabaugh Mill Road
Westminster, MD 21157

WESTOVER

Somerset County Health Department Addiction Services
7920 Crisfield Highway
Westover, MD 21871

WHEATON

Guide Program Montgomery County, Inc. Adult Treatment Services
11141 Georgia Avenue
Suite 420
Wheaton, MD 20902

WOOLFORD

Addiction Treatment Centers of Maryland, Inc. New Beginnings at White Oak
1441 Taylors Island Road
Woolford, MD 21677

MASSACHUSETTS

ASHBURNHAM

Naukeag Hospital, Inc.
Alcohol and Drug Abuse Program/Detox
Alcohol and Drug Abuse Program/INPT
216 Lake Road
Ashburnham, MA 01430

ATTLEBORO

Versacare
Outpatient Substance Abuse Services
Youth Assistance Program
140 Park Street
Attleboro, MA 02703

The Road Back
7 Forest Street
Attleboro, MA 02703

BEDFORD

Veterans' Affairs Medical Center Substance Abuse Rehabilitation Program
200 Springs Road
Building 7 B and C
Bedford, MA 01730

BELMONT

McLean Hospital Alcohol and Drug Abuse Treatment Center
115 Mill Street
Appleton Building
Belmont, MA 02178

BEVERLY

North Shore Counseling Center Outpatient Substance Abuse Services
23 Broadway
Beverly, MA 01915

Project Rap, Inc.
Project Rap Outpatient Program
Youth Assistance Program
202 Rantoul Street
Beverly, MA 01915

BOSTON

Bay Cove Human Services
Methadone Services
Outpatient Substance Abuse Services
104 Lincoln Street
Boston, MA 02111

Boston Alcohol and Substance Abuse Program
30 Winter Street
Boston, MA 02108

Boston Alcohol Detox Project, Inc. Boston Detoxification
19 Father Francis J Gilday Street
Boston, MA 02118

Boston Asian Youth Essential Services, Inc./YAP
199 Harrison Avenue
Boston, MA 02111

Bridge Over Troubled Waters, Inc.
Outpatient Program
Youth Intervention Program
47 West Street
Boston, MA 02111

Fenway Community Health Center
Acupuncture Detoxification Clinic
Outpatient Substance Abuse Services
7 Haviland Street
Boston, MA 02115

Habit Management Institute/ Boston Methadone Services
648 Beacon Street
Boston, MA 02215

Harvard Community Health Plan Substance Abuse Day Treatment
23 Miner Street
Boston, MA 02215

Hope House, Inc. Recovery Home
42 Upton Street and
24 Hanson Street
Boston, MA 02118

Mass General Hospital Addiction Services/Outpatient
15 Parkman Street
Ambulatory Care Center Suite 812
Boston, MA 02114

Medical Foundation Prevention Center
95 Berkeley Street
Boston, MA 02116

MSPCC Family Counseling Center Outpatient Substance Abuse Program
95 Berkeley Street
Boston, MA 02116

Salvation Army Harbor Light Center
407 Shawmut Avenue
Boston, MA 02118

Spaulding Rehabilitation Hospital Alcohol and Chemical Dependency Rehabilitation Program
125 Nashua Street
Boston, MA 02114

Trustees of Health and Hospitals
Alternative Recovery Program/AC
723 Massachusetts Avenue
Boston, MA 02118

BCH Alcohol and Drug Clinic/ Outpatient
818 Harrison Avenue
Boston, MA 02118

Narcotic Addiction Clinic/ Methadone Services
400 Frontage Road
Boston, MA 02118

Veterans' Affairs Medical Center Substance Abuse Treatment Program
150 South Huntington Avenue
Boston, MA 02130

Victory House, Inc.
Women's Hope
226 South Huntington Avenue
Boston, MA 02130

Recovery Home
566 Massachusetts Avenue
Boston, MA 02118

Volunteers of America Hello House
686 Massachusetts Avenue
Boston, MA 02118

BOSTON (BRIGHTON)

Addiction Treatment Center of New England, Inc. Methadone Services
77 Warren Street
F
Boston (Brighton), MA 02135

Granada House Recovery House
77 D Warren Street
Boston (Brighton), MA 02135

Saint Elizabeth's Hospital Comprehensive Alcoholism Program (SECAP)
736 Cambridge Street
Cardinal Cushing Building
Boston (Brighton), MA 02135

BOSTON (CHARLESTOWN)

Boys'/Girls' Club of Boston Youth Intervention Program
15 Green Street
Boston (Charlestown), MA 02129

John F. Kennedy Family Service Center, Inc. Outpatient Substance Abuse Services
27 Winthrop Street
Boston (Charlestown), MA 02129

BOSTON (DORCHESTER)

Boston Hamilton House, Inc. Hamilton Recovery Home
25 Mount Ida Road
Boston (Dorchester), MA 02122

Carney Hospital Drug and Alcohol Program/Outpatient Psychiatry
2100 Dorchester Avenue
Boston (Dorchester), MA 02124

Federal Dorchester Neighborhood Houses
Little House/Outpatient Youth Assistance
275 East Cottage Street
Boston (Dorchester), MA 02125

First, Inc.
Algonquin House Therapeutic Community
22 Algonquin Street
Boston (Dorchester), MA 02125

First Step Outpatient Services
34 Intervale Street
Boston (Dorchester), MA 02121

Hispanic Academy Therapeutic Community
80 Hamilton Street
Boston (Dorchester), MA 02125

Flynn Christian Fellowship House, Inc. Flynn House/ Recovery Home
10 Chamblet Street
Boston (Dorchester), MA 02125

Harbor Health Services, Inc. Neponset Alcoholism Counseling Program
398 Neponset Avenue
Boston (Dorchester), MA 02122

Interim House, Inc. Recovery Home
62 Waldeck Street
Boston (Dorchester), MA 02124

La Alianza Hispana, Inc. Outpatient Substance Abuse Services
112 Stoughton Street
Boston (Dorchester), MA 02125

Peaceful Movement Committee, Inc.
Outpatient Substance Abuse Services
Youth Assistance Program
879 Blue Hill Avenue
Boston (Dorchester), MA 02124

Shepherd House, Inc. Recovery Home
22 and 24 Windermere Road
Boston (Dorchester), MA 02125

Women, Inc.
Outpatient Substance Abuse Services
Residential Program
244 Townsend Street
Boston (Dorchester), MA 02121

BOSTON (EAST BOSTON)

After Care Services Outpatient Substance Abuse Services
1A Monmouth Square
Boston (East Boston), MA 02128

Noddles Island Multi-Services Agency, Inc.
Meridian House
Boston (East Boston), MA 02128

Outpatient Substance Abuse Services
14 Porter Street
Boston (East Boston), MA 02128

Rehabilitation and Health, Inc. Recovery Home
52 White Street
Boston (East Boston), MA 02128

BOSTON (JAMAICA PLAIN)

Boston Alcohol Detox Project, Inc. Boston Alcohol Transitional Care
170 Morton Street
Boston (Jamaica Plain), MA 02130

Brigham and Women's Hospital
Brookside Community Health Center
3297 Washington Street
Boston (Jamaica Plain), MA 02130

50 Jamaica Plain Health Center/ Facts
687 Centre Street
Boston (Jamaica Plain), MA 02130

Faulkner Hospital Addiction Recovery Program
1153 Centre Street
Boston (Jamaica Plain), MA 02130

The Arbour Inpatient Substance Abuse Program/STIIT
49 Robinwood Avenue
Boston (Jamaica Plain), MA 02130

BOSTON (MATTAPAN)

Positive Lifestyles, Inc.
River Street Detoxification Center
249 River Street
Boston (Mattapan), MA 02126

Stair Program
Boston (Mattapan), MA 02126

BOSTON (ROSLINDALE)

Boston Community Services Outpatient Substance Abuse Services
780 American Legion Highway
Boston (Roslindale), MA 02131

BOSTON (ROXBURY)

Casa Esperanza, Inc. Men's Recovery Home
291 Eustis Street
Boston (Roxbury), MA 02119

Dimock Community Health Center
Acupuncture Detox
Alcohol and Drug Detox Program
Ambulatory Substance Abuse Services
John Flowers Recovery Home
Youth Assistance Program
55 Dimock Street
Boston (Roxbury), MA 02119

Roxbury Court Clinic
85 Warren Street
Boston (Roxbury), MA 02119

First Inc./First Academy Therapeutic Community
167 Centre Street
Boston (Roxbury), MA 02119

La Alianza Hispana, Inc. Youth Assistance Program
409 Dudley Street
Boston (Roxbury), MA 02119

Sociedad Latina, Inc. Youth Assistance Program
1530 Tremont Street
Boston (Roxbury), MA 02120

BOSTON (SOUTH BOSTON)

Arch Foundation, Inc. James F. Gavin House
675 East 4 Street
Boston (South Boston), MA 02127

New Pathways to Life Foundation, Inc. Answer House Recovery Home
5 G. Street
Boston (South Boston), MA 02127

South Boston Action Council, Inc.
Turning Point/Outpatient
Turning Point/Youth Assistance
424 West Broadway
Boston (South Boston), MA 02127

BROCKTON

Catholic Charities
Edwina Martin Recovery House
678 North Main Street
Brockton, MA 02401

Outpatient Substance Abuse Services
Youth Assistance Program
686 North Main Street
Brockton, MA 02401

Resurrection House
40 Bartlett Street
Brockton, MA 02401

Life Resources, Inc./Phaneuf Center Outpatient Substance Abuse Services
792 North Main Street
Brockton, MA 02401

MSPCC Family Counseling Center Outpatient Substance Abuse Services
130 Liberty Street
Brockton, MA 02401

South Bay Mental Health Center Outpatient Substance Abuse Services
37 Belmont Street
Brockton, MA 02401

Teen Challenge New England, Inc. Drug/Alcohol Abuse Treatment and Rehab
1315 Main Street
Brockton, MA 02401

Veterans' Affairs Medical Center Alcohol and Drug Dependence Program
940 Belmont Street
Brockton, MA 02401

BROOKLINE

First Psychiatric Planners Bournewood Hospital/Detox
300 South Street
Brookline, MA 02167

CAMBRIDGE

Cambridge Hospital Outpatient Addiction Services
1493 Cambridge Street
Cambridge, MA 02139

Caspar, Inc.
Womanplace Halfway House for Women
11 Russell Street
Cambridge, MA 02140

Women's Program/Outpatient
6 Camelia Avenue
Cambridge, MA 02139

Concilio Hispano, Inc.
Addictions Program/Outpatient
Youth Assistance Program
105 Windsor Street
3rd Floor
Cambridge, MA 02139

Mount Auburn Hospital Outpatient Substance Abuse Services
330 Mount Auburn Street
Cambridge, MA 02238

CHELSEA

Chelsea ASAP, Inc.
Chelsea Substance Abuse Clinic
Youth Assistance Program
100 Everett Avenue
Unit 4
Chelsea, MA 02150

CHICOPEE

Providence Hospital Chicopee Counseling Center
317 Maple Street
Chicopee, MA 01040

CONCORD

Assabet Human Services, Inc. Outpatient Substance Abuse Services
Damonmill Square
Suite 2A
Concord, MA 01742

Emerson Hospital Aftercare Addiction Services/Outpatient
Old Road to Nine Acre Corner
Concord, MA 01742

DANVERS

Center for Addictive Behaviors, Inc.
Detoxification Unit
Methadone Services
450 Maple Street
Danvers, MA 01923

EVERETT

Whidden Memorial Hospital Addiction Treatment Center
96 Garland Street
Everett, MA 02149

FALL RIVER

Family Service Assoc. of Greater Fall River Outpatient Substance Abuse Services
151 Rock Street
Fall River, MA 02720

Portuguese Youth Cultural Organization
Outpatient Substance Abuse Services
Youth Assistance Program
339 Spring Street
Fall River, MA 02721

Saint Anne's Hospital Lifeline/ Methadone Services
795 Middle Street
Fall River, MA 02721

Stanley Street Treatment and Resources
Alcoholism/Drug Detox Program
Chemical Dependency Services/ Outpatient
Women's Rehab Program/Section 35
Youth Assistance Program
386 Stanley Street
Fall River, MA 02720

Steppingstone, Inc.
Halfway House
466 North Main Street
Fall River, MA 02720

Outpatient Substance Abuse Services
101 Rock Street
Fall River, MA 02720

Therapeutic Community
522 North Main Street
Fall River, MA 02720

FALMOUTH

CCAIRU Gosnold Counseling Center
Outpatient Substance Abuse Services
Youth Assistance Program
196 Ter Heun Drive
Falmouth, MA 02540

CCAIRU Gosnold on Cape Cod Detoxification Center
200 Ter Heun Drive
Falmouth, MA 02540

CCAIRU Stephen Miller House Recovery Home
165 Woods Hole Road
Falmouth, MA 02540

FITCHBURG

Luk Crisis Center, Inc. Youth Assistance Program
99 Day Street
Fitchburg, MA 01420

FRAMINGHAM

Metro West Youth Guidance Center
88 Lincoln Street
Framingham, MA 01701

Metrowest Medical Center Synthesis Women's Chemical Dependency Program
115 Lincoln Street
Framingham, MA 01701

New England Aftercare Ministries, Inc. The Bridge House/Halfway House
18–20 Summit Street
Framingham, MA 01701

South Middlesex Addiction Services
Framingham Detoxification Program
3 Merchant Road
Framingham, MA 01701

Outpatient Services
63 Fountain Street
4th Floor
Framingham, MA 01701

GARDNER

Gardner Athol Area Mental Health Association, Inc. Pathway House
171 Graham Street
Gardner, MA 01440

GEORGETOWN

Baldpate Hospital Detoxification Unit
Baldpate Road
Georgetown, MA 01833

GLOUCESTER

Greater Cape Ann Human Services Mental Health Center
Addison Gilbert Hospital/Detox
Outpatient Substance Abuse Services
298 Washington Street
Gloucester, MA 01930

Nuva, Inc.
Methadone Services
Outpatient Services
100 Main Street
Gloucester, MA 01930

GREENFIELD

Beacon Programs of Franklin Medical
Beacon Detoxification Center
164 High Street
Greenfield, MA 01301

Beacon House for Men/Recovery House
53 Beacon Street
Greenfield, MA 01301

Beacon House for Women/ Recovery House
153 High Street
Greenfield, MA 01301

Beacon Outpatient Clinic
Beacon Youth Assistance Program
60 Wells Street
Greenfield, MA 01301

Franklin Community Action Corp. Youth Assistance Program
86 Washington Street
Greenfield, MA 01301

HAVERHILL

Team Coordinating Agency, Inc.
Outpatient Substance Abuse Services
Youth Assistance Program
350 Main Street
Haverhill, MA 01831

Phoenix East
20 Newcomb Street
Haverhill, MA 01831

HINGHAM

Spectrum Addiction Services, Inc. Project Turnabout/Therapeutic Community
224 Beal Street
Hingham, MA 02043

HOLYOKE

MSPCC Family Counseling Center Outpatient Substance Abuse Services
1727 Northampton Street
Holyoke, MA 01040

Providence Hospital
Chemical Dependency Unit/First Step
1233 Main Street
Holyoke, MA 01040

Honor House
40 Brightside Drive
Holyoke, MA 01040

Methadone Services
210 Elm Street
Holyoke, MA 01040

Substance Abuse Outpatient Programs
Youth Assistance Program
317 Maple Street
Holyoke, MA 01040

HYANNIS

Cape Cod Human Services
Outpatient Substance Abuse Services
175 West Main Street
Hyannis, MA 02601

Youth Assistance Program
120 Yarmouth Road
Hyannis, MA 02601

CCAIRU Education and Intervention Center Non-Medical Detoxification
71 Pleasant Street
Hyannis, MA 02601

LAKEVILLE

Catholic Charities Substance Abuse Treatment Center/ Alcohol Detox
33 Main Street
Human Service Building Suite 1
Lakeville, MA 02346

Center for Human Services, Inc. Pathways Prevention Center
109 Rhode Island Road
Route 79
Lakeville, MA 02347

LAWRENCE

Centro Panamericano, Inc. Substance Abuse Outpatient Services
90 Broadway Street
Lawrence, MA 01841

Family Service Assoc. of Greater Lawrence Outpatient Substance Abuse Services
430 North Canal Street
Lawrence, MA 01840

MSPCC Family Counseling Center Outpatient Substance Abuse Services
439 South Union Street
Lawrence, MA 01843

Psychological Center
Foundation House Recovery Home
10 Haverhill Street
Lawrence, MA 01841

Outpatient Substance Abuse Services
Prevention Center
488 Essex Street
Lawrence, MA 01840

Pegasus Youth Residence
482 Lowell Street
Lawrence, MA 01840

LEOMINSTER

North Central Alcoholism Commission
Detoxification Center
17 Orchard Street
Leominster, MA 01453

Outpatient Counseling
71 Pleasant Street
Leominster, MA 01453

LEXINGTON

Choate Health Systems Addiction Recovery Corp./Day Treatment
1050 Waltham Street
Lexington, MA 02173

LOWELL

Family Service of Greater Lowell Outpatient Substance Abuse Services
97 Central Street
Suite 400
Lowell, MA 01852

Habit Management Institute/ Lowell Methadone Services
660 Suffolk Street
Lowell, MA 01854

Lowell General Hospital
Community Health Initiatives/ Detox
295 Varnum Avenue
Service Building
Lowell, MA 01854

Community Health Initiatives/ OUTPT
15 Hurd Street
Lowell, MA 01852

Lowell House, Inc.
Outpatient Substance Abuse Services
Youth Assistance Program
555 Merrimack Street
Lowell, MA 01854

Residential Services
102 Appleton Street
Lowell, MA 01852

LYNN

Atlanticare Hospital Substance Abuse Unit R 1
212 Boston Street
Lynn, MA 01904

Center for Addictive Behaviors, Inc.
Residential Intermediate Care Facility
110 Green Street
Lynn, MA 01902

Ryan Rehabilitation Center
100 Green Street
Lynn, MA 01902

Project Cope
Outpatient Substance Abuse Services
The Women's Program/Recovery Home
117 North Common Street
Lynn, MA 01902

MALDEN

Adult/Adolescent Counseling, Inc. Outpatient Substance Abuse Services
110 Pleasant Street
Malden, MA 02148

Eastern Middlesex Alcoholism Services
Outpatient Substance Abuse Services
557 Salem Street
Malden, MA 02148

Recovery House
12 Cedar Street
Malden, MA 02148

MARLBOROUGH

Marlborough Hospital MARCAP
57 Union Street
Marlborough, MA 01752

Together, Inc.
Outpatient Substance Abuse Services
Youth Assistance Program
133 East Main Street
Marlborough, MA 01752

MEDFIELD

Bay State Community Services, Inc. Odyssey Family Treatment Program
54 Hospital Road
Medfield, MA 02052

MIDDLEBORO

Middleboro/Lakeville Mental Health Services Outpatient Substance Abuse Services
94 South Main Street
Middleboro, MA 02346

MILFORD

Community Counseling Center
Outpatient Substance Abuse Services
Youth Assistance Program
10 Asylum Street
Milford, MA 01757

Urban/Suburban Ministry Hospitality House/Recovery Home
245 West Street
Milford, MA 01757

NANTUCKET

Family and Children's Service/ Nantucket Outpatient Substance Abuse Services
Off Vesper Lane
Nantucket, MA 02554

Nantucket Cottage Hospital Outpatient Substance Abuse Services
57 Prospect Street
Nantucket, MA 02554

NATICK

South Middlesex Addiction Services Serenity House
20 Cottage Street
Natick, MA 01760

Valle Management Associates, Inc.
Leonard Morse Hosp Start Addict Serv
Start Out Prog
67 Union Street
Natick, MA 01760

NEW BEDFORD

Center for Health and Human Services
Outpatient Alcohol and Drug Program
848 Pleasant Street
Melville Plaza Suite 6
New Bedford, MA 02741

Drug Treatment Program/ Methadone Services
848 Pleasant Street
Suite 5
New Bedford, MA 02741

Youth Assistance Program
800 Purchase Street
Suite 6
New Bedford, MA 02741

New Bedford Council on Alcoholism, Inc. Harmony House
234 Earle Street
New Bedford, MA 02746

NEWBURYPORT

Link House Recovery Home
37 Washington Street
Newburyport, MA 01950

North Essex Community Mental Health Outpatient Substance Abuse Services
21 Water Street
Newburyport, MA 01950

Turning Point, Inc. Outpatient Substance Abuse Counseling
21 Storey Avenue
Newburyport, MA 01950

NEWTON

Family Counseling Region West Family Substance Abuse/ Outpatient
74 Walnut Park
Newton, MA 02158

NEWTON CENTER

Riverside Comm. Mental Health and Retardation Center Multi Service Center, Inc./Outpatient
1301 Centre Street
Newton Center, MA 02159

NORTH ADAMS

Northern Berkshire Mental Health Assoc. Substance Abuse Services Program
85 Main Street
Suite 500
North Adams, MA 01247

NORTHAMPTON

Cooley Dickinson Hospital The Dickinson Programs
Outpatient
Prevention One
Youth Intervention
76 Pleasant Street
Northampton, MA 01060

Veterans' Affairs Medical Center Substance Abuse Treatment Program
Northampton, MA 01060

NORTON

North Cottage Program, Inc. Halfway House
69 East Main Street
Norton, MA 02766

NORWOOD

Billings Human Services, Inc. Outpatient Substance Abuse Services
275 Prospect Street
Norwood, MA 02062

Norfolk Mental Health Association Cutler Counseling Center
886 Washington Street
Norwood, MA 02062

OAK BLUFFS

Martha's Vineyard Community Services Island Counseling Center/Outpatient
Off Edgartown/Vineyard Haven Road
Oak Bluffs, MA 02557

PALMER

Wing Memorial Hospital Griswold Center Substance Abuse Services
Wright Street
Palmer, MA 01069

PITTSFIELD

Berkshire Council on Alcohol Addictions
Doyle Treatment Center/Detox
793 North Street
Pittsfield, MA 01201

Keenan House Recovery Home
206 Francis Avenue
Pittsfield, MA 01201

Outpatient Clinic
Youth Assistance Program
131 Bradford Street
Pittsfield, MA 01202

Hillcrest Hospital Thomas W. McGee Unit
165 Tor Court
Pittsfield, MA 01201

PLYMOUTH

Anchor House Recovery Home
60 1/2 Cherry Street
Plymouth, MA 02362

Center for Health and Human Services/AFR
Outpatient Substance Abuse Services
Youth Assistance Program
71 Christa McAuliffe Boulevard
Plymouth, MA 02360

National Alcoholism Programs
High Point
Detox
Outpatient
STIT
1233 State Road
Plymouth, MA 02360

QUINCY

Bay Cove Human Services Andrew House Detoxification Center
Long Island Hospital Ground
Morris Building
Quincy, MA 02169

Bay State Community Services, Inc.
Outpatient Substance Abuse Service
Prevention Resources
Youth Assistance Program
15 Cottage Avenue
Quincy, MA 02169

Quincy Detoxification Center, Inc. DBA Faxon Recovery Service
120 Whitwell Street
Quincy, MA 02169

South Shore Halfway House
10 Dysart Street
Quincy, MA 02169

Valle Management Associates, Inc. Right Turn
7 Foster Street
Quincy, MA 02169

Volunteers of America Rebound
Long Island Hospital
Quincy, MA 02269

SALEM

Center for Addictive Behaviors, Inc.
Ambulatory Service
Prevention Services
Youth Assistance Program
27 Congress Street
Salem, MA 01970

Essex County District Attorney Juvenile Diversion Program/ YIP
1 East India Square
Museum Place
Salem, MA 01970

Health and Education Services Outpatient Substance Abuse Program
162 Federal Street
Salem, MA 01970

North Shore Medical Center Addictive Disease Unit
81 Highland Avenue
Salem, MA 01970

Salem Hospital Addictive Disease Program/Outpatient
172 Lafayette Street
Salem Hospital/Professional Services Building
Salem, MA 01970

SOMERVILLE

Caspar, Inc.
Alcohol and Drug Education Youth Assistance
226 Highland Avenue
Somerville, MA 02143

Caspar Men's Residences
16 Highland Avenue
Somerville, MA 02143

New Day
242 Highland Avenue
Somerville, MA 02143

Heritage Hospital Adult Addictions Program
26 Central Street
Somerville, MA 02143

North Charles Institute
Methadone Services
Outpatient Substance Abuse Services
260 Beacon Street
Somerville, MA 02143

Somerville Portuguese American League
Acupuncture Services
Outpatient Substance Abuse Services
Youth Assistance Program
92 Union Square
Somerville, MA 02143

SOUTH YARMOUTH

Habit Management Institute/ Yarmouth Methadone Services
20 Forsyth Street
South Yarmouth, MA 02664

SOUTHBRIDGE

Harrington Memorial Hospital Harrington Substance Abuse Center
29 Pine Street
Southbridge, MA 01550

Tri Link Counseling and Family Services Youth Assistance Program
51 Everett Street
Southbridge, MA 01550

SPRINGFIELD

Alcohol/Drug Services of Western Mass, Inc.
Carlson Detox Center
Sloan Clinic
1400 State Street
Springfield, MA 01109

Opportunity House
59–61 Saint James Avenue
Springfield, MA 01109

Women's Division/My Sisters House
89 Belmont Avenue
Springfield, MA 01108

Youth Assistance Program
20 Maple Street
Springfield, MA 01105

Child/Family Service of Pioneer Valley Outpatient Substance Abuse Services
367 Pine Street
Springfield, MA 01105

Habit Management Institute/ Springfield Methadone Services
2257 Main Street
Springfield, MA 01107

Marathon House, Inc. Residential Unit
5 Madison Avenue
Springfield, MA 01105

Northern Educational Services, Inc.
Ethos I/Recovery Home
56 Temple Street
Springfield, MA 01105

Ethos III Outpatient Services
756 State Street
Springfield, MA 01109

Youth Assistance Program
622 State Street
Springfield, MA 01109

Providence Hospital Insights of Providence Hospital
209 Carew Street
Springfield, MA 01104

STONEHAM

New England Memorial Hospital Addictions Treatment Services/ Outpatient
5 Woodland Road
Stoneham, MA 02180

TAUNTON

Community Counseling of Bristol County Outpatient Substance Abuse Services
19 Cedar Street
Taunton, MA 02780

Greater Taunton Council on Alcoholism
Alcoholism Counseling Center/ Outpatient
Youth Intervention Program
4 Cohannet Street
Taunton, MA 02780

TEWKSBURY

National Alcoholism Programs
High Point
Detox
Outpatient
STIT
2580 Main Street
Tewksbury, MA 01876

VINEYARD HAVEN

Martha's Vineyard Community Services Youth Assistance Program
Vineyard Haven, MA 02568

WAKEFIELD

Eastern Middlesex Human Services Outpatient Substance Abuse Services
338 Main Street
Wakefield, MA 01880

Eastern Middlesex Human Services Youth Assistance Program
338 Main Street
Wakefield, MA 01880

WALTHAM

Hurley House Recovery Home
12–14 Lowell Street
Waltham, MA 02154

Middlesex County Hospital Driving Under the Influence of Liquor
775 Trapelo Road
Waltham, MA 02154

Middlesex Human Service Agency, Inc.
Regional Addictions Treatment Center, Inc.
Detox/Outpatient/TCF
775 Trapelo Road
Waltham, MA 02154

Youth Assistance Program
205 Bacon Street
Waltham, MA 02154

Mount Auburn Hospital Prevention and Training Center
24 Crescent Street
Suite 301
Waltham, MA 02154

Waltham/Weston Hospital Addictions Treatment Program
Hope Avenue
Waltham, MA 02254

WATERTOWN

Center for Mental Health and Retardation Services Outpatient Substance Abuse Services
372 Main Street
Watertown, MA 02172

Watertown Multi-Service Center Youth Assistance Program
127 North Beacon Street
Watertown, MA 02172

WELLESLEY

Charles River Hospital Dual Diagnosis Program
203 Grove Street
Wellesley, MA 02181

WEST FALMOUTH

CCAIRU Emerson House
Women's Recovery Home
Youth Residence
554 West Falmouth Highway
West Falmouth, MA 02574

WESTBORO

Spectrum Addiction Services, Inc. Spectrum Primary Care/STIIT
154 Oak Street
Westboro, MA 01581

WESTBOROUGH

Spectrum Addiction Services, Inc.
Detoxification Center
154 Oak Street
Westborough, MA 01581

Spectrum Residential Program
155 Oak Street
Westborough, MA 01581

WESTFIELD

Alcohol/Drug Services of Western Mass, Inc. Quarry Hill Treatment Center
137 East Mountain Road
Westfield, MA 01085

Community Health Care Methadone Services
138 East Mountain Road
Westfield, MA 01085

Providence Hospital Westfield Counseling Center
41 Church Street
Westfield, MA 01085

WESTWOOD

Westwood Lodge Hospital Dual Diagnosis Program
45 Clapboardtree Street
Westwood, MA 02090

WOBURN

Bay Colony Health Services Outpatient Substance Abuse Services
800 West Cummings Park
Woburn, MA 01801

WORCESTER

Adcare Hospital Substance Abuse Treatment Program
107 Lincoln Street
Worcester, MA 01605

Catholic Charities/Worcester Crozier House
15 Ripley Street
Worcester, MA 01610

Community Healthlink
Thayer Institute/Detoxification
Thayer Institute/DUI
26 Queen Street
Worcester, MA 01610

Faith, Inc. Recovery Home
142 Burncoat Street
Worcester, MA 01606

Henry Lee Willis Community Center
Channing House Recovery Home
18 Channing Street
Worcester, MA 01605

Channing II/Linda F. Griffin House
15 Northampton Street
Worcester, MA 01605

Outpatient Substance Abuse Services
Youth Assistance Program
110 Lancaster Street
Worcester, MA 01609

MSPCC Family Counseling Center Outpatient Substance Abuse Services
286 Lincoln Street
Worcester, MA 01605

North Central Alcoholism Commission Tri Prevention First
100 Grove Street
Worcester, MA 01605

Saint Vincent's Hospital Dept. of Alcohol and Drug Services/ Outpatients
25 Winthrop Street
Worcester, MA 01604

Spectrum Addiction Services, Inc.
Methadone Services
Outpatient Services/Drug Free
105 Merrick Street
Worcester, MA 01609

MICHIGAN

ADRIAN

Emma L. Bixby Medical Center Sage Center for Substance Abuse Treatment
818 Riverside Avenue
Adrian, MI 49221

Family Service and Children's Aid
405 Mill Street
Adrian, MI 49221

Lenawee Intermediate School District Prevention First
4107 North Adrian Highway
Adrian, MI 49221

McCullough Vargas and Associates
127 South Winter Street
Adrian, MI 49221

ALBION

Chemical Dependency Resources
112 South Superior Street
Albion, MI 49224

Minority Program/South Central Michigan Substance Abuse Prevention Program
112 South Superior Street
Albion, MI 49224

ALGONAC

Downriver Community Services, Inc. Substance Abuse Services
329 Columbia Street
Algonac, MI 48001

ALLEGAN

Allegan County Substance Abuse Agency Vintage Program
120 Cutler Street
Allegan, MI 49010

Family Recovery Center of Allegan County
138 B Hubbard Street
Allegan, MI 49010

ALLEN PARK

Evergreen Counseling Centers
7445 Allen Road
Suite 108
Allen Park, MI 48101

Josephine Sheehy Program
10501 Allen Road
Suite 207
Allen Park, MI 48107

Veterans' Affairs Medical Center Chemical Dependence Treatment Services
Southfield and Outer Drive
116A
Allen Park, MI 48101

ALLENDALE

Grand Valley State University Office of Alcohol and Wellness Education
163 Fieldhouse
Allendale, MI 49401

ALMA

Adapt/A Program Service of Mount Plesant Counseling Services
409 Gratiot Avenue
Alma, MI 48801

Gratiot Community Hospital Pine River Recovery Center
300 Warwick Drive
C-2
Alma, MI 48801

ALPENA

Birchwood Center for Chemical Dependency
1501 West Chisholm Street
Alpena Hospital
Alpena, MI 49707

Boys' and Girls' Club of Alpena
601 River Street
Alpena, MI 49707

Catholic Charities Community Family/Children's Services (CFCS)
228 South 3 Avenue
Suite C
Alpena, MI 49707

Sunrise Centre
630 Walnut Street
Alpena, MI 49707

ANN ARBOR

Alpha House
4290 Jackson Road
Ann Arbor, MI 48103

Ann Arbor Community Center Substance Abuse Treatment and Prevention Services
625 North Main Street
Ann Arbor, MI 48104

Ann Arbor Consultation Services, Inc.
5331 Plymouth Road
Ann Arbor, MI 48105

Catholic Social Services Substance Abuse Services
117 North Division Street
Ann Arbor, MI 48104

Chelsea Community Hospital Older Adult Recovery Program
955 West Eisenhower Circle
Suite E
Ann Arbor, MI 48103

Child and Family Service of Washtenaw Family Counseling
2301 Platt Road
Ann Arbor, MI 48104

Clear House Chemical Dependency Program
704 Spring Street
Ann Arbor, MI 48103

Community Education and Training Service Prevention and Education
2008 Hogback Road
Ann Arbor, MI 48105

Dawn, Inc. Dawn Re-Entry
544 North Division Street
Ann Arbor, MI 48104

Eastwood Clinics
1829 West Stadium Boulevard
Suite 100
Ann Arbor, MI 48103

Employee Assistance Associates, Inc.
1580 Eisenhower Place
Ann Arbor, MI 48108

Green Road Counseling Center
2000 Green Road
Suite 250
Ann Arbor, MI 48105

Huron Valley Consultation Center
Carpenter Outpatient
1750 Carpenter Road
Ann Arbor, MI 48108

Eisenhower Outpatient
955 West Eisenhower Parkway
Suite B
Ann Arbor, MI 48103

Institute for Psychology and Medicine
2010 Hogback Road
Suite 6
Ann Arbor, MI 48105

McAuley Chemical Dependency Center
2006 Hogback Road
Ann Arbor, MI 48105

Occupational Health Centers of America Substance Abuse Services
3800 Packard Street
Suite 130
Ann Arbor, MI 48108

Soundings/A Center for Continuing Growth, Inc.
117 North First Street
Suite 100
Ann Arbor, MI 48104

The Domestic Violence Project, Inc.
2301 Platt Road
Ann Arbor, MI 48107

Washtenaw Council on Alcoholism
2301 Platt Road
Ann Arbor, MI 48104

Washtenaw County
Alcohol Highway Safety Education
4133 Washtenaw Road
Ann Arbor, MI 48107

Correctional Services Program
2201 Hogback Road
Ann Arbor, MI 48107

ATLANTIC MINE

New Creation Center
Highway 26 and Erickson Drive
Atlantic Mine, MI 49905

AUBURN HILLS

Boys' and Girls' Club of Auburn Hills
220 South Squirrel Road
Auburn Hills, MI 48326

Havenwyck Hospital Substance Abuse Services
1525 University Drive
Auburn Hills, MI 48326

BAD AXE

Catholic Family Service Family Counseling Services
123 Scott Street
Bad Axe, MI 48413

Huron County Substance Abuse Program Common Ground
1108 South Van Dyke Road
Bad Axe, MI 48413

BALDWIN

Family Health Care Substance Abuse Center
1090 North Michigan Avenue
Baldwin, MI 49304

BARAGA

New Day Treatment Center Bedabin Family Home
Route 1
Baraga, MI 49908

BARK RIVER

Three Fires Halfway House Substance Abuse Program
3093 D Road
Bark River, MI 49807

BATTLE CREEK

Central Diagnostic and Referral Services
67 West Michigan Avenue
Suite 411
Battle Creek, MI 49017

Chemical Dependency Resources
67 West Michigan Mall
300 Old Kent Bank Building
Battle Creek, MI 49017

Helpnet Employee Assistance Program
131 East Columbia Street
Suite 112
Battle Creek, MI 49015

New Day Center of Battle Creek Adventist Hospital
165 North Washington Avenue
Battle Creek, MI 49016

New Day Center of Battle Creek
330 East Columbia Street
Battle Creek, MI 49016

Occupational Health Centers of America, Inc.
395 South Shore Drive
Suite 204
Battle Creek, MI 49015

Peer Listening Program and South Central Michigan Substance Abuse Prevention
131 East Colombia Street
Suite 205
Battle Creek, MI 49015

Substance Abuse Council of Greater Battle Creek
80 North 20 Street
Battle Creek, MI 49015

Substance Abuse Prevention Group of Greater Battle Creek
233 South Kendall Street
Battle Creek, MI 49016

Veterans' Affairs Medical Center Substance Abuse Treatment Unit
Unit 116E
Battle Creek, MI 49015

BAY CITY

Bay Haven Chemical Dependency and Mental Health Programs
713 9 Street
Bay City, MI 48708

Boys' and Girls' Club of Bay County Life Planning System
300 West Lafayette Street
Bay City, MI 48706

Catholic Family Services Family Counseling Services
1008 South Wenona Avenue
Bay City, MI 48706

New Friendship House of Bay County Basis
700 North Van Buren Street
Bay City, MI 48708

Riverside Center for Substance Abuse Treatment/Basis
1110 North Washington Avenue
Bay City, MI 48708

SCHSC Pathways Outpatient Program
2303 East Amelith Road
Bay City, MI 48706

Tri City/Service Jobs for Progress
301 North Farragut Street
Bay City, MI 48708

BELLAIRE

Charles Allen Ransom Counseling Center
7053 M-88 Highway South
Bellaire, MI 49615

BELLEVILLE

Community Care Services Substance Abuse Service
25 Owen Street
Belleville, MI 48111

Eastwood Clinics
418 Main Street
Belleville, MI 48111

BENTON HARBOR

Berrien County Health Department Alcohol/Drug Abuse Program
769 Pipestone Street
Benton Harbor, MI 49022

Horizon Recovery Center at Mercy Memorial Medical Center
960 Agard Street
Benton Harbor, MI 49022

Southwestern Michigan Commission Community Assessment Services
185 East Main Street
Suite 701
Benton Harbor, MI 49022

BERKLEY

Berkley Schools Substance Abuse Program
2077 Oxford Road
Berkley, MI 48072

Oakland Family Services Berkley Substance Abuse Services
2351 West 12 Mile Road
Berkley, MI 48072

BIG RAPIDS

Program for Alcohol and Substance Treatment, Inc.
110 Sanborn Avenue
Big Rapids, MI 49307

BIRMINGHAM

Alcohol and Alcoholism Info Program
30700 Telegraph Road
Suite 4680
Birmingham, MI 48025

Beacon Hill Clinic PC
31000 Lahser Road
Suite 1
Birmingham, MI 48025

Evergreen Counseling Centers Substance Abuse Program
999 Haynes Street
Suite 245
Birmingham, MI 48009

Joyce K. Frazho, MSW
111 South Woodward Avenue
Suite 256
Birmingham, MI 48009

Lakewood Clinic/North Point
280 North Woodward Avenue
Suite 213
Birmingham, MI 48009

Michael F. Abramsky PhD.PC/ Associates
111 South Woodward Avenue
Birmingham, MI 48009

Michigan Communities in Action for Drug Free Youth
925 East Maple Road
Suite 103
Birmingham, MI 48009

Montgomery and Associates
770 South Adams Road
Suite 210
Birmingham, MI 48009

Psychological Evaluation Service Substance Abuse Services
700 East Maple Street
Suite 303
Birmingham, MI 48011

South Woodward Clinic
111 South Woodward Avenue
Suite 250
Birmingham, MI 48009

Special Tree Rehabilitation System Substance Abuse Services
2100 East Maple Street
Suite 300
Birmingham, MI 48009

BLOOMFIELD HILLS

Auro Medical Center Substance Abuse Services
2515 Woodward Avenue
Suite 250
Bloomfield Hills, MI 48304

Center for Contemporary Psychology PC Outpatient Substance Abuse
950 North Hunter Street
Bloomfield Hills, MI 48304

Counseling Alternatives Bloomfield Hills
10 West Square Lake Road
Suite 300
Bloomfield Hills, MI 48302

Evergreen Counseling Centers
1760 South Telegraph Road
Bloomfield Hills, MI 48302

LSSM Family Counseling Bloomfield Hills
5631 North Adams Street
Bloomfield Hills, MI 48013

Oakland Psychological Clinic PC Substance Abuse Services
2000 South Woodward Avenue
Suite 102
Bloomfield Hills, MI 48013

The Counseling Center PC
1411 South Woodward Avenue
Suite 101
Bloomfield Hills, MI 48302

BLOOMFIELD VILLAGE

Sis Wenger and Associates Substance Abuse Services
3355 Bradway Boulevard
Bloomfield Village, MI 48301

BOYNE CITY

Boyne Valley Counseling Service
624 State Street
Boyne City, MI 49712

BRIGHTON

Brighton Hospital Alcoholism Treatment Services
12851 East Grand River Street
Brighton, MI 48116

Evergreen Counseling Centers Substance Abuse Services
7600 Grand River Avenue
Brighton, MI 48116

Let Us Heal Our Youth (LUHOY)
6815 West Grand River Avenue
Brighton, MI 48116

Livingston County Catholic Social Services Substance Abuse Prevention
8619 West Grand River
Suite I
Brighton, MI 48116

Livingston/Ingham Counseling Services
7400 West Grand River
Suite B
Brighton, MI 48116

BRIMLEY

Bay Mills Indian Community Substance Abuse Program
Route 1
Brimley, MI 49715

CADILLAC

Community Family and Children's Services Alcohol and Drug Services
140 West River Street
Suite 7
Cadillac, MI 49601

Staircase Runaway and Youth Services, Inc.
124 1/2 North Mitchell Street
Cadillac, MI 49601

Wexford/Missaukee Intermediate School District
9905 East 13th Street
Cadillac, MI 49601

CALUMET

Phoenix House, Inc.
422 Pine Street
Calumet, MI 49913

Willow Hall, Inc.
210 8 Street
Calumet, MI 49913

CANTON

Evergreen Counseling Centers
5840 Canton Center Road
Suite 290
Canton, MI 48187

Hegira Programs, Inc. Oakdale Recovery Center
43825 Michigan Avenue
Canton, MI 48188

CARO

Cass River Services, Inc.
337 Montague Avenue
Caro, MI 48723

Catholic Family Service Family Counseling Services
758 North State Street
Caro, MI 48733

List Psychological Services
443 North State Street
Caro, MI 48723

Tuscola Substance Abuse Services
1309 Cleaver Road
Caro, MI 48723

CEDAR SPRINGS

New Day Center of Cedar Springs
73 South Main Street
Cedar Springs, MI 49319

CENTER LINE

Medical Resource Center, Inc.
26328 Van Dyke Street
Center Line, MI 48015

Options Counseling Services, Inc.
25529 Van Dyke Street
Center Line, MI 48015

CHARLEVOIX

Charles Allen Ransom Counseling Center
14695 Park Avenue
Charlevoix, MI 49720

CHARLOTTE

Eaton Substance Abuse Program, Inc.
138 South Washington Street
Suite C
Charlotte, MI 48813

Region 13 Substance Abuse Prevention Education (SAPE)
1790 East Packard Highway
Charlotte, MI 48813

CHEBOYGAN

Charles Allen Ransom Counseling Center CHIP
520 North Main Street
Suite 202
Cheboygan, MI 49721

Prevention Education Specialties
520 North Main Street
Suite 200
Cheboygan, MI 49721

CHELSEA

Chelsea Arbor Treatment Center Substance Abuse Program
775 South Main Street
Chelsea, MI 48118

CLARE

Human Aid, Inc.
1416 North McEwan Street
Clare, MI 48617

CLARKSTON

Consortium for Human Development, Inc.
5645 Sashabaw Road
Clarkston, MI 48346

Insight Recovery Center/ Clarkston
9075 Big Lake Road
Clarkston, MI 48347

Psychological Evaluation Service Substance Abuse Services
5850 Lorac Drive
Suite A
Clarkston, MI 48347

CLINTON TOWNSHIP

Community Assessment Referral and Education (CARE) Employee Assistance Center
36358 Garfield Street
Clinton Office Plaza South
Clinton Township, MI 48035

Eastwood Community Clinics
35455 Garfield Road
Suite C
Clinton Township, MI 48035

Life Skills Counseling Clinic Substance Abuse Services
23823 15 Mile Road
Clinton Township, MI 48035

North Star Counseling
43900 Garfield Street
Suite 222
Clinton Township, MI 48038

Saint Joseph Hospital Outpatient Behavioral Medicine
42524 Hayes Road
Suite 800
Clinton Township, MI 48044

COLDWATER

Community Health Center of Branch County Substance Treatment and Referral Service
316 East Chicago Street
Coldwater, MI 49036

COMMERCE

Substance Abuse Prevention Coalition of Southeast Michigan (PREVCO)
2822 Lauryl Drive
Commerce, MI 48382

COMSTOCK PARK

Coordinated Health Services, Inc.
3655 Alpine Street NW
Suite 230
Comstock Park, MI 49321

CORUNNA

Shiawassee County Probate Court In-Home Supervision Program
110 East Mack Street
Corunna, MI 48817

DAVISON

Gap
1035 Dayton Street
Davison, MI 48423

DEARBORN

Arab Community Center for Economic and Social Services (ACCESS)
2651 Saulino Court
Dearborn, MI 48120

Bazini Oak Clinic PC Alcoholism and Substance Abuse Program
18181 Oakwood Boulevard
Oakwood Medical Building Suite 401
Dearborn, MI 48124

Eastwood Clinics
19855 West Outer Drive
Suite 204W
Dearborn, MI 48124

Employee Assistance Associates, Inc.
One Parklane Boulevard
1217 East Parklane Towers
Dearborn, MI 48126

Family Services of Detroit/Wayne County Dearborn Office
19855 West Outer Drive
Suite 104
Dearborn, MI 48124

Henry Ford Health Systems Maplegrove Fairlane Center
19401 Hubbard Drive
Dearborn, MI 48126

Insight
22190 Garrison Street
Suite 302
Dearborn, MI 48124

Midwest Mental Health and Employee Assistance Clinic
5050 Schaefer Road
Dearborn, MI 48126

North Point Mental Health Associates of Dearborn
19855 West Outer Drive
Suite 202W
Dearborn, MI 48124

Occupational Health Centers of America
5495 Schaeffer Road
Dearborn, MI 48126

Personal Dynamics Center Substance Abuse Program
23810 Michigan Avenue
Dearborn, MI 48124

Regional Mental Health Clinic PC Substance Abuse Services
23100 Cherry Hill Road
Dearborn, MI 48124

Serenity Manor, Inc.
1637 Ferney Street
Dearborn, MI 48120

Vonschwarz Associates Community Resource Services
22361 Michigan Avenue
Dearborn, MI 48126

DEARBORN HEIGHTS

Dearborn Heights Human Service Center
25639 Ford Road
Dearborn Heights, MI 48127

Westside Mental Health Services
24548 West Warren Avenue
Dearborn Heights, MI 48127

DETROIT

Adult Well Being Services Well Being Program
2111 Woodward Avenue
410 Palms Building
Detroit, MI 48201

Alternatives for Girls
1950 Trumbull Street
Detroit, MI 48216

BAPCO Substance Abuse Treatment and Prevention Program
17357 Klinger Street
First Community Baptist Church
Detroit, MI 48212

Black Family Development, Inc.
15231 West McNichols Street
Detroit, MI 48235

Boniface Community Action Corporation
Boniface Youth and Aftercare Service
7737 Kercheval Street
Room 105
Detroit, MI 48214

Boniface Youth Services
1025 East Forest Street
Room 315
Detroit, MI 48201

Fort Street Clinic
5886 West Fort Street
Detroit, MI 48209

Catholic Social Services of Wayne County
9851 Hamilton Avenue
Detroit, MI 48202

Catholic Youth Organization Comp. Substance Abuse Prevention Services
6900 McGraw Avenue
Detroit, MI 48210

Center of Behavioral Therapy PC
24453 Grand River Avenue
Detroit, MI 48219

Children's Center of Wayne County
101 East Alexandrine Street
Detroit, MI 48201

Christian Guidance Center
3684 Trumbull Avenue
Detroit, MI 48208

5801 Woodward Street
Detroit, MI 48202

Community Corrections Center Monica House
15380 Monica Street
Detroit, MI 48238

Comprehensive Services, Inc.
4630 Oakman Boulevard
Detroit, MI 48204

Detroit American Indian Health Center Substance Abuse/ Prevention/Treatment
4400 Livernois Street
Detroit, MI 48210

Detroit Central City Community Mental Health, Inc.
10 Peterboro Street
Detroit, MI 48201

Detroit City Rescue Mission
3535 3 Street
Detroit, MI 48201

Detroit Counseling Center, Inc.
3800 Woodward Avenue
Suite 1010
Detroit, MI 48201

Detroit East Dual Diagnosis Program
9141 East Jefferson Avenue
Detroit, MI 48214

Detroit Health Department Central Diagnostic and Referral Service
1151 Taylor Street
Building 1
Detroit, MI 48202

Detroit Public Library Drug Abuse and Family Living Program
5201 Woodward Avenue
Detroit, MI 48202

Detroit Recreation Department Leisure Education and Sports Program
735 Randolph Street
Room 1707
Detroit, MI 48226

Detroit Riverview Hospital
Beacon Center for Behavioral Medicine
7733 East Jefferson Avenue
Detroit, MI 48214

Detroit Urban League
Black Family and Substance Abuse
15770 James Couzens Freeway
Detroit, MI 48238

Donald Tate and Associates PC Psychological Services
17320 Livernois Street
Detroit, MI 48221

Eastwood Community Clinics
Beginnings
15085 East 7 Mile Road
Detroit, MI 48205

Conner House
11542 Conner Street
Detroit, MI 48205

Outpatient
15125 Gratiot Avenue
Detroit, MI 48205

Eleonore Hutzel Recovery Center
301 East Hancock Street
Detroit, MI 48201

13301 Mound Road
Detroit, MI 48213

Elmhurst Home, Inc.
12007 Linwood Street
Detroit, MI 48206

Elrose Health Services, Inc. Outer Drive Services
1475 East Outer Drive
Detroit, MI 48234

Employee Assistance Associates, Inc.
Penobscot Building Suite 1866
Detroit, MI 48236

Family Services of Detroit/Wayne County
Downtown Detroit Office
220 Bagley Street
Michigan Building Suite 700
Detroit, MI 48226

Mack/Warren Office
18585 Mack Street
Detroit, MI 48236

Northwest District/Family Trouble Clinic
11000 West McNichols Road
Suite 320-S
Detroit, MI 48221

Fathers and Mothers in League with Youth (FAMILY) Substance Abuse Prevention Program
18954 James Couzens Highway
Detroit, MI 48235

Franklin Wright Settlements, Inc. Project Kujichagulia
3360 Charlevois Street
Detroit, MI 48207

Ganesh Clinic
13011 West McNichols Street
Detroit, MI 48238

Genesis Community Treatment Center
3875 Lillibridge Street
Detroit, MI 48214

Genesis House I
131 Stimson Street
Detroit, MI 48201

Grateful Home, Inc.
335 East Grand Boulevard
Detroit, MI 48207

Dream Weaver Project/Fairview
3840 Fairview Street
Detroit, MI 48214

Dreamweaver Power Project
11031 Mack Avenue
Detroit, MI 48214

Greenfield Services Agency, Inc. Drug Free Program
13200 Fenelon Street
Detroit, MI 48212

Grosse Pointe Center for Individual and Family Therapy
15224 Kercheval Street
Detroit, MI 48236

Harper Hospital Chemical Dependence Program
3990 John R Street
Detroit, MI 48201

Health Service Technical Assistance, Inc. Addiction Treatment Services
1545 East Lafayette Street
Detroit, MI 48207

Henry Ford Health System/ Maplegrove Detroit Center for Chemical Dependency
3011 West Grand Boulevard
200 Fisher Building
Detroit, MI 48202

Hope, Unity and Growth, Inc.
4875 Coplin Street
Detroit, MI 48215

Insight
7430 2 Avenue
Detroit, MI 48202

J and J Youth Services Center
12919 East 7 Mile Road
Detroit, MI 48205

Jefferson House
8311 East Jefferson Avenue
Detroit, MI 48214

Latino Family Services, Inc.
3815 West Fort Street
Detroit, MI 48216

Pretreatment/Aftercare Center
4748 West Vernor Street
Detroit, MI 48209

Lutheran Child and Family Services Choices/Substance Abuse Prevention for High Risk Youth
10811 Puritan Avenue
Detroit, MI 48238

Magic Intervention
8401 Woodward Avenue
A and D Building
Detroit, MI 48202

Mariners Inn Transitional Living Community
445 Ledyard Street
Detroit, MI 48201

Medical Center Psychiatric Associates PC
4727 Saint Antoine Street
Detroit, MI 48201

Mercy Hospital Chemical Dependency Services
5555 Conner Avenue
Detroit, MI 48213

Metro East Substance Abuse Treatment Corporation
8047 East Harper Avenue
2nd Floor
Detroit, MI 48213

13627 Gratiot Avenue
Detroit, MI 48205

Metro Medical Group East Chemical Dependency Program
4401 Conner Street
Detroit, MI 48215

Metro Medical Group/Detroit Northwest Health Alliance Plan Substance Abuse Services
1800 Tuxedo Avenue
Tuxedo Medical Center
Detroit, MI 48206

Metropolitan Arts Complex, Inc. Metro Arts Therapy Services
11000 West McNichols Road
Detroit, MI 48221

Michigan Health Center Lifeline Program
2700 Martin Luther King Jr. Boulevard
Detroit, MI 48208

Nardin Park Substance Abuse Center
9605 West Grand River Avenue
Detroit, MI 48204

National Council on Alcohol and Drug Dependency (Greater Detroit Area)/Vantage Point
10601 West 7 Mile Road
Detroit, MI 48221

Neighborhood Service Organization (NSO)
Concord
22041 Grand River Avenue
Detroit, MI 48219

24-Hour Walk-In Center
3430 3 Street
Detroit, MI 48201

Calvin Wells Treatment Center
8600 Woodward Street
Detroit, MI 48202

Drug Abuse Center Area I
8809 John C. Lodge
Herman Keifer Hospital Building 5
Detroit, MI 48202

Gratiot Services Center
3506 Gratiot Avenue
Detroit, MI 48207

New Center Community Mental Health Services
2051 West Grand Boulevard
Grand Dex Plaza
Detroit, MI 48208

North Park
801 Virginia Park Street
Detroit, MI 48202

Northeast Guidance Center
Central Screening Unit
13340 East Warren Street
Detroit, MI 48215

Community Support Program
2670 Chalmers Street
Detroit, MI 48215

Intensive Family Services
17000 East Warren Street
Detroit, MI 48224

Northwest Treatment Center Drug Treatment Program
14602 Greenfield Road
Detroit, MI 48227

NPL, Inc.
18641 West 7 Mile Road
Detroit, MI 48219

Parkside Mental Health Services, Inc. Substance Abuse Services
567 Van Dyke Street
Detroit, MI 48214

People on an Elected Mission (POEM)
5700 30 Street
Detroit, MI 48210

Project Life Substance Abuse Center
18609 West 7 Mile Road
Detroit, MI 48219

Renaissance Education and Training Center
19042 West McNichols Road
Detroit, MI 48219

Renaissance West Community Health Service Community Mental Health Chemical Dependency Service
8097 Decatur Street
Detroit, MI 48228

Sacred Heart Rehabilitation Center, Inc. Alcohol and Drug Treatment Services
2203 Saint Antoine Street
Detroit, MI 48201

Salvation Army Evangeline Center for Women/Children
130 West Grand Boulevard
Detroit, MI 48216

Salvation Army Harbor Light Substance Abuse Center
2643 Park Avenue
Detroit, MI 48201

Self-Help Addiction Rehab (SHAR)
Aftercare
5675 Maybury Grand Avenue
Detroit, MI 48208

Day Treatment
14301 Longview Street
Detroit, MI 48213

East
4216 McDougall Street
Detroit, MI 48207

SHAR I
1852 West Grand Boulevard
Detroit, MI 48208

Sobriety House, Inc.
2081 West Grand Boulevard
Detroit, MI 48208

SW Detroit Community Mental Health Services, Inc. Substance Abuse Services
1700 Waterman Street
Detroit, MI 48209

University of Detroit Mercy Addiction Studies Inst. Alternatives Project
8200 West Outer Drive
Detroit, MI 48219

Urge Chemical Abuse/Dependent Prevention Program
19190 Schaefer Highway
Detroit, MI 48235

Wayne State University
Addiction Research Institute
4201 Saint Antoine Street
University Health Center Suite 9A
Detroit, MI 48201

College of Lifelong Learning
6001 Cass Street
Detroit, MI 48202

Center for Urban Studies/Detroit Family Project
656 West Kirby Street
Faculty Administration Building 3049
Detroit, MI 48202

Wendie D. Lee Institute of Life Management, Inc.
11000 West McNichols Street
Suite 212
Detroit, MI 48221

Wolverine Human Services Drug Education and Prevention
2629 Lenox Street
Detroit, MI 48215

2nd Chance, Inc. Int. Resocialization and A/C Program (IRAP)
1249 Washington Boulevard
Suite 1200
Detroit, MI 48226

DRAYTON PLAINS

Oakland Family Services
4440 West Walton Boulevard
Unit E
Drayton Plains, MI 48020

EAST DETROIT

Eastwood Community Clinics
20811 Kelly Street
Suite 103
East Detroit, MI 48021

EAST LANSING

Central Diagnostic and Referral Services, Inc.
2875 Northwind Drive
Suite 211
East Lansing, MI 48823

Gateway Community Services Family Services
910 Abbott Road
Suite 100
East Lansing, MI 48823

Horizon Center
610 Abbott Road
East Lansing, MI 48823

Inst. for Public Policy/Social Research Addiction Relapse Prevention Project
Michigan State University
129 Psychology Research Building
East Lansing, MI 48824

Meridian Prof. Psychological Consultants PC
5031 Park Lake Road
East Lansing, MI 48823

Tamarack Substance Abuse Treatment Component
2775 East Lansing Drive
East Lansing, MI 48823

ECORSE

Tri-City Renaissance, Inc.
3890 West Jefferson Street
Ecorse, MI 48229

ESCANABA

Delta County Substance Abuse Center
2920 College Avenue
Delta County Service Center
Escanaba, MI 49829

FARMINGTON

Adult/Youth Developmental Services PC Substance Abuse Program
23133 Orchard Lake Road
Suite 104
Farmington, MI 48336

Botsford General Hospital Eastwood at Botsford General Hospital
28050 Grand River Avenue
Farmington, MI 48336

Catholic Social Services of Oakland County
26105 Orchard Lake Road
Suite 303
Farmington, MI 48334

Chemical Dependency Awareness Program
32795 West Ten Mile Road
Farmington, MI 48336

Evergreen Counseling/Orchard Lake Road
29226 Orchard Lake Road
Suite 250
Farmington, MI 48334

North Point Mental Health Associates Substance Abuse Services
28595 Orchard Lake Road
Suite 301
Farmington, MI 48334

Professional Psychotherapy and Counseling Center
28521 Orchard Lake Road
Suite A
Farmington, MI 48334

FARMINGTON HILLS

Evergreen Counseling Centers
31400 Northwestern Highway
Suite E
Farmington Hills, MI 48334

Farmington Area Advisory Council, Inc. Middlebelt Road Outpatient Services
23450 Middlebelt Road
Farmington Hills, MI 48336

Total Therapy Management Farmington Hills Center Substance Abuse Program
28511 Orchard Lake Road
Suite A
Farmington Hills, MI 48334

FERNDALE

Community Services of Oakland
345 East 9 Mile Road
Ferndale, MI 48220

Kingswood Hospital Second Step Program
10300 West 8 Mile Road
Ferndale, MI 48220

FLINT

Auburn Counseling Associates
400 North Saginaw Street
Suite 300
Flint, MI 48502

Care Inc./Community Alliance Resource Environment
G-5171 North Saginaw Street
Flint, MI 48505

Catholic Social Services
202 East Boulevard Drive
Suite 210
Flint, MI 48503

Community Recovery Services
119 North Grand Traverse Street
Market Fire Station
Flint, MI 48503

Connexion, Inc.
310 East 3 Street
Flint, MI 48502

CRS at Flint County Jail
1002 South Saginaw Street
Flint, MI 48502

CRS at Flint Dept. of Corrections
411 East 3 Street
Flint, MI 48503

CRS at Flint New Paths
765 East Hamilton Avenue
Flint, MI 48505

Flint Genesee County Community Coordinated Child Care Association (4C)
310 East 3 Street
5th Floor
Flint, MI 48502

Flint Odyssey House, Inc.
1225 Martin Luther King Avenue
Flint, MI 48503

Genesis Health System Center for Addiction Treatment
2811 East Court Street
Flint, MI 48506

Hurley Medical Center Substance Abuse Services Department
One Hurley Plaza
Flint, MI 48502

Insight Recovery Center at Miller Road
G-3426 Miller Road
Flint, MI 48507

Insight Recovery Center, Inc.
1110 Eldon Baker Drive
Flint, MI 48507

Intake Assessment and Referral Center
G-3500 Flushing Road
Suite 4100
Flint, MI 48504

McLaren Regional Medical Center Substance Abuse Services
401 South Ballenger Highway
Flint, MI 48532

National Council on Alcoholism and Addictions/Greater Flint Area
202 East Boulevard Drive
Suite 310
Flint, MI 48503

Oakland Psychological Clinic Substance Abuse Services
2360 Linden Road
Suite 300
Flint, MI 48532

Occupational Health Centers of America
2503 South Linden Road
Flint, MI 48532

Transition House, Inc.
939 Martin Luther King Boulevard
Flint, MI 48503

FLUSHING

Special Family Services of Genesee County
3158 McKinley Road
Flushing, MI 48433

FRASER

Alcohol Highway Safety Program
31900 Utica Road
Suite 201
Fraser, MI 48026

FREMONT

Fremont Medical Center PROST
230 West Oak Street
Suite 106
Fremont, MI 49412

GARDEN CITY

Garden City Osteopathic Hospital
Brookfield Clinic Substance Abuse Services
6245 North Inkster Road
Garden City, MI 48135

Chemical Dependency Program
6701 Harrison Street
Garden City, MI 48135

Northwestern Community Services
6012 Merriman Road
Garden City, MI 48135

GAYLORD

Central Diagnostic and Referral Services, Inc.
1665 West M32 Baraga Building
Gaylord, MI 49735

Community Family and Children's Services Alcohol and Drug Services
111 South Michigan Avenue
Gaylord, MI 49735

Northern Michigan Counseling Services
209 West First Street
Gaylord, MI 49735

87th District Court Alcohol Awareness and Information Program
800 Livingston Boulevard
Alpine Center
Gaylord, MI 49735

GLADWIN

Human Aid, Inc. Substance Abuse Services
1302 Chatterton Street
Gladwin, MI 48624

GRAND BLANC

Oakland Psychological Clinic PC
8341 Office Park Drive
Grand Blanc, MI 48439

TFA/Fuller and Associates
4033 South Dort Highway
Grand Blanc, MI 48439

GRAND HAVEN

Child/Family Services of Western Michigan, Inc.
321 South Beechtree Street
Grand Haven, MI 49417

North Ottawa Community Hospital Chemical Dependency Clinic
436 Ottawa Street
Grand Haven, MI 49417

Ottagan Addictions Rehab, Inc.
120 South 5 Street
Grand Haven, MI 49417

GRAND RAPIDS

Advanced Therapeutics Corporation Solutions
2020 Nelson Street SE
Grand Rapids, MI 49507

Advisory Center for Teens Neighborhood Service Center
1115 Ball Avenue NE
Grand Rapids, MI 49505

Alcohol and Chemical Abuse Consultants
2020 Raybrook Street SE
Suite 102
Grand Rapids, MI 49506

Alcohol Outpatient Services, Inc. (AOS Inc)
1331 Lake Drive SE
Grand Rapids, MI 49506

Assessment Unit
700 Fuller Street NE
Grand Rapids, MI 49503

Brief Therapy Center
2504 Ardmore Street SE
Grand Rapids, MI 49506

Butterworths Resolve
21 Michigan Avenue NE
Suite 370
Grand Rapids, MI 49503

Care Unit of Grand Rapids
1931 Boston Street SE
Grand Rapids, MI 49506

Center for Family Recovery
4467 Cascade Road
Suite 4481
Grand Rapids, MI 49546

Christian Counseling
3300 36 Street SE
Grand Rapids, MI 49518

Community Alternatives Program Project Rehab
801 College Street SE
Grand Rapids, MI 49507

Community Services Project Rehab
822 Cherry Street SE
Grand Rapids, MI 49506

Dakotah Family Treatment Center Project Rehab
315 Richard Terrace SE
Grand Rapids, MI 49506

Employee Assistance Center Substance Abuse Services
161 Ottawa Avenue NW
Waters Building Suite 301
Grand Rapids, MI 49503

Family Day Care Program
200 Eastern Street SE
Grand Rapids, MI 49503

Family Outreach Center Outpatient Substance Abuse Counseling
1922 South Division Avenue
Grand Rapids, MI 49507

Family Service
1400 Leonard Street NE
Grand Rapids, MI 49505

Gerontology Network Services of Kent County
516 Cherry Street SE
Grand Rapids, MI 49503

Glenbeigh of Kent Community Hospital
750 Fuller Avenue NE
Grand Rapids, MI 49503

Grand Rapids Inter-Tribal Council Substance Abuse Services
45 Lexington Avenue NW
Grand Rapids, MI 49504

Harbinger of Grand Rapids/ Team IV
1155 Front Avenue NW
Grand Rapids, MI 49504

Hope Network Disabilities Substance Abuse Services
1490 East Beltline Street SE
Grand Rapids, MI 49506

Jellema House
523 Lyon Street NE
Grand Rapids, MI 49503

Alpha Omega
242 College Avenue NE
Grand Rapids, MI 49503

Kootstra/Jansma/Elders/Teitsma/ Dejonge
3330 Claystone Street SE
Grand Rapids, MI 49546

Lake Drive Recovery Program
1230 Lake Drive SE
Grand Rapids, MI 49506

Occupational Health Centers of America, Inc.
400 Ann Street
Suite 210A
Grand Rapids, MI 49506

Our Hope Association
324 Lyon Street NE
Grand Rapids, MI 49503

Pine Rest Christian Hospital Outpatient Division/Campus
300 68 Street SE
Grand Rapids, MI 49501

Project Rehab
Adult Residential Services
200 Eastern Avenue SE
Grand Rapids, MI 49503

Family Support Center
3637 Clyde Park Street SW
Grand Rapids, MI 49507

Shiloh Family I
750 Cherry Street SE
Grand Rapids, MI 49503

Prospectives
2755 Birchcrest Street SE
Grand Rapids, MI 49506

Salvation Army Substance Abuse Services
1215 East Fulton Street
Grand Rapids, MI 49503

Substance Abuse Prevention Education Regions 7 and 8
2930 Knapp Street NE
Grand Rapids, MI 49505

The American People Against Drugs
419 Leonard Street NW
Grand Rapids, MI 49504

West Michigan Addiction Consultants PC Professional Recovery System
1514 Wealthy Street SE
Suite 292
Grand Rapids, MI 49506

Women's Service Network
124 East Fulton Street
Suite 301
Grand Rapids, MI 49503

61st District Court Substance Abuse Services
333 Monroe Street NW
Hall of Justice Building Room 101
Grand Rapids, MI 49503

GRANDVILLE

New Day Center of Grand Rapids
2990 Franklin Street
Grandville, MI 49418

GRAYLING

New Life Community Service
6441 West M-72
Grayling, MI 49738

GREENVILLE

Occupational Health Center of America, Inc.
1810 West Washington Street
Suite 1
Greenville, MI 48838

GROSSE POINTE

Eastwood Clinics Grosse Point Woods
19251 Mack Avenue
Mack Office Building Suite 300
Grosse Pointe, MI 48236

Employee Assistance Associates, Inc.
16845 Kercheval Street
Suite 5
Grosse Pointe, MI 48236

Grosse Pointe Probation Substance Abuse Services
15115 East Jefferson Street
Grosse Pointe, MI 48230

GROSSE POINTE WOODS

Catholic Social Services of Wayne County Substance Abuse Services
19653 Mack Avenue
Grosse Pointe Woods, MI 48236

HANCOCK

Western Up Assessment Service, Inc.
1100 1/2 Quincy Street
Hancock, MI 49930

Western Up District Health Department Substance Abuse Services
540 Depot Street
Hancock, MI 49930

HART

New Life Recovery and Prevention Services, Inc.
208 Washington Street
Hart, MI 49420

HARTFORD

Van Buren County Health Department Substance Abuse Services
57418 County Road 681
Hartford, MI 49057

HASTINGS

Barry County Substance Abuse Services
220 West Court Street
Hastings, MI 49058

HIGHLAND PARK

Black Family Development, Inc. Family Abstinence Commitment to Empower (FACE)
16041 Woodward Avenue
Highland Park, MI 48203

New Center Community Mental Health Services
13700 Woodward Avenue
Highland Park, MI 48203

HILLSDALE

Nielsen Center
3300 Beck Road
Hillsdale, MI 49242

Seniors Prevention and Education Services
1580 South Hillsdale Road
Hillsdale, MI 49242

HOLLAND

Center for Women in Transition
304 Garden Avenue
Holland, MI 49424

Chester A. Ray Center
118 East 9 Street
Holland, MI 49423

Child/Family Services of Western Michigan, Inc. Substance Abuse Services
412 Century Lane
Holland, MI 49423

Employee Assistance Center
426 Century Lane
Holland, MI 49423

Mercy Glen Family Recovery Center Substance Abuse Services
603 East 16 Street
Holland, MI 49423

Ottagan Addictions Rehab, Inc.
483 Century Lane
Holland, MI 49423

Pine Rest Christian Hospital Substance Abuse Services
525 Michigan Avenue
Holland, MI 49423

HOLLY

Highland Waterford Center, Inc. Holly Gardens
4501 Grange Hall Road
Holly, MI 48442

HOLT

Child and Family Services of Michigan, Inc. Capitol Area Substance Abuse Services
4801 Willoughby Street
Suite 1
Holt, MI 48842

HOUGHTON

Dial Help, Inc.
616 Sheldon Avenue
Houghton, MI 49931

HOWELL

Livingston Counseling and Assessment Services, Inc.
3744 East Grand River Avenue
Howell, MI 48843

McPherson Hospital McPherson Treatment Center
620 Byron Road
Ground Floor
Howell, MI 48843

Women's Resource Center
207 East Grand River Street
Howell, MI 48843

HUNTINGTON WOODS

Recovery Consultants, Inc. Larry Smaller NCAC II
26711 Woodward Avenue
Suite 204
Huntington Woods, MI 48070

INKSTER

Youth Living Centers, Inc.
30000 Hively Street
Inkster, MI 48141

IONIA

Ionia Subtance Abuse/Addiction Counseling (ISAAC)
848 East Lincoln Avenue
Ionia, MI 48846

Personal Counseling Services
2300 West Main Street
Ionia, MI 48846

IRON MOUNTAIN

Dickinson/Iron Substance Abuse Services, Inc. Outpatient
427 South Stephenson Avenue
Iron Mountain, MI 49801

Guardian Consultants
218 East Ludington Street
Iron Mountain, MI 49801

Marquette Medical Center Family Addiction Services
800 East Boulevard
Iron Mountain, MI 49801

Veterans' Affairs Medical Center Substance Abuse Treatment Program
East H Street
Iron Mountain, MI 79801

IRON RIVER

Dickinson/Iron Substance Abuse Services, Inc.
117 West Genesee Street
Iron River, MI 49935

IRONWOOD

Lutheran Social Services of Wisconsin and Upper Michigan, Inc.
333 East Aurora Street
Salem Lutheran Church
Ironwood, MI 49938

Western Up Assessment Services, Inc.
237 East Cloverland Drive
Ironwood, MI 49938

JACKSON

Breakout Drug Education Program, Inc.
1115 East White Oak Street
Jackson, MI 49201

Bridgeway Center, Inc.
301 Francis Street
Suite 100A
Jackson, MI 49201

Central Diagnostic and Referral Service, Inc.
301 Francis Street
Suite 101A
Jackson, MI 49201

Chelsea Community Hospital Jackson Outpatient Substance Abuse
One Jackson Square
Suite 111
Jackson, MI 49201

Double Talk
5020 Ann Arbor Road
Jackson, MI 49204

Family Service and Children's Aid
115 West Michigan Street
Jackson, MI 49204

Jackson County Intermediate School District Substance Abuse Services
6700 Browns Lake Road
Jackson, MI 49201

Michigan Therapeutic Consultants PC
2212 4 Street
Suite A
Jackson, MI 49203

Occupational Health Centers of America Substance Abuse Services
123 North West Avenue
Suite E
Jackson, MI 49201

Student Assistance Program Substance Abuse Prevention Program
209 East Washington Street
Suite 301D
Jackson, MI 49201

Washington Way Recovery Center
2424 West Washington Street
Jackson, MI 49203

JENISON

Child and Family Services of Michigan Western Michigan Branch
1384 Baldwin Street
Jenison, MI 49428

KALAMAZOO

Catholic Family Services The Ark/Kalamazoo
990 West Kilgore Road
Kalamazoo, MI 49001

Community Assessment and Screening
629 Pioneer Street
Kalamazoo, MI 49008

Gateway Northside Outreach Services
129 Roberson Street
Kalamazoo, MI 49007

Gateway Outpatient Services
333 Turwill Lane
Kalamazoo, MI 49006

Gateway Residential Services
1910 Shaffer Road
Kalamazoo, MI 49001

Gryphon Place
1104 South Westnedge Avenue
Kalamazoo, MI 49008

Health Psychology and Medicine PC Substance Abuse Services
3503 Greenleaf Boulevard
Suite 102
Kalamazoo, MI 49008

Human Services Dept./Nazareth Campus Substance Abuse Prevention Services
3299 Gull Road
Kalamazoo, MI 49001

Kalamazoo Psychology PC
122 West South Street
Suite 207
Kalamazoo, MI 49007

Mid-America Psychological Services
8036 Moorsbridge Road
Suite 2
Kalamazoo, MI 49002

Occupational Health Centers of America Substance Abuse Services
5360 Holiday Terrace
Kalamazoo, MI 49009

Senior Services, Inc. Older Adult Recovery Program
918 Jasper Street
Kalamazoo, MI 49001

The Guidance Clinic
2615 Stadium Drive
Kalamazoo, MI 49008

University Substance Abuse Clinic SPADA
112 West South Street
Kalamazoo, MI 49008

Western Michigan University
Campus Substance Abuse Services
1921 West Michigan Street
1150 Kanley Chapel
Kalamazoo, MI 49008

Specialty Prog in Alcohol/Drug Abuse
1111 Oliver Street
Kalamazoo Corrections Center
Kalamazoo, MI 49008

Substance Abuse Services
Sindecuse Health Center
Room 3235
Kalamazoo, MI 49008

Womancare, Inc.
2836 West Main Street
Kalamazoo, MI 49006

KALKASKA

Substance Abuse Education Service
556 Cedar Street
Kalkaska, MI 49646

KINGSFORD

Community Substance Abuse Services, Inc.
373 Woodward Avenue
Kingsford, MI 49801

L'ANSE

Keweenaw Bay Tribal Council Bedabin Family Group Home
Brewry Road
Route 2
L'Anse, MI 49946

LAKE ORION

Education Intervention Programs
1520 South Lapeer Road
Suite 208
Lake Orion, MI 48360

Guest House
1840 West Scripps Road
Lake Orion, MI 48361

Oakland Psychological Clinic PC Substance Abuse Services
1520 South Lapeer Road
Suite 111
Lake Orion, MI 48360

LANSING

Cristo Rey Counseling Services Substance Abuse Program
1717 North High Street
Lansing, MI 48906

Dimensions of Life
3320 West Saginaw Street
Lansing, MI 48917

Glass House
419 North Logan Street
Lansing, MI 48915

Holden House
3300 South Pennsylvania Avenue
Lansing, MI 48910

House of Commons
517 North Walnut Street
Lansing, MI 48933

LSSM Family Counseling/ Lansing
801 South Waverly Road
Suite 202
Lansing, MI 48917

Michigan Department of State Police Drug Abuse Resistance Education (DARE)
2722 East Michigan Avenue
Suite 100
Lansing, MI 48912

Michigan Substance Abuse and Traffic Safety Information Center
2409 East Michigan Avenue
Lansing, MI 48912

Midwest Hispanic Institute
1135 North Washington Avenue
Lansing, MI 48906

National Council on Alcoholism and Drug Dependence of Michigan
913 West Holmes Road
Suite 225
Lansing, MI 48910

National Council on Alcoholism Lansing Regional Area
3400 South Cedar Street
Suite 200
Lansing, MI 48910

Occupational Health Centers of America, Inc.
801 South Waverly Road
Suite 305
Lansing, MI 48917

Older Adult Substance Abuse Prevention and Treatment
808 Southland Street
Suite A
Lansing, MI 48910

Saint Lawrence Hospital Outpatient Mental Health and Substance Abuse Services
1210 West Saginaw Street
Lansing, MI 48915

Southland Counseling Center Comp. Substance Abuse Treatment Program
808 Southland Street
Suite C
Lansing, MI 48910

Total Health Education, Inc.
3400 South Cedar Street
Suite 100
Lansing, MI 48911

LAPEER

Alcohol Information and Counseling Center
1575 Suncrest Drive
Lapeer County Health Department
Lapeer, MI 48446

Christian Family Services of Lapeer County
441 Clay Street
Lapeer, MI 48446

Lapeer Regional Hospital Vail Center
1375 North Main Street
Lapeer, MI 48446

LATHRUP VILLAGE

Tri-County Drug/Substance Abuse Prevention Arab/ Chaldean Youth
28551 Southfield Road
Lathrup Village, MI 48076

LINCOLN PARK

Boniface Community Action Corporation
25050 West Outer Drive
Suite 201
Lincoln Park, MI 48146

Community Care Services
Counseling and Resource Center
1673 Fort Street
Lincoln Park, MI 48146

Substance Abuse Services
1174 Fort Street
Lincoln Park, MI 48146

Counseling Alternatives/Lincoln Park
1530 John A. Papalas Drive
Lincoln Park, MI 48183

Oakwood Downriver Medical Center Substance Abuse Services
25750 West Outer Drive
Lincoln Park, MI 48146

Oxford Institute Outpatient Clinic Lincoln Park
25050 West Outer Drive
Lincoln Park, MI 48146

25th District Court Probation Dept. Substance Abuse Services
1475 Cleophus Street
Lincoln Park, MI 48146

LIVONIA

Butterfly Center The Recovery Corporation
29586 5 Mile Road
Livonia, MI 48154

Catholic Social Service of Wayne County
17332 Farmington Road
Livonia, MI 48152

Counseling Alternatives/Livonia
37650 Professional Center Drive
Suite 145A
Livonia, MI 48154

Eastwood Clinics
17250 Farmington Road
Livonia, MI 48154

Education Training Research Services Alcohol/Substance Insight and Assessment
37895 Ann Arbor Road
Livonia, MI 48150

Employee Assistance Associates, Inc.
38705 7 Mile Road
Suite 130
Livonia, MI 48152

Family Services of Detroit/Wayne County Livonia Office
16755 Middlebelt Road
Livonia, MI 48154

Insight Services of Livonia
14800 Farmington Road
Suite 107
Livonia, MI 48152

Livonia Counseling Center
13325 Farmington Road
Livonia, MI 48150

Medical Center Psychiatric Assoc. PC
16832 Newburgh Road
Livonia, MI 48154

Metro Medical Group Rehabilitation and Fitness Center
29105 Buckingham Street
Livonia, MI 48154

Oakland Psychological Clinic PC Substance Abuse Services
29865 6 Mile Road
Suite 112
Livonia, MI 48152

Saint Mary Hospital Brookfield Clinic Substance Abuse Services
36475 5 Mile Road
Livonia, MI 48154

LUDINGTON

Mason County Comm. Mental Health Services Substance Abuse Services
920 Diana Street
Ludington, MI 49431

New Life Recovery and Prevention Services
1105 South Washington Street
Ludington, MI 49431

MADISON HEIGHTS

Brookfield Clinics/East
28091 Dequindre Road
Suite 308
Madison Heights, MI 48071

Gateway Counseling Center
30785 Stephenson Highway
Madison Heights, MI 48071

Medical Resource Center, Inc.
1400 East 12 Mile Road
Madison Heights, MI 48071

MANISTEE

Manistee/Benzie Comm. Mental Health Substance Abuse Services
395 3 Street
Manistee, MI 49660

Paradigm Counseling Center of West Michigan, Inc.
59 Maple Street
Manistee, MI 49660

Staircase Runaway and Youth Services
225 9 Street
Manistee, MI 49660

MANISTIQUE

LMAS District Health Department Substance Abuse Program/ Schoolcraft
129 1/2 South Cedar Street
Manistique, MI 49854

MARQUETTE

Catholic Social Services New Visions/Youth Substance Abuse Services
347 Rock Street
Marquette, MI 49855

Great Lakes Recovery Center, Inc.
241 Wright Street
Marquette, MI 49855

Lutheran Social Services of Wisconsin and Upper Michigan
135 West Washington Street
Marquette, MI 49855

Marquette General Hospital Upper Michigan Addiction Rehabilitation
420 West Magnetic Street
Marquette, MI 49855

Martin Psychological PC Substance Abuse Services
6044 U.S. 41 South
Marquette, MI 49855

Substance Abuse Prevention Education/Upper Peninsula Marquette/Alger Intermediate School District
427 West College Avenue
Marquette, MI 49855

96th District Court Alcohol Awareness Program
234 West Baraga Street
Marquette, MI 49855

MARSHALL

Alcohol/Drug Highway Safety and Highway Safety Seminars
111 North Jefferson Street
Suite 109
Marshall, MI 49068

Oaklawn Life Improvement Center Psychological Services
13697 15 Mile Road
Marshall, MI 49068

Personal Counseling Service
113 North Jefferson Street
Suite 107
Marshall, MI 49068

Region 12 Substance Abuse Prevention Education
17111 G Drive North
Marshall, MI 49068

MASON

Correctional Assessment and Treatment Services Comp. Substance Abuse Treatment Program
630 North Cedar Street
Ingham County Jail
Mason, MI 48854

MEMPHIS

Sacred Heart Rehabilitation Center, Inc.
400 Stoddard Road
Memphis, MI 48041

Tom Foley Associates
11206 Gilbert Road
Memphis, MI 48041

MENOMINEE

Bay Area Medical Center Substance Abuse Services
1110 10 Avenue
Menominee, MI 49858

Menominee County CMH Alcohol/Drug Center
2608 10 Street
Menominee, MI 49858

MIDLAND

Catholic Family Service Family Counseling Services
220 West Main Street
Midland, MI 48640

CRW Associates
5114 Eastman Street
Midland, MI 48640

Family and Children's Services of Midland
1714 Eastman Avenue
Midland, MI 48641

Focus Substance Abuse Counseling and Information Service
4604 North Saginaw Road
Suite C
Midland, MI 48640

Occupational Health Centers of America, Inc.
5103 Eastman Avenue
Suite 124
Midland, MI 48640

Ten Sixteen Home
1016 Eastman Avenue
Midland, MI 48640

MILAN

Federal Correctional Institutional Substance Abuse Services
Milan, MI 48160

MILFORD

Oakland Psychological Clinic PC Substance Abuse Services
1203 North Milford Road
Suite A
Milford, MI 48381

Palmer Drug Abuse Program Southeast Michigan, Inc./ Milford
238 North Main Street
Milford, MI 48381

MIO

Ausable Valley Community Mental Health Substance Abuse Services
325 North Mount Tom Road
Mio, MI 48647

MONROE

Caknipe/Kovach Association PC
214 East Elm Street
Suite 112
Monroe, MI 48161

Catholic Social Services of Monroe County
16 East 5 Street
Monroe, MI 48161

Mercy Memorial Hospital
Family Center
700 Stewart Road
Monroe, MI 48161

Substance Abuse Services
740 North Macomb Street
Monroe, MI 48161

Monroe County Community Mental Health Substance Abuse Services
1001 South Raisinville Road
Monroe, MI 48161

Monroe County Intermediate School District Student Assistance Program
1101 South Raisinville Road
Monroe, MI 48161

Monroe County Jail Substance Abuse Education and Counseling Program
100 East 2 Street
Monroe, MI 48163

Monroe Public Schools Prevention Program
901 Herr Street
Monroe, MI 48161

Occupational Health Centers of America Substance Abuse Services
214 East Elm Street
Suite 110
Monroe, MI 48161

Sacred Heart Rehabilitation Center, Inc.
214 Elm Street
Suite 107
Monroe, MI 48161

Saint Vincent Tennyson Center Outpatient
901 North Macomb Street
Suite 104
Monroe, MI 48161

Salvation Army Harbor Light
Monroe County Alcohol Center
3580 South Custer Road
Monroe, MI 48161

Monroe County Center
502 West Elm Street
Monroe, MI 48161

MOUNT CLEMENS

Clinton Counseling Center
Comprehensive Youth Services
2 Crocker Boulevard
Suite 101
Mount Clemens, MI 48043

Macomb County Jail Substance Abuse Program
43565 Elizabeth Road
Mount Clemens, MI 48043

Genesis Counseling Center
308 North Gratiot Street
Mount Clemens, MI 48043

Macomb Family Services Inc I
2 Crocker Boulevard
Suite 202
Mount Clemens, MI 48043

Macomb Intermediate School District Region 18 Substance Abuse Prevention
44001 Garfield Road
Mount Clemens, MI 48044

Options Counseling Services, Inc.
15985 Canal Road
Suite 2
Mount Clemens, MI 48044

Oxford Institute at Saint John Hospital Macomb Center
26755 Ballard Road
Mount Clemens, MI 48045

Total Therapy Management Groesbeck Place
279 North Groesbeck Highway
Mount Clemens, MI 48043

MOUNT PLEASANT

Catholic Family Services Family Counseling Service
120 South University Street
Mount Pleasant, MI 48858

Central Michigan University Alcohol and Other Drug Abuse Prevention Program
205 Foust Hall
Mount Pleasant, MI 48859

Choices of Mount Pleasant, Inc.
5805 Pickard Street
Suite 135
Mount Pleasant, MI 48858

Mount Pleasant Counseling Services
3480 South Isabella Road
Mount Pleasant, MI 48858

OJIBWE Substance Abuse Program
7363 East Broadway
Mount Pleasant, MI 48855

Youth Group Home
3548 South Shepherd Street
Mount Pleasant, MI 48858

MUNISING

LMAS District Health Department Substance Abuse Program/ Alger County
202 Elm Avenue
Munising, MI 49862

MUSKEGON

Child and Family Services and Muskegon Children's Home
1352 Terrace Street
Muskegon, MI 49442

East Side Substance Abuse Clinic
445 East Sherman Boulevard
Muskegon, MI 49444

Every Woman's Place, Inc.
425 West Western Street
Suite 204
Muskegon, MI 49440

Muskegon County Comprehensive Drug Program
1611 East Oak Avenue
Muskegon, MI 49442

New Day Center of Muskegon
90–92 Seaway Drive
Muskegon, MI 49444

West Michigan Therapy, Inc.
130 East Apple Avenue
Muskegon, MI 49442

NEW BALTIMORE

Psychiatric Center of Michigan Hospital Great Lakes Recovery Center
35031 23 Mile Road
New Baltimore, MI 48047

NEW HAVEN

Community Human Services, Inc.
57737 Gratiot Avenue
New Haven, MI 48048

NEWBERRY

LMAS District Health Department Substance Abuse Program/ Luce County
County Road 428
Hamilton Lake Road
Newberry, MI 49868

NILES

Charter Counseling Center at Niles
1956 U.S. 31 South
Niles, MI 49120

NORTHVILLE

Community Commission on Drug Abuse Northville Counseling Center
115 North Center Street
Suite 204
Northville, MI 48167

Psychotherapy and Counseling Services, Inc.
670 Griswold Street
Suite 4
Northville, MI 48167

NOVI

Consortium for Human Development, Inc. Substance Abuse Services
24230 Karim Boulevard
Suite 160
Novi, MI 48375

Insight
39555 West 10 Mile Road
Suite 303
Novi, MI 48375

Janice Cotter/Albert Dicken Substance Abuse Services
22935 Brook Forest Street
Novi, MI 48375

Orchard Hills Psychiatric Center Substance Abuse Services
42450 West 12 Mile Road
Suite 305
Novi, MI 48337

OAK PARK

Metropolitan Rehabilitation Clinics
21700 Greenfield Street
Suite 130
Oak Park, MI 48237

OKEMOS

Highway Safety Associates
2187 Jolly Street
Suite 100
Okemos, MI 48864

OWOSSO

Alcohol Driving Education (ADE)
606 West Main Street
Owosso, MI 48867

Families Against Substance Abuse
280 East Riley Road
Owosso, MI 48867

PACE, Inc. (Positive Alternatives Counseling/Education)
120 West Exchange Street
Suite 300
Owosso, MI 48867

PETOSKEY

Charles Allen Ramson Counseling Center
1310 Spring Street
Petoskey, MI 49770

Harbor Hall
704 Emmet Street
Petoskey, MI 49770

Women's Resource Center of Northern Michigan, Inc.
423 Porter Street
Petoskey, MI 49770

PLYMOUTH

Adult/Youth Developmental Services PC
199 North Main Street
Suite 202
Plymouth, MI 48170

Growth Works Counseling and Intervention Services
271 South Main Street
Plymouth, MI 48170

Knopf Institute
1126 South Main Street
Plymouth, MI 48170

Personalized Nursing Light House, Inc.
575 Main Street
Suite 6
Plymouth, MI 48170

Plymouth Family Service
880 Wing Street
Plymouth, MI 48170

The Knopf Company, Inc.
1126 South Main Street
Plymouth, MI 48170

PONTIAC

Catholic Social Services of Oakland County
53 Franklin Boulevard
Pontiac, MI 48341

Highland Waterford Center, Inc.
377 South Telegraph Road
Pontiac, MI 48341

NPL, Inc.
989 University Drive
Suite 102
Pontiac, MI 48058

Oakland County Youth Assistance
1200 North Telegraph Road
Department 452
Pontiac, MI 48341

Oakland Family Services Substance Abuse Services
114 Orchard Lake Road
Pontiac, MI 48341

Pontiac Area Urban League, Inc. Hispanic Outreach Services
295 West Huron Street
Pontiac, MI 48341

Procare at Pontiac Osteopathic Hospital Chemical Dependency Unit
50 North Perry Street
Pontiac, MI 48058

Providential Health Care Unlimited, Inc. at Pontiac Osteopathic Hospital
24 East Huron Street
Pontiac, MI 48342

Saint Joseph Mercy Hospital Outpatient Psychiatric Clinic
900 Woodward Avenue
Pontiac, MI 48341

Telcom Care Network Personal Consulting Service
367 South Telegraph Street
Suite 105
Pontiac, MI 48341

The Alternatives Center
210 North Perry Street
Pontiac, MI 48343

Turning Point Recovery Center Prall Street Unit
121 Prall Street
Pontiac, MI 48341

Whitlow Center at Pontiac Osteopathic Hospital
16 1/2 East Huron Street
Pontiac, MI 48058

Woodward Counseling, Inc.
35 South Johnson Street
Suite 3D
Pontiac, MI 48341

PORT HURON

Blue Water Mental Health and Child Guidance Clinic
1501 Krafft Road
Port Huron, MI 48059

Catholic Social Services of Saint Clair County/Substance Abuse Services
2601 13 Street
Port Huron, MI 48060

Center for Human Resources Military Street
1001 Military Street
Port Huron, MI 48060

Center for Personal Growth Substance Abuse Services
817 10 Avenue
Port Huron, MI 48060

Comprehensive Youth Services (CYS) The Harbor
3061 Commerce Drive
Suite 2
Port Huron, MI 48060

Port Huron Hospital Outpatient Chemical Dependence Services
1205 Pine Grove Street
Port Huron, MI 48060

Professional Counseling Center PC
520 Superior Street
Port Huron, MI 48060

PRUDENVILLE

Human Aid, Inc. Substance Abuse Services
3000 West Houghton Lake Drive
Prudenville, MI 48651

REDFORD

Aurora Corporation Aurora Community Programs/Redford
14157 Telegraph Street
Redford, MI 48239

Botsford Family Service Center Substance Abuse Services
26905 Grand River Avenue
Redford, MI 48240

Michigan Alcohol and Drug Education
26075 West 6 Mile Road
Redford, MI 48240

Redford Counseling Center
25945 West 7 Mile Road
Redford, MI 48240

17th District Court Probation Department Substance Abuse Services
15111 Beech Daly Street
Redford, MI 48239

REED CITY

Human Aid, Inc. Substance Abuse Services
8625 South Mackinaw Trail
Suite 2
Reed City, MI 49677

RIVER ROUGE

Downriver Guidance Clinic River Rouge/Ecorse Substance Abuse Prevention Project
421 Beechwood Street
Beechwood Resource Center
River Rouge, MI 48218

RIVERVIEW

Evergreen Counseling Centers
14600 King Road
Suite A-1
Riverview, MI 48192

27-2 District Court Substance Abuse Services
14100 Civic Park Drive
Riverview, MI 48192

ROCHESTER

Individual/Group Psychological Services Substance Abuse Services
139 Walnut Street
Rochester, MI 48307

OU Substance Abuse Prevention East Wing Graham Health Center
Rochester, MI 48309

Vincam Human Resources of Michigan, Inc. Employee Assistance Program (EAP)
134 West University Street
Suite C
Rochester, MI 48307

ROCHESTER HILLS

Oxford Institute Outpatient Clinic
455 Barclay Circle Drive
Rochester Hills, MI 48307

ROMEO

Community Human Services, Inc.
332 South Main Street
Romeo, MI 48065

ROMULUS

Romulus Help Center A Division of Matrix Associates, Inc.
9340 Wayne Road
Suite A
Romulus, MI 48174

ROSEVILLE

Metro Medical Group/Roseville
18223 East 10 Mile Road
Suite B
Roseville, MI 48066

NPI., Inc.
27115 Gratiot Street
Roseville, MI 48066

ROYAL OAK

Boys' and Girls' Club of South Oakland County
1545 East Lincoln Street
Royal Oak, MI 48067

Catholic Social Services of Oakland County/Talbott Center
1424 East 11 Mile Road
Royal Oak, MI 48067

Common Ground
751 Hendrie Boulevard
Royal Oak, MI 48067

Eastwood Community Clinics
2801 North Woodward Avenue
Suite 200
Royal Oak, MI 48073

Residential Substance Abuse Treatment Program
1515 North Stephenson Highway
Royal Oak, MI 48067

Judson Center Recovering/A Substance Abuse Program
4410 West 13 Mile Road
Royal Oak, MI 48073

Smith Counseling Services Substance Abuse Services
1719 Crooks Road
Royal Oak, MI 48067

The Sanctuary, Inc.
1222 South Washington Street
Royal Oak, MI 48067

William Beaumont Hospital Alcoholism and Drug Abuse Services
3601 West 13 Mile Road
Royal Oak, MI 48072

SAGINAW

Alcoholism Information Center Prevention and Youth Services
1600 North Michigan Avenue
Suite 501
Saginaw, MI 48602

American Comp. Treatment Services, Inc.
1230 South Washington Street
Suite 105B
Saginaw, MI 48601

Bay Area Consultants
765 North Center Street
Saginaw, MI 48703

Catholic Family Service Family Counseling Services
710 North Michigan Avenue
Saginaw, MI 48602

Child and Family Service of Saginaw County
1226 North Michigan Avenue
Saginaw, MI 48602

Dot Caring Centers, Inc.
Halfway House/Residential Center
1915 Fordney Street
Saginaw, MI 48601

Outpatient
3190 Hallmark Court
Saginaw, MI 48603

First Ward Community Center
1410 North 12 Street
Saginaw, MI 48601

Insight Intl. Inc./Saginaw Division Substance Abuse Services
1230 South Washington Avenue
Suite 305
Saginaw, MI 48601

Insight Recovery Center
420 North Michigan Avenue
Saginaw, MI 48602

Intervention and Rehab Associates, Inc.
1616 Court Street
Saginaw, MI 48602

LSSM Neighborhood House Substance Abuse Prevention
3145 Russell Street
Saginaw, MI 48601

New Perspectives Center, Inc.
Residential and Day Treatment
1321 South Fayette Street
Saginaw, MI 48602

Residential Treatment/Women
613 Wayne Street
Saginaw, MI 48602

Occupational Health Centers of America Substance Abuse Services
1 Tuscola Street
Morley Building
Saginaw, MI 48607

Prof. Psychological and Psychiatric Service Substance Abuse Services
1502 Court Street
Saginaw, MI 48602

Saginaw Community Hospital Pathways Alcohol and Drug Treatment Center
3340 Hospital Road
Saginaw, MI 48603

Saginaw Inter Tribal Substance Abuse Outreach
3239 Christy Way
Saginaw, MI 48603

Saginaw Psychological Services, Inc.
2100 Hemmeter Street
Saginaw, MI 48603

Samaritan Counseling Center of Saginaw Valley
2405 Bay Street
Faith Lutheran Church
Saginaw, MI 48602

Tri-City/SER Jobs for Progress, Inc.
620 Thompson Street
Saginaw, MI 48607

SAINT CLAIR

Counseling Center at River District Hospital Substance Abuse Program
4100 South River Road
Saint Clair, MI 48079

SAINT CLAIR SHORES

Evergreen Counseling Centers
19900 10 Mile Road
Saint Clair Shores, MI 48081

Oxford Institute Saint Clair Shores
23411 Jefferson Avenue
Saint Clair Shores, MI 48080

40th District Court Alcohol Prevention/Traffic Awareness Program
27701 Jefferson Street
Saint Clair Shores, MI 48081

SAINT IGNACE

LMAS Health Department Substance Abuse Program/ Mackinac County
749 Hombach Street
Saint Ignace, MI 49781

SAINT JOHNS

Clinton County Counseling Center
1000 East Sturgis Street
Saint Johns, MI 48879

SAINT JOSEPH

Occupational Health Centers of America, Inc.
815 Main Street
Saint Joseph, MI 49085

SALINE

Saline Community Hospital Greenbrook Recovery Center
400 West Russell Street
Saline, MI 48176

SANDUSKY

Catholic Family Service Family Counseling Services
119 East Sanilac Plaza
Suite 1
Sandusky, MI 48471

Sanilac County Health Department Alcohol and Drug Program
37 Austin Street
Sandusky, MI 48471

SAULT SAINTE MARIE

American Indian Substance Abuse Program
2154 Shunk Road
Sault Sainte Marie, MI 49783

Great Lakes Recovery Center, Inc.
New Hope House/Men
1139 East Portage Avenue
Sault Sainte Marie, MI 49783

New Hope House/Women
1111 Minneapolis Street
Sault Sainte Marie, MI 49783

Marquette Medical Clinic Addiction Treatment Services
500 Osborne Boulevard
Sault Sainte Marie, MI 49783

SHELBY TOWNSHIP

Evergreen Counseling Centers
53950 Van Dyke Street
Shelby Township, MI 48087

Phoenix Center, Inc. North Branch Office
52935 Mound Road
Shelby Township, MI 48316

SOUTHFIELD

Burdette and Doss Associates Psychological Services
17336 West 12 Mile Road
Suite 100
Southfield, MI 48076

Caknipe/Kovach Associates PC Substance Abuse Services
4400 Town Center
Suite 280
Southfield, MI 48175

Central Therapeutic Services, Inc.
17600 West 8 Mile Road
Suite 7
Southfield, MI 48075

Clark and Associates Psychological Services
16250 Northland Drive
Suite 245
Southfield, MI 48075

Counseling Associates
26699 West 12 Mile Road
Suite 100
Southfield, MI 48034

Courage Drug Prevention and Education Program
24300 Southfield Road
Suite 100
Southfield, MI 48075

Employee Assistance Associates, Inc.
21415 Civic Center Drive
Suite 315
Southfield, MI 48076

Evergreen Counseling Centers
20755 Greenfield Road
Suite 1001
Southfield, MI 48075

LSSM Family Counseling/ Southfield
21700 Northwestern Highway
Suite 801
Southfield, MI 48075

Lutheran Child and Family Service of Michigan Choices/ Substance Abuse Prevention for High Risk Youth
20830 Rutland Drive
Suite 201
Southfield, MI 48076

Michigan Alcohol and Addictions Assoc. (MAAA)
21711 West 10 Mile Road
Southfield, MI 48075

National Council on Alcohol and Drug Dependency (Greater Detroit Area)/Vantage Point
17330 Northland Park Court
Southfield, MI 48075

Northland Clinic
17117 West 9 Mile Road
Suite 1221
Southfield, MI 48075

Oakland Psychological Clinic PC Substance Abuse Services
21700 Northwestern Highway
Suite 750
Southfield, MI 48075

Providence Hospital Division of Substance Abuse Services
16001 West 9 Mile Road
Southfield, MI 48037

Relapse Prevention Services
21415 South Civic Center Drive
Suite 211
Southfield, MI 48076

Southfield Lathrup Counseling Services
26080 Berg Road
Southfield, MI 48034

46th District Court Alcohol Awareness Program
26000 Evergreen Road
Southfield, MI 48037

SOUTHGATE

Downriver Community Alliance Central Diagnostic and Referral
15100 Northline Road
Room 144
Southgate, MI 48195

Downriver Guidance Clinic
14665 Northline Road
Southgate, MI 48195

Adult and Family Services
13645 Northline Road
Southgate, MI 48195

NPL, Inc.
13309 Reeck Road
Southgate, MI 48195

28th District Court Substance Abuse Services
14720 Reaume Parkway
Southgate, MI 48195

SPRING LAKE

Mercy Glen Family Recovery Center Substance Abuse Services
17160 130 Avenue
Spring Lake, MI 49456

STANTON

Substance Abuse Consultation Services
308 East Main Street
Stanton, MI 48888

STERLING

Sterling Area Health Center
725 East State Street
Sterling, MI 48659

STERLING HEIGHTS

Evergreen Counseling Centers
33200 Dequindre Road
Suite 200
Sterling Heights, MI 48310

North Point Mental Health Assoc. of Sterling Heights
36250 Dequindre Road
Suite 310
Sterling Heights, MI 48310

Oakland Psychological Clinic PC Substance Abuse Services
39880 Van Dyke Street
Sterling Heights, MI 48313

Phoenix Center, Inc.
4455 Metropolitan Parkway
Sterling Heights, MI 48310

STURGIS

Myrtle Treatment Services
900 Myrtle Avenue
Sturgis, MI 49091

Ruster Foundation
903 C and D North Clay Street
Sturgis, MI 49091

Sturgis Outpatient Center
300 West Chicago Street
Suite 1212
Sturgis, MI 49091

SUTTONS BAY

Grand Traverse Band of Ottawa Chippewa Indians Substance Abuse Services
2807 NE Manitou Street
Route 1 Box 135
Suttons Bay, MI 49682

SYLVAN LAKE

Evergreen Counseling Centers
2360 Orchard Lake Road
Suite 104
Sylvan Lake, MI 48053

TAWAS CITY

Ausable Valley Community Mental Health Substance Abuse Services
1199 West Harris Avenue
Tawas City, MI 48763

TAYLOR

ACT Specialists
20600 Eureka Road
Suite 604
Taylor, MI 48180

Catholic Social Services of Wayne County Substance Abuse Services
24331 Van Born Road
Taylor, MI 48180

Community Care Services Substance Abuse Service
8750 Telegraph Road
Suite 420
Taylor, MI 48180

Downriver Mental Health/ Advanced Psychiatric Services Chemical Dependency Program
20600 Eureka Road
Suite 819
Taylor, MI 48180

Employee Assistance Associates, Inc.
20600 Eureka Road
Suite 500
Taylor, MI 48180

Heritage Hospital Chemical Dependency Rehabilitation Program
1000 Telegraph Road
Taylor, MI 48180

Taylor Teen Health Center
21115 Eureka Road
Taylor, MI 48180

Wayne County Alcohol Highway Safety Education
20600 Eureka Street
Suite 401
Taylor, MI 48180

TECUMSEH

Exceptional Training Systems
9960 Matthews Highway
Tecumseh, MI 49286

Herrick Outpatient Substance Abuse Services
500 East Pottawatamie Street
Tecumseh, MI 49286

TEMPERANCE

Bedford Public Schools Student Assistance Program
1623 Sterns Road
Temperance, MI 48182

Catholic Social Services of Monroe County
8330 Lewis Street
Temperance, MI 48182

THREE RIVERS

Substance Abuse Council of St. Joseph County Counseling Center
222 South Main Street
Three Rivers, MI 49093

TRAVERSE CITY

Central Diagnostic and Referral Services, Inc.
808A South Garfield Avenue
Traverse City, MI 49684

Charles Bethea Associates
10850 Traverse Highway
Grandview Plaza Suite 3304
Traverse City, MI 49684

Community Family and Children's Services
1000 Hastings Street
Traverse City, MI 49684

Munson Medical Center Alcohol and Drug Treatment Center
1105 6 Street
Traverse City, MI 49684

Northern Michigan Alcoholism and Addiction Treatment Services, Inc.
116 East 8 Street
Traverse City, MI 49684

Northwest Michigan Private Industry Council, Inc./CJA
2200 Dendrinos Drive
Suite 201
Traverse City, MI 49685

Phoenix Hall
445 East State Street
Traverse City, MI 49684

Third Level Crisis Intervention Center, Inc.
908 West Front Street
Traverse City, MI 49684

Traverse Bay Area Intermediate School District
2325 North Garfield Road
Traverse City, MI 49684

Wedgewood Christian Counseling Center
3301 Veterans Drive
Suite 125
Traverse City, MI 49684

TRENTON

Downriver Guidance Clinic
1651 Kingsway Court
Trenton, MI 48183

Eastwood Clinics
1650 Fort Street
Suite A
Trenton, MI 48183

Family Services of Detroit/Wayne County Trenton Office
19366 Allen Road
Trenton, MI 48183

Seaway Hospital Chemical Dependency Rehab Services
5450 Fort Street
Trenton, MI 48183

TROY

Assessment Center
2820 West Maple Road
Troy, MI 48084

Consortium for Human Development, Inc.
755 West Big Beaver Road
14th Floor Suite 1401
Troy, MI 48084

Eastwood Clinics Heritage Square
1771 West Big Beaver Road
Troy, MI 48084

Employee Assistance Associates, Inc.
3155 West Big Beaver Road
Suite 121
Troy, MI 48084

Insight
631 East Big Beaver Road
Suite 211
Troy, MI 48083

Maplegrove Troy Center
1475 West Big Beaver Road
Suite 310
Troy, MI 48084

Perspectives of Troy PC
1000 West Maple Road
Suite 1E
Troy, MI 48084

UTICA

Bi-County Outpatient Counseling Center Substance Abuse Services
49310 Van Dyke Street
Utica, MI 48317

Catholic Social Services of Macomb County Substance Abuse Program
45100 Sterritt Street
Suite 300
Utica, MI 48317

Employee Assistance Associates, Inc.
52188 Van Dyke Street
Suite 200
Utica, MI 48087

Ganesh Clinic
7877 Stead Street
Utica, MI 48317

Macomb Family Services, Inc. II
45445 Mound Road
Utica, MI 48317

VANDALLIA

Cass County Drug and Alcohol Program
17321 M-60 East
Vandallia, MI 49095

WALKER

Northwest Counseling Center
3755 Remembrance Road NW
Walker, MI 49504

WALLED LAKE

First Step Mental Health and Substance Abuse Center
2346 South Commerce Street
Walled Lake, MI 48390

Oakland Family Services
2045 West Maple Road
Suite D 405
Walled Lake, MI 48088

WARREN

Alcohol Evaluation Services
29200 Hoover Road
Suite 104
Warren, MI 48093

Bi-County Outpatient Counseling Center Substance Abuse Services
26091 Sherwood Street
Suite 4A
Warren, MI 48091

Catholic Social Services of Macomb Substance Abuse Program
12434 East 12 Mile Road
Suite 201
Warren, MI 48093

Growth Associates
27440 Hoover Road
Suite B
Warren, MI 48093

Harper/Warren Chemical Dependency Program
4050 East 12 Mile Road
Warren, MI 48092

Kern Hospital Eastmore Center at Kern
21230 Dequindre Street
Warren, MI 48091

Lakewood Substance Abuse Center, Inc.
29600 Hoover Street
Warren, MI 48089

Medical Center Psychiatric Associates PC
28800 Ryan Road
Warren, MI 48092

Occupational Health Centers of America
31201 Chicago Road
Suite 202A
Warren, MI 48093

Salvation Army Harbor Light Center Macomb County Satellite
23700 Van Dyke Avenue
Warren, MI 48089

WATERFORD

Catholic Social Services of Oakland County/Waterford
5770 Highland Road
Waterford, MI 48327

Community Programs, Inc.
1435 North Oakland Boulevard
Waterford, MI 48327

Ronald K. Dudley
3950 Lotus Drive
Waterford, MI 48329

WATERSMEET

Lac Vieux Desert Substance Abuse Program
Choate Road
Watersmeet, MI 49969

WAYNE

Caknipe/Kovach Associates PC
36040 West Michigan Avenue
Wayne, MI 48184

Michigan Elite Teens, Inc.
35609 West Michigan Avenue
Wayne, MI 48184

Wayne County Regional Education Services Agency Substance Abuse Prevention Education (SAPE)
33500 Van Born Road
Wayne, MI 48184

WEST BLOOMFIELD

Affordable Counseling
5640 West Maple Road
Suite 205
West Bloomfield, MI 48322

Gerger Spivack and Associates
5793 Maple Road
Suite 155
West Bloomfield, MI 48322

Henry Ford Maplegrove Center Chemical Dependency Center
6773 West Maple Road
Maplegrove Center
West Bloomfield, MI 48322

Living Life Drug Free
2276 Shorehill Drive
West Bloomfield, MI 48323

New Start, Inc.
5839 West Maple Road
Suite 112
West Bloomfield, MI 48322

Palmer Drug Abuse Program Southeast Michigan, Inc./ Orchard Lake
5171 Commerce Road
West Bloomfield, MI 48324

WEST BRANCH

Au Sable Valley Comm. Mental Health Center Substance Abuse Program
511 Griffin Road
West Branch, MI 48661

WESTLAND

Evergreen Counseling Centers
35180 Nankin Boulevard
Suite 200
Westland, MI 48185

Oakwood United Hospitals Annapolis Hospital Westland Center
2345 Merriman Road
Westland, MI 48185

Occupational Health Centers of America
35360 Nankin Boulevard
Suite 801
Westland, MI 48185

Parent to Parent for Prevention
8137 August Street
Westland, MI 48185

Westland Counseling Center
8623 North Wayne Street
Suite 156
Westland, MI 48185

WILLIAMSBURG

Grand Traverse Band of Ottawa Chippewa Indians Substance Abuse Services
7741 M72 East
Williamsburg, MI 49690

WILSON

Hannahvil Indian Community Substance Abuse Project
N 14925 Hannahville B-1 Road
Wilson, MI 49896

WYANDOTTE

Wyandotte Health Center Substance Abuse Services
2411 Fort Street
Wyandotte, MI 48192

YPSILANTI

Beyer Hospital Chemical Dependency Services
135 South Prospect Street
Ypsilanti, MI 48198

Catholic Social Services of Washtenaw County
101 South Huron Street
Ypsilanti, MI 48197

Clear Day Women's Treatment Program
118 South Washington Street
Ypsilanti, MI 48197

Corner Health Center Substance Abuse Services/Prevention
47 North Huron Street
Ypsilanti, MI 48197

Dawn, Inc. Dawn Farm
6633 Stony Creek Road
Ypsilanti, MI 48197

Huron Services for Youth, Inc. Student-Parent Center
633 Harriet Street
Ypsilanti, MI 48197

Parents Together
448 South Huron Street
Ypsilanti, MI 48197

Share House, Inc.
411 Ballard Street
Ypsilanti, MI 48197

SOS Community Crisis Center
Prospect Place Family Shelter
11 North Prospect Street
Ypsilanti, MI 48198

114 North River Street
Ypsilanti, MI 48198

ZILWAUKEE

J/C Consultants
859 Waukee Lane
Zilwaukee, MI 48604

MINNESOTA

AH-GWAH-CHING

Lakeside Center Chemical Dependency Services
Ah-Gwah-Ching, MN 56430

ALBERT LEA

Fountain Lake Treatment Center
408 Fountain Street
Albert Lea, MN 56007

ALEXANDRIA

Douglas County Hospital Detox Services
111 17 Avenue East
Alexandria, MN 56308

Lakeview Chemical Dependency Unit of Douglas County Hospital
700 Cedar Street
Marian Building Suite 154
Alexandria, MN 56308

ANOKA

Anoka/Metro Regional Treatment Center
3300 4 Avenue North
Anoka, MN 55303

Minnesota Prevention Resource Center
2829 Verndale Avenue
Anoka, MN 55303

Riverplace Counseling Center
1806 South Ferry Street
Anoka, MN 55303

Transformation House
1410 South Ferry Street
Anoka, MN 55303

Transformation House II
2532 North Ferry Street
Anoka, MN 55303

ARDEN HILLS

Group Health Mental Health Center Outpatient Chemical Dependency Program
4105 North Lexington Avenue
Arden Hills, MN 55126

AUSTIN

Agape Halfway House, Inc.
200 5 Street SW
Austin, MN 55912

BARNESVILLE

Red River Serenity Manor, Inc.
123 2 Street NE
Barnesville, MN 56514

BEMIDJI

Counseling Associates of Bemidji
1615 5 Street NW
Bemidji, MN 56601

Lakes Region Chemical Dependency
1411 Bemidji Avenue
Bemidji, MN 56601

Upper MS Mental Health Center Program for Addictions Recovery
722 15 Street
Bemidji, MN 56601

BLAINE

Anthony Louis Center/Blaine
1000 Paul Parkway
Blaine, MN 55434

Ramsey/New Connection Programs Blaine Outpatient Treatment
10267 University Avenue NE
Blaine, MN 55434

BRAHAM

Danewood Board and Lodging
Route 2
Braham, MN 55006

BRAINERD

Adapt of Minnesota
510 Bluff Avenue
Brainerd, MN 56401

Brainerd Regional Human Services Center Aurora Chemical Dependency Program
1777 Highway 18 East
Brainerd, MN 56401

Break Free Adolescent Outpatient
521 Charles Street
Brainerd, MN 56401

Northern Pines Mental Health Center Chemical Dependency Treatment Program
520 NW 5 Street
Brainerd, MN 56401

Saint Joseph's Medical Center Focus Unit
523 North 3 Street
Brainerd, MN 56401

BRECKENRIDGE

Saint Francis Medical Center Hope Unit
401 Oak Street
Breckenridge, MN 56520

BROOKLYN CENTER

Behavioral Health Services, Inc.
3300 County Road 10
Suite 306
Brooklyn Center, MN 55429

BUFFALO

Health One Buffalo Hospital Chemical Dependency Unit
111 Highway 25 North
Plaza Professional Building Suite 203
Buffalo, MN 55313

BURNSVILLE

Fairview Ridges Hospital
Adult Inpatient Chemical Dependency
201 East Nicollet Boulevard
Burnsville, MN 55337

Adult Outpatient Chemical Dependency Program
156 Cobblestone Lane
Burnsville, MN 55337

Illusions
1000 East 146 Street
Suite 221
Burnsville, MN 55337

River Ridge Nonresidential Treatment Center
1515 East Highway 13
Burnsville, MN 55337

Riverside Medical Center/ Fairview Deaconess Adolescent Outpatient Chemical Dependency Program
14569 Grand Avenue South
Burnsville, MN 55337

CARLTON

Lake Venoah Community Relapse Treatment Program
310 East Chub Lake Road
Carlton, MN 55718

CASS LAKE

Leech Lake Outpatient Drug Abuse Program
Route 3
Cass Lake, MN 56633

CENTER CITY

Hazelden Foundation
15245 Pleasant Valley Road
Center City, MN 55012

CHASKA

Stafford CD Treatment Center, Inc.
302 Walnut Street
Chaska, MN 55318

CLOQUET

Fond Du Lac Chemical Dependency Program
927 Trettel Lane
Cloquet, MN 55720

Phoenix Treatment Center Center for Alcohol and Drug Treatment
40 11 Street
Cloquet, MN 55720

COON RAPIDS

Mercy Medical Center Chemical Dependency Services
4050 Coon Rapids Boulevard
Coon Rapids, MN 55433

COTTAGE GROVE

Hawthorn Institute, Inc.
7064 West Point Douglas Road
Suite 102
Cottage Grove, MN 55016

CROOKSTON

Glenmore Recovery Center
323 South Minnesota Street
Crookston, MN 56716

CRYSTAL

Fairview Deaconess Adolescent CD Program Crystal Outpatient
2960 Winnetka Avenue North
Suite 101
Crystal, MN 55427

Fairview Southdale/Crystal Adult Chemical Dependency Treatment Programs
2960 Winnetka Avenue North
Suite 206
Crystal, MN 55427

DELANO

Professional Counseling Center of Delano
500 Babcock Boulevard
Delano, MN 55328

DETROIT LAKES

Lakes Counseling Center
928 8 Street SE
Detroit Lakes, MN 56501

DULUTH

Center for Alcohol and Drug Treatment
Mental Illness/Chemical Dependency
1402 East 2 Street
Duluth, MN 55805

Outpatient Treatment Program
314 West Superior Street
400 Torrey Building
Duluth, MN 55802

Duluth Detoxification Center
1001 East First Street
Duluth, MN 55805

Federal Prison Camp Substance Abuse Services
Duluth, MN 55814

Howard Friese Memorial Halfway House
1520 East 2 Street
Duluth, MN 55812

Marty Mann Halfway House
714 North 11 Avenue East
Duluth, MN 55805

Messabi Chemical Abuse and Awareness Service
Port Rehab Center
Work Release Program
23 Mesaba Avenue
Duluth, MN 55806

Messabi Treatment/St. Louis County Jail
521 West 2 Street
Duluth, MN 55802

Thunderbird Halfway House
229 North 4 Avenue West
Duluth, MN 55806

Wren Halfway House
1731 West First Street
Duluth, MN 55806

EAST GRAND FORKS

Douglas Place, Inc.
322 5 Street NW
East Grand Forks, MN 56721

Glenmore Recovery Center Outpatient Clinic
1620 Central Avenue NE
East Grand Forks, MN 56721

EDEN PRAIRIE

Pride Institute
14400 Martin Drive
Eden Prairie, MN 55344

Ramsey/New Connection Programs Eden Prairie Outpatient Treatment
8110 Eden Road
Suite 205
Eden Prairie, MN 55344

EDINA

Fairview Southdale
Chemical Dependency Treatment Program
3101 West 69 Street
Edina, MN 55435

Inpatient Chemical Dependency Services
6401 France Avenue South
Edina, MN 55435

ELK RIVER

Gateway Center
9816 NW Highway 10
Elk River, MN 55330

ELY

Arrowhead Center, Inc.
118 South 4 Avenue East
Ely, MN 55731

Ely Community Resource, Inc.
103 East Chapman Street
Ely, MN 55731

FARIBAULT

Faribault Family Focus
201 South Lyndale Avenue
Suite F
Faribault, MN 55021

Southern Minnesota Chemical Dependency Services
127A NW 4 Street
Faribault, MN 55021

FARMINGTON

Journey Counseling Services
209 Oak Street
Suite 2
Farmington, MN 55024

FERGUS FALLS

Fergus Falls Regional Treatment Center Chemical Dependency Services
Fergus Falls, MN 56537

Saint Francis Medical Center New Awareness
125 West Lincoln Avenue
Fergus Falls, MN 56537

FOREST LAKE

District Memorial Hospital Family Services
246 11 Avenue SE
Forest Lake, MN 55025

FRIDLEY

Clinical Associates
7260 University Avenue NE
Suite 320
Fridley, MN 55432

Unity Medical Center Chemical Dependency Program
550 Osborne Road
Station 2-East
Fridley, MN 55432

GRAND MARAIS

Cook County Social Services North Shore Chemical Dependency Outpatient Program
Arrowhead Professional Building
Grand Marais, MN 55604

GRAND PORTAGE

Grand Portage Chemical Dependency Services
Grand Portage, MN 55605

GRAND RAPIDS

Hope House of Itasca County
604 South Pokegama Avenue
Grand Rapids, MN 55744

Northland Recovery Center Substance Abuse Services
1215 7 Avenue SE
Grand Rapids, MN 55744

GRANITE FALLS

Project Turnabout
660 18 Street
Granite Falls, MN 56241

HASTINGS

Dakota County Jail Treatment Program
1560 West Highway 55
Law Enforcement
Hastings, MN 55033

Dakota County Program Dakota County Receiving Center
1200 East 18 Street
Building 1
Hastings, MN 55033

Ramsey/Hastings Alcohol and Drug Abuse Program
1320 South Frontage Road
Hastings, MN 55033

HOPKINS

New Beginnings at Waverly/ Hopkins Outpatient Program
810 First Street South
Johnson Building Suite 260
Hopkins, MN 55343

West Suburban Counseling Clinic
13815 Ridgedale Drive
Hopkins, MN 55343

HUGO

Northern Lights Hope Center Lino Lakes Program
2209 Phelps Road
Hugo, MN 55038

HUTCHINSON

Hutchinson Community Hospital
Outpatient Chemical Dependency Program
Hutchinson Receiving Center
1095 Highway 15 South
Hutchinson, MN 55350

INTERNATIONAL FALLS

Koochiching Counseling Center
1804 3 Street
International Falls, MN 56649

Pineview Regional Recovery Center
1800 West 3 Street
International Falls, MN 56649

Rational Alternatives, Inc.
206 14 Street East
International Falls, MN 56649

JORDAN

Scott County Jail Chem Dependency Program
17706 Valley View Drive
Jordan, MN 55352

KARLSTAD

M/K Recovery Center
400 Northland Square
Karlstad, MN 56732

LA CRESCENT

Counseling Clinic/La Crescent
33 South Walnut Street
La Crescent, MN 55947

LE CENTER

Choices
203 West Lanesburg Street
Le Center, MN 56057

LITCHFIELD

J L Counseling Service
230 North Sibley Avenue
Litchfield, MN 55355

Recovery Plus/Litchfield
201 South Selby Street
Library Square Building
Litchfield, MN 55355

LITTLE FALLS

Effective Living Center, Inc.
111 5 Street SE
Little Falls, MN 56345

Saint Gabriel's Hospital Chemical Dependency Unit
815 SE 2 Street
Little Falls, MN 56345

LONG PRAIRIE

Freedom Plus Recovery Center
127 Central Avenue
Long Prairie, MN 56347

LORETTO

Vinland National Center
Lake Independence
Loretto, MN 55357

LUVERNE

Southwestern Mental Health Center Alcoholism and Drug Abuse Program
2 Round Wind Road
Luverne, MN 56156

MAHNOMEN

Center for Human Environment
Mahnomen, MN 56557

Mahnomen County Human Services Outpatient Chemical Dependency Treatment
Mahnomen County Courthouse
Mahnomen, MN 56557

MANKATO

Addictions Recovery Technologies
423 South Broad Street
Mankato, MN 56001

House of Hope
119 Fulton Street
Mankato, MN 56001

Immanuel/Saint Joseph's Hospital Family Recovery Program
325 Garden Boulevard
5th Floor
Mankato, MN 56001

MARSHALL

Project Turnabout/Marshall Halfway House and Outpatient Center
1220 Birch Street
Marshall, MN 56258

MILACA

Fairview Milaca Hospital Outpatient Chemical Dependency Services
150 10 Street NW
Milaca, MN 56353

MINNEAPOLIS

Affirmation Place, Ltd.
127 West Grant Street
Minneapolis, MN 55403

Bridgeway Treatment Center
22 27 Avenue SE
Minneapolis, MN 55414

Chrysalis/A Center for Women
2650 Nicollet Avenue South
Minneapolis, MN 55408

Create, Inc.
1911 Pleasant Avenue
Minneapolis, MN 55403

Eden Programs, Inc.
Eden Day Progrma
Eden House
1025 Portland Avenue South
Minneapolis, MN 55404

Eden Youth Program
1035 East Franklin Avenue South
Minneapolis, MN 55404

Fairview Deaconess Center
Adolescent Chemical Dependency Program
Hearing Impaired Chemical Dependency Program
2450 Riverside Avenue
Minneapolis, MN 55454

Hazelden Women's Outpatient Program
1400 Park Avenue South
Minneapolis, MN 55404

Health Recovery Center, Inc.
3255 Hennepin Avenue South
Minneapolis, MN 55408

Hennepin County Methadone Program
Hennepin County Outpatient Programs
Hennepin County Shelter
1800 Chicago Avenue South
Minneapolis, MN 55404

Kelly Institute
1111 3 Avenue South
Suite 30
Minneapolis, MN 55404

Lifestyle Counseling of Richfield/Bloomington
9801 Dupont Avenue South
Suite 110
Minneapolis, MN 55431

Minnesota Indian Women's Resource Center
2300 15 Avenue South
Minneapolis, MN 55404

Mustangs Chemical Dependency Outpatient Treatment Center
904 West Broadway
Minneapolis, MN 55411

Nuway House II
2518 First Avenue South
Minneapolis, MN 55404

Park Avenue Center
2525 Park Avenue
Minneapolis, MN 55404

Prodigal House
5103 Minnehaha Avenue South
Minnesota Veterans Home Bldg 1
Minneapolis, MN 55417

Progress Valley I
3033 Garfield Avenue South
Minneapolis, MN 55408

Resource, Inc. Recovery Resource Center
1900 Chicago Avenue
Minneapolis, MN 55404

Saint Mary's Adult Chemical Dependency Services at Fairview Riverside Medical Center
2450 Riverside Avenue
Minneapolis, MN 55454

Salvation Army Harbor Light Beacon Program
1010 Currie Avenue North
Minneapolis, MN 55405

The Wayside House, Inc.
3705 Park Center Boulevard
Minneapolis, MN 55416

Triumph Services
3735 Lakeland Avenue North
Suite 200
Minneapolis, MN 55422

Turning Point, Inc. Demand, Inc.
1015 Olson Memorial Highway
Minneapolis, MN 55405

Turning Point Primary Care
1105 16 Avenue North
Minneapolis, MN 55411

Veterans' Affairs Medical Center Addictive Disorders Section
1 Veterans Drive
Highway 55 and County 62
Minneapolis, MN 55417

Vinland Outpatient Chemical Dependency Program
1313 SE 5 Street
Suite 105
Minneapolis, MN 55414

West Suburban Counseling Clinic
111 3 Avenue South
Suite 110
Minneapolis, MN 55401

3 RS Counseling Center
1304 University Avenue NE
Minneapolis, MN 55413

MINNEAPOLIS (SHOREWOOD)

West Metro Recovery Services
6140 Lake Linden Drive
Minneapolis (Shorewood), MN 55331

MINNETONKA

Family Therapy and Recovery Center
10801 Wayzata Boulevard
Suite 320
Minnetonka, MN 55305

Omegon, Inc.
2000 Hopkins Crossroads
Minnetonka, MN 55343

On Belay House
1502 Archwood Road
Minnetonka, MN 55343

River Ridge Nonresidential Treatment Center
15612 West Highway 7
Highwood Office Center Suite 150
Minnetonka, MN 55345

MONTICELLO

Counseling Center Chemical Dependency Unit
407 Washington Street
Monticello, MN 55362

MOORHEAD

Clay County Outpatient Treatment Program
123 1/2 21 Street South
Moorhead, MN 56560

Clay County Receiving Center
715 North 11 Street
Moorhead, MN 56560

Wellness Center of Fargo/ Moorhead
403 Center Avenue
Suite 409
Moorhead, MN 56560

MOOSE LAKE

Moose Lake Regional Treatment Center Chemical Dependency Services
1000 Lakeshore Drive
Moose Lake, MN 55767

MORRIS

Stevens Community Memorial Hospital New Beginning Center
400 East First Street
Morris, MN 56267

MORTON

Lower Sioux Alcoholism Program
Route 1
Morton, MN 56270

MOUND

Lifestyle Counseling and Training, Inc. Lifestyle Counseling of Mound
5600 Lynwood Boulevard
Suite C-8
Mound, MN 55364

NETT LAKE

Bois Forte Reservation/Health Division Chemical Dependency Programs
Saint John Road
Health Services Building
Nett Lake, MN 55772

NEVIS

Pine Manor, Inc. Chemical Dependency Services
Route 2
Nevis, MN 56467

NEW PRAGUE

Queen of Peace Alcohol and Substance Abuse Program
301 2 Street NE
New Prague, MN 56071

NEW ULM

Brown County Detox and Evaluation Center
510 North Front Street
New Ulm, MN 56073

Sioux Valley Hospital Substance Abuse Services
1324 North 5 Street
New Ulm, MN 56073

ODIN

Freedom Bound/Odin
203 2 Street
Odin, MN 56160

OWATONNA

Owatonna Family Focus
110 North Cedar Street
Owatonna, MN 55060

West Hills Lodge, Inc.
545 Florence Avenue
Owatonna, MN 55060

PINE CITY

Meadow Creek
Route 4
Pine City, MN 55063

Pine Shores Chemical Dependency Services
Route 2
Pine City, MN 55063

PIPESTONE

Southwest Mental Health Center Substance Abuse Services
210 South Hiawatha Street
Pipestone, MN 56164

PLYMOUTH

Anthony Louis Center/Plymouth
115 Forestview Lane North
Plymouth, MN 55441

Ark Counseling of Plymouth
3025 Harbor Lane
Suite 125
Plymouth, MN 55447

Create/Telesis
1345 Shenandoah Lane
Hennepin County Workhouse
Plymouth, MN 55447

Hazelden Center for Youth and Family
11505 36 Avenue North
Plymouth, MN 55441

Mission Care Center
3409 East Medicine Lake Boulevard
Plymouth, MN 55441

PRIOR LAKE

Lifestyle Counseling Services
14033 Commerce Avenue NE
Suite 140
Prior Lake, MN 55372

RED LAKE

Indian and Free Drug Abuse Program
Red Lake, MN 56671

RED WING

Dawn Counseling Services
416 Bush Street
Red Wing, MN 55066

Saint John's Regional Health Center Alcohol and Drug Abuse Services
1407 West 4 Street
Red Wing, MN 55066

REDBY

Red Lake Halfway House Oosh Kii Mii Kah Nah
Redby, MN 56670

REDWOOD FALLS

Project Turnabout/Redwood Falls Outpatient Program
600 South Gould Drive
Redwood Falls, MN 56283

RICHFIELD

Progress Valley II
308 East 78 Street
Richfield, MN 55423

ROCHESTER

Aiimsonion Clinic Chemical Dependency Program
300 3 Avenue SE
Ironwood Square Suite 206
Rochester, MN 55904

Cronin Homes, Inc.
622 North Broadway
Rochester, MN 55906

Federal Medical Center Chemical Abuse Program
Rochester, MN 55903

Fountain Center
4104 18 Street NW
Cedarwood Mall
Rochester, MN 55901

Guest House
4800 48 Street NE
Rochester, MN 55903

Mayo Clinic Adolescent Substance Abuse Service
Baldwin Desk 3A
Rochester, MN 55905

Mayo Clinic/Rochester Methodist Hospital Alcoholism and Drug Dependence Unit
201 West Center Street
Rochester, MN 55902

New Beginnings at Rochester
333 16 Avenue NE
Rochester, MN 55901

Pathway House
103 6 Avenue SW
Rochester, MN 55902

Saint Francis Community Programs, Inc. Carillion House
1623 4 Street NW
Rochester, MN 55901

The Gables
604 5 Street SW
Rochester, MN 55902

Zumbro Valley Mental Health Center, Inc.
Chemical Health Services
2116 Campus Drive SE
Rochester, MN 55904

Right to Recovery Program
423 3 Avenue SE
Rochester, MN 55904

Crisis Receiving Unit
2116 SE Campus Drive
Suite 105
Rochester, MN 55904

ROSEAU

Glenmore Recovery Center Outpatient Clinic
101 South Main Street
Roseau, MN 56751

ROSEVILLE

First Step Professional Services, Inc.
2085 Rice Street
Roseville, MN 55113

SAGINAW

Northeast Regional Correction Center Phase I Outpatient Treatment Program
6102 Abrahamson Road
Saginaw, MN 55779

SAINT CLOUD

Focus 12 Halfway House
3220 North 8 Street
Saint Cloud, MN 56303

Journey Home
210 5 Avenue NE
Saint Cloud, MN 56304

Passage Home
1003 South 8 Avenue
Saint Cloud, MN 56301

Saint Cloud Hospital Recovery Plus
1406 North 6 Avenue
Saint Cloud, MN 56301

Veterans' Affairs Medical Center Alcohol/Drug Dependence Treatment Program
4801 North 8 Street
Unit 116C
Saint Cloud, MN 56303

SAINT LOUIS PARK

Changing Lifestyle Counseling of Saint Louis Park
7515 Wayzata Boulevard
Suite 202
Saint Louis Park, MN 55426

SAINT PAUL

Children Are People Support Groups, Inc.
2489 Rice Street
Suite 275
Saint Paul, MN 55113

Conceptual Counseling, Inc.
245 East 6 Street
Suite 435
Saint Paul, MN 55101

Genesis House
1060 Greenbriar Street
Saint Paul, MN 55106

Hazelden/Fellowship Club
680 Stewart Avenue
Saint Paul, MN 55102

Hispanos en Minnesota
155 South Wabasha Street
Suite 128
Saint Paul, MN 55107

Minnesota Prevention Resource Center
417 University Avenue
Saint Paul, MN 55103

People, Inc.
Dayton House
565 Dayton Avenue
Saint Paul, MN 55102

Ramsey New Connection Programs Primary Outpatient Chemical Dependency Services
570 Asbury Street
Suite 301
Saint Paul, MN 55104

Saint Joseph Hospital/Health East Chemical Dependency Program
69 West Exchange Street
Saint Paul, MN 55102

Saint Paul Ramsey Medical Center Alcohol and Drug Abuse Program
445 Etna Street
Suite 55
Saint Paul, MN 55106

Sherburne House
550 Galtier Street
Saint Paul, MN 55103

Triumph Services Straight Friends of America
2147 University Avenue
Suite 202
Saint Paul, MN 55114

Twin Town Treatment Center
1706 University Avenue
Saint Paul, MN 55104

United Hospital Chemical Dependency Services
333 North Smith Avenue
Suite 4900
Saint Paul, MN 55102

SAINT PAUL (MAPLEWOOD)

Fairview Deaconess/Maplewood Adolescent Chemical Dependency Program
1707 Cope Avenue
Saint Paul (Maplewood), MN 55109

SAINT PAUL (NEW BRIGHTON)

Amethyst Counseling Services Outpatient Chemical Dependency Treatment Services
1405 Silver Lake Road
Saint Paul (New Brighton), MN 55112

SAINT PAUL (OAKDALE)

We Care Counseling Center
6060 50 Street North
Suite 1
Saint Paul (Oakdale), MN 55128

SAINT PETER

Saint Peter Regional Treatment Center Johnson Chemical Dependency Center
100 Freeman Drive
Johnson Hall
Saint Peter, MN 56082

SANDSTONE

Federal Correctional Institution Drug Abuse Program
Sandstone, MN 55072

SAUK CENTRE

Saint Michael's Chemical Dependency Unit
425 North Elm Street
Sauk Centre, MN 56378

SAWYER

Mash Ka Wisen Treatment Center
Sawyer, MN 55780

SHAKOPEE

Stafford CD Treatment Center, Inc.
1100 East 4 Avenue
Suite 60
Shakopee, MN 55379

SOUTH SAINT PAUL

Healtheast Adolescent Behavioral Health Services
724 19 Avenue North
South Saint Paul, MN 55075

STAPLES

Greater Staples Hospital Recovery Plus
401 Prairie Avenue East
Staples, MN 56479

STILLWATER

Health East/Stillwater
6381 Osgood Avenue
Stillwater, MN 55082

THIEF RIVER FALLS

Glenmore Recovery Center Outpatient Clinic
City Hall
Thief River Falls, MN 56701

Trident Services
Extended Care
621 North Labree Avenue
Thief River Falls, MN 56701

Halfway House
504 North Knight Avenue
Thief River Falls, MN 56701

Inpatient/Outpatient/Detox
1902 East Greenwood Street
Thief River Falls, MN 56701

TWO HARBORS

Lake View Memorial Hospital Outpatient Chemical Dependency Unit
325 11 Avenue
Two Harbors, MN 55616

VICTORIA

Cornerstone Recovery Center
1600 Arboretum Boulevard
Victoria, MN 55386

VIRGINIA

Arrowhead Center, Inc.
505 12 Avenue West
Virginia, MN 55792

Range Mental Health Center, Inc. Detoxification Service
901 9 Avenue
Virginia, MN 55792

Twelfth Step House, Inc.
512 2 Street North
Virginia, MN 55792

WACONIA

Counseling Center of Waconia
24 South Olive Street
Waconia, MN 55387

WADENA

Bell Hill Recovery Center
Wadena, MN 56482

Neighborhood Counseling Center
209 South Jefferson Street
Wadena, MN 56482

WARROAD

Glenmore Recovery Center
611 East Lake Street
Warroad, MN 56763

WASECA

Waseca Family Focus
203 South State Street
Waseca, MN 56093

WAVERLY

New Beginnings at Waverly
109 North Shore Drive
Waverly, MN 55390

WAYZATA

Way 12 Halfway House
645 East Wayzata Boulevard
Wayzata, MN 55391

WELCH

Prairie Island Chemical Dependency Program
5750 Sturgeon Lake Road
Welch, MN 55089

WEST SAINT PAUL

Fountain Center/West Saint Paul
1276 South Robert Street
West Saint Paul, MN 55118

WILLMAR

Bradley Center
1550 Highway 71 NE
Willmar Regional Treatment Center
Willmar, MN 56201

Cardinal Recovery Center/ Willmar Regional Treatment Center
316 Becker Avenue SW
Cardinal Square Suite 323
Willmar, MN 56201

West Central Community Services Center, Inc.
1125 6 Street SE
Willmar, MN 56201

WINDOM

Southwest Mental Health Center Substance Abuse Services
Windom Industrial Park
Windom, MN 56101

WINNEBAGO

Adolescent Treatment Center of Winnebago
550 Cleveland Avenue West
Winnebago, MN 56098

WINONA

Community Memorial Hospital First Step Center
825 Mankato Avenue
Suite 214
Winona, MN 55987

Hiawatha Valley Mental Health Center Outpatient Chemical Dependency Program
111 Riverfront
Winona, MN 55987

Saint Francis Community Programs, Inc.
Aamethyst House
428 West Broadway
Winona, MN 55987

Winona Counseling Clinic Chemical Dependency Services
256 East Sarnia Street
Winona, MN 55987

WINSTED

Health One/Saint Mary's
551 4 Street North
Winsted, MN 55395

WOODBURY

Fairview Deaconess/Woodbury
1665 Woodbury Drive
Woodbury, MN 55125

WOODSTOCK

New Life Treatment Center County Road
Woodstock, MN 56186

WORTHINGTON

Southwest Mental Health Center Challenges
1024 7 Avenue
Worthington, MN 56187

ZUMBROTA

Zumbrota Health Services, Inc.
The Center
383 West 5 Street
Zumbrota, MN 55992

MISSISSIPPI

BILOXI

Gulf Oaks Hospital Chemical Dependency Unit
180-C De Buys Road
Biloxi, MS 39535

Veterans' Affairs Medical Center Substance Abuse Treatment Program/Inpatient
400 Veterans Boulevard
Biloxi, MS 39531

BOONEVILLE

Timber Hills Region IV Alcohol and Drug Abuse Services
Highway 30 By Pass
Booneville, MS 38829

BROOKHAVEN

Southwest Mississippi MH/MR Complex
Opportunity House 3/4 House
310 West Cherokee Street
Brookhaven, MS 39601

Newhaven Recovery Center
Route 1
Brookhaven, MS 39601

CHARLESTON

Region I MH/MR Center Tallahatchie County Satellite Center
North Market Street
Charleston, MS 38921

CLARKSDALE

Region I Mental Health Center
Alcohol and Drug Services
1742 Cheryl Street
Health Services Building
Clarksdale, MS 38614

Fairland House
200 Fairland Place
Clarksdale, MS 38614

COLUMBUS

Community Counseling Services
Lowndes County Office
1001 Main Street
Columbus, MS 39701

Residential Alcohol/Drug Program
108 10 Street North
Columbus, MS 39701

Recovery House, Inc.
770 Golding Road
Columbus, MS 39704

CORINTH

Timber Hills Region IV
Alcohol and Drug Abuse Services
U.S. Highway 45 South
Corinth, MS 38834

Haven House/Residential Services
1402 East 4 Street
Corinth, MS 38834

GREENVILLE

Delta Community Mental Health Services Substance Abuse Services
1654 East Union Street
Greenville, MS 38701

GREENWOOD

Denton House/Female
Highway 82 East
Greenwood, MS 38930

Denton House/Male
Highway 82 East
Greenwood, MS 38930

GULFPORT

CPC Sand Hill Hospital
11150 Highway 49 North
Gulfport, MS 39503

Gulf Coast Mental Health Center Substance Abuse Services
4514 Old Pass Road
Gulfport, MS 39501

Live Oaks Treatment Center (LOTC)
15120 County Barn Road
Gulfport, MS 39503

HATTIESBURG

Pine Belt Mental Healthcare Resources
Programs for Chemical Dependency/IOP
Programs for Chemical Dependency/Outpatient
820 South 28 Avenue
Hattiesburg, MS 39401

Pine Grove Recovery Center
2255 Broadway Drive
Hattiesburg, MS 39401

IUKA

Timber Hills Region IV MH/MR Comm., Inc. Alcohol and Drug Abuse Services
Maria Lane
Iuka, MS 38852

JACKSON

Alcohol Services Center, Inc. Drug Treatment Unit
950 North West Street
Jackson, MS 39202

Developing Resources for Education in America, Inc. (DREAM)
817 East River Place
Jackson, MS 39202

Harbor Houses of Jackson, Inc.
Men's Division Alcoholism Treatment
1019 West Capitol Street
Jackson, MS 39203

Women's Division
3588 Flowood Drive
Jackson, MS 39208

Jackson Recovery Center
5354 I-55 South Frontage Road
Jackson, MS 39212

Mississippi Baptist Medical Center Chemical Dependency Center
1225 North State Street
Jackson, MS 39201

Mississippi Children's Home Society and Family Services Association/The Ark
1801 North West Street
Jackson, MS 39205

National Council on Alcoholism of the Central Mississippi Area, Inc.
333 North Mart Plaza
Jackson, MS 39206

New Hope Foundation, Inc.
Day Treatment Service Program
Outreach/Aftercare
516 North Mill Street
Jackson, MS 39202

Three Quarterway House/Female
2326 Saint Charles Street
Jackson, MS 39209

Three Quarterway House/Male
428 Hooker Street
Jackson, MS 39204

New Roads Alcohol/Drug Services Center
200 Park Circle Drive
Suite 6
Jackson, MS 39208

Veterans' Affairs Medical Center Chemical Dependence Treatment Program
1500 East Woodrow Wilson
Unit 116B1
Jackson, MS 39216

LAUREL

Pine Belt Recovery House/ Females
712 Royal Street
Laurel, MS 39440

Pine Belt Recovery House/Males
527 South Magnolia Street
Laurel, MS 39440

Serenity House
711 Royal Street
Laurel, MS 39440

Serenity House (Female)
712 Royal Street
Laurel, MS 39440

MARKS

Region I MH/MR Center Quitman County Satellite Center
Locust Street
Marks, MS 38646

MCCOMB

Southwest Mississippi MH/MR Complex Pike County Center
1701 White Street
McComb, MS 39648

MENDENHALL

New Roads Residential Treatment Center
1060 Smith Road
Mendenhall, MS 39114

MERIDIAN

East Mississippi State Hospital Alcohol and Drug Services
4555 Highland Park Drive
Meridian, MS 39302

Laurel Wood Center
5000 Highway 39 North
Meridian, MS 39305

Weems Community Mental Health Center
Weems Lifecare Medical Dependency Center
2806 7 Street
Meridian, MS 39301

Alcohol and Drug Program/ Lauderdale County
1415 Junior College Road
Meridian, MS 39304

MOOREVILLE

Region III Mental Health Center Drug and Alcohol Program
Route 1
Mooreville, MS 38857

NATCHEZ

Humana Hospital/Natchez Chemical Dependency Unit
129 Jefferson Davis Boulevard
Natchez, MS 39120

OXFORD

Communicare Alcohol and Drug Program Haven House
1908 Highway 7 Bypass South
Oxford, MS 38655

PARCHMAN

Mississippi State Penitentiary Department of Alcohol and Drug Rehab
Parchman, MS 38738

PASCAGOULA

McClamroch Center
2602 West Criswell Avenue
Pascagoula, MS 39567

Singing River Mental Health Center Alcohol and Drug Program
4507 McArthur Street
Pascagoula, MS 39567

Stevens Center
4905 Telephone Road
Pascagoula, MS 39567

Teen Challenge of Mississippi
3219 Nathan Hale Street
Pascagoula, MS 39581

3/4 Women Program
Stevens Center
4905 Telephone Road
Pascagoula, MS 39567

PHILADELPHIA

Choctaw Community Mental Health Center Mississippi Band of Choctaw Indians
Route 7
Choctaw Health Center
Philadelphia, MS 39350

POCAHONTAS

Friends of Alcoholics
F O A Road
Pocahontas, MS 39072

RIPLEY

Timber Hills Region IV Alcohol and Drug Abuse Services
Industrial Park
Ripley, MS 38663

TUNICA

Region I MH/MR Center Tunica County Satellite Center
North Edwards Street
Tunica, MS 38676

TUPELO

North Mississippi Medical Center Alcohol and Drug Recovery Services
830 South Gloster Street
Tupelo, MS 38801

VANCLEAVE

Home of Grace
14200 Jericho Road
Vancleave, MS 39564

VICKSBURG

Belmont House
809 Belmont Street
Vicksburg, MS 39180

Marian Hill Chemical Dependency Center
100 McAuley Drive
Vicksburg, MS 39180

Warren/Yazoo Mental Health Service Vicksburg Satellite Center
1315 Adams Street
Vicksburg, MS 39180

WHITFIELD

Mississippi State Hospital Chemical Dependency Unit
Building 84
Whitfield, MS 39193

YAZOO CITY

Warren/Yazoo Mental Health Service Yazoo City Satellite Center
129 South Main Street
Yazoo City, MS 39194

MISSOURI

BOONVILLE

Boonville Valley Hope
1415 Ashley Road
Boonville, MO 65233

Family Counseling Center of Missouri, Inc. Outpatient Clinic
214 1/2 Main Street
Boonville, MO 65233

BRAGGADOCIO

Alpha for Adolescents
Highway J West
Braggadocio, MO 63826

BRANSON

Sigma House, Inc. Branson Branch
HCR 7
Branson, MO 65616

CAPE GIRARDEAU

Family Counseling Center, Inc. Women's CSTAR
20 South Sprig Street
Suite 2
Cape Girardeau, MO 63701

Saint Francis Center for Recovery Alcohol/Chemical Dependency Treatment Program
211 Saint Francis Drive
Cape Girardeau, MO 63701

CLINTON

Community Counseling Consultants
Business Highway 13 South
Clinton, MO 64735

COLUMBIA

Comprehensive Human Service
409 Vandiver Drive
Building 7 Suite 101
Columbia, MO 65202

Family Counseling Center of Missouri, Inc.
Administration/Prevention Resource Center
Alcohol/Drug Treatment Services
117 North Garth Street
Columbia, MO 65203

CSTAR McCambridge Center
201 North Garth Street
Columbia, MO 65203

EXCELSIOR SPRINGS

Kansas City Community Center (KCCC) Northland Community Center
106 Elizabeth Street
Excelsior Springs, MO 64024

FARMINGTON

Mineral Area Regional Medical Center Mineral Area Program
1212 Weber Road
Farmington, MO 63640

Southeastern Missouri Treatment Center
Administrative Unit
Aquinas Center
CSTAR
Highway 32 East
Farmington, MO 63640

FESTUS

Community Treatment, Inc. (COMTREA)
227 Main Street
Festus, MO 63028

FLORISSANT

Christian Hospital Northwest
1225 Graham Road
Florissant, MO 63031

FULTON

Fulton State Hospital
Alcohol and Drug Treatment Unit
Fulton, MO 65251

Serve, Inc.
Administrative Unit
302 Market Street
Fulton, MO 65251

Recovery Home
1411 Airport Road
Route 4
Fulton, MO 65251

HANNIBAL

Hannibal Council on Alcohol/ Drug Abuse, Inc.
Administrative Unit
Men's Primary Treatment Program
Women's Comprehensive Care Center
146 Communications Drive
Hannibal, MO 63401

HAYTI

Family Counseling Center, Inc. Stapleton Center
Highway J North
Hayti, MO 63851

INDEPENDENCE

Comprehensive Mental Health Services
Administrative Unit
CSTAR
Substance Abuse Services
10901 Winner Road
Independence, MO 64052

Independence Regional Health Center Step Up Unit
1509 West Truman Road
Independence, MO 64050

JEFFERSON CITY

Family Counseling Center of Missouri, Inc. Capitol Unit
502 East McCarty Street
Jefferson City, MO 65101

JOPLIN

Family Self Help Center, Inc. DBA Lafayette House/CSTAR
1809 Connor Avenue
Joplin, MO 64804

Scott Greening Center for Youth Dependency, Inc.
1315 East 20 Street
Joplin, MO 64804

KANSAS CITY

Ad Hoc Group Against Crime
2360 East Linwood Boulevard
Kansas City, MO 64109

Baptist Medical Center Adult Chemical Dependency Unit
6601 Rockhill Road
8th Floor
Kansas City, MO 64131

Catholic Charities Drug and Alcohol Aftercare Program
1112 Broadway
Kansas City, MO 64105

Community Addictions Programs
616 East 63 Street
Kansas City, MO 64110

De La Salle Education Center CSTAR
3740 Forest Avenue
Kansas City, MO 64109

DRD Kansas City Medical Clinic
723 East 18 Street
Kansas City, MO 64108

Ewing Marion Krauffman Foundation Project Star
4900 Oak Street
Kansas City, MO 64112

Greater Kansas City Mental Health Foundation Administrative Unit
2055 Holmes Street
Kansas City, MO 64108

Greater Kansas City MH Foundation Outpatient Component (NARA/DPSP)
2055 Holmes Street
Kansas City, MO 64108

Kansas City Community Center (KCCC)
Administrative Unit
Detox Unit
1514 Campbell Street
Kansas City, MO 64108

South Unit
2751 Charlotte Street
Kansas City, MO 64109

National Council on Alcoholism and Drug Dependence/Kansas City Area
601 East 63 Street
Suite 511
Kansas City, MO 64110

Renaissance West, Inc.
Administrative Unit
CSTAR
406 West 34 Street
Kansas City, MO 64130

Central
Women's Place
5840 Swope Parkway
Kansas City, MO 64130

North Clinic
5756 North Broadway
Kansas City, MO 64118

Residential Unit
3220 East 23 Street
Kansas City, MO 64127

Research Mental Health Services North Star Recovery CSTAR
2801 Wyandotte Street
6th Floor
Kansas City, MO 64108

Rodgers South
2701 East 31 Street
Kansas City, MO 64128

Safety and Health Council of Western Missouri and Kansas
901 Charlotte Street
Kansas City, MO 64106

Speas Resource Center on Alcoholism and Substance Abuse
616 East 63 Street
Kansas City, MO 64110

Veterans' Affairs Medical Center Substance Abuse Treatment Unit
4801 Linwood Boulevard
Kansas City, MO 64128

Western Missouri Mental Health Center
Alcohol and Drug Abuse Detox
Outpatient Drug Program/Admin.
600 East 22 Street
Kansas City, MO 64108

Chemical Free Program
Paseo Comprehensive Rehab Clinic
2211 Charlotte Street
Kansas City, MO 64108

Wise Council House
3005 Benton Boulevard
Kansas City, MO 64128

KENNETT

Family Counseling Center, Inc.
Administrative Unit
Adolescent CSTAR
CSTAR
925 Highway VV
Kennett, MO 63857

KIRKSVILLE

Professional Counseling Services
403 North Elson Street
Kirksville, MO 63501

LEES SUMMIT

Research Mental Health Services CSTAR
1001 NE Independence Avenue
Lees Summit, MO 64063

LINN CREEK

Family Counseling Center of Missouri, Inc. Cedar Ridge Treatment Center
Route 1
Linn Creek, MO 65052

MARYVILLE

Family Guidance Center/ Maryville
114 East South Hills Drive
South Hills Medical Building
Maryville, MO 64468

MEXICO

Hannibal Council on Alcohol/ Drug Abuse, Inc.
Mexico Area Recovery Center
1130 South Elmwood Street
Mexico, MO 65265

NEVADA

Community Mental Health Consultants
815 South Ash Street
Nevada, MO 64772

POPLAR BLUFF

Diversified Treatment Services New Era/Westwood Center
Route 11
Poplar Bluff, MO 63901

Doctors Regional Medical Center Substance Abuse Services
621 Pine Boulevard
Poplar Bluff, MO 63901

Veterans' Affairs Medical Center Substance Abuse Treatment Unit
1500 North Westwood Boulevard
Poplar Bluff, MO 63901

REEDS SPRING

Teen Challenge of the Ozarks
Route 6
Reeds Spring, MO 65737

ROLLA

Phelps County Regional Medical Center Stress Center
1000 West 10 Street
Rolla, MO 65401

SAINT CHARLES

Alternative Health Management
2747 West Clay Street
Saint Charles, MO 63301

Bridgeway Counseling Services, Inc.
Administrative Unit
Bridgeway Center/CSTAR
CSTAR/Women
1601 Old South River Road
Saint Charles, MO 63303

SAINT JOSEPH

Family Guidance Center CMHC Alcohol and Drug Services
Admin. Unit
CSTAR
910 Edmond Street
Suite 100
Saint Joseph, MO 64501

Heartland Health System Addiction Recovery Services
801 Faraon Street
Heartland Hospital West/5 West
Saint Joseph, MO 64501

North Central Missouri Mental Health Center Sunrise Center
909 Felix Street
Saint Joseph, MO 64502

SAINT LOUIS

Alexian Brothers Hospital Pathways Program/CSTAR
3933 South Broadway
Saint Louis, MO 63118

Archway Communities, Inc.
5652 Pershing Avenue
Saint Louis, MO 63112

Black Alcohol/Drug Service Info. Center (BASIC)
CSTAR
Men's CSTAR
1501 Locust Street
Suite 1100
Saint Louis, MO 63103

Central Medical Center Hospital CSTAR Family Center
4411 North Newstead Street
Saint Louis, MO 63115

Cope and Twin Town of Saint Louis
777 South New Ballas Road
Suite 230 West
Saint Louis, MO 63141

Counseling Center of Clayton
225 South Meramec Avenue
Room 205
Saint Louis, MO 63105

Dart, Inc.
Administrative Unit
Medication Unit
Outpatient Unit
1307 Lindbergh Plaza Center
Saint Louis, MO 63132

East
1207 South Vandeventer Street
Saint Louis, MO 63110

Deaconess Health Systems Behavioral Institutes/ Deaconess West
530 Des Peres Road
Saint Louis, MO 63131

Harris House Foundation
8327 South Broadway
Saint Louis, MO 63111

Hyland Center
10020 Kennerly Road
Saint Louis, MO 63128

Malcolm Bliss Mental Health Center Substance Abuse Referral Services
5400 Arsenal Street
Saint Louis, MO 63139

NASCO Central Clinic
2305 Saint Louis Avenue
Saint Louis, MO 63106

National Council on Alcoholism and Drug Abuse/Saint Louis Area
8790 Manchester Road
Saint Louis, MO 63144

Provident Counseling Family Care Program
9109 Watson Street
Saint Louis, MO 63126

Queen of Peace CSTAR Center
325 North Newstead Street
Saint Louis, MO 63108

Salvation Army Harbor Light Center
3010 Washington Avenue
Saint Louis, MO 63103

SSM Recovery Centers at Saint Mary's
6420 Clayton Road
Saint Louis, MO 63117

West End Clinic
5917 Martin Luther King Drive
Saint Louis, MO 63112

SALEM

SE Missouri Community Treatment Center Salem Center
203 North Grand Street
Salem, MO 65560

SPRINGFIELD

Bridgeway Substance Abuse Program
2828 North National Avenue
Springfield, MO 65803

Burrell Community Mental Health Center
CSTAR
201 South Campbell Street
Springfield, MO 65806

Substance Abuse Program
1300 Bradford Parkway
Springfield, MO 65804

Checkmate, Inc.
5335 South Campbell Street
Suite A
Springfield, MO 65810

Cox Care Center
1423 North Jefferson Avenue
Springfield, MO 65802

Lakeland Regional Hospital Dual Diagnosis Track
440 South Market Street
Springfield, MO 65806

Lakes Country Rehabilitation
Administrative Unit
2626 West College Road
Springfield, MO 65802

427 South Grant Street
Springfield, MO 65802

Ozarks National Council on Alcoholism and Drug Dependence
205 Saint Louis Street
Suite 407
Springfield, MO 65806

Sigma House, Inc. Springfield Unit
800 South Park Avenue
Springfield, MO 65802

TRENTON

North Central Missouri Mental Health Center
Administrative Unit
Substance Abuse Program
1601 East 28 Street
Trenton, MO 64683

TROY

Bridgeway Counseling Services, Inc. Troy Recovery Center
1011 East Cherry Street
Troy, MO 63379

WARRENSBURG

West Central Missouri MHC Substance Abuse Program
520-C Burkarth Road
Warrensburg, MO 64093

WASHINGTON

Clayton Concepts, Inc. New Hope
1321 East 5 Street
Washington, MO 63090

WENTZVILLE

Bridgway Counseling Services, Inc. Bridgeway Longterm
300 North Linn Street
Wentzville, MO 63385

WEST PLAINS

South Central Missouri Rehab Center
1015 Lanton Road
West Plains, MO 65775

MONTANA

ANACONDA

Deer Lodge County Alcohol and Drug Services of Anaconda
100 West Park Street
Anaconda, MT 59711

BILLINGS

Rimrock Foundation
1231 North 29 Street
Billings, MT 59101

South Central Montana Mental Health Center Chemical Dependency Program
1245 North 29 Street
Billings, MT 59103

BOZEMAN

Alcohol/Drug Services of Gallatin County
502 South 19 Street
Suite 302
Bozeman, MT 59715

BROWNING

Blackfeet Chemical Dependency Program
Blackfoot Indian Reservation
Browning, MT 59417

BUTTE

Butte/Silver Bow Chemical Dependency Services
125 West Granite Street
Suite 102
Butte, MT 59701

DEER LODGE

Chemical Dependency and Family Counseling, Inc.
304 Milwaukee Avenue
Deer Lodge, MT 59722

Galen Alcoholism Services Center Alcoholism Treatment and Rehab Program
Montana State Hospital
Deer Lodge, MT 59722

Montana State Prison Chemical Dependency Program
500 Conley Lake Road
Deer Lodge, MT 59722

GLENDIVE

District II Alcohol and Drug Program
119 South Kendrick Street
Glendive, MT 59330

GREAT FALLS

Montana Deaconess Medical Center Chemical Dependency Unit
1101 26 Street South
Great Falls, MT 59405

Providence Center
401 3 Avenue North
Great Falls, MT 59401

Rocky Mountain Treatment Center
920 4 Avenue North
Great Falls, MT 59401

HAMILTON

Crossroads/Ravalli County
824 First Avenue South
Hamilton, MT 59840

HAVRE

Northern Montana Chemical Dependency Program
1410 First Avenue
Havre, MT 59501

HELENA

Boyd Andrew Chemical Dependency Care Center
Arcade Building Unit 1-E
Helena, MT 59601

KALISPELL

Flathead Valley Chemical Dependency Clinic, Inc.
1312 North Meridian Road
Kalispell, MT 59901

Glacier View Hospital Chemical Dependency Unit
200 Heritage Way
Kalispell, MT 59901

LEWISTOWN

Alcohol and Drug Services of Central Montana
300 First Avenue North
Centennial Plaza Suite 204
Lewistown, MT 59457

LIBBY

Recovery Northwest/Lincoln County Main Office
418 Main Avenue
Libby, MT 59923

LIVINGSTON

Southwest Chemical Dependency Program
414 East Callendar Street
Livingston, MT 59047

MARION

Wilderness Treatment Center
200 Hubbart Dam Road
Marion, MT 59925

MILES CITY

Chemical Dependency Services, Inc.
108 North Haynes Avenue
Miles City, MT 59301

Pine Hills School for Boys Chemical Dependency Program
Miles City, MT 59301

Veterans' Affairs Medical Center Substance Abuse Services
210 South Winchester Street
Miles City, MT 59301

MISSOULA

Missoula Indian Alcohol and Drug Services
2010 South Avenue West
Missoula, MT 59801

Saint Patrick Hospital Addiction Treatment Program
500 West Broadway
Missoula, MT 59802

Western Montana Regional Mental Health Turning Point
500 North Higgins Street
Suite 101
Missoula, MT 59802

POLSON

Lake County Chemical Dependency Program
12 5 Avenue East
Polson, MT 59860

POPLAR

Spotted Bull Treatment Center
603 1/2 Court Avenue
Poplar, MT 59255

RONAN

Confederated Salish/Kootenai Tribes Tribal Human Services
26 Round Butte Road/West
Ronan, MT 59864

SCOBEY

High Plains Chemical Dependency Services Daniels County Satellite
205 Smith Avenue
Scobey, MT 59263

SWAN LAKE

Swan River Forest Camp Chemical Dependency Program
Swan Lake, MT 59911

WILSALL

Wilderness Treatment Center II
RR 1
Wilsall, MT 59086

NEBRASKA

AINSWORTH

Sandhills Mental Health and Substance Abuse Services, Inc.
312 North Main Street
Ainsworth, NE 69210

ALLIANCE

Human Services, Inc.
419 West 25 Street
Alliance, NE 69301

Valley Hope Alcoholism Treatment Center
2101 Box Butte
Alliance, NE 69301

AUBURN

Blue Valley Mental Health Center Southeastern Nebraska Comm. Alcohol/Drug Abuse Program/ Beatrice
1719 Courthouse Avenue
Auburn, NE 68305

BEATRICE

Blue Valley Mental Health Center Southeastern Nebraska Comm. Alcohol/Drug Abuse Program/ Beatrice
1200 South 9 Street
Beatrice, NE 68310

BELLEVUE

Lutheran Family Services/ Bellevue
1318 Federal Square Drive
Bellevue, NE 68005

Renaissance Program
703 West 24 Avenue
Bellevue, NE 68005

BLAIR

Immanuel Family Counseling Center/Blair
753 North 21 Street
Blair, NE 68008

BROKEN BOW

North Central Alcoholism Outpatient Counseling Service/ Broken Bow
145 Memorial Drive
Broken Bow, NE 68822

CHADRON

Northwest Nebraska Alcohol and Drug Abuse Services
245 East 10 Street
Chadron, NE 69337

COLUMBUS

Mid-East Nebraska Mental Health Clinic, Inc.
3314 26 Street
Columbus, NE 68601

Transitional Living Center, Inc.
4807 29 Street
Columbus, NE 68601

CRETE

Pioneer Mental Health Center Substance Abuse Program
1240 Ivy Street
Crete, NE 68333

DAVID CITY

Pioneer Mental Health Center Substance Abuse Program
367 E Street
David City, NE 68632

FAIRBURY

Blue Valley Mental Health Center Southeastern Nebraska Alcohol/Drug Abuse Program/ Beatrice
521 E Street
Fairbury, NE 68352

FALLS CITY

Blue Valley Mental Health Center Southeastern Nebraska Alcohol/Drug Abuse Program/ Beatrice
116 West 19 Street
Falls City, NE 68355

FREMONT

Immanuel Family Counseling Center/Fremont
1627 East Military Street
Fremont, NE 68025

Pathfinder Alcohol/Drug Outpatient Clinic
630 North H Street
Fremont, NE 68025

United Family Services, Inc. U Turn
640 North H Street
Fremont, NE 68025

GENEVA

Blue Valley Mental Health Center Alcohol and Drug Abuse Services
831 F Street
Geneva, NE 68361

GORDON

Northeast Panhandle Substance Abuse Center
305 Foch Street
Gordon, NE 69343

Northwest Nebraska Alcohol and Drug Abuse Services/Gordon
Gordon Community Hospital
Gordon, NE 69343

GRAND ISLAND

Central Nebraska Council on Alcoholism
219 West 2 Street
Grand Island, NE 68801

Friendship House, Inc.
406 West Koenig Street
Grand Island, NE 68801

Lutheran Family Services of Nebraska, Inc.
2121 North Webb Road
Suite 206
Grand Island, NE 68803

Mid-Plains Center for Professional Services
914 Bauman Street
Grand Island, NE 68801

Milne Detoxification Center
406 West Koenig Street
Grand Island, NE 68801

North Central Alcoholism Outpatient Counseling Services
2112 West Faidley Street
Memorial Health Center
Grand Island, NE 68803

Region III School Community Intervention Prevention Program
424 West 3 Street
Grand Island, NE 68802

Saint Francis Alcoholism/Drug Treatment Center
2116 West Faidley Avenue
Grand Island, NE 68803

Veterans' Affairs Medical Center Alcohol Treatment Program
2201 North Broadwell Street
Grand Island, NE 68803

HASTINGS

Hastings Area Council on Alcoholism and Drugs
432 North Minnesota Street
Hastings, NE 68901

Hastings Regional Center Chemical Dependency Unit
Hastings, NE 68901

Mary Lanning Memorial Hospital Behavioral Services
715 North Saint Joseph Street
Hastings, NE 68901

South Central Counseling
Hastings Clinic
612 West 6 Street
Hastings, NE 68901

The Bridge, Inc.
922 North Denver Street
Hastings, NE 68901

HEBRON

Blue Valley Mental Health Center SE Nebraska Comm. Alcohol/ Drug Abuse Program/Beatrice
Thayer County Courthouse
Hebron, NE 68370

HOLDREGE

South Central Counseling Holdrege Clinic
Holdrege, NE 68949

IMPERIAL

Heartland Counseling and Consulting Clinic
839 Douglas Street
Imperial, NE 69033

KEARNEY

Richard Young Hospital Chemical Dependency Unit
4600 17 Avenue
Kearney, NE 68847

South Central Counseling Substance Abuse Treatment Program
3810 Central Avenue
Kearney, NE 68847

LEXINGTON

Heartland Counseling and Consulting Clinic
307 East 5 Street
Lexington, NE 68850

LINCOLN

Alcoholism/Drug Abuse Council of Nebraska Alcohol and Drug Info Clearinghouse
650 J Street
Suite 215
Lincoln, NE 68508

Care/Life
129 North 10 Street
Suite 222
Lincoln, NE 68508

Cornhusker Place
Detoxification Program
Long Term Care Program
721 K Street
Lincoln, NE 68508

Houses of Hope of Nebraska, Inc.
2015 South 16 Street
Lincoln, NE 68502

Lincoln Correctional Center Substance Abuse Program
3210 West Van Dorn Street
Lincoln, NE 68502

Lincoln Council Alcohol and Drugs, Inc.
914 L Street
Lincoln, NE 68508

Lincoln Employee Assistance Program
201 North 8 Street
Suite 101
Lincoln, NE 68508

Lincoln General Hospital
Independence Center
Youth Treatment Services
1650 Lake Street
Lincoln, NE 68502

Lincoln Indian Center Counseling Services
1100 Military Road
Lincoln, NE 68508

Lincoln Lancaster Drug Projects, Inc.
Administrative/Outpatient Offices
Adult Residential Program
Youth Residential Program
610 J Street
Lincoln, NE 68508

Lincoln Medical Education Foundation School Community Intervention Program
4600 Valley Road
Lincoln, NE 68510

Lincoln Valley Hope Alcohol and Drug Counseling and Referral Center
4600 Valley Road
Suite D
Lincoln, NE 68510

Lincoln/Lancaster Attention Center
2220 South 10 Street
Lincoln, NE 68502

Lincoln/Lancaster County Child Guidance Adolescent Substance Abuse Program
215 Centennial Mall South
312 Lincoln Center Building
Lincoln, NE 68508

Lutheran Family Social Services Substance Abuse Program
4620 Randolph Street
Lincoln, NE 68510

Methodist Richard Young Addictions Service/Lincoln
4535 Normal Boulevard
Suite 165
Lincoln, NE 68506

Nebraska Council on Alcohol and Drug Education, Inc.
4547 Calvert Street
Suite 6
Lincoln, NE 68506

Nebraska Prevention Center for Alcohol and Drug Abuse
125 Mabel Lee Hall
University of Nebraska/Lincoln
Lincoln, NE 68588

Nebraska State Penitentiary Mental Health Unit Substance Abuse Services
14 Street and Pioneer Boulevard
Lincoln, NE 68502

Saint Monica's Halfway House and Outpatients Counseling
2130 North 64 Street
Lincoln, NE 68505

Veterans' Affairs Medical Center Chemical Abuse Services
600 South 70 Street
Unit 116A
Lincoln, NE 68510

Youth Service System
2202 South 11 Street
Lincoln, NE 68502

MACY

Macy Alcoholism Counseling Clinic Outpatient Program
Macy, NE 68039

Macy Alcoholism Counseling Services
Men's Halfway House
Women's Halfway House
Macy, NE 68039

Macy Youth and Family Services
Macy, NE 68039

MCCOOK

Heartland Counseling and Consulting Clinic
302 West B Street
McCook, NE 69001

NEBRASKA CITY

Blue Valley Mental Health Center SE Nebraska Comm. Alcohol/ Drug Abuse Program/Beatrice
1903 4 Corso
Nebraska City, NE 68410

Saint Mary's Hospital Oak Arbor Recovery Center
1314 3 Avenue
Nebraska City, NE 68410

NIOBRARA

Santee Sioux Tribe of Nebraska/ Health Education Addictions Recovery Training (HEART)
Route 2
Niobrara, NE 68760

NORFOLK

Methodist Richard Young Lutheran Community Hospital
2700 Norfolk Avenue
Norfolk, NE 68702

Monroe Mental Health Center, Inc. Substance Abuse Services
201 Miller Avenue
Norfolk, NE 68701

The Aware Program
103 North 4 Street
Norfolk, NE 68701

The Link, Inc. Halfway House
1001 Norfolk Avenue
Norfolk, NE 68701

Well Link, Inc.
305 North 9 Street
Norfolk, NE 68702

NORTH PLATTE

Center for Substance Abuse Prevention
311 North Bailey Street
North Platte, NE 69103

Heartland Counseling and Consulting Clinic
110 North Bailey Street
North Platte, NE 69101

Lutheran Family Services
1300 East 4 Street
North Platte, NE 69101

New Horizon Halfway House
New Horizons Detoxification Unit
110 North Bailey Street
North Platte, NE 69101

Parents Action in Nebraska for Drug Free Youth, Inc.
3014 Cedar Berry Road
North Platte, NE 69101

OFFUTT AIR FORCE BASE

Offutt Air Force Base Social Action Office
109 Grant Circle
Suite 101
Offutt Air Force Base, NE 68113

OGALLALA

Heartland Counseling and Consulting Clinic
103 East 10 Street
Ogallala, NE 69153

OMAHA

Alcoholics Resocialization Conditioning Help (ARCH Inc)
604 South 37 Street
Omaha, NE 68105

Chicano Awareness Center
4821 South 24 Street
Omaha, NE 68107

Equilibria Medical Center Substance Abuse Services
544 South 24 Avenue
Omaha, NE 68105

Immanuel Medical Center Addictions Treatment Center
6901 North 72 Street
Omaha, NE 68122

Mercy Chemical Dependency Services
6910 Pacific Street
Suite 204
Omaha, NE 68106

Methodist Richard Young
Chemical Dependency Unit
Dual Diagnosis Units
515 South 26 Street
Omaha, NE 68105

National Council on Alcoholism and Drug Dependence/Omaha
115 North 49 Street
Omaha, NE 68132

Native American Comm. Development Corp. Substance Abuse Program
2451 Saint Mary's Avenue
Omaha, NE 68105

Nebraska Urban Indian Coalition Inter Tribal Treatment Center
2301 South 15 Street
Omaha, NE 68108

North Omaha Alcoholism Counseling Program
Extended Care Program
1924 Lothrop Street
Omaha, NE 68110

Outpatient Services
3014 North 24 Street
Omaha, NE 68111

Reaching Out/Outreach and Youth Services
3805 North 16 Street
Omaha, NE 68110

Nova Therapeutic Community
Outpatient
Partial Care/North
Residential
3473 Larimore Avenue
Omaha, NE 68111

Partial Care/South
4932 South 24 Street
Omaha, NE 68107

Omaha Correctional Center Substance Abuse Program
2323 East Avenue J
Omaha, NE 68110

Operation Bridge, Inc.
114th Street Unit
701 North Street
Miracle Hills Square
Omaha, NE 68154

42nd Center Street Unit
42 and Center Streets
122 Applewood Mall/The Center Mall
Omaha, NE 68105

Pride Omaha, Inc.
3534 South 108 Street
Omaha, NE 68144

Saint Joseph Center for Mental Health
Addiction Recovery Center/Outpatient
14748 West Center Road
Suite 301
Omaha, NE 68144

Addiction Recovery Center/Residential
819 Dorcas Street
Omaha, NE 68108

Santa Monica, Inc.
103 North 39 Street
Omaha, NE 68131

South Omaha Alcoholism Counseling Agency
2900 O Street
Livestock Exchange Building Suite 521
Omaha, NE 68107

United Catholic Social Services
Christus House
Outpatient Chemical Dependency Services
3300 North 60 Street
Omaha, NE 68104

Saint Gabriel's Center
3483 Larimore Avenue
Omaha, NE 68111

Saint Raphael Transitional Living
4727 Hamilton Street
Omaha, NE 68132

The Shelter
Omaha, NE 68104

Veterans' Affairs Medical Center Substance Abuse Treatment Center
4101 Woolworth Avenue
Omaha, NE 68105

O'NEILL

Sandhills Mental Health and Substance Abuse Services, Inc.
204 East Everett Street
O'Neill, NE 68763

Valley Hope Alcoholism Treatment Center
North 10 Street
O'Neill, NE 68763

ORD

North Central Alcoholism Outpatient Counseling Services
314 South 14 Street
Ord, NE 68862

OSCEOLA

Pioneer Mental Health Center Substance Abuse Program
121 North State Street
Osceola, NE 68651

PAPILLION

Lutheran Family Services Papillion Clinic
1246 Golden Gate Drive
Papillion, NE 68046

Midlands Community Hospital Recovery Services
Highway 370 at 84 Street
Papillion, NE 68046

Sarpy County Alcohol Diversion Program
1210 Golden Gate Drive
Sarpy County Courthouse Suite 2101
Papillion, NE 68046

PAWNEE CITY

Blue Valley Mental Health Center Southeastern Nebraska Comm. Alcohol/Drug Abuse Program/ Beatrice
701 I Street
Pawnee City, NE 68420

PLATTSMOUTH

Lutheran Family Services Cass Family Clinic
2302 West 8 Avenue
Suite 4
Plattsmouth, NE 68048

SCOTTSBLUFF

Addiction Treatment Center at Regional West Medical Center
3700 Avenue B
Scottsbluff, NE 69361

Homestead Halfway House, Inc.
Men's Program
1624 Avenue A
Scottsbluff, NE 69361

Women's Program
513 West 24 Street
Scottsbluff, NE 69361

Panhandle Mental Health Center Substance Abuse Program
4110 Avenue D
Scottsbluff, NE 69361

Panhandle Substance Abuse Council
1517 Broadway
Suite 124
Scottsbluff, NE 69361

SEWARD

Pioneer Mental Health Center Substance Abuse Program
729 Seward Street
Seward, NE 68434

Rivendell Psychiatric Center Substance Abuse Programs
3770 Rivendell Drive
Seward, NE 68434

SIDNEY

Memorial Health Center Addiction Center
908 11 Avenue
Sidney, NE 69162

SOUTH SIOUX CITY

Heartland Counseling Services, Inc.
105 East 28 Street
South Sioux City, NE 68776

New Hope Alcoholism and Addiction Center
South Sioux City, NE 68776

TECUMSEH

Blue Valley Mental Health Center Southeastern Nebraska Comm. Alcohol/Drug Abuse Program/ Beatrice
202 High Street
Tecumseh, NE 68450

VALENTINE

Sandhills Mental Health and Substance Abuse Services, Inc.
325 North Victoria Street
Presbyterian Church
Valentine, NE 69201

WAHOO

Pioneer Mental Health Center Substance Abuse Program
543 North Linden Street
Wahoo, NE 68066

WAYNE

Heartland Counseling Services, Inc. Wayne Satellite Clinic
421 Pearl Street
Saint Paul Lutheran Church
Wayne, NE 68787

WEST POINT

Monroe Mental Health Center Substance Abuse Services
900 East Decatur Street
Trinity Church
West Point, NE 68788

WINNEBAGO

American Indian Human Resource Center
Halfway House
Outpatient Clinic
Winnebago, NE 68071

Winnebago Public Health Hospital Drug Dependency Unit
Winnebago, NE 68071

YORK

Pioneer Mental Health Center Substance Abuse Program
727 Lincoln Avenue
York, NE 68467

NEVADA

AUSTIN

Churchill Council on Alcohol Yomba
Route 1
Austin, NV 89310

Yomba Shoshone Tribe Substance Abuse Prevention
HC61
Austin, NV 89310

BATTLE MOUNTAIN

Battle Mountain Alcohol Program
35 Mountain View Drive
Suite 205
Battle Mountain, NV 89802

CARSON CITY

Carson City Community Counseling Center
625 Fairview Drive
Suite 111
Carson City, NV 89701

Carson Detox Center
105 North Roop Street
Carson City, NV 89701

Rural Clinics CMHC Substance Abuse Services
Stewart Facility
Building 107
Carson City, NV 89710

State of Nevada Department of Prisons Substance Abuse Program Admin. Office
5500 Snyder Avenue
Carson City, NV 89702

Young Volunteers of Nevada
1937 North Carson Street
Suite 101
Carson City, NV 89706

DUCKWATER

Duckwater Tribal Alcohol and Drug Program
Duckwater, NV 89314

ELKO

Nevada Youth Training Center
Highway 40
Elko, NV 89801

Ruby View Counseling Center Outpatient
401 Railroad Street
Suite 301
Elko, NV 89801

Vitality Center Residential Treatment
3740 East Idaho Street
Elko, NV 89801

Vitality Center Teen Discovery
1297 Idaho Street
Elko, NV 89801

ELY

Bristlecone Counseling Service Outpatient
2200 Aultman Street
Ely, NV 89301

FALLON

Churchill Council
Day Treatment
90 North Main Street
Fallon, NV 89406

New Frontier Treatment Center
165 North Carson Street
Fallon, NV 89406

HAWTHORNE

Mineral County Council on Alcohol and Drug Abuse
407 E Street
Hawthorne, NV 89415

INDIAN SPRINGS

Nevada Department of Prisons Men's DUI Center
Cold Creek Road
Indian Springs, NV 89702

LAS VEGAS

Bridge Counseling Associates Outpatient
550 East Charleston Boulevard
South Building
Las Vegas, NV 89104

Center for Behavioral Health/ Nevada Methadone Outpatient Treatment Center
3050 East Desert Inn Road
Suite 117
Las Vegas, NV 89121

Clark County Health District Addiction Treatment Clinic/ Methadone
625 Shadow Lane
Las Vegas, NV 89127

Clark County Juvenile Court Services Family Based Drug Treatment Program
3401 East Bonanza Road
Las Vegas, NV 89101

Community Counseling Center AIDS and Substance Abuse Program
1006 East Sahara Avenue
Las Vegas, NV 89104

Economic Opportunity Board of Clark County
EOB Treatment Center
522 West Washington Street
Las Vegas, NV 89106

EOB Women's Transitional Residential
428 North 15 Street
Las Vegas, NV 89030

Las Vegas Indian Center
2300 West Bonanza Road
Las Vegas, NV 89106

Las Vegas Paiute Tribal Council Substance Abuse Prevention/ Mental Health Program
1321 Ken Street
Las Vegas, NV 89106

Nevada Assoc. of Latin Americans (NALA) High on Life Youth Program
2629 East Searles Street
Las Vegas, NV 89101

Nevada Treatment Center
1721 East Charleston Boulevard
Las Vegas, NV 89104

Salvation Army Adult Rehabilitation Program
211 Judson Street
Las Vegas, NV 89030

Samaritan House, Inc.
1001 North 4 Street
Las Vegas, NV 89101

Temporary Assistance for Domestic Crisis (TADC) Shelter
Las Vegas, NV 89116

Veterans' Affairs Outpatient Clinic
1703 West Charleston Boulevard
Las Vegas, NV 89102

Vista Professional Services, Inc.
314 East Charleston Boulevard
Las Vegas, NV 89104

We Care Foundation
2216 South 6 Street
Las Vegas, NV 89104

Westcare, Inc.
Adult Detox/Residential/Aftercare
930 North 4 Street
Las Vegas, NV 89101

Boys' Residential
Harris Springs Ranch
Prevention Project
401 South Martin Luther King Boulevard
Las Vegas, NV 89106

Outpatient/Adult Intensive OPC/ Hispanic Outreach
2325 West Charleston Street
Las Vegas, NV 89102

LAUGHLIN

Bridge Counseling Associates Laughlin Outpatient Program
3650 South Point Circle
Laughlin Professional Plaza Suite 13
Laughlin, NV 89029

LOVELOCK

Lovelock Counseling Clinic
1055 Cornell Avenue
Lovelock, NV 89419

NORTH LAS VEGAS

Lake Mead Hospital and Medical Center Substance Abuse Services
1409 East Lake Mead Boulevard
North Las Vegas, NV 89030

RENO

Dickson O'Bryan and Associates Substance Abuse Program
729 Evans Avenue
Reno, NV 89512

Family Counseling Service of Northern Nevada
777 Sinclair Street
Suite 100
Reno, NV 89501

Northern Area Substance Abuse Council Outpatient/Central Intake
320 Flint Street
Reno, NV 89501

Oikos, Inc. Adolescent Care and Treatment
3000 Dickerson Road
Reno, NV 89503

Reno Sparks Drug and Alcohol Program
34 Reservation Road
Reno, NV 89502

Ridge House
57 Vine Street
Reno, NV 89503

Saint Mary's Regional Medical Center Saint Mary's Adolescent Chemical Dependency
235 West 6 Street
3 East
Reno, NV 89520

Salvation Army Alcoholism Rehab Program
835 East 2 Street
Reno, NV 89505

Step Two
78 Keystone Avenue
Reno, NV 89503

Transition House Residential and Aftercare
52 West 9 Street
Reno, NV 89503

Truckee Meadows Hospital Alcoholism and Drug Abuse Program
1240 East 9 Street
Reno, NV 89512

University Health Professionals
941 North Virginia Street
Reno, NV 89503

SILVER SPRINGS

Nevada Department of Prisons Women's DUI Center
Silver Springs Conservation Camp
Silver Springs, NV 89429

SPARKS

Northern Area Substance Abuse Council Chemical Dependency Unit/Detox
480 Galletti Way
Buildings 3 and 4 Second Floor
Sparks, NV 89431

WENDOVER

Great Basin Counseling Service Outpatient Services
391 Aria Boulevard
Wendover, NV 89883

WINNEMUCCA

Silver Sage Counseling Service Outpatient Services
530 Melarkey Street
Winnemucca, NV 89445

YERINGTON

Lyon Council Alcohol and Drugs Yerington Project
26 Nevin Way
Yerington, NV 89447

Yerington Paiute Tribe Substance Abuse Program
171 Campbell Lane
Yerington, NV 89447

NEW HAMPSHIRE

BEDFORD

New Life Center
169 South River Road
Bedford, NH 03110

BERLIN

Berlin/Gorham Drug Free Schools and Communities
220 Main Street
Berlin, NH 03570

BETHLEHEM

Friendship House
Route 302
Bethlehem, NH 03574

CLAREMONT

Bailey House
18 Bailey Avenue
Claremont, NH 03743

Community Youth Advocates
36 Tremont Square
Claremont, NH 03743

Counseling Center of Claremont
241 Elm Street
Claremont, NH 03743

CONCORD

Community Alcohol Information Program, Inc.
10 Ferry Street
The Concord Center
Concord, NH 03301

Merrimack County Alcohol and Drug Intervention Program
48 Branch Turnpike
Concord, NH 03301

New Hampshire State Prison Dept. of Corrections/Substance Abuse Prevention
281 North State Street
Concord, NH 03301

New Hampshire Teen Institute for Prevention of Alcohol and Other Drug Abuse, Inc.
46 South Main Street
Concord, NH 03301

Second Start
17 Knight Street
Concord, NH 03301

Southeastern New Hampshire Services at Concord
105 Pleasant Street
State Office Park Square
Concord, NH 03301

State of New Hampshire Department of Education/ Substance Abuse Services
101 Pleasant Street
State Office Park South
Concord, NH 03301

CONWAY

Carroll County Mental Health Service Drug Abuse Treatment Program
West Main Street
Kennett House
Conway, NH 03818

DOVER

Southeastern New Hampshire Services
Crisis Site
Halfway House for Men and Women
Country Farm Crossroad
Dover, NH 03820

DUBLIN

Beech Hill Hospital Substance Abuse Services
New Harrisville Road
Dublin, NH 03444

Marathon House
Alcohol Crisis Intervention Program
Long Term Residential
1 Pierce Road
Dublin, NH 03444

The Lodge at Dublin
1 Pierce Road
Dublin, NH 03444

EPPING

Amethyst Foundation, Inc. Alcohol Awareness Program
120 Hedding Road
Route 87
Epping, NH 03042

EXETER

Rockingham Counseling Center
9 Hampton Road
Exeter, NH 03833

GROVETON

Northumberland Citizens Committee Eagle's Nest
3C State Street
Groveton, NH 03582

HAMPTON

Odyssey House, Inc.
30 Winnacunnet Road
Hampton, NH 03842

KEENE

Cheshire Medical Center Mental Health Unit
580 Court Street
Keene, NH 03431

Monadnock Region Substance Abuse Services, Inc.
Swanzey Homestead Office
RR 32/Old Homestead Highway
Keene, NH 03431

LACONIA

Horizons Counseling Center
390 Union Avenue
Laconia, NH 03246

Lakes Region General Hospital Nathan Brody Chemical Dependency Program
80 Highland Street
Laconia, NH 03246

Multiple DWI Offender Program
Parade Road
Laconia, NH 03246

New Hampshire Dept. of Corrections Lakes Region Facility Summit House
1 Right Way Path
Laconia, NH 03246

LEBANON

Counseling Center of Lebanon
23 Old Etna Road
Wheelock Office Park
Lebanon, NH 03766

Headrest
14 Church Street
Lebanon, NH 03766

West Central Community Support Services
9 Commerce Building
Lebanon, NH 03766

Administrative Offices/IDIP
2 Whipple Place
Suite 202
Lebanon, NH 03766

LITTLETON

White Mountain Mental Health Center Substance Abuse Services
16 Maple Street
Littleton, NH 03561

MANCHESTER

Boy Scouts of America Daniel Webster Council Education Program Services
571 Holt Avenue
Manchester, NH 03103

Catholic Medical Center Riverway Center for Recovery
100 McGregor Street
Manchester, NH 03102

Farnum Center
235 Hanover Street
Manchester, NH 03104

Lake Shore Hospital Substance Abuse Services
200 Zachary Road
Manchester, NH 03109

National Council on Alcoholism Greater Manchester Area/Crisis Site
93–101 Manchester Street
Manchester, NH 03103

Office of Youth Services
66 Hanover Street
Suite 106
Manchester, NH 03101

Tirrell House
15–17 Brook Street
Manchester, NH 03104

Veterans' Affairs Medical Center Substance Abuse Treatment Program (SATP)
718 Smyth Road Building 5
Manchester, NH 03104

NASHUA

Brookside Hospital Substance Abuse Services
29 Northwest Boulevard
Nashua, NH 03063

Greater Nashua Council on Alcoholism Pine Street Extension
Keystone Hall
Nashua, NH 03060

Nashua Youth Council
63 Temple Street
Nashua, NH 03060

Saint Joseph's Hospital New Start
172 Kinsley Street
Nashua, NH 03061

NEWPORT

Counseling Center of Newport
92 South Main Street
Newport, NH 03773

PORTSMOUTH

New Heights
135 Daniel Street
Portsmouth, NH 03801

Southeastern New Hampshire Services at Portsmouth
140 Congress Street
Portsmouth, NH 03801

SUNAPEE

Seminole Point Hospital Substance Abuse Services
Woodland Road
Sunapee, NH 03782

WHITEFIELD

White Mountain Regional High School Student Assistance Program
Route 3
Whitefield, NH 03598

WOLFEBORO

Carroll County Mental Health Substance Abuse Services
Huggins Hospital
Wolfeboro, NH 03894

WOLFEBORO FALLS

FXH Consulting, Inc.
18 Center Street
Wolfeboro Falls, NH 03896

NEW JERSEY

ABSECON

Family Service Association of Atlantic County/Alcoholism Services
312 East Whitehorse Pike
Absecon, NJ 08201

IDRC Counseling Associates/ Thomas E. Hand Professional Associates
283 East Jimmie Leeds Road
Absecon, NJ 08201

William A. Miller, PhD., CAC
18 Bayview Drive
Absecon, NJ 08201

ASBURY PARK

Center of Love, Inc.
115 Dewitt Avenue
Asbury Park, NJ 07712

Jersey Shore Addiction Services, Inc. T/A Asbury Park Drug Treatment Center
1200 Memorial Drive
Asbury Park, NJ 07712

ATLANTIC CITY

Institute for Human Development (IHD)
1315 Pacific Avenue
Atlantic City, NJ 08401

ATLANTIC HIGHLANDS

Alan E. Matonti BSW CADC NCAC II
30 Briarwood Avenue
Atlantic Highlands, NJ 07716

BASKING RIDGE

Jeremiah Bresnahan ACSW CAC and Maureen Bresnahan MS CADC
36 Manchester Drive
Basking Ridge, NJ 07920

BAYONNE

Alternatives Group Counseling Services of Bayonne, Inc.
11 West 42 Street
Bayonne, NJ 07002

Army Alcohol and Drug Abuse Program HQMTMC-EA MTEA-PE-ADCO
Bayonne, NJ 07002

Bayonne Community Mental Health Center Alcoholism Services
601 Broadway
Bayonne, NJ 07002

Community Psychotherapy Associates
479 Avenue C
Bayonne, NJ 07002

BELLE MEAD

Carrier Foundation Addiction Recovery Unit/Russell Hall
Belle Mead, NJ 08502

BELLEVILLE

CMHC of Belleville/Bloomfield/ Nutley Alcoholism Services/ Partial Care Program
570 Belleville Avenue
Belleville, NJ 07109

Marriage and Family Counseling Center
387 Union Avenue
Belleville, NJ 07109

BELMAR

Wall Youth Center
1824 South M Street
Belmar, NJ 07719

BELVIDERE

Warren County Council on Alcoholism and Drug Abuse
311 Front Street
Belvidere, NJ 07823

BERKELEY HEIGHTS

Runnells Specialized Hospital of Union County/Addiction Recovery Unit
40 Watchung Way
Berkeley Heights, NJ 07922

BERLIN

Bowling Green Adolescent Center Residential Program
109–111 Jackson Road
Berlin, NJ 08009

BLAIRSTOWN

Little Hill/Alina Lodge
Paulinskill River and Squires Road
Blairstown, NJ 07825

BOONTON

Saint Clare's/Riverside Medical Center Acute Alcohol/Chemical Dependency Program
Powerville Road
Boonton, NJ 07005

BORDENTOWN

Laurie Eberle Snyder ACSW CDC
15 Lyons Lane
Bordentown, NJ 08505

BRICK

Bernadette Brewer Counseling Associates, Inc.
515 Route 70
Brick, NJ 08723

Counseling and Referral Services of Brick Substance Abuse Services/Outpatient
35 Beaverson Boulevard
Lion's Head Office Park, Building 6B
Brick, NJ 08723

BRIDGETON

Counseling Associates
143 East Commerce Street
Bridgeton, NJ 08302

Crystal Associates
12 Franklin Street
1st Floor
Bridgeton, NJ 08302

Cumberland County Alcoholism and Drug Treatment/ Outpatient
Cumberland Drive/Route 22
Cumberland County Medical Center
Bridgeton, NJ 08302

Faith Farm, Inc.
21 Stretch Road
Bridgeton, NJ 08302

South Jersey Drug Treatment Center Cumberland Drive
Bridgeton, NJ 08302

BRIDGEWATER

Catholic Charities/Somerset CFTAAA Substance Abuse Services/Adult Outpatient
540–550 Route 22 East
Bridgewater, NJ 08807

Crawford House/Cedar House
520 North Bridge Street
Bridgewater, NJ 08807

Richard Hall CMHC Outpatient Substance Abuse Services
500 North Bridge Street
Bridgewater, NJ 08807

BURLINGTON

Catholic Charities of Burlington
13 East Brand Street
Burlington, NJ 08016

Edmund S. Bienkowski and Associates Alcoholism Services
206 East Union Street
Burlington, NJ 08016

CALDWELL

Rehabilitation Counseling Associates
15 Johnson Avenue
Caldwell, NJ 07006

The Bridge, Inc.
14 Park Avenue
Caldwell, NJ 07006

CAMDEN

Alcove Drug and Alcohol Treatment Program
1000 Atlantic Avenue
Camden, NJ 08104

Cooper Hospital/University Medical Center Cooper House Residential and Outpatient Services
600 Benson Street
First Floor
Camden, NJ 08103

Hispanic Family Center of Southern New Jersey La Esperanza
425 Broadway
Camden, NJ 08103

Nigeria American Institute Substance Abuse Unit
509 Cooper Street
Camden, NJ 08102

Substance Abuse Center of Southern Jersey, Inc.
417 Broadway
Segaloff Treatment Center
Camden, NJ 08103

The Starting Point of New Jersey, Inc.
216 Haddon Avenue
Sentry Office Plaza Suite 608
Camden, NJ 08108

CAPE MAY

Richard P. Koeppel Associates, Inc.
Bank Street Commons
Suite 130
Cape May, NJ 08204

CAPE MAY COURT HOUSE

Burdette Tomlin Hospital Detox Unit
Stone Harbor Boulevard
Route 9
Cape May Court House, NJ 08210

Cape May Council on Alcoholism and Drug Abuse, Inc.
6 Moore Road
Crest Haven Complex
Cape May Court House, NJ 08210

Cape May County Juvenile/ Family Crisis Intervention/ Youth Drug and Alcohol Services
Cape May Court House, NJ 08210

Cape May County Youth Shelter Substance Abuse Services
151 Crest Haven Road
Cape May Court House, NJ 08210

CARTERET

James P. Dunlap Associates
75 Willow Street
Carteret, NJ 07008

CEDAR GROVE

Essex County Employee Assistance Program
125 Fairview Avenue
Essex County Hospital Center
Cedar Grove, NJ 07009

George Snider MA CADC CEAP
102 Brunswick Road
Cedar Grove, NJ 07009

Turning Point, Inc.
125 Fairview Avenue
Cedar Grove, NJ 07009

CHERRY HILL

Kennedy Memorial Hospital/ Cherry Hill Division Substance Abuse Services/Detox and Outpatient
Chapel Avenue and Cooperlanding Road
Cherry Hill, NJ 08034

Steininger Center of Camden County, Inc. MICA Program
19 East Ormond Avenue
Cherry Hill, NJ 08034

West Jersey Recovery Network
2301 West Marlton Park
Cherry Hill, NJ 08002

CINNAMINSON

Healthmark Counseling
101 Route 130
Madison Building Suite 321
Cinnaminson, NJ 08077

CLIFTON

Clifton Counseling Services
60 Hadley Avenue
Suite A
Clifton, NJ 07011

Counseling Center
716 Broad Street
Suite 2
Clifton, NJ 07013

Passaic County Council on Alcoholism and Drug Abuse Prevention, Inc.
880 Clifton Avenue
Clifton, NJ 07015

COLLINGSWOOD

Genesis Counseling Center Alcoholism Outpatient Services
636 Haddon Avenue
Collingswood, NJ 08108

DENVILLE

OPT Counseling Services
61 Broadway
Denville, NJ 07834

DOVER

Hope House Outpatient Services
19–21 Belmont Avenue
Dover, NJ 07802

EAST BRUNSWICK

Middlesex Council on Alcoholism and Drug Abuse
330 Milltown Road
East Brunswick, NJ 08816

EAST ORANGE

East Orange Substance Abuse Treatment Program Administrative Unit
160 Halsted Street
East Orange, NJ 07018

Veterans' Affairs Medical Center
Alcohol Dependence Treatment Program
Drug Dependency Treatment Program
385 Tremont Avenue
Unit 116E
East Orange, NJ 07019

EATONTOWN

CPC/Chemical Dependency Services
270 Route 35
Eatontown, NJ 07724

ELIZABETH

Elizabeth General Medical Center Substance Abuse Services
655 East Jersey Street
3rd Floor
Elizabeth, NJ 07201

Flynn House/Elizabeth Halfway House/Male
1089/1091 East Jersey Street
Elizabeth, NJ 07201

Proceed, Inc. Addiction Services
815 Elizabeth Avenue
Elizabeth, NJ 07201

Seton Center for Chemical Dependency
225 Williamson Street
Saint Elizabeth Hospital, Unit 3X
Elizabeth, NJ 07207

YWCA of Eastern Union County Cooperative Program for Addicted Women
1131 East Jersey Street
Elizabeth, NJ 07201

ENGLEWOOD

Community Centers for Mental Health, Inc. Substance Abuse Services
93 West Palisade Avenue
Englewood, NJ 07631

The Van Ost Institute for Family Living, Inc./Alcohol Drug Outpatient Services
113 Engle Street
Englewood, NJ 07631

ENGLISHTOWN

Manalapan Community and Family Services
120 Route 522
Englishtown, NJ 07726

FAIRTON

Federal Correctional Institution Substance Abuse Services
Fairton, NJ 08320

FLEMINGTON

Catholic Charities Substance Abuse Services Care Program
84 Park Avenue
Flemington, NJ 08822

Good News Home for Women
33 Bartles Corner Road
Flemington, NJ 08822

Hunterdon Drug Awareness Program
8 Main Street
Suite 7
Flemington, NJ 08822

Hunterdon Medical Center Addictions Treatment Services
1200 Wescott Drive
Mental Health Center
Flemington, NJ 08822

Hunterdon Youth Services Inside Out Program
Rural Route 2
322 Highway 12
Flemington, NJ 08822

National Council on Alcoholism and Drug Dependence/ Hunterdon County
153 Broad Street
Flemington, NJ 08822

FORT DIX

Fort Dix Counseling Center U.S. Army Garrison
Attention AFTZ-CFA-DA
Fort Dix, NJ 08640

FORT LEE

Behavioral Counseling Associates
1580 Lemoine Avenue
Suite 8
Fort Lee, NJ 07024

Mark Dannemfelfer MA CAC
1 Bridge Plaza
Suite 400
Fort Lee, NJ 07024

FRANKLIN

Wallkill Valley Hospital Counseling Center
19 Hospital Road
Franklin, NJ 07416

FREEHOLD

Freehold Community Counseling Service
30 Jackson Mills Road
Freehold, NJ 07728

Monmouth County Office on Aging Alcoholism Program
Freehold, NJ 07728

GARFIELD

Rehabilitation Counseling Associates
33 Outwater Lane
Garfield, NJ 07026

GLASSBORO

Together, Inc. Drug Treatment Program
7 State Street
Glassboro, NJ 08028

GLEN GARDNER

Freedom House Adult Male Halfway House
3 Pavillion Road
Glen Gardner, NJ 08826

HACKENSACK

Alternatives to Domestic Violence Chemical Dependency Unit
21 Main Street
Room 111W
Hackensack, NJ 07601

Bergen County Community Action Program Addictions Program
214 State Street
Hackensack, NJ 07601

Bergen County Division of Family Guidance Jupiter Program/Adolescent Substance Abuse Program
296 East Ridgewood Avenue
Hackensack, NJ 07625

Department of Health Services Alcohol Recovery Program Unit
21 Main Street
Court Plaza South Room 104E
Hackensack, NJ 07601

HACKETTSTOWN

Hackettstown Community Hospital Substance Abuse Department
651 Willow Grove Street
Hackettstown, NJ 07840

HADDONFIELD

Miriam Mundy Broderick CADC NCAC II
15 South Haddon Avenue
2nd Floor
Haddonfield, NJ 08033

HAMMONTON

Ancora Psychiatric Hospital Addiction Services
202 Spring Garden Road
Hammonton, NJ 08037

HAZLET

Women's Center of Monmouth County, Inc. Outpatient Alcohol Counseling
1 Bethany Road
Building 3 Suite 42
Hazlet, NJ 07730

HIGHLAND PARK

Ellen Hulme ED D
308 Raritan Avenue
Highland Park, NJ 08904

Human Resources Associates
324 Raritan Avenue
Suite 105
Highland Park, NJ 08904

HOBOKEN

Saint Mary's CMHC Alcoholism Treatment Unit
314 Clinton Street
Hoboken, NJ 07030

Saint Mary's Hospital Giant Steps/Adolescent Substance Abuse Program
527 Clinton Street
Hoboken, NJ 07030

The Harbor Residential Services
1405 Clinton Street
Hoboken, NJ 07030

HOLMDEL

Bayshore Community Hospital Substance Abuse Program/ Outpatient
719 North Beers Street
Holmdel, NJ 07733

HOWELL

Howell Township Youth and Family Counseling Services
Preventorium Road
Howell, NJ 07731

IRVINGTON

L and L Clinics, Inc. Methadone Maintenance and Detox
57–59 New Street
Irvington, NJ 07111

Underwood Counseling
1354 Clinton Avenue
Irvington, NJ 07111

JERSEY CITY

Catholic Community Services MICA Drop-In Center
308 Academy Street
Jersey City, NJ 07306

Christ Hospital Substance Abuse Services
176 Palisade Avenue
Jersey City, NJ 07306

Hudson Catholic Community Services
249 Virginia Avenue
Jersey City, NJ 07304

Hudson County Council on Alcoholism and Drug Abuse
83 Wayne Street
Barrow Mansion Suite 105
Jersey City, NJ 07302

Salvation Army Adult Rehab Center/Inpatient and Outpatient
248 Erie Street
Jersey City, NJ 07302

Spectrum Health Care, Inc.
74–80 Pacific Avenue
Jersey City, NJ 07304

KEARNY

Inter County Council on Drug/ Alcohol Abuse Administration/ Drug Free Counseling
480 Kearny Avenue
Kearny, NJ 07032

KEYPORT

Endeavor House
Keyport, NJ 07735

LAFAYETTE

Sunrise House Foundation, Inc. Alcohol Residential Program
Sunset Inn Road
Intersection of Routes 15 and 94
Lafayette, NJ 07848

LAKELAND

Camden County Division of Alcohol and Substance Abuse Administrative Unit
Collier Drive
Blackwood Post Office
Lakeland, NJ 08012

LAKEWOOD

Alcoholism and Drug Abuse Council of Ocean County, Inc.
117 East County Line Road
Lakewood, NJ 08701

Counseling Center for Self Discovery
222 River Avenue
Route 9 South
Lakewood, NJ 08701

Kimball Medical Center/The Greenhouse Substance Abuse Services/Adolescent
528 River Avenue
Lakewood, NJ 08701

Shore Mental Health Center Drug and Alcohol Rehabilitation (DARE)
700 Airport Road
Lakewood, NJ 08701

LEDGEWOOD

Professional Guidance Firm, Inc.
1451 Route 46W
Ledgewood, NJ 07853

LEONARDO

Middletown Office of Substance Abuse Services
20 Leonardville Road
Croydon Hall
Leonardo, NJ 07737

LINWOOD

Behavioral Health Services PA
2106 New Road
Linwood Commons Suite C-10
Linwood, NJ 08221

LONG BRANCH

Epiphany House, Inc.
373 Brighton Avenue
Long Branch, NJ 07740

LYNDHURST

South Bergen Mental Health Center Alcoholism Services
516 Valley Brook Avenue
Lyndhurst, NJ 07071

LYONS

Veterans' Affairs medical Center Alcohol Dependence Treatment Program
Building 57 A South
Lyons, NJ 07939

MARGATE CITY

AB Counseling
210 North Rumson Avenue
Margate City, NJ 08402

MARLBORO

Discovery Institute for Addictive Disorders
Marlboro State Hospital
Marlboro, NJ 07746

Marlboro Psychiatric Hospital Atlas Residential MICA Program
Cottage 17 Station A
Marlboro, NJ 07746

New Hope Foundation, Inc. Substance Abuse Services
Route 520
Marlboro, NJ 07746

MATAWAN

Bay Shore Youth and Family Services
166 Main Street
Matawan, NJ 07747

MAYS LANDING

Lighthouse/Recovery Services of New Jersey
5034 Atlantic Avenue
Mays Landing, NJ 08330

MEDFORD

Levin and Fox Counseling Associates
1 North Main Street
Suite 3B
Medford, NJ 08055

MENDHAM

Daytop Village Adolescent Services
80 West Main Street
Mendham, NJ 07945

METUCHEN

First Step Counseling
15 Calvin Place
Metuchen, NJ 08840

MILLTOWN

Brunswick Group
134 North Main Street
Milltown, NJ 08850

MONTCLAIR

Mountainside Hospital Alcoholism Treatment Unit/ Counsellor Coord.
Bay and Highland Avenues
Montclair, NJ 07042

National Council on Alcohol and Drug Dependency North Jersey Area, Inc.
60 South Fullerton Avenue
Montclair, NJ 07042

North Essex Development and Action Council (NEDAC)
104 Bloomfield Avenue
Montclair, NJ 07042

MOORESTOWN

Parkside Lodge of New Jersey
611 East 2 Street
Moorestown, NJ 08057

MORRISTOWN

Center for Addictive Illnesses Alcoholism Residential and Outpatient
95 Mount Kemble Avenue
Morristown, NJ 07962

Market Street Mission New Reality of Recovery Program
9 Market Street
Morristown, NJ 07960

Mrs. Wilson's Halfway House
56 Mount Kemble Avenue
Morristown, NJ 07960

Saint Clare's/Esther T. Dutton Counseling Center
3 Schuyler Place
Morristown, NJ 07960

MOUNT HOLLY

Amity House, Inc.
211 Garden Street
Mount Holly, NJ 08060

Burlington Comp. Counseling, Inc.
75 Washington Street
Mount Holly, NJ 08060

Drenk Mental Health Center Path Substance Abuse Services
205 High Street
Mount Holly, NJ 08060

NEPTUNE

Jersey Shore Medical Center CMHC Outpatient
1945 Corlies Avenue
Neptune, NJ 07753

NEW BRUNSWICK

Damon House, Inc. Residential and Outpatient
105 Joyce Kilmer Avenue
New Brunswick, NJ 08901

New Brunswick Counseling Center
84 New Street
New Brunswick, NJ 08901

Rutgers Univ. Student Health Center Alcohol/Other Drug Assistance for Students
11 Bishop Place
Hurtado Health Center
New Brunswick, NJ 08903

Rutgers University Personnel Counseling Service
88 College Avenue
New Brunswick, NJ 08901

Saint Peter's Medical Center Center for Treatment of Pregnancy and Addiction
240 Easton Avenue
New Brunswick, NJ 08903

The Open Door Alcoholism Treatment Program
2–4 Kirkpatrick and New Street
New Brunswick, NJ 08901

Univ. of MDNJ CMHC at Piscataway The Club/ Habilitation Services
195 New Street
New Brunswick, NJ 08901

NEW LISBON

Burlington County Health Department Post House
600 Pemberton/Browns Mills Road
New Lisbon, NJ 08064

NEWARK

Choices, Inc. United Community Corp.
169 Roseville Avenue
Newark, NJ 07107

Community United for Rehab of Addiction CURA, Inc.
35 Lincoln Park
Newark, NJ 07102

Essex Substance Abuse Treatment Center, Inc.
164 Blanchard Street
Newark, NJ 07105

Integrity House, Inc.
103 Lincoln Park
Newark, NJ 07102

La Casa de Don Pedro Alcoholism and Drug Services
23 Broadway
Newark, NJ 07104

Mount Carmel Guild
Halfway House
56 Freeman Street
Newark, NJ 07105

Narcotic/Rehab Center Substance Abuse Therapy
17 Mulberry Street
Newark, NJ 07102

Mount Carmel Social Service Center of Catholic Community Services
56 Freeman Street
Newark, NJ 07105

Spectrum Health Care Inc./ Newark
461 Frelinghuysen Avenue
Newark, NJ 07114

United Community Alcoholism Network/I UCAN/I
493 Clinton Avenue
Newark, NJ 07108

NEWTON

Domestic Abuse of Sussex County Decide Program for Men
Newton, NJ 07860

Newton Memorial Hospital Alcohol and Substance Abuse Program
175 High Street
Newton, NJ 07860

Professional Counseling Associates
11 High Street
Newton, NJ 07860

NORTH BERGEN

Palisades Counseling Center Counseling Center Alcoholism MICA
7101 Kennedy Boulevard
North Bergen, NJ 07047

NORTHFIELD

Atlantic County Council on Alcoholism and Substance Abuse, Inc.
101 South Shore Road
Shoreview Complex
Northfield, NJ 08225

OAKHURST

Ocean Township Dept. of Human Services Substance Abuse Services
Oakhurst, NJ 07755

ORADELL

Professional Counseling Associates
370 Kinderkamack Road
Oradell, NJ 07649

ORANGE

City of Orange Drug/Alcohol Abuse Program
301 Main Street
Room 1
Orange, NJ 07050

Family Service and Child Guidance Center
395 South Center Street
Orange, NJ 07050

OXFORD

Domestic Abuse and Rape Crisis Center Decide Program
Rural Route 1
Oxford, NJ 07863

Northwest New Jersey Community Action Program, Inc. (NORWESCAP)
67 James Burns Drive
Oxford, NJ 07863

PARAMUS

Bergen County Council on Alcoholism and Drug Abuse, Inc.
Bergen Pines Hospital Complex
Building 7/Spring House
Paramus, NJ 07652

Bergen Pines County Hospital Evergreen Treatment Center
East Ridgewood Avenue
Paramus, NJ 07652

Family Recovery Centers, Inc.
11 Mackay Avenue
Paramus, NJ 07652

Hispanic Institute for Research/ Development Services for Hispanics with Alcohol/Drug Problems
17 Arcadian Avenue
Suite 202
Paramus, NJ 07652

Mid-Bergen Mental Health Center Alcoholism Services/Outpatient
610 Industrial Avenue
Paramus, NJ 07652

Spring House/Halfway House East Ridgewood Avenue
Bergen Pines Complex Bldg 7
Paramus, NJ 07652

PARLIN

South Amboy Memorial Hospital Chemical Dependency Recovery Center/Outpatient
1145 Bordentown Avenue
Claremont Building Suite 4
Parlin, NJ 08859

PARSIPPANY

Alcohol Recovery Center
620 West Hanover Avenue
Parsippany, NJ 07054

PASSAIC

Hispanic Information Center Alcohol Outreach Program for Minorities
270 Passaic Street
Passaic, NJ 07055

Northeast Life Skills Associates, Inc.
121 Howe Avenue
Passaic, NJ 07055

PATERSON

Barnert Memorial Hospital Center Substance Abuse Program
680 Broadway
Paterson, NJ 07514

Eva's Shelter Halfway House for Men
96 Hamilton Street and Saint John's St.
Paterson, NJ 07505

Paterson Counseling Center, Inc.
321 Main Street
Paterson, NJ 07505

Saint Joseph's Hospital and Medical Center
Adolescent MICA Treatment Program
Harbor House
703 Main Street
Paterson, NJ 07503

The New Beginning Is Now, Inc. A Community Family Care Services Program
833 Madison Avenue
Paterson, NJ 07501

PENNINGTON

The Gabrielson Group
65 South Main Street
Pennington, NJ 08534

PEQUANNOCK

Glen Burke Sr NCAC II CADC CAC
Pequannock, NJ 07440

PERTH AMBOY

Addiction Management Treatment Program
182 Jefferson Street
YMCA Building
Perth Amboy, NJ 08861

Raritan Bay Medical Center
Addiction Treatment Services
530 New Brunswick Avenue
Perth Amboy, NJ 08861

Alcoholism Treatment Unit/ Outpatient
570 Lee Street
Perth Amboy, NJ 08861

PHILLIPSBURG

Catholic Charities/Warren ADAPT
700 Sayre Avenue
Phillipsburg, NJ 08865

Warren Hospital Alcohol/Drug Recovery Center/Detox
185 Roseberry Street
Phillipsburg, NJ 08865

Warren Hospital MICA Program/ Inpatient
185 Roseberry Street
Mental Health Unit 2 South
Phillipsburg, NJ 08865

PISCATAWAY

UMDNJ/CMHC at Piscataway Addiction Recovery Outpatient Services
667 Hoes Lane
Piscataway, NJ 08854

PLAINFIELD

Plainfield Red Cross Options Program
332 West Front Street
Plainfield, NJ 07060

Plainfield Treatment Center
519 North Avenue
Plainfield, NJ 07060

Project Alert/Alertop/Dudley House
930 Putnam Avenue
Plainfield, NJ 07060

Union County Psychiatric Clinic Adolescent Alcohol Program
117–119 Roosevelt Avenue
Plainfield, NJ 07062

POMPTON LAKES

Matthew E. Collins CDC CRPS Counseling and Relapse Prevention Services
109 Beech Avenue
Pompton Lakes, NJ 07442

PRINCETON

Family Service of Princeton Outpatient Alcoholism Counseling and Education
120 John Street
Princeton, NJ 08542

Princeton House Addiction Recovery Program
905 Herrontown Road
Princeton, NJ 08540

Township of Princeton Cornerhouse
369 Witherspoon Street
Valley Road Building
Princeton, NJ 08540

PRINCETON JUNCTION

Institute for Family Therapy
14 Washington Road
Princeton Junction, NJ 08500

RANCOCAS

Hampton Hospital Alcohol and Drug Abuse Services
1530 Beverly/Rancocas Road
Rancocas, NJ 08073

RANDOLPH

Morris County Aftercare Center Outpatient/Drug Free and Methadone
1574 Sussex Turnpike
Randolph, NJ 07869

RED BANK

Mary Anne Ruane MSW CAC
16 Spring Street
Suite 1607
Red Bank, NJ 07701

Monmouth Council on Alcoholism and Drug Abuse, Inc.
208 Maple Avenue
Suite 4
Red Bank, NJ 07701

Riverview Medical Center/ Century House Alcoholism Services
87 East Front Street
Red Bank, NJ 07701

RIDGEWOOD

Furman Clinic, Inc.
487 Goffle Road
Ridgewood, NJ 07450

Henry Seligson PhD./Bryan Granelli PhD.
112 Prospect Street
Ridgewood, NJ 07452

RINGWOOD

Sandra A. Carlson Counseling
11 Sunset Road
Ringwood, NJ 07456

RIVERSIDE

Zurbrugg Memorial Hospital Riverside Division Alcohol Services/OATS Program
Hospital Plaza
Riverside Division
Riverside, NJ 08075

ROCKLEIGH

Bergen County Community Action Program Ladder Project
35 Piermont Road
Rockleigh, NJ 07647

Phoenix House Academy
35 Piermont Road
Rockleigh, NJ 07647

SECAUCUS

Hudson County Meadowview Hospital Meadowview Detox
595 County Avenue
Secaucus, NJ 07094

Kids of North Jersey, Inc.
200 Seaview Drive
Secaucus, NJ 07094

New Jersey Treatment Campus Demonstration Project
Meadowview Hospital
Secaucus, NJ 07094

SHREWSBURY

Counseling Care
55 White Road
Shrewsbury, NJ 07702

SKILLMAN

Crawford House, Inc. Halfway House for Women Alcoholics
362 Sunset Road
Skillman, NJ 08558

Princeton Associates for Total Health The Path to Wellness
100–102 Tamarack Circle
Skillman, NJ 08558

SOMERS POINT

Alpha Care
320 New Road
Somers Point, NJ 08244

Amethyst Addictions Services
1409 Roberts Avenue
Somers Point, NJ 08244

SOMERVILLE

Guided Life Structures
19 Davenport Street
Somerville, NJ 08876

Somerset Council on Alcoholism and Drug Dependency, Inc.
112 Rehill Avenue
Somerville, NJ 08876

Somerset Medical Center/ Specialized Treatment for Addictions Recovery Program (STAR)
128 Rehill Avenue
Somerville, NJ 08876

Somerset Treatment Services
256 East Main Street
Somerville, NJ 08876

SOUTH AMBOY

Jack Hynes MA
South Amboy, NJ 08879

South Amboy Memorial Hospital
Center for Behavioral Medicine
Starting Point Program
540 Bordentown Avenue
South Amboy, NJ 08879

169 North Stevens, Inc.
169 North Stevens Avenue
South Amboy, NJ 08879

SOUTH RIVER

South Amboy Center for Behavioral Medicine MICA Counseling
77-B Water Street
South River, NJ 08882

STRATFORD

YWCA of Camden County and Vicinity Solace/Dove Program
710 West Laurel Road
Stratford, NJ 08084

SUMMIT

Blakes Associates
43 Ashwood Avenue
Summit, NJ 07901

Fair Oaks Hospital Outpatient Recovery Center
2 Broad Street
Summit, NJ 07901

Overlook Hospital Drug Treatment Program
99 Beauvoir Avenue
Summit, NJ 07902

TEANECK

Center for Addiction Rehab and Education Care Program Substance Abuse Services
718 Teaneck Road
1 Marion
Teaneck, NJ 07666

TOM'S RIVER

Allan J. Horen ACSW
24 Pine Fork Drive
Tom's River, NJ 08755

Alternatives Counseling Center, Inc.
96 East Water Street
Tom's River, NJ 08754

Healy Counseling Associates
1108 Hooper Avenue
Tom's River, NJ 08753

Ocean County Board of Social Services
1027 Hooper Avenue
Tom's River, NJ 08754

Schaffer Treatment Center
1594 Route 9
Tom's River, NJ 08755

Tom's River Outreach Center Student Team Training
247 Main Street
Tom's River, NJ 08753

TRENTON

Catholic Charities Alcoholism/ Addictions Program
47 North Clinton Avenue
Trenton, NJ 08607

Comm. Guidance Center of Mercer County Substance Abuse Recovery Program
2300 Hamilton Avenue
Trenton, NJ 08619

Family Services of Trenton Adolescent Aftercare/Day Evening
941 Whitehorse Mercerville Road
Suite 21
Trenton, NJ 08610

Helene Fuld Medical Center First Step Chemical Dependency Unit/Detox
750 Brunswick Avenue
Trenton, NJ 08638

J102335 MICA Project
132 North Warren Street
Trenton, NJ 08607

Mercer Council on Alcoholism and Drug Addiction
408 Bellevue Avenue
Trenton, NJ 08618

New Horizon Treatment Services, Inc.
132 Perry Street
2nd Floor
Trenton, NJ 08618

Rescue Mission of Trenton Vince's Place
98 Carroll Street
Trenton, NJ 08604

United Progress, Inc. Detoxification Center
541 East State Street
Trenton, NJ 08609

UNION

Catholic Community Services
438 Clermont Terrace
Union, NJ 07083

Genesis Program of Union Hospital Drug and Alcohol Services/Detox
1090 Morris Avenue
Union, NJ 07083

VAUXHALL

Suburban Clinic
1 Roselyn Place
Vauxhall, NJ 07088

VENTNOR CITY

Jewish Family Services Addiction Services
3 South Weymouth Avenue
Ventnor City, NJ 08406

VINELAND

Lloyd Reynolds and Mary Reynolds
733 Elmer Street
Vineland, NJ 08360

VOORHEES

Reality House, Inc.
Ashland Office Center
1 Alpha Avenue
Suite 43
Voorhees, NJ 08043

WALDWICK

Valley Hospital Comp. Outpatient Addiction Program (COAP)
20 Franklin Turnpike
Waldwick, NJ 07463

WASHINGTON

Family Guidance Center of Warren Outpatient Substance Abuse Treatment Program
492 Route 57 West
Washington, NJ 07882

WAYNE

Wayne TWP Counseling Center
475 Valley Road
Wayne, NJ 07470

WEST ORANGE

West Orange Family Youth Service
4 Charles Street
West Orange, NJ 07052

WEST TRENTON

Trenton Psychiatric Hospital Mentally Ill Chemical Abuse (MICA) Program
West Trenton, NJ 08628

WESTFIELD

National Council on Alcoholism and Drug Dependence/Union County (NCADD)
300 North Avenue East
Westfield, NJ 07090

WHITING

America's Keswick Keswick Colony Division
601 Route 530
Whiting, NJ 08759

WILDWOOD

Cape Counseling Services, Inc. Drug and Alcohol Unit
2604 Pacific Avenue
Wildwood, NJ 08260

WILLIAMSTOWN

Maryville, Inc.
Grant Avenue
RD 2
Williamstown, NJ 08094

WOODBRIDGE

J.F.K. Center for Drug and Alcohol Problems Prevention and Treatment
73 Green Street
Woodbridge, NJ 07095

WOODBURY

Services to Overcome Drug Abuse Among Teenagers of New Jersey, Inc. (SODAT, Inc.)
124 North Broad Street
Woodbury, NJ 08096

NEW MEXICO

ALAMOGORDO

La Placita Reintegration Center Substance Abuse Program
3102 North Florida Street
Route 7
Alamogordo, NM 88310

Otero County Council on Alcohol Abuse and Alcoholism
909 East 10 Street
Alamogordo, NM 88310

ALBUQUERQUE

Albuquerque Health Care for the Homeless
1001 Gold Avenue SW
Albuquerque, NM 87102

Albuquerque Public Schools TASA/Instructional Services Support
220 Monroe Street SE
Albuquerque, NM 87108

Albuquerque Substance Abuse Clinic (ASAC)
117 Quincy Street NE
Albuquerque, NM 87108

Bernalillo County MH/MR Center Center for Alcohol/Substance Abuse and Addictions
2350 Alamo Drive SE
Albuquerque, NM 87106

Charter Hospital of Albuquerque Substance Abuse Services
5901 Zuni Road SE
Albuquerque, NM 87108

Citizens' Council on Alcoholism and Drug Abuse
7711 Zuni Road SE
Albuquerque, NM 87108

Family Recovery, Inc.
11805 Menaul Boulevard NE
Albuquerque, NM 87112

Hogares, Inc.
1218 Griegos Road NW
Albuquerque, NM 87107

Mesa Mental Health Professionals Substance Abuse Services
4640 Jefferson Lane NE
Suite A
Albuquerque, NM 87109

New Mexico Children Youth and Families Juvenile Reintegration Centers
3505 Pan American Highway
Albuquerque, NM 87125

New Mexico Families in Action
5321 A Menaul Boulevard NE
Albuquerque, NM 87110

Peanut Butter and Jelly Therapeutic Preschool and Family Center
1101 Lopez Street SW
Albuquerque, NM 87105

Presbyterian Alcohol and Drug Treatment Center
5901 Harper Drive NE
Albuquerque, NM 87109

Saint Martin's Hospitality Center
1201 3 Street
Albuquerque, NM 87125

Selby Consultants Alcoholism Treatment Education and Training
4233 Montgomery Boulevard NE
Suite 200W
Albuquerque, NM 87109

Staying Straight
3900 Georgia Street NE
Albuquerque, NM 87110

Turquoise Lodge
6000 Isleta Boulevard SW
Albuquerque, NM 87105

Veterans' Affairs Medical Center Substance Abuse Treatment Program
2100 Ridgecrest Drive SE
Albuquerque, NM 87108

Western Clinical Health Services of New Mexico/Monroe Clinic
223 Monroe Street SE
Albuquerque, NM 87108

Youth Development, Inc.
6031 Central Avenue NW
Albuquerque, NM 87105

ANTHONY

Southern New Mexico Human Development, Inc.
880 North Main Street
Suite H
Anthony, NM 88047

ANTHONY (TEXAS)

Federal Correctional Institution La Tuna Drug Abuse Program
Anthony (Texas), NM 88021

BERNALILLO

La Buena Vida, Inc.
720 Camino Del Pueblo
Bernalillo, NM 87004

Life Options
2000 Camino Del Pueblo
Bernalillo, NM 87004

CARLSBAD

Carlsbad Mental Health Association Villa de Esperanza
914 North Canal Street
Carlsbad, NM 88220

Carlsbad Municipal Schools Substance Abuse Services
408 North Canyon Street
Carlsbad, NM 88220

Life Management Program at Guadalupe Medical Center
2430 West Pierce Street
Carlsbad, NM 88220

Southeastern United Family Services The Spring
402 East Wood Street
Carlsbad, NM 88221

CLOVIS

Curry Roosevelt County Educational Services/DWI Center
5104 North Prince Street
Clovis, NM 88101

EMBUDO

Rio Grande Alcoholism Treatment Program, Inc.
Embudo, NM 87531

ESPANOLA

Santa Clara Rehabilitation Center Santa Clara Pueblo
Espanola, NM 87532

FARMINGTON

ABC Counseling Services
219 North Orchard Street
Farmington, NM 87401

Meyers Counseling Services
2124 North Sullivan Avenue
Farmington, NM 87401

Presbyterian Medical Services Main Office Outpatient Alcoholism Services
1001 West Broadway
Farmington, NM 87410

San Juan Detoxification Services Four Winds Addiction Recovery Center
1313 Mission Avenue
Farmington, NM 87401

FORT BAYARD

Fort Bayard Medical Center
Yucca Lodge
Fort Bayard, NM 88036

GALLUP

Behavioral Health Services
650 Vanden Bosch Parkway
Gallup, NM 87301

Gallup/McKinley County Schools Substance Abuse Prevention Program
700 South Boardman Street
Gallup, NM 87301

KDM and Associates, Ltd. Adolescent Outpatient Counseling Program
101 South Clark Street
Gallup, NM 87301

Turquoise Club
218 East Highway 66
Gallup, NM 87301

HOBBS

Guidance Center of Lea County Treatment Center
920 West Broadway
Hobbs, NM 88240

New Mexico Junior College Southeast New Mexico DWI School
5317 Lovington Highway
Hobbs, NM 88240

Palmer Drug Abuse Program of Lea County
200 East Snyder Street
Hobbs, NM 88241

ISLETA

Pueblo of Isleta Alcoholism and Drug Program
Isleta, NM 87022

LAS CRUCES

Associates for Counseling and Recovery
1990 East Lohman Street
Las Cruces, NM 88001

Families and Youth, Inc.
Las Cruces, NM 88004

Mesilla Valley Hospital Chemical Dependency Unit
3751 Del Rey Boulevard
Las Cruces, NM 88005

Southwest Counseling Center, Inc.
1480 North Main Street
Las Cruces, NM 88005

LAS VEGAS

New Mexico Highlands Univ. Northeast Regional Center Coop. 4 Drug Free Schools Consortium
Department of Education
Central Receiving
Las Vegas, NM 87701

West Las Vegas Schools Substance Abuse Prevention Program (SAPP)
179 Bridge Street
Las Vegas, NM 87701

LOS ALAMOS

Los Alamos Council on Alcoholism/Substance Abuse
2132 Central Avenue
Los Alamos, NM 87544

Los Alamos Public Schools Alcohol/Drug Abuse Prevention
751 Trinity Drive
Los Alamos, NM 87544

LOS LUNAS

Cottonwood de Albuquerque
804 Blythe Road
Los Lunas, NM 87031

Los Lunas Public Schools Substance Abuse Services
Los Lunas, NM 87031

Valencia Counseling Services
Los Lunas, NM 87301

MORA

Helping Hands, Inc. North Highway
Mora, NM 87732

POJOAQUE

Pojoaque Valley Schools Substance Abuse Prevention/ Education
Route 11
Pojoaque, NM 87501

PORTALES

Mental Health Resources, Inc. Substance Abuse Services/ Outpatient
300 East First Street
Portales, NM 88130

RATON

Raton Alcohol and Drug Program
207 South 2 Street
Raton, NM 87740

ROSWELL

Counseling Associates, Inc.
109 West Bland Street
Roswell, NM 88201

New Mexico Rehabilitation Center Chemical Dependency Unit
31 Gail Harris Avenue
Roswell, NM 88201

SAN JUAN PUEBLO

Delancey Street/New Mexico, Inc. Old Alcalde Road
Swan Lake Ranch
San Juan Pueblo, NM 87566

Eight Northern Indian Pueblos Council Drug Abuse Program
San Juan Pueblo, NM 87566

SANTA FE

Ayudantes, Inc. Santa Fe Northern Clinic
1316 Apache Street
Santa Fe, NM 87504

Pinon Hills Hospital Substance Abuse Services
313 Camino Alire
Santa Fe, NM 87501

Recovery of Alcoholics Program, Inc.
4100 Lucia Lane
Santa Fe, NM 87505

Saint Vincent Hospital Substance Abuse Services
455 Saint Michael's Drive
Santa Fe, NM 87501

Santa Fe Public Schools Drug Prevention Program
1300 Sierra Vista Annex
Santa Fe, NM 87501

SANTA ROSA

Greater Santa Rosa COA, Inc. The Sure House
130 4 Street
Santa Rosa, NM 88435

SILVER CITY

Border Area Mental Health Services Substance Abuse Services
315 South Hudson Street
Silver City, NM 88061

SOCORRO

Socorro Mental Health Foundation
204-B Heel Avenue
Socorro, NM 87801

TAOS

Taos Alcohol and Drug Program
141 East Kit Carson Road
Taos, NM 87571

ZUNI

Division of Alcoholism Substance Abuse Program
Zuni, NM 87327

NEW YORK

AKRON

American Indian Child And Family Services Drug Abuse Prevention Services
29 Main Street
Akron, NY 14001

ALBANY

Aids Council of Northeastern New York
Alcoholism Prevention Education
Drug Abuse Prevention
750 Broadway
Albany, NY 12207

AL Care
Alcohol Abuse Treatment/ Outpatient
Drug Abuse Treatment
445 New Karner Road
Albany, NY 12205

Albany Citizens Council on Alcohol and Other Chemical Dependence, Inc.
Alcohol Crisis Center
75 New Scotland Avenue Unit G
Capital District Psychiatric Center
Albany, NY 12208

ALC Outpatient Clinic
283 Central Avenue
3rd Floor
Albany, NY 12206

Community Education
283 Central Avenue
Albany, NY 12206

McCarty Ave HH
90 McCarty Avenue
Albany, NY 12202

Albany County
Substance Abuse Prevention Program
845 Central Avenue
Suite LL 101
Albany, NY 12206

Substance Abuse Clinic
845 Central Avenue
Albany, NY 12206

Vocational and Educational Rehab of Substance Abusers
845 Central Avenue
East 1
Albany, NY 12206

Albany Diocesan Drug Education Ministry
40 North Main Avenue
Albany, NY 12203

Arbor Hill Alcoholism Program (AHAP) Supportive Living Facility
295 Clinton Avenue
Albany, NY 12206

City of Albany Department of Human Resources
88 North Lake Avenue
Albany, NY 12206

Colonie Youth Center, Inc.
1653 Central Avenue
Albany, NY 12205

Equinox, Inc. Equinox Counseling Center
214 Lark Street
Albany, NY 12210

Homer Perkins Center, Inc. Residential Drug Free
76–82 2nd Street
Albany, NY 12210

Hope House, Inc.
Hill House Drug Abuse Treatment
Residential Chemical Dependency for Youth/Long Term
1 Hope Lane
Albany, NY 12212

Outpatient Clinic
1500 Western Avenue
Albany, NY 12203

Residential
261 North Pearl Street
Albany, NY 12207

Hospitality House Therapeutic Community, Inc./Residential
271 Central Avenue
Albany, NY 12206

Ironworkers EAP, Inc. Workplace Intervention/EAP
900 North Manning Boulevard
Albany, NY 12207

La Salle School, Inc.
391 Western Avenue
Albany, NY 12203

Saint Anne Institute
160 North Main Avenue
Albany, NY 12206

Saint John's Project Lift, Inc.
Alcoholism Community Residence
37 South Ferry Street
Albany, NY 12202

Drug Abuse Services
45 South Ferry Street
Albany, NY 12202

Saint Peter's Hospital Addiction Recovery Center (SPARC)
Acute Care Unit
315 South Manning Boulevard
Cusack Pavilion
Albany, NY 12208

Community Residence
Day Rehabilitation Program
64 2 Avenue
Albany, NY 12202

Outpatient Alcoholism Clinic
875 Central Avenue
Albany, NY 12206

Admissions/Evaluation Unit
315 South Manning Boulevard
Albany, NY 12208

Hospital Intervention Services
315 South Manning Boulevard
Albany, NY 12208

The Next Step, Inc. Recovery Home for Women
276 Sherman Street
Albany, NY 12206

Visiting Nurse Assoc. of Albany, Inc. Geriatric Alcohol Program
35 Colvin Avenue
Albany, NY 12206

Whitney M. Young, Jr. Health Center, Inc. Family Alcoholism Treatment Services
900 Lark Drive
Albany, NY 12207

ALBION

Park Ridge Hospital Chemical Dependency Services
Outpatient Clinic
Prevention/Treatment
168 South Main Street
Medical Arts Center
Albion, NY 14411

ALDEN

Brylin Hospitals Chemical Dependency Treatment
11438 Genesee Street
Alden, NY 14004

AMHERST

Sisters of Charity Hospital Star/ Amherst
4512 Main Street
Amherst, NY 14226

AMITYVILLE

The Long Island Home, LTD. at South Oaks Hospital
Alcoholism Outpatient and Drug Clinic
Bailey House Alcohol Inpatient Detox Unit
Robbins Inpatient Alcoholism Rehab Center
400 Sunrise Highway
Amityville, NY 11701

Town of Babylon Division of Drug And Alcohol Services
400 Broadway
Amityville, NY 11701

AMSTERDAM

Catholic Family and Community Services Montgomery County Administrative Unit
1 Kimball Street
Amsterdam, NY 12010

Saint Mary's Hospital
Alcoholism Inpatient Rehab Program
427 Guy Park Avenue
Amsterdam, NY 12010

Comprehensive Alcohol Outpatient Clinic
Hospital Alcohol Intervention Services
76 Guy Park Avenue
Amsterdam, NY 12010

APPLETON

Fellowship House, Inc. Somerset House/Alcohol Halfway House
7397 Lake Road
Appleton, NY 14008

ARDSLEY

Student Assistance Services Alcohol Employee Assistance Program Services
300 Farm Road
Ardsley High School
Ardsley, NY 10502

ASTORIA

Hellenic American Neighborhood Action Committee
St. Demetrios HS
30-03 30 Drive
Astoria, NY 11102

Treatment Satellite 1
31-14 30 Avenue
Astoria, NY 11102

Treatment Satellite 2
23-16 30 Avenue
Astoria, NY 11102

AUBURN

Confidential Help for Alcohol Drugs (CHAD)
Alcoholism Outpatient Clinic
Drug Abuse Unit
31 Market Street
Auburn, NY 13021

BABYLON

Crossings Recovery Program, Inc. Crossings Alcoholism Outpatient Clinic
133 East Main Street
Berger Professional Plaza Suite 1B
Babylon, NY 11702

Suffolk County Dept. of Alcohol Substance Abuse Babylon Chemotherapeutic Program
1121 Deer Park Avenue
Babylon, NY 11703

BALDWIN

Baldwin Council Against Drug Abuse (BCADA) Outpatient Drug Free
1914 Grand Avenue
Baldwin, NY 11510

Baldwin Union Free School District Drug Abuse Prevention
Hastings Street
Administrative Building
Baldwin, NY 11510

BALLSTON SPA

Alcohol/Substance Abuse Council of Saratoga County, Inc. Alcoholism Information and Referral
433 Geyser Road
Suite 3
Ballston Spa, NY 12020

Clinical Services and Consult, Inc. Ballston SPA Drug Abuse Alcohol Outpatient Clinic
433 Geyser Road
Ballston Spa, NY 12020

Transitional Services Association, Inc.
Hedgerow Apartment Program
RD 3 Route 67
Ballston Spa, NY 12020

BARRYVILLE

New Hope Manor, Inc. Residential Unit
35 Hillside Road
Barryville, NY 12719

Veritas Therapeutic Community, Inc. Lucy Rudd House
375 Route 55
Barryville, NY 12719

BATAVIA

Genessee Council on Alcohol/ Substance Abuse, Inc.
Alcohol Occupational/Industrial Program
Alcoholism Info. and Referral Program
Drug Abuse Services
Rural Alcoholism Emergency Care Program
Substance Abuse Outpatient
Workplace Intervention/EAP
30 Bank Street
Batavia, NY 14020

Satellite
Student Assistance Program
1 Mill Street
Batavia, NY 14020

Saint Jerome Hospital
Hospital Alcohol Intervention Services
Mercy Hall Alcoholism Inpatient Program
16 Bank Street
Batavia, NY 14020

BATH

Bath Area Hope for Youth, Inc.
117 East Steuben Street
Bath, NY 14810

Steuben County Alcoholism and Substance Abuse Services
3 East Pulteney Square
County Office Building
Bath, NY 14810

Veterans' Affairs Medical Center Chemical Dependency Services
Bath, NY 14810

BAY SHORE

Family Consultation Service, Inc. Family Alcoholism Treatment Center
38 Park Avenue
Bay Shore, NY 11706

Mid-Island Rehabilitation Center Alcoholism Outpatient Clinic
1322 5 Avenue
Bay Shore, NY 11706

Suffolk County Dept. of Alcohol Substance Abuse Bayshore Chemotherapeutic Program
5 Shore Lane
Bay Shore, NY 11706

BAYSIDE

Long Island Jewish Hillside Medical Center Family Treatment Program Outpatient Alcohol Clinic
212-02 41 Avenue
Bayside, NY 11364

BEACON

Craig House Hospital Corporation Drug Abuse Services
Howland Avenue
Beacon, NY 12508

Dutchess County Dept. of Mental Hygiene Beacon Mental Health Clinic/Alcohol Outpatient Clinic
223 Main Street
Beacon, NY 12508

Saint Francis Hospital Poughkeepsie/Beacon
Turning Point/Acute Care
Turning Point/Inpatient Rehabilitation
60 Delavan Avenue
Beacon, NY 12058

BEDFORD HILLS

Renaissance Project, Inc. Bedford Hills Unit
524–26 North Bedford Road
Bedford Hills, NY 10507

BELLMORE

Bellmore Union Free School District Saint Mark's Avenue South
Reinhard School
Bellmore, NY 11710

North Bellmore Union Free School Dist. Drug Abuse Program
2616 Martin Avenue
Bellmore, NY 11710

Rapport of the Merricks & Bellmore, Inc.
217 Bedford Avenue
Bellmore, NY 11710

BETHPAGE

Bethpage Adolescent Development Assoc. (BADA) Outpatient Drug Free
936 Stewart Avenue
Bethpage, NY 11714

BINGHAMTON

Alcoholism Center of Broome County, Inc. Alcoholism Outpatient Clinic
455 State Street
Binghamton, NY 13901

Broome County Community Mental Health Youth Education Intervention I and R Program Yescap
168 Water Street
Binghamton, NY 13901

Broome County Council on Alcoholism Alcoholism Information/Referral Center
25 Main Street
Binghamton, NY 13905

Fairview Halfway House, Inc.
Alcohol Crisis Center
38 Carrol Street
Binghamton, NY 13901

Merrick House Supportive Living Residence
1 Merrick Street
Binghamton, NY 13904

United Health Services, Inc.
Binghamton General Hospital
New Horizons Alcohol Inpatient Rehab Unit
New Horizons Detox Program
New Horizons Outpatient Alcoholism Clinic
Mitchell Avenue
Binghamton, NY 13903

Wilson Memorial Hospital
Alcoholism Hospital Intervention Services
Mitchell Avenue
Binghamton, NY 13903

Workplace Intervention/Alcohol EAP
24–42 Mitchell Avenue
Binghamton, NY 13903

YWCA Clear Visions for Women Halfway House
80 Hawley Street
Binghamton, NY 13901

BLAUVELT

Daytop Village, Inc. Rockland Outreach Center
523 Route 303
First Floor
Blauvelt, NY 10913

South Orangetown School District Drug Abuse Program/ Admin.
Van Wyck Road
Blauvelt, NY 10913

BOHEMIA

Catholic Charities (Talbot House) Alcohol Crisis Center
30-C Carlough Road
Bohemia, NY 11716

Connetquot CSD of Islip Youth and Family Services Program
780 Ocean Avenue
Bohemia, NY 11716

Institute for Rational Counseling, Inc.
Alcoholism Outpatient Clinic
Bohemia Drug Abuse Treatment
150 Knickerbocker Avenue
Bohemia, NY 11716

BRENTWOOD

A Program Planned For Life Enrichment, Inc. (APPLE)
Drug Abuse Treatment
Pilgrim Psychiatric Center
Building 5
Brentwood, NY 11717

Linkage Suffolk
600 Suffolk Avenue
Brentwood, NY 11717

Brentwood Union Free School District Drug Abuse Prevention Program
3 Avenue
Admin. Bldg. Office of Funded Programs
Brentwood, NY 11717

Outreach Development Corporation Outreach II
400 Crooked Hill Road
Brentwood, NY 11717

Town of Islip Dept. of Human Services Access/Accesso Alcoholism Outpatient Clinic
452 Suffolk Avenue
Brentwood, NY 11717

BREWSTER

Community Addiction Treatment Services (CATS) Lakeview Outpatient Alcohol Clinic
Route 6
Middlebranch Offices
Brewster, NY 10509

BRONX

Albert Einstein College of Medicine, Division of Substance Abuse
Soundview 1
1764 Randall Avenue
Bronx, NY 10473

HUB 1/2/3
368 East 149 Street
Bronx, NY 10455

Melrose
260 East 161 Street
Bronx, NY 10451

Pharmacy/Trailers 1 and 2
1500 Waters Place
Betty Parker Building
Bronx, NY 10461

Van Etten Hospital Clinic 2
Morris Park and Seminole Avenue
Wing A
Bronx, NY 10461

Archdiocese of New York Drug Abuse Prevention Program
1725 Castle Hill Avenue
Bronx, NY 10462

Argus Community Inc. Unit 1/ Unit 2
760 East 160 Street
Bronx, NY 10456

Beth Jacob/Beth Miriam School Drug Intervention Program
2126 Barnes Avenue
Bronx, NY 10462

Bronx AIDS Service, Inc. Alcoholism Prevention/ Intervention Program
1 Fordham Plaza
Suite 903
Bronx, NY 10458

Bronx Alcoholism Treatment Center Alcoholism Rehabilitation Unit
1500 Waters Place
Bronx PC Building 1
Bronx, NY 10461

Bronx Committee For the Betterment of Alcoholism Services in the Bronx
3164 3 Avenue
Bronx, NY 10451

Bronx Municipal Hospital Center
CATC/Inpatient Acute Care
CATC/Outpatient Clinic
Morris Park Avenue and Seminole Avenue
Van Etten Hospital 4A Inpatient
Bronx, NY 10461

Bronx/Lebanon Hospital Center
Alcoholism Halfway House
Alcoholism Outpatient Clinic
321 East Tremont Street
Bronx, NY 10457

Alcoholism Inpatient Rehab
Dept. of Psychiatry Detox Unit
1276 Fulton Avenue
Bronx, NY 10456

Methadone Maintenance Treatment Program/KEEP
3100 3 Avenue
Bronx, NY 10451

Cardinal McCloskey School and Home Drug Abuse Treatment Unit
953 Southern Boulevard
Bronx, NY 10459

Daytop Village, Inc. Bronx Outreach Center
16 Westchester Square
Bronx, NY 10461

Featherbed Lane Education Improvement Association I
1705 Andrews Avenue
Bronx, NY 10453

Genesis Medical Professional Services Corp/Outpatient
880 Morris Avenue
Bronx, NY 10451

Hunt's Point Multi-Service
Substance Abuse Treatment Prog/ MMTP 1/MMTP 2/MMTP 3
785 Westchester Avenue
Bronx, NY 10455

Alcoholism Outpatient Clinic
630 Jackson Avenue
Bronx, NY 10455

Lincoln Medical and Mental Health Center Alcoholism Outpatient Clinic
349 East 140 Street
Bronx, NY 10454

Metropolitan Assistance Corp. Drug Abuse Treatment Unit/ Bronx
2530 Grand Concourse
4th and 5th Floors
Bronx, NY 10458

Montefiore Medical Center
SATP Unit II
SATP Unit III
2005 Jerome Avenue
Bronx, NY 10453

Narco Freedom, Inc.
Children and Families Together
391 East 149 Street
Suite 520
Bronx, NY 10455

Key Extended Entry Program
2780 3 Avenue
Bronx, NY 10455

Medically Supervised Ambulatory
477–481 Willis Avenue
Bronx, NY 10455

Methadone Maintenance Treatment Program
477–479 Willis Avenue
Bronx, NY 10455

Neighborhood Youth Diversion Program
4137 3 Avenue
Bronx, NY 10457

New York City School District 10 Community Involvement Program
1 Fordham Plaza
Room 821
Bronx, NY 10458

New York City School District 11 Project Esteem
1250 Arnow Avenue
Bronx, NY 10469

New York City School District 12
Education Prevention/Referral Program
TFIP Prevention Program
1000 Jennings Street
Bronx, NY 10460

New York City School District 7 Substance Abuse Prevention New Directions
501 Courtlandt Avenue
Bronx, NY 10451

New York City School District 8 ADAPT Program
650 White Plains Road
Bronx, NY 10473

New York City School District 9 Spins 9
1020 Anderson Avenue
Room 314
Bronx, NY 10452

Osborne Association Treatment Services
807 Westchester Avenue
Bronx, NY 10455

Our Lady of Mercy Medical Center Alcoholism Outpatient Clinic
4401 Bronx Boulevard
Bronx, NY 10470

Phoenix House
Homeless Induction Unit
480 East 185 Street
Bronx, NY 10458

Phelan Place
1851 Phelan Place
Bronx, NY 10453

Planned Parenthood Project Street Beat
349 East 149 Street
6th Floor Room 601
Bronx, NY 10451

Project Return Foundation, Inc.
Exodus
Exdous House Homeless Unit
Olympus House
Residential
1600 Macombs Road
Bronx, NY 10452

Promesa, Inc.
Drug Free Residential/DFAR/Re-Entry
Homeless/MTA Outpatient/MTA Residential
1776 Clay Avenue
Bronx, NY 10457

Drug Free Residential/Youth Component
2064 Boston Road
Bronx, NY 10460

Riverdale Mental Health Association Riverdale Mental Health Clinic/Outpatient
5676 Riverdale Avenue
Bronx, NY 10471

Saint Barnabas Hospital
Alcohol Detox Program
Alcoholism Outpatient Rehab Program
3 Avenue and East 183 Street
Bronx, NY 10457

Samaritan Village, Inc. Residential Drug Free Program
1381 University Avenue
Bronx, NY 10452

Scan New York Volunteer Parent Aides Assoc. Family Renewal Center Drug Abuse Treatment
1075 Grand Concourse
Bronx, NY 10452

Seneca Center, Inc. Bronx Unit/ Admin
1241 Lafayette Avenue
Bronx, NY 10474

Soundview Throgs Neck CMHC Yeshiva Univ. A. Einstein College OPC
1967 Turnbull Avenue
Suite 34
Bronx, NY 10473

South Bronx Mental Health Council, Inc. CMHC Alcoholism Outpatient Clinic
1241 Lafayette Street
Bronx, NY 10474

South Fordham Organization, Inc. Substance Prevention Education and Recreation (SPEAR)
2385 Valentine Avenue
Suite 1
Bronx, NY 10458

Sports Foundation, Inc.
391 East 149 Street
Room 317
Bronx, NY 10455

TRI Center, Inc. Drug Abuse Treatment/Bronx
400 East Fordham Road
Bronx, NY 10458

Unitas Therapeutic Community, Inc.
928 Simpson Street
Bronx, NY 10459

United Bronx Parents, Inc.
La Casita
834–836 East 156 Street
Bronx, NY 10455

United Youth Center
773 Prospect Avenue
Bronx, NY 10455

Veterans' Affairs Medical Center
Alcohol Dependence Treatment Program
Drug Dependency Treatment Program
130 West Kingsbridge Road
Unit 151G
Bronx, NY 10468

Women In Need, Inc. Casa Rita Alcoholism Outpatient Clinic
284 East 151 Street
Bronx, NY 10451

BROOKLYN

Anchor House
976 Park Place
Brooklyn, NY 11213

Bedford/Stuyvesant Comp. Alcoholism Treatment Center
Alcoholism Outpatient Clinic
Holloman Halfway House
1121 Bedford Avenue
Brooklyn, NY 11216

Bensonhurst Mental Health Clinic, Inc.
Drug Abuse Services
Outpatient/Prevention
86-20 18 Avenue
Brooklyn, NY 11214

Beth Israel Medical Center MMTP
Coney Island Clinic
2601 Ocean Parkway
Brooklyn, NY 11235

Cumberland Clinic
98 Flatbush Avenue
Brooklyn, NY 11217

Methodist Clinic
502 8 Avenue
Brooklyn, NY 11215

Brooklyn Addiction Task Force, Inc. Innovative Alcholism Prevention and Intervention Program
465 Dean Street
Brooklyn, NY 11217

Brooklyn USA Athletic Association, Inc. Jackie Robinson Center for Physical Culture
1424 Fulton Street
Brooklyn, NY 11216

Canarsie Aware, Inc.
Day Service
Outpatient/Prevention
1310 Rockaway Parkway
Brooklyn, NY 11236

Church Avenue Merchants' Block Assoc., Inc. Drug Abuse Prevention Services
1720 Church Avenue
Brooklyn, NY 11226

Coney Island Hospital
Alcoholism Treatment Program
Hospital Intervention Services
2601 Ocean Parkway
Hammett Pavilion
Brooklyn, NY 11235

Congress of Italian Americans Org., Inc.
5901 New Utreche Avenue
Brooklyn, NY 11219

Counseling Service of Eastern District New York, Inc.
186 Montague Street
Brooklyn, NY 11201

Crown Heights Services Center, Inc. Prevention Program
786 Nostrand Avenue
Brooklyn, NY 11216

Cumberland Diagnostic and Treatment Center Alcoholism Treatment Program
100 North Portland Avenue
Brooklyn, NY 11205

Daytop Village, Inc. Brooklyn Outreach Center
401 State Street
Brooklyn, NY 11201

Diocese of Brooklyn Dept. of Education Drug Abuse Prevention Program
6025 6 Avenue
Brooklyn, NY 11220

Discipleship Outreach Ministries
Discipleship Outreach Center Prevention
5711 4 Avenue
Brooklyn, NY 11220

District 3 Youth and Adult, Inc. Outpatient Drug Free
271 Melrose Street
Brooklyn, NY 11206

Dynamic Youth Community, Inc.
Daycare Unit
Outpatient Unit
1830 Coney Island Avenue
Brooklyn, NY 11230

El Puente
211 South 4 Street
Brooklyn, NY 11211

EL Regreso Foundation, Inc.
Referral/Information Services Center
215 Roebling Street
Brooklyn, NY 11211

Drug Abuse Treatment
189–191 South 2 Street
Brooklyn, NY 11211

Federation of Italian American Organizations of Brooklyn, Ltd.
6209 20 Avenue
Brooklyn, NY 11204

Health Science Center Brooklyn/ Kings County
Polydrug Unit 1/Unit 2
The Family Youth Center
600 Albany Avenue
Building K Box 9 Code 26
Brooklyn, NY 11203

The Loft
157–159 Atlantic Avenue
Brooklyn, NY 11201

Hebrew Educational Society
9502 Seaview Avenue
Brooklyn, NY 11236

HHC New York City Kings County Hospital MTA
600 Albany Avenue
Brooklyn, NY 11204

HHC/Woodhill Medical CMH Center Chemical Dependency Services Drug Detox Unit
760 Broadway
Brooklyn, NY 11206

Interfaith Medical Center
Alcoholism Detoxification Unit
Drug Detoxification Unit
1545 Atlantic Avenue
Brooklyn, NY 11213

Jewish Board of Family/Children Services Midwood Adolescent Project
1118 Avenue J
Brooklyn, NY 11230

Kings County Hospital Center
Acute Alcohol Withdrawal Unit
Short Term Alcohol Treatment Unit
600 Albany Avenue
Brooklyn, NY 11203

Comprehensive Alcoholism Outpatient Clinic
606 Winthrop Street
Brooklyn, NY 11203

Kingsboro Alcoholism Treatment Center Methodist Hospital/ Alcoholism Inpatient Rehab Program
506 6 Street
Buckley 5
Brooklyn, NY 11215

Long Island College Hospital
Alcoholism Acute Care Program
70 Atlantic Avenue
Brooklyn, NY 11201

MMTP/KEEP
132–34 Van Dyke Street
Brooklyn, NY 11231

NU Prospect Halfway House
566 Sterling Place
Brooklyn, NY 11238

Outpatient Clinic
255 Duffield Street
Brooklyn, NY 11201

Court Street Clinic
217 Court Street
Brooklyn, NY 11201

Red Hook Clinic
132–34 Van Dyke Street
Brooklyn, NY 11231

Lutheran Medical Center
Alcoholism Outpatient Clinic
514 49 Street
Brooklyn, NY 11220

Acute Care Alcoholism Program
Drug Abuse Treatment
Hospital Intervention Services
150 55 Street
Brooklyn, NY 11220

Metropolitan Assistance Corporation Family Alliance
3021 Atlantic Avenue
Brooklyn, NY 11208

Mid-Brooklyn Health Society, Inc. Alcohol Crisis Center
599 Ralph Avenue
Brooklyn, NY 11233

National Society for Hebrew Day Schools
601 Ocean Parkway
Brooklyn, NY 11218

New Directions Alcoholism and Substance Abuse Treatment Program
206 Flatbush Avenue
Brooklyn, NY 11217

New York City School District 13 Positive Directions for Youth
355 Park Place
Community School District 13
Brooklyn, NY 11238

New York City School District 15 Alternative 15
360 Smith Street
Room 15
Brooklyn, NY 11231

New York City School District 16 Community Counseling Service Program
1010 Lafayette Avenue
Brooklyn, NY 11221

New York City School District 17 Project 17 Drug Prevention Program
402 Eastern Parkway
Room 125
Brooklyn, NY 11225

New York City School District 18 Spins 18/Project Concern
755 East 100 Street
Brooklyn, NY 11236

New York City School District 19 Reach Out/TFIP Prevention Program
557 Pennsylvania Avenue
Brooklyn, NY 11207

New York City School District 20 Drug/Alcohol Abuse Prevention Program
1031 59 Street
PS 105K
Brooklyn, NY 11219

New York City School District 21 Directions 21
345 Van Sicklen Street
Room 335
Brooklyn, NY 11223

New York City School District 22
Alternatives/Spins
Drug Abuse Prevention Council
2525 Haring Street
Room 208
Brooklyn, NY 11235

New York City School District 23 Spins Program
2240 Dean Street
Room 314
Brooklyn, NY 11233

New York City School District 32 Spins 32/Substance Prevention and Intervention Network
797 Bushwick Avenue
Brooklyn, NY 11221

Paul J. Cooper Center for Human Services Outpatient Alcoholism Clinic
106 New Lots Avenue
Brooklyn, NY 11212

Phoenix House Re-Entry
174 Prospect Place
Brooklyn, NY 11238

Ridge Associates Alcoholism Outpatient Clinic
375 76 Street
Brooklyn, NY 11209

Ridgewood/Bushwick Senior Citizens Project Ex TTD/Drug Abuse Prevention Services
217 Wyckoff Avenue
6th Floor
Brooklyn, NY 11237

Saint Martin de Porres Alabama 31 MMTP
480 Alabama Avenue
Brooklyn, NY 11207

Saint Mary's Hospital
Brownsville Clinic 30 MMTP
229 Powell Street
Brooklyn, NY 11212

Classon Unit 29 MMTP
635 Classon Avenue
Brooklyn, NY 11238

South Brooklyn Medical Associates Methadone Maintenance
685 3 Avenue
Brooklyn, NY 11232

Spins 14 Narcotics Education and Prevention
215 Heyward Street
Brooklyn, NY 11206

Urban Center for Alcoholism Services Alcoholism Outpatient Clinic
937 Fulton Street
Brooklyn, NY 11238

Veterans' Affairs Medical Center
Alcohol Dependence Treatment
Drug Dependence Treatment Program
800 Poly Place
Unit 116B
Brooklyn, NY 11209

Woodhull Medical and Mental Center
Chemical Dependency Services
Alcoholism Outpatient Treatment
Alcohol Detox Unit
760 Broadway
Brooklyn, NY 11206

YMCA of Greater New York Kids in Control (KIC)
224 Berry Street
Brooklyn, NY 11211

BUFFALO

AIDS Community Services of Western New York Alcoholism Prevention/Education Program
121 West Tupper Street
Buffalo, NY 14201

Alcohol and Drug Dependency Services Inc.
Alcohol Crisis Program
Inpatient Rehabilitation Services
291 Elm Street
Buffalo, NY 14203

Casa de Vita Halfway House/ Women
200 Albany Street
Buffalo, NY 14213

Chemical Dependency Program for Youth/LT
920 Harlem Road
Buffalo, NY 14224

Family Addiction Outpatient Services
210 Franklin Street
Buffalo, NY 14202

Men's Halfway House
2025 Broadway
Buffalo, NY 14213

Beacon Center Alcoholism Outpatient Clinic
2440 Sheridan Drive
Buffalo, NY 14150

Brylin Hospital, Inc.
Rush Hall/Inpatient Rehab Unit
1263 Delaware Avenue
Buffalo, NY 14209

Chemical Dependency Outpatient Program Alcohol Outpatient
2625 Delaware Avenue
Buffalo, NY 14216

Drug Abuse Treatment Unit
5830 Main Street
Buffalo, NY 14221

Buffalo Columbus Hospital Acute Care Alcoholism Program
300 Niagara Street
Buffalo, NY 14201

Buffalo General Hospital Hospital Alcohol Intervention Services
112 Goodrich Street
Room 106
Buffalo, NY 14203

CAO/Dart Drug Abuse Research and Treatment Program
1237 Main Street
Buffalo, NY 14209

City of Buffalo Board of Education Preventive Education
City Hall
Room 709-C
Buffalo, NY 14202

City of Buffalo DDAS Fillmore/ Leroy Counseling Center
2255 Fillmore Avenue
Buffalo, NY 14214

City of Buffalo DHR Riverside Counseling Clinic
155 Lawn Avenue
Buffalo, NY 14207

City of Buffalo DSAS
Ellicott/Masten Counseling Clinic
425 Michigan Avenue
Sheehan Memorial Hospital
Buffalo, NY 14203

Elmwood Counseling Clinic
656 Elmwood Avenue
Suite 100
Buffalo, NY 14222

Genesee/Moselle Clinic
1532 Genesee Street
Buffalo, NY 14211

Community Prevention Network of Western New York
4255 Harlem Road
Buffalo, NY 14226

Effective Parenting Information (EPIC) Drug Abuse Prevention Services
1300 Elmwood Avenue
Cassety Hall Room 317
Buffalo, NY 14222

Erie County Lakeshore Corp. VT El Comienzo Hispanic Alcoholism Outpatient Clinic
508 Niagara Street
Buffalo, NY 14201

Erie County Medical Center
Alcoholism Acute Care Program
Alcoholism Inpatient Rehabilitation Program
Detoxification Unit
Emergency Room Alcoholism Program
462 Grider Street
Buffalo, NY 14215

Alcoholism Downtown Clinic
1280 Main Street
Buffalo, NY 14209

Northern Erie Clinical Services
2282 Elmwood Avenue
Buffalo, NY 14217

Erie County Sheriff's Department Public Awareness/ Narcotics
134 West Eagle Street
4th Floor
Buffalo, NY 14202

Erie Niagara Counseling Associates Alcoholism Outpatient Clinic
6245 Sheridan Drive
Buffalo, NY 14221

Everywoman Opportunity Center Women's Alcohol Education Program/I and R
237 Main Street
Suite 330
Buffalo, NY 14203

Friends of Cazenovia Manor, Inc.
Cazenovia Manor Community Residence
486 North Legion Drive
Buffalo, NY 14210

Supportive Living Program
923 Sycamore Street
Buffalo, NY 14212

Greater Buffalo Council on Alcohol and Substance Abuse
Alcohol Public Education Info and Referral
Task Force on Integrated Projects (TFIP)
220 Delaware Avenue
Jackson Building, Suite 509
Buffalo, NY 14202

Hispanics United of Buffalo Substance Abuse Prevention/ Education Program
254 Virginia Street
Buffalo, NY 14201

Horizon Human Services, Inc.
Addictions Outpatient/Bailey
3297 Bailey Avenue
Buffalo, NY 14215

Addictions Outpatient/Black Rock
655 Hertel Avenue
Buffalo, NY 14207

Addictions Outpatient/Central Park
60 East Amherst Street
Buffalo, NY 14214

Mental Health Services Corporation VI Drug Abuse Services/Lower West Side
485 Niagara Street
Buffalo, NY 14201

Mid-Erie Alcoholism Outpatient Clinic
463 William Street
Buffalo, NY 14204

Mid-Erie Mental Health Services, Inc. Chemical Dependency Program
1520 Walden Avenue
Buffalo, NY 14225

Monsignor Carr Institute Chemical Dependency Program
76 West Humboldt Parkway
Buffalo, NY 14214

Native American Community Services of Erie and Niagara Counties, Inc.
1047 Grant Street
Buffalo, NY 14207

Northwest Community Mental Health Center
Elmwood Avenue Unit
2495 Elmwood Avenue
Buffalo, NY 14217

Niagara Street Unit
1300 Niagara Street
Buffalo, NY 14213

Parents Anonymous of Buffalo and Erie County
27 Jewett Parkway
Buffalo, NY 14214

Prevention Focus
656 Elmwood Avenue
Suite 300
Buffalo, NY 14222

Research Institute on Addictions Clinical Research Center/ Clinic/Rehab Unit
1021 Main Street
Buffalo, NY 14203

Sheeham Memorial Hospital Chemical Dependency Treatment
425 Michigan Avenue
Buffalo, NY 14203

Sisters of Charity Hospital Star Alcoholism Outpatient Clinic/ West Seneca
1500 Union Road
Buffalo, NY 14224

Spectrum Human Services
New Alternatives
1235 Main Street
Buffalo, NY 14201

South Buffalo Counseling Center
Alcoholism Outpatient Clinic
2040 Seneca Street
Buffalo, NY 14210

Stutzman Alcoholism Treatment Center Alcoholism Inpatient Rehab Unit
360 Forest Avenue
Buffalo, NY 14213

United Way of Buffalo and Erie County, Inc. Western New York United Against Drug and Alcohol Abuse
1170 Niagara Street
Buffalo, NY 14213

Veterans' Affairs Medical Center Ambulatory Substance Abuse Program
3495 Bailey Avenue
Unit 116A
Buffalo, NY 14215

CAIRO

Catholic Family and Community Services Greene County Schools Substance Abuse Prevention Program
Main Street
Cairo, NY 12413

Greene County Community Services Board Drug Abuse Services
Route 3
County Building
Cairo, NY 12413

CAMILLUS

Professional Counseling Services Alcoholism Outpatient Clinic
5099 West Genesee Street
Camillus, NY 13031

CANADAIGUA

Clifton Springs Hospital Alcoholism Outpatient Clinic
11 North Street
Canandaigua, NY 14424

Finger Lakes Alcohol Counseling Referral Agency FLACRA Supportive Living Facilities
Camelot Square Apartments
Canandaigua, NY 14424

Ontario County Division of Health/Medical Services
Prevention
Turning Point/Drug Free Outpatient
3907 County Road 46
County Office Building 2
Canandaigua, NY 14424

Park Ridge Chemical Dependency, Inc. Alcoholism Outpatient Clinic
468 South Pearl Street
Lemac Professional Building
Canandaigua, NY 14424

Veterans' Affairs Medical Center Substance Abuse Treatment Program
400 Fort Hill Avenue
Building 6
Canandaigua, NY 14424

CANASTOTA

Madison County Council Alcohol Substance Abuse
Information and Referral Program
Student Assistance Program
RR 5
Birchwood Office Building
Canastota, NY 13032

CANTON

Alcohol Substance Abuse Council of St. Lawrence County Inc./ Alcoholism Info and Referral Program
7 Main Street
Canton, NY 13617

North Country Freedom Homes The Canton House/Halfway House
25 Dies Street
Canton, NY 13617

CARLE PLACE

Carle Place Union Free School District Substance Prevention Education Program/Admin.
Cherry Lane
Carle Place, NY 11514

Nassau TASC Alcoholism Info. Referral Unit
1 Old Country Road
Suite 420
Carle Place, NY 11514

CARMEL

Arms Acres, Inc.
Alcoholism Inpatient Rehab Unit
Alcoholism Primary Care Program
Residential Chemical Dependency for Youth/Short-Term
Seminary Hill Road
Carmel, NY 10512

Natl. Council on Alcoholism and Other Drug Dependencies Community Education Information
2 Church Street
Carmel, NY 10512

Putnam County Alcohol/ Substance Abuse Services
Common Sense Outpatient Drug Free Program
Common Sense Prevention Program
Talbot House Alcoholism Outpatient Clinic
17 Brewster Avenue
Carmel, NY 10512

CATSKILL

Alcoholism Center of Columbia County Greene County Alcohol Services Outpatient Clinic
66 William Street
Catskill, NY 12414

CENTER MORICHES

Hampton Council of Churches Family Counseling Service
529 Main Street
School Special Education Admin. Bldg.
Center Moriches, NY 11934

CENTEREACH

YMCA of Long Island, Inc. Family Services Outpatient Drug Free Center
11 Unity Drive
Unity Drive School
Centereach, NY 11720

CENTRAL ISLIP

La Union Hispanica en Suffolk County Drug Free Day Service
45 West Suffolk Avenue
Central Islip, NY 11722

CLIFTON PARK

Lifestart Health Services Alcoholism Outpatient Clinic
1745 Route 9
Clifton Park, NY 12065

Shenendehowa Central School Substance Abuse Program
1 Fairchild Square
Clifton Park, NY 12065

CLIFTON SPRINGS

Clifton Springs Hospital Alcohol Intervention Services
2 Coulter Road
Clifton Springs, NY 14432

Finger Lakes Alcohol Counseling Referral Agency
Alcoholism Crisis Center/ Outpatient Clinic
Community Residence I
28 East Main Street
Clifton Springs, NY 14432

COBLESKILL

New Directions Schoharie County Substance Abuse Program/OP DF
150 East Main Street
Cobleskill, NY 12043

Schoharie County Community Services Program for Alcoholism Recovery
150 East Main Street
Cobleskill, NY 12043

Schoharie County Council on Alcohol and Substance Abuse, Inc./Drug Unit
150 East Main Street
Suite 102
Cobleskill, NY 12043

COHOES

Saint Peter's Addiction Recovery Center (SPARC) Cohoes Outpatient Clinic
50 Remsen Street
Cohoes, NY 12047

COLD SPRING HARBOR

Huntington Youth Bureau/Drug and Alcohol Cold Spring Harbor YDA
Goose Hill Road
Goose Hill School
Cold Spring Harbor, NY 11724

COMMACK

Huntington Youth Bureau/Drug and Alcohol Commack YDA/ Long Acre School
Sarina Drive and Betty Lane
Commack, NY 11725

COOPERSTOWN

Leatherstocking Council on Alcoholism Addictions Foundation, Inc./TFIP
45 Pioneer Street
Cooperstown, NY 13320

CORAM

Passages Counseling Center Alcoholism Outpatient Clinic/ Montauk
3680 Route 112
Coram, NY 11727

CORNING

Family Service Society of Corning Youth Services Program/Admin.
254 Denison Parkway East
Corning, NY 14830

Southern Tier Office of Social Ministries Counseling Services of the Southern Tier
65 East First Street
Corning, NY 14830

Steuben Council on Alcoholism
Prevention/Info. and Referral
Student Assistance Program
Youth Assistance Program
27 Denison Parkway East
Corning, NY 14830

CORONA

Elmcor Youth and Adult Activities, Inc.
Day Service
Prevention Unit
107-20 Northern Boulevard
Corona, NY 11368

Residential Homeward Bound
Residential Unit 2
107-10 Northern Boulevard
Corona, NY 11368

CORTLAND

Alcohol Services Inc. Cortland Alcoholism Outpatient Clinic
17 Main Street
Cortland, NY 13045

Catholic Charities of Cortland County The Charles Street Halfway House
29 Charles Street
Cortland, NY 13045

Cortland City School District Comp. School-Based Prevention Program
1 Valley View Drive
Cortland, NY 13045

Family Counseling Services of Cortland County
Alcoholism Outpatient Clinic
Medically Supervised Amb. Substance Abuse Program
10 North Main Street
Cortland, NY 13045

Seven Valleys Council on Alcohol Substance Abuse Alcoholism Comm. Education and Intervention I and R
15 Central Avenue
Cortland, NY 13045

DANSVILLE

Livingston County Council/ Alcoholism Alcoholism Outpatient Clinic
Red Jacket Street
Dansville, NY 14437

DELMAR

Addiction Counseling Center of Bethlehem Cross Road Alcoholism Outpatient Clinic
4 Normanskill Boulevard
Delmar, NY 12054

DIX HILLS

Huntington Youth Bureau/Drug and Alcohol Half Hollow Hills YDA
525 Half Hollow Road
Dix Hills, NY 11746

DUNKIRK

Chautauqua Alcohol and Substance Abuse Council Alcoholism Public Education Info. and Referral
314 Central Avenue
Room 308
Dunkirk, NY 14048

Chautauqua County Dept. Of Mental Health Alcohol and Substance Abuse Clinic/ Alcohol Unit
55 East 4 Street
Dunkirk, NY 14048

EAST ELMHURST

Montefiore Medical Center Rikers Island
Health Services Alcohol AT1
15-15 Hazen/Units 1 and 2
18-18 Hazen Unit
15-15 Hazen Street
East Elmhurst, NY 11370

EAST GREENBUSH

Leonard Hospital Alcohol Substance Abuse Services
Drug Abuse Unit
Chemical Dependency Outpatient Clinic
743 Columbia Turnpike
East Greenbush, NY 12061

EAST HAMPTON

A Program Planned for Life
Enrichment, Inc. (APPLE)/East
End of Industrial Road
East Hampton, NY 11937

W-L
43 Main Street
East Hampton, NY 11937

EAST MEADOW

Nassau County Coalition Against Domestic Violence/Drug Abuse Prevention
Nassau County Medical Center
East Meadow, NY 11554

Nassau County Substance Alternative Clinic
MMTP Clinic I
2201 Hempstead Turnpike
Nassau County Medical Center Building Z
East Meadow, NY 11554

MMTP Clinic II
NCDDAA Drug Free Outpatient Support Services
2201 Hempstead Turnpike
Nassau County Medical Center Building K
East Meadow, NY 11554

EAST NORTHPORT

Huntington Youth Bureau/Drug and Alcohol Northport/East Northport YDA
7 Diane Court
East Northport, NY 11731

EAST ROCKAWAY

East Rockaway Union Free School District 19 Alcohol and Drug Abuse Program Admin.
Ocean Avenue
East Rockaway High School
East Rockaway, NY 11518

EDEN

Friends of Cazenovia Manor, Inc. Turning Point House Recovery Home
9136 Sandrock Road
Eden, NY 14057

ELIZABETHTOWN

North Country Council on Alcoholism (NCCA) Comm. Education and Intervention/ Information and Referral
Church Street
Elizabethtown, NY 12932

Saint Joseph's Rehabilitation Center Inc. Alcoholism Outpatient Cinic
Maple Avenue
Elizabethtown, NY 12932

ELLENVILLE

Ellenville Community Hospital Acute Care Program
Route 209
Ellenville, NY 12428

Renaissance Project, Inc. Ellenville Residential Facility
767 Cape Road
Ellenville, NY 12428

Samaritan Village, Inc. Ellenville Residential
751 Briggs Highway
Ellenville, NY 12428

Ulster County Mental Health Services Ellenville Alcohol Abuse Outpatient Clinic
50 Center Street
Trudy Resnick Farber Center
Ellenville, NY 12428

ELMHURST

Elmhurst Hospital Center
Alcoholism Inpatient Detoxification Unit
79-01 Broadway
Ward A-9
Elmhurst, NY 11373

Community Residential Program
81-30 Baxter Avenue
Elmhurst, NY 11373

Human Service Centers, Inc. Alcoholism Outpatient Clinic
87-08 Justice Avenue
Suite 1G
Elmhurst, NY 11373

ELMIRA

Alcohol Drug Abuse Council of Chemung County
Alcohol and Drug Information Center
Student Assistance Program
380 West Gray Street
Elmira, NY 14905

Economic Opportunity Program, Inc.
TFIP Prevention Program
318 Madison Avenue
Elmira, NY 14901

Alcoholism and Drug Rehab Clinic
310 West 3 Street
Elmira, NY 14901

Family Services of Chemung County, Inc. Workplace Intervention/Alcoholism EAP
1019 East Water Street
Elmira, NY 14901

Saint Joseph's Hospital S Tier Alcoholism Rehab Services (STARS)
Alcoholism Outpatient
Inpatient
555 East Market Street
Elmira, NY 14902

Salvation Army/Citadel Lormore House Community Residence
314 Lormore Street
Elmira, NY 14904

Schuyler/Chemung/TIOGA Boces Workplace Intervention/ Alcoholism EAP
431 Philo Road
Elmira, NY 14903

ELMONT

Elmont Union Free School District 16 Alcohol and Drug Abuse Prevention Program
Elmont Road School
Elmont, NY 11003

Long Island Jewish Medical Center
Elmont Treatment Center/ Alcoholism Treatment
Drug Abuse TRT
40 Elmont Road
Elmont, NY 11003

West Nassau Counseling Center Drug Abuse Services
90 Meachem Avenue
Elmont, NY 11003

ELMSFORD

Mid-Hudson Valley AIDS Task Force (ARCS)
AIDS Prevention and Management Program
Alcoholism Prevention Program
2269 Saw Mill River Road
Executive Park 1
Elmsford, NY 10523

EVANS MILLS

Credo Foundation, Inc. Drug Free/Residential Facility
RD 1
Evans Mills, NY 13637

FAR ROCKAWAY

Daytop Village, Inc. Far Rockaway Entry and Re-Entry
316 Beach 65 Street
Far Rockaway, NY 11692

Episcopal Health Services, Inc.
South Shore Alcoholism Inpatient Detox Unit
327 Beach 19 Street
Far Rockaway, NY 11691

South Shore Alcoholism Outpatient Program
1815 Cornaga Avenue
Far Rockaway, NY 11691

Saint John's Episcopal Hospital Alcoholism Outpatient Clinic
1908 Brookhaven Avenue
Far Rockaway, NY 11691

FARMINGDALE

Price, Inc. Counseling Center/ Outpatient DF
399 Conklin Street
Farmingdale, NY 11735

FISHKILL

Mid-Hudson Alcoholism Recovery Center Clove Manor Community Residence
Route 9 South
RD 3
Fishkill, NY 12524

FLUSHING

AIDS Center of Queens County, Inc. Innovative Alcoholism Prevention Intervention Program
97-45 Queens Boulevard
Suite 1220
Flushing, NY 11374

Aurora Concept, Inc.
Medically Supervised Ambulatory Program
78-39 Parsons Boulevard
Flushing, NY 11366

Residential Units 1 and 2
160-40 78 Road
Flushing, NY 11366

Booth Memorial Medical Center Inpatient Detox Unit
56-45 Main Street
Flushing, NY 11355

Elmhurst Hospital Center Methadone Maintenance Treatment Program
79-01 Broadway
Room DI-71
Flushing, NY 11373

Jewish Comm. Services of Long Island, Inc. Living Free Drug Program/Rego Park
97-45 Queens Boulevard
Flushing, NY 11374

Long Island Consultation Center, Inc. Alcoholism Outpatient Clinic
97-29 64 Road
Flushing, NY 11374

Mental Health Providers of Western Queens, Inc. Alcoholism Services
62-07 Woodside Avenue
Flushing, NY 11377

New York City School District 24 Project Friend
8000 Cooper Avenue
Flushing, NY 11385

New York City School District 25
Project 25/Drug Abuse Program
Early Intervention Alcohol Program
34-65 192 Street Room 226
Flushing, NY 11358

New York City School District 26 Drug Abuse Program
174-10 67 Avenue
Flushing, NY 11365

Psychiatric and Addictions Recovery Services (PARS) PC
92-29 Queens Boulevard
Suite 2E
Flushing, NY 11374

Queens Child Guidance Center, Inc.
41-25 Kissena Boulevard
Suite 118
Flushing, NY 11355

FOREST HILLS

New York City School District 28 Spins 28
108-55 69 Avenue
Forest Hills, NY 11375

FRANKLIN SQUARE

Comm. Counseling Services of West Nassau
Alcoholism Outpatient Clinic
Outpatient Drug Free
1200-A Hempstead Turnpike
Franklin Square, NY 11010

FREEPORT

Freeport Citizens' Comm. on Drug Abuse Operation Pride/ Outpatient DE
33 Guy Lombardo Avenue
Freeport, NY 11520

Freeport Hospital
Alcoholism Inpatient Rehab Program
Alcoholism Primary Care Program
267 South Ocean Avenue
Freeport, NY 11520

Freeport Public Schools Youth Counselor Services
South Brookside Avenue
Freeport, NY 11520

Mercy Hospital Association
Mercy Hill HWH/Women
95 Pine Street
Freeport, NY 11520

Women's Day Rehabilitation Services
90 Mill Road
Freeport, NY 11520

South Shore Child Guidance Center Care Alcoholism Program
87 Church Street
Freeport, NY 11520

FULTON

Oswego County Council on Alcoholism Alcoholism Information and Referral
4 Tower Drive
Suite B
Fulton, NY 13069

GARDEN CITY

Garden City Public School
56 Cathedral Avenue
Garden City, NY 11530

Mercy Hospital Association Family Counseling Alcoholism Outpatient Clinic
385 Oak Street
Garden City, NY 11530

Mercy Hospital New Hope Primary Care Program
8 Street Avenue P
Mitchel Field Complex
Garden City, NY 11530

GARNERVILLE

Haverstraw/Stony Point Central School District/Drug Abuse Prevention
65 Chapel Street
Garnerville, NY 10923

GENEVA

Council on Alcoholism and Other Chemical Dependencies of the Finger Lakes/Ontario Council Office
620 West Washington Street
Geneva, NY 14456

Finger Lakes Alcoholism Counseling Referral Agency
Alcoholism Outpatient Clinic
246 Castle Street
Geneva, NY 14456

Hobart and William Smith Colleges Alcohol Prevention Program
119 Saint Clair Street
Geneva, NY 14456

Planned Parenthood of Finger Lakes Alcoholism Prevention/ Education Program
601 West Washington Street
Geneva, NY 14456

GLEN COVE

Angelo J. Melillo Center for Mental Health, Inc. Alcoholism Counseling Services
30A Glen Street
Glen Cove, NY 11542

Glen Cove City School District How To Win
Dosoris Lane
Glen Cove, NY 11542

North Shore University Hospital at Glen Cove
Adolescent Day Service
Women's/Children's Program
Detox
Outpatient
Saint Andrew's Lane
Glen Cove, NY 11542

GLENS FALLS

Center for Recovery Alcoholism Outpatient Rehabilitation
300 Bay Road
Glens Falls, NY 12804

Family Treatment Center For Alcoholism of Glens Falls Hospital Alcoholism Outpatient Clinic
126 South Street
Glens Falls, NY 12801

First Step Counseling Center Alcoholism Outpatient Clinic
16 Exchange Street
Glens Falls, NY 12801

Glens Falls Hospital CMHC Human Resource Center
46 Elm Street
Glens Falls, NY 12801

GLOVERSVILLE

Alcoholism Council of Hamilton/ Fulton/Montgomery Counties/ Alcoholism Prevention
40 North Main Street
Gloversville, NY 12078

Fulton County Comm. Services Board Fulton County Alcoholism Services
34 West Fulton Street
Gloversville, NY 12078

GOSHEN

Alcoholism and Drug Abuse Council of Orange County Inc.
Alcoholism Prevention and Education
Drug Prevention and Education
224A Main Street
Goshen, NY 10924

Forensic Mental Health Clinic Drug Abuse Services/DF
40 Erie Street
Goshen, NY 10924

Pius XII Chemical Dependency Program Goshen Cocaine Clinic
224 Main Street
Goshen, NY 10924

Warwick Area Migrant Committee, Inc. Community Center/Info. and Referral
Pulaski Highway
RD 2
Goshen, NY 10924

GOWANDA

Tri-County Memorial Hospital
Alcoholism Inpatient Rehab Program
Chemical Dependency Programs
Drug Abuse Treatment
100 Memorial Drive
Gowanda, NY 14070

GREAT NECK

Great Neck Union Free School District 7
345 Lakeville Road
Great Neck, NY 11020

Great Neck Community Organization for Parents and Youth (COPAY)/Outpatient DF
21 North Station Plaza
2nd Floor
Great Neck, NY 11021

GREEN ISLAND

Albany Citizens' Council on Alcoholism and Other Chemical Dependencies, Inc. Green Island HH
123 George Street
Green Island, NY 12183

GREENLAWN

Huntington Youth Bureau/Drug and Alcohol Harbor Fields Elwood YDA
8 Gates Street
Greenlawn, NY 11740

GREENPORT

Eastern Long Island Hospital Quannacut Alcoholism Inpatient Rehab Program
201 Manor Place
Greenport, NY 11944

GROTON

Ithaca Alpha House Center, Inc. Medically Supervised
101 Cayuga Street
Groton, NY 13073

GUILDERLAND

Saint Peter's Addiction Recovery Center (SPARC) Inpatient Rehabilitation Program
2232 Western Avenue
Guilderland, NY 12084

HAMBURG

Erie County Medical Center Southern Erie Clinical Services
517 Sunset Drive
Hamburg, NY 14075

Mental Health Services Corporation VI Drug Abuse Services/Hamburg Counseling
91 Union Street
Hamburg, NY 14075

HAMDEN

Delaware County Alcohol and Drug Abuse Services
Route 10
Hamden, NY 13782

Delaware County Comm. Services Board De County Alcohol Drug Abuse Services/Hamden
RD 1
Hamden, NY 13782

HAMPTON BAYS

Hampton Council of Churches Family Counseling Service
Ponquoque Avenue
Hampton Bays, NY 11946

HANNIBAL

Oswego County Opportunities, Inc. Arbor House Alcoholism Halfway House
RD 1 Hall Road
Hannibal, NY 13074

HARRIS

Comm. General Hospital of Sullivan County Biochemical Dependency Unit
Bushville Road
Harris, NY 12742

HARRISON

Saint Vincent's Hospital and Medical Center Alcoholism Outpatient Program
275 North Street
Harrison, NY 10528

HARTSDALE

Center for Human Options, Inc.
Central Avenue and Jane Street
Hartsdale, NY 10530

Daytop Village, Inc. Westchester Outreach Center
246 Central Park Avenue
Hartsdale, NY 10530

HASTINGS-ON-HUDSON

Youth Advocate Program Drug Abuse Prevention Council/ Admin.
44 Main Street
Hastings-on-Hudson, NY 10706

HAUPPAUGE

A Program Planned for Life Enrichment, Inc. (APPLE)
Homeless/Residential
220 Veterans Highway
Hauppauge, NY 11788
Outpatient/Satellite
1373-42 Veterans Highway
Hauppauge, NY 11788

Suffolk County Coalition on Alcoholism and Other Drug Dependency Alcohol Comm. Education and Intervention/ Info. and Referral
900 Wheeler Road
Suite 260
Hauppauge, NY 11788

Suffolk County Dept of Alcohol/ Substance Abuse
Keep Program
Suffolk County Drug Free Unit
1330 Motor Parkway
Hauppauge, NY 11788

HAVERSTRAW

Open Arms, Inc. Halfway House
57 Sharp Street
Haverstraw, NY 10927

Village of Haverstraw Counseling Center/Reachout
220A Route 9W
Haverstraw, NY 10927

HAWTHORNE

Cortland, Inc. Cortland Medical Alcoholism Outpatient Clinic
4 Skyline Drive
Hawthorne, NY 10532

HEMPSTEAD

Alliance Counseling Center, Inc.
497 South Franklin Street
Hempstead, NY 11550

Counseling Service of Eastern District New York, Inc. Drug Abuse Treatment
175 Fulton Avenue
Suite 301C
Hempstead, NY 11501

East Meadow Union Free School Dist 3 Substance Prevention/ Education Program
101 Carman Avenue
East Meadow High School
Hempstead, NY 11554

EOC Nassau County, Inc.
106 Main Street
Hempstead, NY 11550

Family Service Association East Meadow Counseling Center
1975 Hempstead Turnpike
Suite 405
Hempstead, NY 11554

Family Services Association of Nassau County
Alcohol Treatment Center
Drug Treatment Center
126 North Franklin Street
Hempstead, NY 11550

Growing Healthy K 7 NCDDAA Education Unit
175 Fulton Avenue
Room 600
Hempstead, NY 11554

Hempstead Union Free School District Drug Abuse Program
265 Peninsula Boulevard
Hempstead, NY 11550

Hispanic Counseling Center (SALSA)
Alcoholism Outpatient Clinic
Outpatient Drug Free Unit
Prevention Unit
250 Fulton Avenue
Hempstead, NY 11500

Jewish Comm. Services of Long Island, Inc. Living Free Drug Program/Hempstead
50 Clinton Street
Hempstead, NY 11550

Long Island Jewish Hillside Medical Center Project Outreach/Intensive Outpatient
600 Hempstead Turnpike
Hempstead, NY 11552

Nassau County Dept. of Drugs and Alcohol
Alcohol Outpatient Clinic/East Meadow
ALTOX
COCAA
2201 Hempstead Turnpike
Nassau County Medical Center
Hempstead, NY 11554

Uniondale High School Drug Abuse Prevention
Goodrich Street
Hempstead, NY 11553

West Hempstead Union Free School District Drug Abuse Program
252 Chestnut Street
Hempstead, NY 11552

Work Evaluation and Resources Center (WERC)
160 North Franklin Street
Hempstead, NY 11550

HERKIMER

Herkimer County Alcoholism Services Alcoholism Outpatient Clinic
119 Mary Street
Herkimer, NY 13350

HICKSVILLE

Central Nassau Guidance and Counseling Services, Inc. Outpatient DE
248 Old Country Road
Hicksville, NY 11801

EAC, Inc. Educational Assistance Center of Long Island
240 A Old Country Road
Hicksville, NY 11801

Family Service Association
Drug Program
Hicksville Alcoholism Outpatient Clinic
23A Jerusalem Avenue
Hicksville, NY 11801

Institute for Rational Counseling, Inc. Hicksville Drug Abuse Treatment
76 North Broadway
Hicksville, NY 11801

HILLBURN

Ramapo Central School District School Community Counselor Program
Mountain Avenue
Hillburn, NY 10931

HOGANSBURG

Saint Regis Mohawk Tribe Health Services
Community Building
Hogansburg, NY 13655

HORNELL

Hornell Area Concern for Youth, Inc. Administrative Unit
30 Seneca Street
Hornell, NY 14843

Saint James Mercy Hospital
Hospital Intervention Service
411 Canisteo Street
Hornell, NY 14843

Mercycare Alcoholism Treatment Center
1 Bethesda Drive
Hornell, NY 14843

HUDSON

Alcoholism Center of Columbia County Alcoholism Outpatient Program
419 Warren Street
Hudson, NY 12534

Catholic Family and Community Services Columbia County/ Outpatient DF
431 East Allen Street
Hudson, NY 12534

Columbia County Schools Community Services Project
Outpatient 1
413 East Allen Street
Hudson, NY 12534

Outpatient 2
71 North 3 Street
Hudson, NY 12534

Outpatient 3
315 Warren Street
Hudson, NY 12534

Recovery Counseling Associates, Inc. Alcoholism Outpatient Clinic
160 Fairview Avenue
Professional Building
Hudson, NY 12534

HUDSON FALLS

Warren/Washington Counties Council on Alcoholism/ Substance Abuse Community Education and Intervention Info. and Referral
56 3/4 Main Street
Hudson Falls, NY 12839

Washington/Warren/Hamilton/ Essex Boces/Substance Abuse Prevention Program
10 La Crosse Street
Hudson Falls, NY 12839

HUNTINGTON

Huntington Youth Bureau/Drug and Alcohol
Counseling Center
423 Park Avenue
Huntington, NY 11743

Huntington Village YDA
Main Street
First Presbyterian Church
Huntington, NY 11743

HUNTINGTON STATION

Daytop Village, Inc.
Huntington Station/Medically Supervised Outpatient
Suffolk Outreach
2075 New York Avenue
Huntington Station, NY 11746

Huntington Youth Bureau/Drug and Alcohol
Huntington Station YDA
4 Railroad Street
Huntington Station, NY 11746

South Huntington YDA
300 West Hills Road
Huntington Station, NY 11746

Long Island Assoc. for AIDS Care, Inc. Alcohol Chemical Dependency Program
1335 New York Avenue
Huntington Station, NY 11746

Long Island Center, Inc. Alcoholism Outpatient Clinic
7 Dawson Street
Huntington Station, NY 11746

Suffolk County Dept. of Alcoholism Substance Abuse Huntington Station MMTP Clinic
689 East Jericho Turnpike
Huntington Station, NY 11746

HYDE PARK

Hyde Park Central School District Partnership in Youth K12
Haviland Road
Hyde Park, NY 12538

ILION

Catholic Family and Community Services Herkimer County/ School Drug Abuse Prevention
61 West Street
Ilion, NY 13357

INDIAN LAKE

Hamilton County Community Services Alcoholism Counseling and Prevention Services
83 White Birch Lane
Indian Lake, NY 12842

ISLIP

Islip Union Free School District Substance Abuse Prevention Program
215 Main Street
Islip, NY 11751

Town of Islip Dept. of Human Services
Access Alcoholism Youth Education and Intervention
Drug Counseling Services/ Outpatient DF
401 Main Street
Islip, NY 11751

ISLIP TERRACE

East Islip School District East Islip Peer Leadership Program
Redmen Street
East Islip High School
Islip Terrace, NY 11752

ITHACA

Alcoholism Council of Tompkins County
Alcoholism Outpatient Clinic
Peer Education Program
201 East Green Street
Suite 500
Ithaca, NY 14850

Alcoholism Council of Tompkins County Community Education and Intervention
201 East Green Street
Suite 500
Ithaca, NY 14850

Ithaca Alpha House Center, Inc. Outreach Center
102 The Commons
Ithaca, NY 14850

Tompkins/Seneca/Tioga Boces Prevention Resources and Education Program
555 Warren Road
Ithaca, NY 14850

JACKSON HEIGHTS

Elmhurst Hospital Center Alcoholism Outpatient Clinic
74-16 Roosevelt Avenue
Jackson Heights, NY 11372

JAMAICA

Beth Israel Medical Center MMTP Queens Clinic
82-68 164 Street
3C1 Bldg. T, Dept. Medicine
Jamaica, NY 11432

Center for Children and Families, Inc. Residential Drug Free
89-12 162 Street
Jamaica, NY 11432

Creedmoor Alcoholism Treatment Center Alcoholism Inpatient Rehab Program
80-45 Winchester Boulevard
Building 19(D)
Jamaica, NY 11427

Daytop Village, Inc. Queens Outreach Center
166-10 91 Avenue
Jamaica, NY 11432

Laurelton/Springfield Gardens C of C Drug Abuse Prevention Services
Jamaica, NY 11413

Outreach Development Corporation
Drug Abuse Services
Outreach Family Services Outpatient Clinic
89-15 Woodhaven Boulevard
3rd Floor
Jamaica, NY 11421

Phoenix House, Inc. Portal
175-15 Rockaway Avenue
Jamaica, NY 11434

Queens County Development Corporation Residential and Aftercare
89-15 Woodhaven Boulevard
Jamaica, NY 11421

Queens Hospital Center Health and Hospital Corp.
Adult Drug Treatment Program/Residential
Adult Drug Detox
Alcoholism Clinic
Alcoholism Consultation Team
Alcoholism Inpatient Detox Unit
Adolescent Drug Treatment Program/Reentry Unit
82-68 164 Street
Jamaica, NY 11432

Stop DWI Program
114-2 Guy Brewer Boulevard
Suite 216
Jamaica, NY 11434

Queens Village Comm. for Mental Health
J Cap Daycare Clinic
J Cap New Spirit II/Outpatient Clinic
162-04 South Road
Jamaica, NY 11433

J Cap Residential Unit 2
177-33 Baisley Boulevard
Jamaica, NY 11434

J Cap Residential Unit 3
156-02 Liberty Avenue
Jamaica, NY 11433

Samaritan Village, Inc.
MTA Ambulatory/Residential
130-15 89 Road
Jamaica, NY 11419

MTA Residential Drug Free Program
88-83 Van Wyck Expressway
Jamaica, NY 11435

Springfield Gardens Comm. Services Agency Drug Abuse Prevention Services
131-29 Farmers Boulevard
Jamaica, NY 11434

JAMESTOWN

Chautauqua Alcohol and Substance Abuse Council
Alcohol Public Education/Info. and Referral
Awareness Theatre
Jamestown Office/Drug Unit
2–6 East 2 Street
Fenton Building, Suite 308
Jamestown, NY 14701

Chautauqua County Dept. of Mental Health Alcohol and Substance Abuse Clinic
Glasgow Street
Jones Hill Professional Building
Jamestown, NY 14901

WCA Hospital Jones Memorial Health Center
Alcoholism Inpatient Rehab Program
51 Glasgow Avenue
Jamestown, NY 14701

The WCA Hospital Alcoholism Outpatient Clinic
73 Forest Avenue
Jamestown, NY 14702

JERICHO

Jericho Union Free School District Drug Abuse Program
Cedar Swamp Road
District Office Jericho High School
Jericho, NY 11753

JOHNSON CITY

Southern Tier Drug Abuse Treatment Center/Outpatient
65 Broad Street
Johnson City, NY 13790

Southern Tier AIDS Program, Inc.
Alcoholism Prevention Program
Drug Abuse Program
122 Baldwin Street
Johnson City, NY 13790

United Health Services Wilson Memorial Hospital Hospital Alcohol Intervention Services
Harrison Street
Johnson City, NY 13790

JOHNSTOWN

Catholic Family and Community Services
Fulton County Administrative Unit
Fulton County Drug Abuse Treatment
208 West State Street
Johnstown, NY 12095

KATONAH

Four Winds Hospital, Inc. Choices Alcoholism Outpatient Clinic
800 Cross River Road
Katonah, NY 10536

KERHONKSON

Veritas Villa, Inc.
Alcoholism Inpatient Rehabilitation
Drug Abuse Inpatient
RR 2
Lower Cherry Town Road
Kerhonkson, NY 12446

KINGSTON

Alcohol Substance Abuse Council/Ulster County Public Education/Info. and Referral
785 Broadway
Kingston, NY 12401

Family of Woodstock, Inc. Prevention
39 John Street
Kingston, NY 12401

Kingston City Schools Consoidated Prevention Activities
61 Crown Street
Kingston, NY 12401

Prevention Connection
785 Broadway
Kingston, NY 12401

Ulster County Mental Health Services
Alcohol Day Rehab/Evening Intensive Program
Drug Free Clinic
Drug Free Jail Program
In-School Prevention Program
Kingston Alcohol Abuse Outpatient Clinic
Methadone Maintenance and Rehab Program/Outpatient
Golden Hill Drive
Kingston, NY 12401

LAKEVILLE

Livingston County Council on Alcoholism/Substance Abuse
Alcoholism and Drug Abuse Prevention Program
Alcoholism Outpatient Clinic
6003 Big Tree Road
Lakeville, NY 14480

LATHAM

Albany/Schoharie/Schenectady Boces/Capit
47 Cornell Road
Latham, NY 12110

Clinical Services and Consult. Inc. Latham Alcoholism Outpatient Clinic/Drug Abuse Treatment
636 New London Road
Latham, NY 12110

LAUREL

Human Understanding and Growth Seminars, Inc. (HUGS)
Laurel, NY 11948

LAWRENCE

Committee on Drug Abuse (CODA) Outpatient Drug Free
270 Lawrence Avenue
Lawrence, NY 11559

Peninsula Counseling Center Alcoholism Counseling Service
270 Lawrence Avenue
5 Towns Community Center
Lawrence, NY 11559

LEVITTOWN

Island Trees Union Free School District
59 Straight Lane
Island Trees High School
Levittown, NY 11756

Levittown Union Free School District Substance Abuse Program/Admin.
Abbey Lane
Memorial Education Center
Levittown, NY 11756

Yours Ours Mine Community Center, Inc.
Adolescent and Family Alcohol Program
Outpatient Ambulatory Drug Free Unit
152 Center Lane Village Green
Building T
Levittown, NY 11756

LEWISTON

Mount Saint Mary's Hospital Alcoholism Inpatient Rehab Program
5300 Military Road
Lewiston, NY 14092

LIBERTY

Inward House Ferndale Drug Abuse Treatment
Upper Ferndale Road
Liberty, NY 12754

Liberty Central School Drug Abuse Prevention Services/ Team 89
115 Buckley Street
Liberty, NY 12754

Sullivan County Boces Substance Abuse Prevention/Intervention Programs
85 Ferndale-Loomis Road
Liberty, NY 12754

Sullivan County Cares Coalition
Liberty, NY 12754

LIVERPOOL

Family Services Associates Alcohol Outpatient Clinic
7445 Morgan Road
Suite 100
Liverpool, NY 13090

LOCKPORT

Alcoholism Council in Niagara County Alcoholism Outpatient Clinic
41 Main Street
Lockview Plaza
Lockport, NY 14094

Lockport Memorial Hospital Reflections Recovery Center
521 East Avenue
Suite 4S
Lockport, NY 14094

LONG BEACH

Long Beach Memorial Hospital
Facts Alcoholism Outpatient Clinic
Methadone Therapy Program/MM OP
455 East Bay Drive
Long Beach, NY 11561

Long Beach Reach
26 West Park Avenue
Long Beach, NY 11561

LONG ISLAND CITY

A Way Out, Inc. II Day Service
10-34 44 Drive
Long Island City, NY 11101

Bridge Plaza Treatment and Rehab Clinic Education and Methadone Treatment Unit
41-15 27 Street
Long Island City, NY 11101

Hellenic American Neighborhood Action Committee, Inc. Project ASAP
27-09 Crescent Street
Long Island City, NY 11102

NYC School District 30 Project Share Drug Education and Prevention
36-25 Crescent Street
Long Island City, NY 11106

LOWVILLE

Lewis County Alcoholism and Substance Abuse Treatment Center Alcoholism Outpatient Clinic
7514 South State Street
Lowville, NY 13367

Lewis County Council on Alcoholism and Substance Abuse
Alcoholism Prevention
Drug Abuse Prevention
7612 North State Street
Lowville, NY 13367

LYNBROOK

Link Counseling Center, Inc. Outpatient Drug Free
21 Langdon Place
Lynbrook, NY 11563

LYONS

Clifton Springs Outpatient Alcoholism Clinic
122 Broad Street
Lyons, NY 14489

Council on Alcoholism and Other Chemical Dependencies of the Finger Lakes/Wayne Council Office
58 Water Street
Lyons, NY 14489

Wayne County Community Counseling Center Drug Abuse Services
7328 Newark-Lyons Road
Lyons, NY 14489

MADRID

North Country Freedom Homes John E. Murphy Community Residence
13 Elm Street
Madrid, NY 13660

MALONE

Citizen Advocates, Inc. North Star Substance Abuse Services
16 4 Street
Malone, NY 12953

Saint Joseph Rehabilitation Center, Inc. Alcoholism Outpatient Clinic/Malone
214 East Main Street
Malone, NY 12953

MALVERNE

Malverne Union Free School District 12 Malverne Substance Abuse Prevention Program Admin.
80 Ocean Avenue
Malverne High School
Malverne, NY 11565

MAMARONECK

Archdiocese of New York
Orange County/School Drug Abuse Prevention
Westchester Drug Abuse Services
145 New Street
Mamaroneck, NY 10543

Larchmont/Mamaroneck Community Counseling Center/ Admin.
234 Stanley Avenue
Mamaroneck, NY 10543

MANHASSET

LIJ/HMC Manhasset Clinic
Daycare Unit
Outpatient Drug Free Unit
1355 Northern Boulevard
Manhasset, NY 11030

Manhasset Union Free School District Substance Abuse Services/Admin.
200 Memorial Place
Manhasset, NY 11030

North Shore University Hospital
Drug Treatment Center/Day Services
400 Community Drive
Manhasset, NY 11030

Family and Maternity Development Program
300 Community Drive
Manhasset, NY 11030

The Long Island Home, LTD./S. Oaks Hospital Alcoholism Outpatient Clinic/Manhasset
535 Plandome Road
Manhasset, NY 11030

MANLIUS

Central New York Services, Inc. Workplace Intervention/ Alcoholism EAP
Manlius, NY 13104

MASSAPEQUA

Massapequa High School Drug Abuse Prevention
4925 Merrick Road
Massapequa, NY 11758

Plainedge Union Free School District
Massapequa, NY 11758

Youth Environmental Services
30 Broadway
Massapequa, NY 11758

MASSENA

Canadian/American Youth Services, Inc. Rose Hill Treatment Center
2 Elizabeth Drive
Massena, NY 13662

Saint Lawrence County Comm. Services Board Alcoholism Outpatient Clinic/Massena
76 North Main Street
Massena, NY 13662

MATTITUCK

Eastern Long Island Hospital Quannacut Outpatient Services
6355 Main Road
Mattituck, NY 11952

Southhampton Drug Abuse Council North Fork Counseling Center
11550 Main Street
Mattituck, NY 11952

MECHANICVILLE

Mechanicville Area Community Services Alcoholism Information and Referral
6 South Main Street
Mechanicville, NY 12118

MEDFORD

Passages Counseling Center, Inc. Drug Abuse Treatment
1741-D North Ocean Avenue
Medford, NY 11763

MELVILLE

Seafield Center, Inc. Melville Drug Abuse Treatment Unit
900 Walt Whitman Road
Suite 102
Melville, NY 11747

MERRICK

Bellmore/Merrick Central High School District
1260 Meadowbrook Road
Merrick, NY 11566

MIDDLE ISLAND

Family Recovery Center Alcoholism Outpatient Clinic
514 Middle Country Road
Middle Island, NY 11953

MIDDLE VILLAGE

NYC School District 24 Community Adolescent Alternatives Program (CAAP)
67-54 80 Street
Middle Village, NY 11379

MIDDLETOWN

Emergency Housing Group, Inc. Middletown Alcohol Crisis Center
Middletown Psychiatric Center
Building 8
Middletown, NY 10940

Horton Outpatient Clinic Horton Family Program for Alcoholism
406 East Main Street
Middletown, NY 10940

Middletown Alcoholism Treatment Center Inpatient Rehabilitation Unit
141 Monhagen Avenue
Middletown, NY 10940

Middletown Community Health Center Pass Program TFIP
14 Grove Street
Middletown, NY 10940

Orange County Dept. of Mental Health Alcoholism Outpatient Clinic
21 Center Street
Middletown, NY 10940

Pius XII Youth and Family Services Outreach/Outpatient DF
10 Orchard Street
Middletown, NY 10940

Regional Economic Comm. Action Program Recap Alcoholism Outpatient Rehab Program
40 Smith Street
Middletown, NY 10940

Riverside House, Inc. Medically Supervised Outpatient
40 1/2 North Street
Middletown, NY 10940

MILFORD

Otsego Northern Catskills Boces Administrative Unit
Route 28
Milford, NY 13807

MILLBROOK

Daytop Village, Inc. Millbrook Residential Center
Route 44
Millbrook, NY 12545

MINEOLA

Herricks Community Life Center EPOD Outpatient Drug Free
450 Jericho Turnpike
2nd Floor
Mineola, NY 11501

Long Island Jewish Medical Center Hillside Hospital Family Consultation Center Alcoholism Outpatient
366 Jericho Turnpike
Mineola, NY 11501

Mineola Union Free School District Drug Abuse Program
200 Emory Road
Mineola, NY 11501

Seafield Center, Inc. Mineola Alcoholism Outpatient Clinic
286 Old Country Road
Mineola, NY 11501

MONROE

Pius XII Chemical Dependency Program
Monroe Clinic/Outpatient Drug Free
11–13 Lake Street
Monroe, NY 10950

Monroe Alcoholism Outpatient Clinic
201 Route 17M
Monroe, NY 10950

MONTICELLO

Monticello Central School District Drug Abuse Program
Port Jervis Road
Monticello Central School
Monticello, NY 12701

Periwinkle Theatre/Halfway There
19 Clinton Avenue
Monticello, NY 12701

Sullivan County Council on Alcohol/Drug Abuse, Inc.
Alcohol and Drug Abuse Outpatient Clinic
Alcohol Public Education/Info. and Referral
Alcohol Crisis Unit
Alcoholism Day Rehab Program
Drug Abuse Treatment
Kids' Klub
Recovery Center Halfway House
17 Hamilton Avenue
Monticello, NY 12701

MONTROSE

Town of Cortlandt PRC Youth Services
129 Albany Post Road
Cortlandt Youth Center
Montrose, NY 10548

Veterans' Affairs Medical Center Drug Dependence Treatment Center
Administrative Building 11D
Montrose, NY 10548

MOUNT KISCO

The Weekend Center, Inc. Alcoholism Outpatient Clinic
120 Kisco Avenue
Mount Kisco, NY 10549

MOUNT VERNON

Mount Vernon Public Schools Drug Abuse Prevention Services
165 North Columbus Avenue
Mount Vernon, NY 10553

Renaissance Project, Inc. Mount Vernon Unit
3 South 6 Street
Mount Vernon, NY 10550

WCMHB Mount Vernon Hospital Methadone Maintenance Treatment Program/Outpatient
3 South 6 Avenue
Mount Vernon, NY 10550

Westchester Community Opportunity Program Mount Vernon Open Door Program
34 South 6 Avenue
Mount Vernon, NY 10550

NANUET

Nanuet Union Free School District School Community Counselor Program
101 Church Street
Nanuet, NY 10954

NEW HARTFORD

Center for Addiction Recovery, Inc. Alcoholism Outpatient Clinic
4299 Middle Settlement Road
New Hartford, NY 13413

NEW HYDE PARK

Herricks Union Free School District Drug Abuse Program
Shelter Rock Road
Administration Building
New Hyde Park, NY 11040

Long Island Jewish Medical Center Daehrs Outpatient Drug Free
270-05 76 Avenue
Building 5
New Hyde Park, NY 11042

NEW PALTZ

Ulster County Mental Health Services New Paltz Alcohol Abuse Outpatient Clinic
40 Main Street
New Paltz, NY 12561

NEW ROCHELLE

Guidance Center, Inc.
Chemical Dependency Treatment Center
403–5 North Avenue
New Rochelle, NY 10801

Vocational Assistance Project
363–369 Huguenot Street
New Rochelle, NY 10801

Renaissance Project, Inc.
New Rochelle Unit
350 North Avenue
New Rochelle, NY 10801

Re-Entry House Apts.
197 Drake Avenue
New Rochelle, NY 10805

Re-Entry Unit/Storefront
350 North Avenue
New Rochelle, NY 10801

United Hospital Alcoholism Outpatient Clinic
3 The Boulevard
New Rochelle, NY 10801

Volunteers of America Shelters, Inc. Crossroads Halfway House
395 Webster Avenue
New Rochelle, NY 10801

Westchester Community Opportunity Program New Rochelle Outreach Center
33 Lincoln Avenue
Suite 2
New Rochelle, NY 10801

NEW YORK

Alcoholism Council of Greater New York
Alcoholism Prevention/Education Program
Information/Referral Unit
49 East 21 Street
New York, NY 10010

Alianza Dominicana, Inc.
2410 Amsterdam Avenue
New York, NY 10033

American Federation of Teachers Work and Family Assistance Program/Alcoholism
48 East 21 Street
12th Floor
New York, NY 10010

American Indian Community Substance Abuse Services
404 Lafayette Street
New York, NY 10003

Apprentice Journeyman Education Fund Workplace Intervention/Alcoholism EAP
49–51 Chambers Street
Suite 220
New York, NY 10007

ARMS Acres, Inc. Alcoholism Outpatient Services of Manhattan
1841 Broadway
Suite 1111
New York, NY 10023

ARTC
Manhattan Clinic 21
Starting Point
136 West 125 Street
6th Floor
New York, NY 10027

Manhattan Clinic 22
Kaleidoscope
136 West 125 Street
New York, NY 10027

Manhattan Clinic 23
Third Horizon
2195 3 Avenue
New York, NY 10035

Barnett and Associates Human Services, Inc.
330 West 58 Street
Suite 612
New York, NY 10019

Betances Health Unit Alternative Resources and Services
34 Gouverneur Street
New York, NY 10002

Beth Israel Medical Center Methadone Maintenance Treatment Program
Avenue A Clinic
26 Avenue A
New York, NY 10009

Clinics 1/2/3/6/7
Interium Clinic
103 East 125 Street
New York, NY 10035

Clinics IE/2F/3G
429 2 Avenue
New York, NY 10003

Clinic 2C
435 2 Avenue
New York, NY 10010

Clinic 3C
433 2 Avenue
New York, NY 10010

Clinic 4, Units 1/2
21 Old Broadway
Basement
New York, NY 10027

Clinics 8/8D
140 West 125 Street
New York, NY 10027

Gouverneur Clinic
109 Delancy Street
New York, NY 10002

Harold L. Trigg Clinic
543 Cathedral Parkway
New York, NY 10025

Keep
567 West 113 Street
New York, NY 10025

Lenox Hill Clinic
1082 Lexington Avenue
New York, NY 10021

Marie Nyswander Clinic
721 9 Avenue
New York, NY 10019

Saint Vincent's Clinic
201 West 13 Street
New York, 10011

Stuyvesant Square Chemical Dependency
380 2 Avenue
10th Floor
New York, NY 10010

Bliss Poston the Second Wind, Inc.
152 Madison Avenue
Suite 505
New York, NY 10016

Boys Harbor, Inc. Prevention Services
1 East 104 Street
New York, NY 10029

BRC Human Services Corporation
Alcohol Crisis Center
324 Lafayette Street
New York, NY 10012

Alcoholism Outpatient Clinic
191 Chrystie Street
New York, NY 10002

Cabrini Medical Center Start Alcoholism Program
137 2 Avenue
New York, NY 10003

Center for Comp. Health Practice, Inc.
CCHP ECP/MAP/PAAM
1900 2 Avenue
12th Floor
New York, NY 10029

CCHP FESP
163 East 97 Street
New York, NY 10029

Central Harlem Emergency Care Services Alcohol Crisis Center
419 West 126 Street
New York, NY 10027

Chinese/American Planning Council, Inc.
1 Orchard Street
2nd Floor
New York, NY 10002

Colonial Park Community Services, Inc.
Day Care Center/Drug Free Outpatient
Drug Abuse Treatment/ Outpatient/Day Services
303 West 145 Street
New York, NY 10039

Cornell University Medical College Midtown Center for Treatment and Research
55 West 44 Street
New York, NY 10036

Create, Inc.
Outpatient Drug Free
2–4 Lenox Avenue
New York, NY 10026

Residence I
121 West 111 Street
New York, NY 10026

Residence II
123 West 111 Street
New York, NY 10026

Daytop Village, Inc.
Adult Outpatient Services
Aftercare
Pregnant Addicted Women's Program
54 West 40 Street
New York, NY 10012

Dr. M. Degiarde MMTP Clinic
233–5 Lafayette Street
New York, NY 10012

East Harlem Block Nursery, Inc. Youth Action Program
1280 5 Avenue
New York, NY 10029

Employment Program For Recovered Alcoholics (EPRA)
225 West 34 Street
4th Floor Room 410
New York, NY 10122

Enter, Inc.
Alcoholism Community Residence
2009 3 Avenue
New York, NY 10029

Alcoholism Outpatient Clinic
302–306 East 111 Street
2nd Floor
New York, NY 10029

Outpatient
254 East 112 Street
New York, NY 10029

Residence
252 East 112 Street
New York, NY 10029

Freedom Institute, Inc. Alcoholism Outpatient Clinic
555 Madison Avenue
6th Floor
New York, NY 10022

Gracie Square Hospital
Alcoholism Inpatient Rehab Unit
Alcoholism Primary Care Program
Drug Rehab Unit
420 East 76 Street
New York, NY 10021

Gramercy Park Medical Group PC Unit 1
253–55 3 Avenue
New York, NY 10010

Greenwich House Counseling Center
80 5 Avenue
10th Floor
New York, NY 10011

Greenwich House, Inc.
Alcohol Treatment Program
55 5 Avenue
New York, NY 10003

Halfway House
312–314 Bowery
New York, NY 10012

Greenwich House MMTP
Cooper Square
50 Cooper Square
New York, NY 10003

Greenwich House West MMTP
24 West 20 Street
New York, NY 10011

Greenwich Village Youth Council
25 Carmine Street
New York, NY 10014

Hamilton Madison House Chinatown Alcoholism Services
146 Elizabeth Street
New York, NY 10012

Harlem Hospital Intervention Service
2238 5 Avenue
Health Building First Floor
New York, NY 10037

HHC Bellevue Hospital Methadone Maintenance Treatment Program
Buildings C and D
New York, NY 10016

HHC Harlem Hospital Center MMTP Clinic 3
500 West 180 Street
New York, NY 10033

Hispanic AIDS Forum, Inc. Innovative Alcohol Prevention and Intervention Program
121 Avenue of Americas
Suite 505
New York, NY 10013

Immigrant Social Services, Inc.
137 Henry Street
New York, NY 10002

Inter Care, Inc. Alcoholism Outpatient Clinic
51 East 25 Street
Suite 400
New York, NY 10010

International Center for the Disabled (TCD) Chemical Dependency Services/ Outpatient Clinic
340 East 24 Street
New York, NY 10010

Inwood Community Services, Inc.
Comprehensive Outpatient Alcoholism Program
Get Centered Mental Health Clinic Drug Prevention Program
651 Academy Street
2nd Floor
New York, NY 10034

John Jay College of Criminal Justice Cuny Substance Abuse Prevention Programs
899 10 Avenue
Room 410
New York, NY 10019

Joint Diseases North General Hospital Alcoholism Community Residence
1887 Madison Avenue
New York, NY 10035

Lesbian and Gay Community Services Center
Project Connect
Youth Enrichment Services (YES)
208 West 13 Street
New York, NY 10011

Lower Eastside International Comm. Schools
Drug Abuse Prevention Services
127 West 22 Street
New York, NY 10011

Homeless
Su Casa Methadone Maintenance Treatment Program
7 Governeur Slip East
New York, NY 10002

Mental Health Clinic/Drug Free Outpatient
Methadone Maintenance Treatment Program Unit 1/ KEEP
46 East Broadway
New York, NY 10002

Methadone Maintenance Treatment Program Unit 3
62 East Broadway
New York, NY 10002

Lucha, Inc. Drug Abuse Treatment/122nd Street
205 East 122 Street
4th Floor
New York, NY 10035

Madison Square Boys' and Girls' Club Youth Empowerment Program
301 East 29 Street
New York, NY 10016

Manhattan Alcoholism Treatment Center Alcoholism Inpatient Rehab Unit/Manhattan PC
600 East 125 Street
Wards Island
New York, NY 10035

Manhattan Bowery Corp.
Alcoholism Outpatient Program
Manhattan Bowery Crisis Center
8 East 3 Street
4th Floor
New York, NY 10003

Shelter Assessment and Referral Program
200 Varick Street
New York, NY 10014

Medical Arts Center Hospital
Alcoholism Acute Care Program
Medically Supervised Outpatient
57 West 57 Street
New York, NY 10019

Medical Arts Sanitarium, Inc. Drug Abuse Treatment
57 West 57 Street
New York, NY 10019

Minority Task Force on AIDS Alcoholism Prevention/ Intervention Program
127 West 127 Street
Room 422
New York, NY 10027

Mount Sinai Hospital
Community Youth Program/AHC
19 East 101 Street
New York, NY 10029

MMTP/KEEP
17 East 102 Street
New York, NY 10029

National Association on Drug Abuse Problems, Inc.
355 Lexington Avenue
2nd Floor
New York, NY 10017

New York Center for Addiction Treatment Services Medically Supervised Outpatient
568 Broadway
8th Floor
New York, NY 10012

New York Council of Smaller Churches, Inc.
69 West 128 Street
New York, NY 10027

New York City Department Juvenile Justice Adolescent Substance Abuse Program
365 Broadway
New York, NY 10013

New York City School Dist 4 School Community Action for a New Tomorrow (SCANT)
176 East 115 Street
C/O PS 57 Room 222
New York, NY 10029

New York City School District 1 Youth Leadership Project
220 Henry Street
JHS 56 Room 134
New York, NY 10002

New York City School District 2 Project Omnibus
80 Catherine Street
PS 126 Room 131
New York, NY 10038

New York City School District 3 Advocate Resource Center
220 West 121 Street
Room 330
New York, NY 10027

New York City School District 4 TFIP Prevention Program
176 East 115 Street
P.S. 57
New York, NY 10029

New York City School District 5 Project Decision Drug Prevention Program
2581 7 Avenue
Fred Doug IS 10 Room 128
New York, NY 10039

New York City School District 6 Project Soar/Substance Abuse Services
c/o School District 6 Office
665 West 182 Street
New York, NY 10033

New York Downtown Hospital Beekman/Trinity MMTP
74 Trinity Place
New York, NY 10006

New York Hospital MMTP Clinic
401 East 71 Street
New York, NY 10021

New York City Spark Program Board of Education
40 Irving Place
Washington Irving High School
New York, NY 10003

New York State Association For Retarded Citizens New York City Chapter/Alcoholism Prevention
200 Park Avenue South
Room 1010
New York, NY 10003

New York Therapeutic Communities, Inc. Stay Out Program
500 8 Avenue
Suite 801
New York, NY 10018

New York Urban Coalition, Inc. Cash/Escort/Vocational Diagnostic Evaluation
356 West 123 Street
New York, NY 10027

NRL Resources, Inc.
450 South Park Avenue
Suite 402
New York, NY 10016

Odyssey House, Inc. of New York
Adolescent Center
309–11 East 6 Street
New York, NY 10003

Homeless
Odyssey House Adult Program
Odyssey House Children
Wards Island
New York, NY 10035

Our Children's Foundation, Inc.
501 West 125 Street
New York, NY 10027

Phase Piggy Back, Inc.
Case Management Outpatient Outpatient
507 West 145 Street
New York, NY 10031

Phoenix House
Drug Free Residential
164 West 74 Street
New York, NY 10023

80th Street Drug Abuse Treatment Unit 1/2
223 West 80 Street
New York, NY 10024

Vernon Boulevard Unit
34-25 Vernon Boulevard
New York, NY 10023

Private Industry Council, Inc. Vocational Preparation Program
17 Battery Place
5th Floor
New York, NY 10004

Project Green Hope Services for Women
Drug Abuse Services/Residential DF
Drug Free Day Service
448 East 119 Street
New York, NY 10035

Project Return Foundation, Inc. Chelsea Tribeca Institute
133 West 21 Street
11th Floor
New York, NY 10011

Reality House, Inc
Drug Free Outpatient
MTA Day Service
637 West 125 Street
New York, NY 10027

Youth Program
Prevention
465 West 162 Street
New York, NY 10032

Realization Center Alcoholism Outpatient Clinic
121 East 18 Street
New York, NY 10003

Regent Hospital, Inc. Drug Abuse Treatment Unit
425 East 61 Street
New York, NY 10021

Richard Koeppel, MD MMTP 1/2
301 West 37 Street
New York, NY 10018

Saint Clare's Hospital Health Center Drug Abuse Services
415 West 51 Street
New York, NY 10019

Saint Luke's/Roosevelt Hospital Center
Alcoholism Halfway House
306 West 102 Street
New York, NY 10025

Alcoholism Inpatient Detox Unit
Amsterdam Avenue at 114 Street
New York, NY 10025

Alcoholism Outpatient Clinic
324 West 108 Street
New York, NY 10025

Smithers Alcoholism Detox Unit
428 West 59 Street
New York, NY 10019

Smithers Alcoholism Outpatient Clinic
17 West 60 Street
New York, NY 10023

Smithers Inpatient Rehabilitation Unit
56 East 93 Street
New York, NY 10028

Samaritan Village, Inc. Samaritan 53rd/Residential Drug Free
225–27 East 53 Street
New York, NY 10022

Single Parent Resource Center
Alcoholism Information and Referral
Student Assistance Program
141 West 28 Street
Room 302
New York, NY 10001

Talbot Perkins Children's Services Drug Abuse Treatment Unit
80 5 Avenue
3rd Floor
New York, NY 10011

Teatro Tiempo, Inc.
220 East 4 Street
New York, NY 10009

The Children's Aid Society Drug Intervention Program/Admin.
105 East 22 Street
New York, NY 10010

The City Kids Foundation Drug Abuse Prevention Services
57 Leonard Street
Street Level
New York, NY 10013

The Door/A Center of Alternatives
Adolescent Treatment Program/Outpatient
Prevention/Outpatient
121 Avenue of the Americas
New York, NY 10013

The Education Alliance Project Contact/Try
Day Treatment
Task Force on Integrated Projects (TFIP)
315 East 10 Street
New York, NY 10009

Pride Site/Homeless
Pride Site/Residential
371 East 10 Street
New York, NY 10009

The Fortune Society, Inc.
39 West 19 Street
New York, NY 10011

Tri-Center, Inc. Drug Abuse Treatment/New York
1776 Broadway
Suites 300–301
New York, NY 10019

Under 21
460 West 41 Street
5th Floor
New York, NY 10036

Upper Manhattan Mental Health Center Alcoholism Outpatient Clinic
1727 Amsterdam Avenue
New York, NY 10031

Veritas Therapeutic Community, Inc. Residential
68 West 106 Street
New York, NY 10025

Veterans' Affairs Medical Center
Alcohol/Drug Dependence Treatment Program
440 First Avenue and 24 Street
New York, NY 10010

MMTP
252 7 Avenue
9th Floor
New York, NY 10001

VNS Home Care First Steps Program
501–505 West 125 Street
3rd Floor
New York, NY 10027

Washington Heights/Inwood Coalition
652 West 187 Street
New York, NY 10033

Washton Institute, Inc.
Alcoholism Outpatient Clinic
Drug Abuse Treatment
4 Park Avenue
Ground Floor
New York, NY 10016

Women in Need Alcoholism Outpatient Clinic
406 West 40 Street
New York, NY 10018

Women's Action Alliance, Inc.
TFIP Prevention Program
Women's Alcohol and Drug Education Project
370 Lexington Avenue
Room 603
New York, NY 10017

YMCA of Greater New York Youth Enrichment Program/Admin.
180 West 135 Street
New York, NY 10030

131st Street Block Association, Inc.
220 West 131 Street
New York, NY 10027

NEWARK

Finger Lakes Alcoholism Counseling Referral Agency Alcoholism Outpatient Clinic
301 West Union Street
Newark, NY 14513

NEWBURGH

Black Community Development, Inc.
245 Liberty Street
Newburgh, NY 12550

City of Newburgh Youth Bureau Drug Abuse Prevention Services
167 Broadway
Newburgh, NY 12550

Pius XII Chemical Dependency Program Newburgh Clinic/Outpatient Drug Free
15 Lake Street
Newburgh, NY 12550

Pius XII Youth and Family Services Workplace Intervention/Alcoholism EAP
372 Fullerton Avenue
Newburgh, NY 12550

Poder, Inc. ADAC
280 Broadway
Newburgh, NY 12550

Riverside House, Inc. Hickory Hill Drug Abuse Treatment
Hickory Hill Road
Newburgh, NY 12550

Saint Luke's Hospital of Newburgh
Hospital Alcohol Intervention Services
70 Dubois Street
Newburgh, NY 12550

Methadone Maintenance Treatment Program
184 First Street
Newburgh, NY 12550

Teen Information Peer Services
290 Broadway
Suite 11
Newburgh, NY 12550

NIAGARA FALLS

Alcohol Council in Niagara County
Information and Referral Program
Student Assistance Program
800 Main Street
Niagara Falls, NY 14301

First Step Alcohol Crisis Center
1560 Buffalo Avenue
Niagara Falls, NY 14303

Fellowship House, Inc. Alcoholism Halfway House
431 Memorial Parkway
Niagara Falls, NY 14303

Milestones Alcoholism Services Alcoholism Outpatient Clinic
501 10 Street
Niagara Falls, NY 14301

Niagara Drug Abuse Program
Drug Free Outpatient
Prevention/Outpatient
1001 11 Street
Trott Access Center
Niagara Falls, NY 14301

Methadone Maintenance Treatment Program
775 3 Street
Niagara County Civic Building
Niagara Falls, NY 14302

NORTH CHILI

Catholic Family Center Union Street
3161 Union Street
North Chili, NY 14514

Restart Alcoholism Outpatient Clinic
3161 Union Street
North Chili, NY 14534

NORTH ROSE

Delphi Drug Abuse Center, Inc. Wayne County/Outpatient/ Prevention
Route 414
North Rose Medical Center
North Rose, NY 14516

NORTH TARRYTOWN

Phelps Mental Health Center Threshold Program/ Alcoholism Outpatient Clinic
38 Beekman Avenue
North Tarrytown, NY 10591

NORTH TONAWANDA

Mount Saint Mary's Hospital Clearview Alcoholism Outpatient Services
66 Mead Street
North Tonawanda, NY 14120

North Tonawanda CSD Project Peer and Project Primary Mental Health
175 Humphrey Street
North Tonawanda, NY 14120

NORTHPORT

Concepts For Narcotics Prevention, Inc. The Place/ Outpatient Drug Free
324 Main Street
Northport, NY 11768

NORWICH

Chenango County Alcohol and Drug Services
Alcoholism Outpatient Clinic
Drug Abuse Treatment
12 Henry Street
Norwich, NY 13815

NYACK

Daytop Village, Inc. Tappan Zee Residential Center
21 Mountainview Avenue
Nyack, NY 10960

Nyack Hospital
Alcoholism Acute Care Program
Alcoholism Inpatient Rehab Program
160 North Midland Avenue
Nyack, NY 10960

Nyack Schools
41 Dickinson Avenue
Nyack, NY 10960

Rockland Council on Alcoholism and Other Drug Dependence, Inc.
11 Division Avenue At Saint Paul's Street
Nyack, NY 10960

Village of Nyack DAPC Reach Out Roy Drug Abuse Treatment
42 Burd Street
Nyack, NY 10960

OCEANSIDE

Oceanside Counseling Center, Inc.
Alcoholism Services
Medically Supervised Outpatient Drug Free
71 Homecrest Court
Oceanside, NY 11572

Oceanside Union Free School District
145 Merle Avenue
Oceanside, NY 11572

South Nassau Communities Hospital YDA/Medical Service Clinic
2445 Oceanside Road
Oceanside, NY 11572

OGDENSBURG

Saint Lawrence Alcoholism Treatment Center Alcoholism Inpatient Rehab Unit
Station A/Hamilton Hall
Ogdensburg, NY 13669

Saint Lawrence County Comm. Services Board Alcoholism Outpatient Clinic/Ogdensburg
206–210 Ford Street
Heritage Center
Ogdensburg, NY 13669

OLEAN

Cattaraugus County Council on Alcoholism and Substance Abuse, Inc.
201 South Union Street
Olean, NY 14760

ONEIDA

Amethyst Chemical Dependency Services Alcoholism Outpatient Clinic
605 Seneca Street
Field Professional Building Suite 202
Oneida, NY 13421

Madison County Substance Abuse Program (ADAPT) Drug Free Outpatient Unit
248 Main Street
Johnson Hall
Oneida, NY 13421

Programs and Domiciles, Inc. Maxwell House Alcoholism Halfway House
239 Broad Street
Oneida, NY 13421

ONEONTA

Otsego County Community Services Otsego Chemical Dependencies Clinic
242 Main Street
County Office Building
Oneonta, NY 13820

ORANGEBURG

Blaisdell Alcoholism Treatment Center Inpatient Rehabilitation Unit
Rockland Psychiatric Center Campus
North Street Building 28
Orangeburg, NY 10962

ORCHARD PARK

Spectrum Human Services Orchard Park Chemical Dependency
Alcoholism Unit
Drug Abuse Unit
227 Thorn Avenue
Orchard Park, NY 14127

OSWEGO

Farnham, Inc. Medically Supervised Drug Free Outpatient
33 East First Street
Oswego, NY 13126

Oswego County Council on Alcoholism
Alcoholism Outpatient Clinic
Comm. Education and Intervention Program
53 East 3 Street
Oswego, NY 13126

OTISVILLE

Federal Correctional Institution Substance Abuse Services
Otisville, NY 10963

OWEGO

TIOGA County Council on Alcohol Substance Abuse Comm. Alcoholism Education Info. and Referral
98 Temple Street
Owego, NY 13827

TIOGA County Alcohol and Drug Services
Medically Supervised Outpatient
Substance Abuse Outpatient Clinic
175 Front Street
Owego, NY 13827

OYSTER BAY

Youth and Family Counseling Agency of Oyster Bay/East Norwich, Inc.
193A South Street
Oyster Bay, NY 11771

OZONE PARK

NYC School District 27 Aspects 27/Drug Prevention Program
82-01 Rockaway Boulevard
Ozone Park, NY 11416

PARKSVILLE

Daytop Village, Inc. Parksville Residential Center
Old Route 17
Parksville, NY 12768

PATCHOGUE

Brookhaven Memorial Hospital Alcoholism Outpatient Clinic/West
365 East Main Street
Patchogue, NY 11772

Patchogue/Medford School District Peer Leadership
241 South Ocean Avenue
Patchogue, NY 11772

PEARL RIVER

Pearl River Senior High School Substance Abuse Services/Admin.
275 East Central Avenue
Pearl River, NY 10965

PEEKSKILL

Hudson Valley Hospital Center Methadone Maintenance Treatment Program
1980 Crompond Road
Peekskill, NY 10566

Peekskill Area Health Center, Inc.
Alcoholism Outpatient Clinic
Peekskill Pathways
1037 Main Street
Peekskill, NY 10566

PELHAM

Pelham Guidance Council
253 Wolf's Lane
Pelham, NY 10803

PENN YAN

Council on Alcoholism and Other Chemical Dependency of the Finger Lakes/Yates Council Office
108 Water Street
Penn Yan, NY 14527

Schuyler Hospital Alcoholism Outpatient Clinic
100 West Lake Road
Penn Yan, NY 14527

Soldiers' and Sailors' Memorial Hospital Youth Counseling Services
165 Main Street
Penn Yan, NY 14527

PLAINVIEW

Nassau County Dept. of Drugs and Alcohol
Alcoholism Inpatient Rehab Unit
Dorothy Young Recovery House
Topic House/Homeless/Residential
1425 Old Country Road
Plainview, NY 11803

Plainview/Old Bethpage CSD
Youth Activities Council/Reflection/Prevention
777 Old Country Road
Plainview, NY 11803

Drug Program
Joyce Road School
Administration Building
Plainview, NY 11803

PLATTSBURGH

Champlain Valley Family Center Drug Treatment Youth Services Inc.
501 North Margaret Street
Plattsburgh, NY 12901

Clinton County Council on Alcoholism Alcoholism Intervention and Education Program
14 Margaret Street
Riverview Mall
Plattsburgh, NY 12901

Clinton County Mental Health Assoc.
Employee Assistance PGM/Consortia Project
159 Margaret Street
Suite 200
Plattsburgh, NY 12901

Twin Oaks Alcoholism Halfway House
79 Oak Street
Plattsburgh, NY 12901

Clinton County Mental Health And Alcoholism Services Clinton County Alcohol Program
6 Ampersand Drive
Plattsburgh, NY 12901

PLEASANTVILLE

Pleasantville Union Free School District Peer Leadership Program
Romer Avenue
Pleasantville, NY 10570

POMONA

Mental Health Assoc. Rockland County Project Rainbow Alcoholism Clinic
Sanatorium Road
Building J
Pomona, NY 10970

Rockland Community Foundation Drug Abuse Prevention Services
Sanitarium Road
Pomona, NY 10970

Rockland County CMHC
MMTP Detox Unit
MMTP Methadone Maintenance Unit
Sanatorium Road
Pomona, NY 10970

Summit Park Hospital
Alcoholism Inpatient Detox Unit
Sobering Up Station
Sanatorium Road
Building D
Pomona, NY 10970

PORT CHESTER

Renaissance Project, Inc. Port Chester Center
4 Poningo Street
Port Chester, NY 10573

WCMHB Saint Vincent's Hospital Methadone Maintenance Treatment Program/Outpatient
350 North Main Street
Port Chester, NY 10573

PORT JEFFERSON

John T. Mather Memorial Hospital Mather Outpatient Alcoholism Clinic
635 Belle Terre Road
Port Jefferson, NY 11777

PORT JEFFERSON STATION

Comsewogue School District Youthful Drug Abuse Program
290 Norwood Avenue
Port Jefferson Station, NY 11776

Crossings of Long Island, Inc.
5225 Route 347/40
Davis Professional Park
Port Jefferson Station, NY 11776

PORT JERVIS

Crossroads At Mercy Community Hospital Crossroads Acute Care Alcoholism Program
160 East Main Street
Port Jervis, NY 12771

Regional Economic Comm. Action Program New Life Manor Halfway House
259 Old Mountain Road
Town of Greenville
Port Jervis, NY 12771

Riverside House, Inc. Residential/Short Term
181 Route 209
Port Jervis, NY 12771

PORT WASHINGTON

Port Counseling Center, Inc.
225 Main Street
Port Washington, NY 11050

Port Washington Union Free School District Substance Abuse Prevention Program
100 Campus Drive
Port Washington, NY 11050

POTSDAM

Canton/Potsdam Hospital
Alcoholism Detoxification Unit
Alcoholism Inpatient Rehab Program
50 Leroy Street
Potsdam, NY 13676

Reachout of Saint Lawrence County, Inc. Alcoholism Information and Referral
Potsdam, NY 13676

Saint Lawrence County Alcohol and Substance Abuse Services Outpatient Clinic
State University of New York at Potsdam
Van Housen Hall
Potsdam, NY 13676

POUGHKEEPSIE

Dutchess County Council on Alcohol and Chemical Dependency, Inc.
Alcohol and Drug Unit
Alcohol Education Intervention
Student Assistance Program
20 Maple Street
Poughkeepsie, NY 12601

Dutchess County Dept. of Mental Hygiene
Alcohol Abuse Clinic
20 Manchester Road
Poughkeepsie, NY 12603

Alcoholism Employee Assistance Program
7 Mansion Street
Poughkeepsie, NY 12601

Alcoholism Youth Clinic Boces TFIPOCDY
350 Dutchess Turnpike
Poughkeepsie, NY 12603

Dutchess County Methadone Clinic Outpatient
230 North Road
Poughkeepsie, NY 12601

Dutchess County Substance Abuse Clinic Medically Supervised
20 Manchester Road
Poughkeepsie, NY 12603

Mid-Hudson Alcoholism Recovery Center
Alcoholism Primary Care Program
Branch B. Ryon Hall
Hudson River PC
Poughkeepsie, NY 12601

Bolger House Community Residence
260 Church Street
Poughkeepsie, NY 12601

New Hope Manor, Inc. Re-Entry House
141 South Avenue
Poughkeepsie, NY 12601

Saint Francis Hospital Workplace Intervention/Alcoholism EAP
68 West Cedar Street
Poughkeepsie, NY 12601

QUEENS VILLAGE

Queensboro Council for Social Welfare Drug Abuse Prevention Services
221-10 Jamaica Avenue
Queens Village, NY 11428

QUEENSBURY

Baywood Center for Treatment of Addictions Alcoholism Outpatient Clinic
386 Bay Road
Queensbury, NY 12804

RAY BROOK

Federal Correctional Institution Drug Abuse Program
Old Ray Brook Road
Ray Brook, NY 12977

RHINEBECK

Almiday Corporation Drug Abuse Treatment
500 Milan Hollow Road
Rhinebeck, NY 12572

Daytop Village, Inc. Rhinebeck Complex
Foxhollow Road
Rhinebeck, NY 12572

Rhinebeck Lodge for Successful Living Alcoholism Inpatient Rehabilitation Unit
500 Milan Hollow Road
Rhinebeck, NY 12572

RIVERHEAD

Hampton Council of Churches Family Counseling Service
Phillips Avenue School
Riverhead, NY 11901

Lighthouse Support Services, Inc. Lighthouse Counseling Center Outpatient Clinic
8 East Main Street
Riverhead, NY 11901

Riverhead Central School District Substance Abuse Prevention Program
700 Osborne Avenue
Riverhead, NY 11901

Riverhead Community Awareness Program, Inc. Drug Abuse Prevention Services
542 East Main Street
Riverhead, NY 11901

Seafield Center, Inc. Riverhead Drug Abuse Treatment Unit
212 West Main Street
Riverhead, NY 11901

Southampton Drug Abuse Council Alternatives Counseling Center
Plaza 540 East Main Street
Riverhead, NY 11901

ROCHESTER

AIDS Rochester, Inc. Alcoholism Prevention/Intervention Program
1350 University Avenue
Suite C
Rochester, NY 14607

Anthony L. Jordan Health Center Alcoholism Outpatient Clinic
82 Holland Street
Rochester, NY 14603

Baden Street Settlement Counseling Center/Outpatient
13 Vienna Street
Rochester, NY 14605

Brylin Hospital, Inc. Alcoholism Outpatient Program
1401 Stone Road
Suite 303
Rochester, NY 14615

Caron Treatment Services, Inc.
Alcoholism and Drug Abuse Outpatient Clinic
1150 University Avenue
Rochester, NY 14607

Drug Abuse Treatment Unit
1501 East Avenue
Rochester, NY 14610

Catholic Charities/Rochester
Outpatient/Intensive Outpatient
Restart Alcoholism Outpatient Clinic
55 Troup Street
Plymouth Park West
Rochester, NY 14608

Catholic Family Center
Joseph Street Unit
1111 Joseph Avenue
Rochester, NY 14621

Center for Youth Services, Inc.
Administrative Unit
Youth Intervention/SAP
258 Alexander Street
Rochester, NY 14607

Citizens Alliance to Prevent Drug Abuse, Inc. (CAPDA)
36 West Main Street
Executive Office Building Suite 690
Rochester, NY 14614

Daybreak/The Alcoholism Treatment Facility
Alcoholism Inpatient Rehab Unit
Alcoholism Primary Care Program
Comprehensive Outpatient Clinic
Supportive Living Facility
435 East Henrietta Road
Rochester, NY 14620

Daybreak Halfway House/Alexander
184 Alexander Street
Rochester, NY 14607

Daybreak Halfway House/Barrington
380 Barrington Street
Rochester, NY 14620

Daybreak House/Wellington
287 Wellington Street
Rochester, NY 14620

Daybreak Halfway House/West
383 West Avenue
Rochester, NY 14611

Entry to Care Primary Care Program
745 West Avenue
Rochester, NY 14611

East House Corporation
Crossroads Apartment Program
259 Monroe Avenue
Rochester, NY 14607

Crossroads House I
269 Alexander Street
Rochester, NY 14607

Crossroads House II
239 Alphonse Street
Rochester, NY 14621

Crossroads III/Cody House
407 Frederick Douglas Street
Rochester, NY 14608

Family Services of Rochester, Inc.
Administrative Unit
Alcoholism Outpatient Clinic
Avon Drug Abuse Prevention Unit
30 North Clinton Avenue
Rochester, NY 14604

Genesee Hospital Dept. of Psychiatry Genesee Alcohol Treatment Center
224 Alexander Street
Rochester, NY 14607

Genesee Valley Group Health Assoc. Drug Abuse Treatment
800 Carter Street
Rochester, NY 14621

Greater Rochester Metro Chamber of Commerce/ Workplace Intervention Alcoholism EAP
55 Saint Paul Street
Rochester, NY 14604

Health Association of Rochester Workplace Intervention/ Alcoholism/EAP
1 Mount Hope Avenue
Rochester, NY 14620

John L. Norris Alcoholism Treatment Center Alcoholism Inpatient Rehab Unit
1600 South Avenue
Rochester Psychiatric Center/ Howard I
Rochester, NY 14620

Men's Service Center Supportive Living Facility
440 Fredrick Douglas Street
Rochester, NY 14608

Monroe County Office of Mental Health Drug and Alcohol Abuse Education Office
375 Westfall Road
Rochester, NY 14620

Native American Cultural Center Alcoholism Comm. Education Intervention/Referral Program
1475 Winton Road North
Suite 12
Rochester, NY 14609

Natl. Council on Alcoholism Drug Dependency Rochester
Hispanic Prevention/Education Program
Prevention Education/Info and Referral
1 Mount Hope Avenue
Rochester, NY 14620

Park Ridge Chemical Dependency, Inc.
Brighton Alcoholism Outpatient Clinic
2000 South Winton Road
Building 2
Rochester, NY 14618

Residential Chemical Dependency for Youth/Long Term
2654 Ridgeway Avenue
Rochester, NY 14626

Women's Community Residence
2650 Ridgeway Avenue
Rochester, NY 14626

Park Ridge Hospital Hospital Intervention Service
1555 Long Pond Road
Rochester, NY 14626

Short Term Rehab Unit
1565 Long Pond Road
Rochester, NY 14626

Rochester City School District Substance Prevention/ Education Program
131 West Broad Street
C O 3
Rochester, NY 14614

Rochester Housing Authority Alcohol and Substance Abuse Program
316 Olean Street
Rochester, NY 14608

Rochester Institute of Technology Substance Alcoholism Intervention Services for the Deaf (SAISD)
Hale Andrews Center
Rochester, NY 14623

Rochester Mental Health Center
Alcoholism Outpatient Clinic
Drug Treatment Services/ MSASATP
490 East Ridge Road
Rochester, NY 14621

Drug Treatment Services/MM
1425 Portland Avenue
Rochester, NY 14621

Rochester Rehabilitation Center Employment Services
46 Mount Hope Avenue
Rochester, NY 14620

Saint Joseph's Villa of Rochester, Inc. Life Program/Residential Chemical Dependency Services/Youth/Long Term
3300 Dewey Avenue
Rochester, NY 14616

Threshold Center for Alternative Youth Services, Inc./Admin.
80 Saint Paul Street
4th Floor
Rochester, NY 14604

University of Rochester/Strong Memorial Hospital
Alcoholism Outpatient Clinic
Methadone Maintenance Treatment Clinic/Outpatient
Drug Dependency Outpatient
300 Crittenden Boulevard
Rochester, NY 14642

Veterans' Affairs Rochester Outpatient Clinic
100 State Street
Rochester, NY 14614

Westfall Associates, Inc.
Alcoholism Outpatient Clinic
Drug Abuse Outpatient Treatment
919 Westfall Road
Suite C-120
Rochester, NY 14618

YWCA of Rochester/Monroe County Steppingstone Drug Program
175 North Clinton Avenue
Rochester, NY 14604

ROCKVILLE CENTRE

Diocese of Rockville Centre Education Dept. Substance Abuse Prevention Program
50 North Park Avenue
Rockville Centre, NY 11570

Mercy Hospital Association Hospital Intervention Services
1000 North Village Avenue
Rockville Centre, NY 11570

Rockville Centre Union Free School District Drug Abuse Program
Shepherd Street
Rockville Centre, NY 11570

Rockville Center Narcotics/Drug Abuse Confide/Outpatient Drug Free
30 Hempstead Avenue
Suite H-6
Rockville Centre, NY 11570

ROME

Rome and Murphy Memorial Hospital Community Recovery Center/Alcoholism Outpatient
1500 North James Street
4th Floor South Wing
Rome, NY 13440

RONKONKOMA

A Program Planned for Life Enrichment, Inc. (APPLE)
161 Lake Shore Road
Ronkonkoma, NY 11779

153 Lake Shore Drive
Ronkonkoma, NY 11779

Board of Cooperative Education Services Boces 2 Student Assistance Service
5018 Express Drive South
Ronkonkoma, NY 11779

Pace Center, Inc. Alcoholism Outpatient Clinic
3555 Veterans Highway
Suite E
Ronkonkoma, NY 11779

ROOSEVELT

NCDDAA Roosevelt Counseling Center
42 East Fulton Avenue
Roosevelt, NY 11575

Roosevelt Ed. Alcoholism Counseling Treatment Center React Alcoholism Outpatient Clinic
27A Washington Place
Roosevelt, NY 11575

Roosevelt Union Free School District Roosevelt Jr./Sr. High School
240 Denton Place
Roosevelt, NY 11575

ROSEDALE

NYC School District 29 Project Youth
1 Cross Island Plaza
PS 136 Mini School
Rosedale, NY 11422

SAG HARBOR

Sag Harbor Union Free School District Substance Abuse Services
Hampton Street
Sag Harbor, NY 11963

SAINT ALBANS

Queens Village Comm. for Mental Health J Cap Residential Unit 1
177-33 Baisley Boulevard
Saint Albans, NY 11434

SALAMANCA

Salamanca District Authority
Changing Seasons Treatment Center
Simtec Alcoholism Outpatient Services
150 Parkway Drive
Salmanca, NY 14779

SARANAC LAKE

Citizen Advocates, Inc. North Star Substance Abuse Services
10 Main Street
Old Village Building
Saranac Lake, NY 12983

Saint Joseph's Rehabilitation Center, Inc.
Alcoholism Inpatient Rehabilitation Program
Glenwood Estates
Saranac Lake, NY 12983

Alcoholism Outpatient Clinic
50 Woodruff Street
Saranac Lake, NY 12983

SARATOGA SPRINGS

Alcoholism/Substance Abuse Council of Saratoga County, Inc. Innovative Alcoholism Prevention Program
19 Maple Avenue
Saratoga Springs, NY 12866

Saratoga County Alcohol and Substance Abuse Prevention Project, Inc. (ASAPP)
10 Lake Avenue
Saratoga Springs, NY 12866

Saratoga County Alcoholism Services Alcoholic Outpatient Clinic
254 Church Street
Saratoga Springs, NY 12866

Saratoga Springs Office of Abused Substances and Intervention Services, Inc.
511 Broadway
Saratoga Springs, NY 12866

SCHENECTADY

Alcohol Council of Schenectady County, Inc.
Alcohol Public Education/Info. and Referral
Alcoholism Outpatient Clinic
302 State Street
Schenectady, NY 12305

Perrin House
575 Lansing Avenue
Schenectady, NY 12303

Purcell House
406–408 Summit Street
Schenectady, NY 12307

Bridge Center of Schenectady, Inc.
Drug Free Ambulatory/Residential
70–72 Union Street
Schenectady, NY 12308

Outreach/El Puente
924 Albany Avenue
Schenectady, NY 12305

Carver Community Counseling Services Medically Supervised Outpatient
949 State Street
Schenectady, NY 12307

Conifer Park, Inc. Intensive Alcoholism Outpatient Clinic
150 Glenridge Road
Schenectady, NY 12302

Leonard Hospital Alcoholism/Substance Abuse Outpatient Clinic
1594 State Street
Schenectady, NY 12304

Lifestart Health Services, Inc. Alcoholism Outpatient Clinic
1356 Union Street
Schenectady, NY 12308

SEA CLIFF

North Shore Central School District
112 Franklin Avenue
Sea Cliff, NY 11579

SEAFORD

Seaford Union Free School District Drug Abuse Program
1590 Washington Avenue
Seaford, NY 11783

SENECA FALLS

Council on Alcoholism and Other Chemical Dependency of the Finger Lakes/Seneca Council Office
49 Fall Street
Seneca Falls, NY 13148

SHRUB OAK

Phoenix House Yorktown/Homeless
Stoney Street
Shrub Oak, NY 10588

SIDNEY CENTER

Delaware County Comm. Services Board Alcoholism Outpatient Clinic
Western Delaware Boces/Rd 1
Sidney Center, NY 13839

SMITHTOWN

Employee Assistance Resource Services, Inc. (EARS)
278 East Main Street
Smithtown, NY 11787

Horizons Counseling Center
Outpatient Drug Free Unit
Prevention Unit
124 West Main Street
Smithtown, NY 11787

North Suffolk Mental Health Center
Alcoholism Outpatient Cinic
Drug Abuse Treatment
11 Route 111
Smithtown, NY 11787

Saint John's Episcopal Hospital Smithtown Alcohol Detoxification Unit
Route 25A
Smithtown, NY 11787

SOUTH KORTRIGHT

Phoenix House South Kortright Drug Abuse Treatment
County Road 513
Old Route 10 Belle Terre
South Kortright, NY 13842

SOUTHAMPTON

Southampton Drug Abuse Council
Outpatient
South Fork Counseling Center
South Fork Prevention Center
291 Hampton Road
Southampton, NY 11968

Southampton Public Schools Peer Assisted Leadership (PAL)
141 Narrow Lane
Southampton, NY 11968

SPRING VALLEY

East Ramapo Central School District School Community Counselor Program
150 South Madison Avenue
Spring Valley, NY 10977

Summit Park Hospital Alcoholism Day Rehabilitation Unit
50A South Main Street
Spring Valley, NY 10977

Summit Park Hospital Alcoholism Outpatient Clinic
50A South Main Street
Spring Valley, NY 10977

Town of Ramapo Youth Counseling Services Outpatient Drug Free
288 North Main Street
Spring Valley, NY 10977

STATEN ISLAND

Amethyst House, Inc. Alcoholism Halfway House
75 Vanderbilt Avenue
Staten Island, NY 10304

Bayley Seton Hospital, Inc.
Alcoholism Acute Care Unit
Alcoholism Crisis Center
Alcoholism Outpatient Clinic
Chemical Dependency Unit
75 Vanderbilt Avenue
Staten Island, NY 10304

Beth Israel Medical Center MMTP Staten Island Clinic
111 Water Street
Staten Island, NY 10304

Camelot of Staten Island, Inc.
Charitable Foundation Inc./Day Service
Outpatient Adult Program
Outpatient/Prevention
263 Port Richmond Avenue
Staten Island, NY 10302

Tier 2
1111 Fr. Capadanno Boulevard
Staten Island, NY 10306

Daytop Village, Inc. Staten Island Outreach Center
1915 Forest Avenue
Staten Island, NY 10303

NYC School District 31 Drug Abuse Prevention Program Project Aware
54 Osborne Street
Room 127
Staten Island, NY 10312

Police Athletic League Staten Island
55 Layton Avenue
Staten Island, NY 10301

Project Hospitality, Inc. Alcohol Outpatient Clinic
100 Central Avenue
Staten Island, NY 10301

Saint Vincent's Hospital Medical Center/N. Richmond
Alcoholism Outpatient Program
427 Forest Avenue
Staten Island, NY 10301

Alcoholism Outpatient Clinic/DWI Program
1794 Richmond Road
Staten Island, NY 10306

South Beach Alcoholism Treatment Center Alcoholism Inpatient Rehab Program
777 Seaview Avenue
South Beach Psychiatric Center Building A
Staten Island, NY 10305

Staten Island Drug Abuse Prevention Council
460 Brielle Avenue
Community Service Building
Staten Island, NY 10314

Staten Island Mental Health Society, Inc.
Project for Academic Student Success (PASS)
30 Nelson Avenue
Staten Island, NY 10308

Teen Alcohol Program
3974 Amboy Road
Staten Island, NY 10308

Staten Island University Hospital
Alcohol Counseling Services
475 Seaview Avenue
Staten Island, NY 10305

Drug Free Services
Key Extended Entry Program
Methadone Maintenance Treatment Program
392 Seguine Avenue
Staten Island, NY 10309

Inpatient Alcohol Detox
Women's Alcoholism Outpatient Clinic
375 Seguine Avenue
Staten Island, NY 10309

United Activities Unlimited, Inc.
485 Clawson Street
Staten Island, NY 10306

YMCA of Greater New York Staten Island YMCA Counseling Services/I & II
3902 Richmond Avenue
Staten Island, NY 10312

SUFFERN

Good Samaritan Hospital of Suffern
Alcoholism Outpatient Clinic
Drug Abuse Treatment Unit
255 Lafayette Avenue
Suffern, NY 10901

Rockland Community College Outreach/Intervention
145 College Road
Suffern, NY 10901

SWAN LAKE

Daytop Village, Inc. Swan Lake Residential Center
Route 55
Swan Lake, NY 12783

SYOSSET

Syosset Central School District Drug Abuse Program
South Woods Road
Syosset High School
Syosset, NY 11791

Syosset's Concern About Its Neighborhood (SCAN)
40 Whitney Avenue
Syosset, NY 11791

SYRACUSE

AIDS Task Force of Central New York, Inc. Alcoholism Prevention/Intervention Program
627 West Genesee Street
Syracuse, NY 13204

Alcohol Services, Inc. Alcoholism Outpatient Clinic
247 West Fayette Street
Syracuse, NY 13202

Benjamin Rush Recovery Center Alcoholism Inpatient Unit
650 South Salina Street
Syracuse, NY 13202

Child and Family Services of Onondago County Alcoholism Outpatient Clinic
450 South Main Street
Syracuse, NY 13212

Clinical Counseling Services Alcoholism Outpatient Clinic
600 East Genesee Street
Suite 208
Syracuse, NY 13202

Crouse/Irving Memorial Hospital
Adolescent Services
Alcoholism Outpatient Clinic
Drug Free Outpatient Unit
Methadone Maintenance Treatment Program
410 South Crouse Avenue
Syracuse, NY 13210

Alcohol Residential Chemical Dependency Program for Youth
Commonwealth Place/Inpatient Rehab Program
Drug Abuse Treatment Unit 2
6010 East Molloy Road
Syracuse, NY 13211

Alcoholism Acute Care Unit
Hospital Intervention Services
736 Irving Avenue
Syracuse, NY 13210

Onondaga Council on Alcoholism Addictions, Inc. Referral Unit/ TFIP
716 James Street
Syracuse, NY 13203

Onondaga County Student Assistance Drug Abuse Prevention
421 Montgomery Street
Civic Center 10th Floor
Syracuse, NY 13202

Onondaga/Cortland/Madison Boces Workplace Intervention/ Alcoholism EAP
6820 Thompson Road
Syracuse, NY 13221

Onondaga/Madison/Cortland Bd. Coop. Educ. Alcohol/Drug Abuse Prevention and Education Program
6075 East Molloy Road
Syracuse, NY 13221

Pelion of Central New York, Inc. Alcoholism Outpatient Clinic
500 South Salina Street
Suite 218
Syracuse, NY 13202

Pelion Prescrip Drug Misuse Program, Inc. Chronic Disorders Outpatient
500 South Salina Street
Suite 218
Syracuse, NY 13202

Rescue Mission Alliance, Inc. Alcohol Crisis Center
120 Gifford Street
Syracuse, NY 13202

Southwest Community Center
401 South Avenue
Syracuse, NY 13204

Syracuse Brick House, Inc.
Alcoholism Outpatient Clinic
716 James Street
Syracuse, NY 13206

Inpatient Rehabilitation
S-1 Van Duyn
Onondaga Hill
Syracuse, NY 13215

Men's Halfway House
121 Green Street
Syracuse, NY 13203

Women's Halfway House
3606 James Street
Syracuse, NY 13206

Syracuse Community Health Center
Alcoholism Outpatient Clinic
Drug Abuse Treatment
819 South Salina Street
Syracuse, NY 13202

Veterans' Affairs Medical Center Chemical Dependency Clinic
800 Irving Avenue
Syracuse, NY 13210

TICONDEROGA

Moses/Ludington Hospital Substance Abuse Prevention Program
Wicker Street
Pavilion Building
Ticonderoga, NY 12883

TONAWANDA

Construction Exchange of Buffalo and Western New York, Inc. EAP
625 Ensminger Road
Tonawanda, NY 14150

Horizon Human Services, Inc.
Addictions Outpatient
1370 Niagara Falls Boulevard
Tonawanda, NY 14150

Addictions Outpatient
36 Delaware Street
Tonawanda, NY 14150

Tonawanda City School Project Team Substance Abuse Prevention Services
202 Broad Street
Tonawanda, NY 14150

TROY

Eight Twenty River Street, Inc.
Alcoholism Halfway House
820 River Street
Troy, NY 12180

Supportive Living
3075 6 Avenue
Troy, NY 12180

Hudson Mohawk Recovery Center Alcoholism Outpatient Clinic
16 First Street
Troy, NY 12180

Leonard Hospital
Alcoholism Inpatient Rehabilitation Unit
Biochemical Dependency Detox Unit
74 New Turnpike Road
Troy, NY 12182

Alcoholism Outpatient Clinic
16 Northern Drive
Troy, NY 12182

Pahl, Inc. Drug Abuse Treatment Services
106–108 9 Street
Troy, NY 12180

Pahl Transitional Apartments
2239–2243 5 Avenue
Troy, NY 12180

Rensselaer County Mental Health Unified Services
Outpatient Drug Free Program
Substance Abuse Prevention
7 Avenue and State Street
County Office Building
Troy, NY 12180

Samaritan Hospital Detoxoification Service
2215 Burdett Avenue
Troy, NY 12180

TRUMANSBURG

Ithaca Alpha House Center, Inc. Residential
RD 1 Route 227
Trumansburg, NY 14886

TUCKAHOE

The Maxwell Institute, Inc. Alcoholism Clinic
167 Scarsdale Road
Tuckahoe, NY 10707

TULLY

Tully Hill Corporation
Alcoholism Inpatient Rehab Program
Alcoholism Primary Care Program
Drug Abuse Treatment Unit
Route 80 and Route 11
Tully, NY 13159

TUPPER LAKE

Citizen Advocates, Inc. North Star Substance Abuse Services Satellite 1
114 Wawbeek Avenue
Tupper Lake, NY 12986

UTICA

Catholic Charities Alcoholism Halfway House
407–409 Rutger Street
Utica, NY 13501

Insight House/Oneida County Substance Abuse Services Agency
Treatment Unit 1
400 Rutger Street
Utica, NY 13501

Treatment Unit 2
320 Rutger Street
Utica, NY 13501

McPike Alcoholism Treatment Center Alcoholism Inpatient Rehab Unit
1213 Court Street
Mohawk Valley Psychiatric Center
Utica, NY 13502

Oneida County Dept. of Mental Health Alcoholism Outpatient Clinic
800 Park Avenue
Utica, NY 13501

VALATIE

Catholic Family and Community Services Columbia County
Route 9
Starkmans Office Building
Valatie, NY 12184

VALHALLA

Westchester County Medical Center Substance Abuse Treatment
Psychiatric Institute
Valhalla Campus
Valhalla, NY 10595

VALLEY STREAM

Friends of Bridge, Inc. Drug Abuse Treatment Program
5–11 Pflug Place
Valley Stream, NY 11580

Valley Stream High School District
1 Kent Road
Valley Stream, NY 11582

WALTON

Delaware County Council on Alcoholism and Other Drug Addictions, Inc.
136 Delaware Street
Walton, NY 13856

Delaware Valley Hospital Alcoholism Inpatient Rehabilitation
1 Titus Place
Walton, NY 13856

WAMPSVILLE

Madison County Substance Abuse Program Alcoholism Outpatient Clinic
North Court Street
Veterans Memorial Building
Wampsville, NY 13163

WANTAGH

South Shore Planning Council Comm. Education/ Intervention/Info. Referral
1742 Old Mill Road
Wantagh, NY 11793

Southeast Nassau Guidance Center (SNG)
Alcoholism Counseling and Treatment
3401 Merrick Road
Wantagh, NY 11793

Drug Abuse Treatment Unit 1
3375 Park Avenue
Suite 2005
Wantagh, NY 11793

Wantagh Union Free School District Substance Abuse Services/Admin.
Beltagh Avenue
Wantagh, NY 11793

WARSAW

Allegany Rehab Associates, Inc. Wyoming County Chemical Abuse Treatment Program
422 North Main Street
Warsaw, NY 14569

Wyoming County Drug Abuse Program/Admin.
338 North Main Street
Warsaw, NY 14569

WARWICK

Sleepy Valley Center Alcoholism Inpatient Rehabilitation Unit
117 Sleepy Valley Road
Warwick, NY 10990

Warwick Community Bandwagon, Inc. Reaching Out
11 Hamilton Avenue
Warwick, NY 10990

WATERLOO

Geneva General Hospital Medical II Detox Unit
369 East Main Street
Waterloo, NY 13165

Seneca County Comm. Counseling Center Alcoholism Outpatient Clinic
31 Thurber Drive
Waterloo, NY 13165

Seneca County Schools Substance Abuse Program
31 Thurber Drive
Waterloo, NY 13165

WATERTOWN

Alcohol and Substance Abuse Council of Jefferson County
Alcohol Info. and Referral
Substance Abuse Prevention
302–314 Court Street
Globe Mall
Watertown, NY 13601

Comm. Center for Alcoholism of Jefferson County
Alcoholism Outpatient Clinic
595 West Main Street
Watertown, NY 13601

Men's Halfway House
1130 State Street
Watertown, NY 13601

Credo Foundation, Inc. Drug Free/Medically Supervised Outpatient
300 Woolworth Building
Watertown, NY 13601

House of Good Samaritan
Mercy Hospital/Alcoholism Inpatient Rehab Program
Alcoholism Acute Care
218 Stone Street
Watertown, NY 13601

WATKINS GLEN

Council on Alcoholism and Other Chemical Dependency of Finger Lakes/Schuyler Council Office
313 North Franklin Street
Watkins Glen, NY 14891

Schuyler Hospital Choices Prevention Program
108 West 4 Street
Watkins Glen, NY 14891

WEBSTER

Delphi Drug Abuse Center, Inc.
Drug Free Outpatient
School Unit/Prevention
55 East Main Street
Webster, NY 14580

Huther/Doyle Memorial Institute
Alcoholism Outpatient Clinic
Drug Abuse Treatment Unit
2112 Empire Boulevard
Webster, NY 14580

WELLSVILLE

Allegany Area Council on Alcohol/Substance Abuse
Alcoholism and Substance Abuse Treatment
Alcoholism Clinic
Trapping Brook Supportive Living
23 Jefferson Street
Wellsville, NY 14895

Trapping Brook House
3084 Trapping Brook Road
Wellsville, NY 14895

WEST BABYLON

West Babylon Schools Drug Abuse Prevention and Education Program Admin.
200 Old Farmingdale Road
West Babylon, NY 11704

WEST BRENTWOOD

Charles K. Post Alcoholism Treatment Center
Alcoholism Independent Living Program
Charles K. Post Inpatient Rehab
Pilgrim Psychiatric Center
Building 1
West Brentwood, NY 11717

WEST COXSACKIE

New York State Coxsackie Correction Facility (ASAT)
West Coxsackie, NY 12192

WEST HEMPSTEAD

Long Island Jewish Hospital Medical Center Project Outreach/Op DF/Prevention
600 Hempstead Turnpike
West Hempstead, NY 11552

WEST NYACK

Clarkstown Awareness Network for a Drugfree Life and Environment, Inc. (CANDLE)
30 Parrott Road
West Nyack, NY 10994

Clarkstown Central School District Substance Abuse Prevention Services
30 Parrott Road
West Nyack, NY 10994

WESTBURY

North Shore Child/Family Guidance Assoc.
The Place/Alcoholism Outpatient Clinic
Drug Abuse Outpatient Clinic
50 Sylvester Street
Westbury, NY 11590

Westbury Union Free School District Drug Abuse Prevention
Jericho Turnpike/Hitchcock Lane
Westbury, NY 11590

WESTHAMPTON BEACH

Hampton Council of Churches Family Counseling Service
Main Street
Beinecke Building
Westhampton Beach, NY 11978

Seafield Center, Inc. Alcoholism Inpatient Rehabilitation Unit
7 Seafield Lane
Westhampton Beach, NY 11978

WESTON MILLS

Cattaraugus County Council on Alcohol and Substance Abuse Program/Weston's Manor
Route 417
Weston Mills, NY 14788

WHITE PLAINS

Cage Teen Center, Inc. Administrative Unit
220 Ferris Avenue
White Plains, NY 10603

College Careers Fund of Westchester Alcoholism Public Education/Info. and Referral
60 Mitchell Place
White Plains, NY 10601

Halfway Houses of Westchester, Inc. Hawthorne House Alcoholism Halfway House
14 Longview Avenue
White Plains, NY 10605

National Council on Alcoholism and Other Drug Addictions/ Westchester
360 Mamaroneck Avenue
White Plains, NY 10605

New York Hospital Cornell Medical Center
Alcoholism Inpatient Rehabilitation Unit
Westchester Division Alcohol Disorders Program
21 Bloomingdale Road
White Plains, NY 10605

The Weekend Center, Inc. Generations Alcoholism Outpatient Clinic
220 Ferris Avenue
White Plains, NY 10601

Treatment Center of Westchester Alcoholism Outpatient Clinic
10 Mitchell Place
White Plains, NY 10601

WCMHB White Plains Hospital Medical Center Methadone Maintenance Treatment Program
Davis Avenue East Post Road
White Plains, NY 10601

Westchester Community Opportunity Program Greenburgh Open Door Drug Program
5 Prospect Avenue
2nd Floor
White Plains, NY 10607

Westchester Putnam Public Employees Employees' Assistance Program
112 East Post Road
Room 243
White Plains, NY 10601

WHITESBORO

Mohawk Valley Council on Alcoholism Addiction, Inc. Public Education/Info. and Referral
210 Oriskany Boulevard
Whitesboro, NY 13492

WILLARD

Dick Van Dyke Alcoholism Treatment Center Alcoholism Inpatient Rehab Unit
Willard Psychiatric Center
Building 112
Willard, NY 14588

WOODMERE

Tempo Group, Inc.
Drug Abuse Treatment/Intensive Program
Outpatient Drug Free Unit
Prevention Unit
112 Franklin Place
Woodmere, NY 11598

WYANDANCH

North Suffolk Mental Health Center Outpatient Alcohol Rehabilitation
240 Long Island Avenue
Wyandanch, NY 11798

YONKERS

Renaissance Project, Inc. Yonkers Unit
42 Warburton Avenue
Yonkers, NY 10701

Saint Joseph's Hospital Yonkers Drug Free Counseling/OP DF
127 South Broadway
2nd Floor
Yonkers, NY 10701

TMB Continuing Care Center, Inc. Alcoholism Day Rehabilitation Program
317 South Broadway
Yonkers, NY 10705

Yonkers General Hospital
Alcoholism Acute Care Program
Hospital Intervention Services
2 Park Avenue
Yonkers, NY 10703

Methadone Maintenance Treatment Program
185 Ashburton Avenue
Yonkers, NY 10701

Yonkers Residential Center
Breakaway Alcoholism Outpatient Clinic
317 South Broadway
Yonkers, NY 10705

NORTH CAROLINA

ABERDEEN

Bethesda, Inc.
204 North Pine Street
Aberdeen, NC 28315

AHOSKIE

Roanoke/Chowan Human Services Center
Route 3
Ahoskie, NC 27910

ALBEMARLE

Albemarle House, Inc.
242 North 2 Street
Albemarle, NC 28001

ASHEBORO

Randolph County Mental Health/ DD and Substance Abuse Services Program
204 East Academy Street
Asheboro, NC 27203

ASHEVILLE

Neil Dobbins Center
277 Biltmore Avenue
Asheville, NC 28801

Veterans' Affairs Medical Center Substance Abuse Treatment Program
1100 Tunnel Road
Asheville, NC 28805

Woodhill Treatment Center Appalachian Hall
60 Caledonia Road
Asheville, NC 28803

BLACK MOUNTAIN

Alcohol and Drug Abuse Treatment Center
301 Tabernacle Road
Black Mountain, NC 28711

Robert Swain Recovery Center
1280 Old U.S. 70
Black Mountain, NC 28711

BOONE

Hebron Colony and Grace Home
Route 3
Boone, NC 28607

New River Substance Abuse Services
885 State Farm Road
Boone, NC 28607

BURLINGTON

Alamance Caswell Area MH/MR and Substance Abuse Program
1946 Martin Street
Burlington, NC 27217

Alamance Memorial Hospital, Inc. Behavioral Medicine Division Inpatient Detox
730 Hermitage Road
Burlington, NC 27215

Residential Treatment Services of Alamance
136 Hall Avenue
Burlington, NC 27215

BUTNER

Alcohol and Drug Abuse Treatment Center
West E Street
Butner, NC 27509

Federal Correctional Institution Drug Abuse Treatment Program
Old North Carolina Highway 75
Butner, NC 27509

CAMP LEJEUNE

Naval Hospital Alcohol Rehabilitation Department
Naval Hospital
Camp Lejeune, NC 28542

CHAPEL HILL

Orange/Person/Chatham Mental Health Center Substance Abuse Services
333 McMasters Street
Chapel Hill, NC 27516

CHARLOTTE

Amethyst Charlotte, Inc.
1715 Sharon Road West
Charlotte, NC 28210

Bethlehem Center Alcohol Education Prevention Program
2705 Baltimore Avenue
Charlotte, NC 28203

Charlotte Council on Alcoholism and Chemical Dependency, Inc.
100 Billingsley Road
Charlotte, NC 28211

Drug Education Center, Inc.
1117 East Morehead Street
Charlotte, NC 28204

Mecklenburg County Area Mental Health Authority Substance Abuse Services
429 Billingsley Road
2nd Floor
Charlotte, NC 28211

Mercy Hospital Horizons
2001 Vail Avenue
Charlotte, NC 28207

Reachline Telephone Counseling Service
1015 Elliott Street
Charlotte, NC 28208

The Randolph Clinic, Inc. Alcohol and Drug Traffic Education School
100 Billingsley Road
Charlotte, NC 28211

CHEROKEE

Cherokee Indian Hospital A Ye Ka Chemical Dependency Unit
Hospital Road
Cherokee, NC 28719

CONCORD

Serenity House, Inc.
172 Spring Street SW
Concord, NC 28025

DILLSBORO

Smoky Mountain Center for MH/MR/Substance Abuse Services Substance Abuse Services Program
U.S. Highway 19A and 441
Dillsboro, NC 28725

DURHAM

Durham Council on Alcoholism and Drug Dependency, Inc. Alcoholism Information Center
3109 University Drive
Suite A
Durham, NC 27707

Oakleigh at Durham
309 Crutchfield Street
Durham, NC 27704

Substance Abuse Services
414 East Main Street
Durham, NC 27701

Veterans' Affairs Medical Center Substance Abuse Program
508 Fulton Street
Durham, NC 27705

ELIZABETH CITY

Albemarle Mental Health Center
111 South Road Street
Elizabeth City, NC 27909

FAYETTEVILLE

Cumberland County Mental Health Center Family Recovery Services
109 Bradford Avenue
4th Floor
Fayetteville, NC 28301

Myrover Reese Fellowship House
613 Quality Road
Fayetteville, NC 28306

Veterans' Affairs Medical Center Substance Abuse Treatment Unit
2300 Ramsey Street
Fayetteville, NC 28301

GASTONIA

Family Service, Inc.
214 East Franklin Boulevard
Gastonia, NC 28052

Gaston/Lincoln Area Mental Health Center Substance Abuse Program
401 North Highland Street
Gastonia, NC 28052

GOLDSBORO

Wayne County Mental Health Center
301 North Herman Street
County Office Building
Goldsboro, NC 27530

GREENSBORO

Fellowship Hall
5140 Dunstan Road
Greensboro, NC 27405

Greater Piedmont Teen Challenge
1912 Boulevard Street
Greensboro, NC 27407

Green Point East
301 East Washington Street
Greensboro, NC 27401

Guilford County Mental Health Center Substance Abuse Program
201 North Eugene Street
Greensboro, NC 27401

Industrial Counseling Service
2302 West Meadowview Road
Suite 209
Greensboro, NC 27407

Links Adolescent Service
714 Huffine Mill Road
Greensboro, NC 27405

Substance Abuse Services of Guilford
312 North Eugene Street
Greensboro, NC 27401

1305 Glenwood Avenue
Greensboro, NC 27403

GREENVILLE

Real Crisis Intervention, Inc.
312 East 10 Street
Greenville, NC 27858

W. B. Jones Alcohol and Drug Abuse Treatment Center
Route 1
Greenville, NC 27834

HENDERSON

Division of Vocational Rehabilitation Alcohol Abuse Program
1924 Ruin Creek Road
Suite 202
Henderson, NC 27536

Franklin/Granville/Vance/Warren Area Mental Health Program
125 Emergency Road
Henderson, NC 27536

HICKORY

Catawba County Mental Health Services
Route 3
Hickory, NC 28602

HIGH POINT

Greenpoint West
119 Chestnut Drive
High Point, NC 27262

Substance Abuse Services of Guilford
118 West Russell Avenue
High Point, NC 27260

JACKSONVILLE

Onslow County Area Mental Health/DD Substance Abuse Services
215 Memorial Drive
Jacksonville, NC 28546

JAMESTOWN

Alcoholics Home, Inc. House of Prayer
5884 Riverdale Road
Route 1
Jamestown, NC 27282

JEFFERSON

New River Mental Health Center Substance Abuse Services
Court Street
Jefferson, NC 28640

KINSTON

Fellowship Homes of Kinston, Inc.
704 Parrot Avenue
Kinston, NC 28501

Lenoir Area MH/MR Substance Abuse Program
2901 North Heritage Street
Kinston, NC 28501

LEXINGTON

Davidson Alcoholic Care, Inc.
1675 East Center Street Extension
Lexington, NC 27292

LUMBERTON

Carolina Manor Treatment Center
1100 Pine Run Drive
Lumberton, NC 28358

Robeson County Mental Health Clinic Non-Hospital Detoxification Center
450 Country Club Road
Lumberton, NC 28359

MORGANTON

Burke Council on Alcoholism and Chemical Dependency
203 White Street
Morganton, NC 28655

Foothills Area Mental Health Center
1001 B East Union Street
Morganton, NC 28655

Burke Youth Alternatives
306 South King Street
Morganton, NC 28655

Foothills Detox/Crisis Program
2130 NC 18/U.S. 64
Morganton, NC 28655

MOUNT AIRY

Crossroads Outpatient Services
113 Gilmer Street
Mount Airy, NC 27030

NEW BERN

Neuse Center for Mental Health/ DD Substance Abuse Services
800 Cardinal Drive
New Bern, NC 28562

NEWLAND

New River Mental Health Center Avery Cares Center
Highway 194
Newland, NC 28657

RALEIGH

Christian Fellowship Home of Raleigh
506 Cutler Street
Raleigh, NC 27603

Drug Action, Inc. Community Treatment Project
2809 Industrial Drive
Raleigh, NC 27609

The Recovery Partnership, Inc.
3900 Barrett Drive
Suite 301
Raleigh, NC 27609

Wake County Alcoholism Treatment Center
3000 Falstaff Road
Raleigh, NC 27610

REIDSVILLE

Rockingham County Mental Health Center Substance Abuse Services
405 NC 65
Reidsville, NC 27320

ROCKINGHAM

Samaritan Colony
Highway 220 North
Rockingham, NC 28379

ROCKY MOUNT

Edgecombe/Nash Mental Health Center Substance Abuse Program
500 Nash Medical Arts Mall
Rocky Mount, NC 27804

ROSE HILL

Duplin/Sampson Area Mental Health Developmental Disabilities and Substance Abuse Services
West Ridge Street
Rose Hill, NC 28458

SALISBURY

Rowan Helping Ministries Dial Help
Salisbury, NC 28144

Rowan Memorial Hospital Careunit
612 Mocksville Avenue
Salisbury, NC 28144

Tri-County Halfway House for Women
1127 South Main Street
Salisbury, NC 28144

Veterans' Affairs Medical Center Substance Abuse Treatment Program
1601 Brenner Avenue
Unit 4-2B (116A3)
Salisbury, NC 28144

SANFORD

Lee/Harnett Mental Health Center Substance Abuse Services
130 Carbonton Road
Sanford, NC 27330

SELMA

Day by Day Treatment Center for Men and Women
1110 River Road
Selma, NC 27576

SHELBY

Cleveland County Mental Health Center Substance Abuse Program
222 Crawford Street
Shelby, NC 28150

SMITHFIELD

Johnston Substance Abuse Program Treatment and Rehabilitation
Bright Leaf Boulevard
Smithfield, NC 27577

SOUTHERN PINES

Bethany House Halfway House
240 East Vermont Avenue
Southern Pines, NC 28387

SPARTA

Alleghany Lodge Halfway House for Substance Abusers
Route 4
Sparta, NC 28675

New River Mental Health Center Alleghany Cares
Highway 18 South
Sparta, NC 28675

STATESVILLE

Bagnal Home Halfway House
1103 West Front Street
Statesville, NC 28677

Tri-County
Detoxification Center
Intensive Outpatient Services
1419 Wilson Lee Boulevard
Statesville, NC 28677

THOMASVILLE

Davidson County MH/DD and Substance Abuse Services
205 Old Lexington Road
Thomasville, NC 27360

WASHINGTON

Tideland Mental Health Center Substance Abuse Division
1308 Highland Drive
Washington, NC 27889

WEST END

Sandhills Mental Health Center Substance Abuse Services
Administrative Offices
West End, NC 27376

WILKESBORO

New River Substance Abuse Services Wilkes County
1216 School Street
Wilkesboro, NC 28697

WILMINGTON

Southeastern MH/MR Substance Abuse Area Program
2023 South 17 Street
Wilmington, NC 28401

Stepping Stone Manor
416 Walnut Street
Wilmington, NC 28401

The Wilmington Treatment Center, Inc.
2520 Troy Drive
Wilmington, NC 28401

WILSON

Wilson Crisis Center
Wilson, NC 27893

Wilson/Greene Substance Abuse Center
208 North Goldsboro Street
Wilson, NC 27895

WINSTON-SALEM

Charter Hospital of Winston-Salem Genesis Recovery Center
3637 Old Vineyard Road
Winston-Salem, NC 27104

Medicorp Recovery Network
200 Charlois Boulevard
Winston-Salem, NC 27103

R. J. Reynolds Tobacco Co. USA Recovery Program
Building 601-1
Winston-Salem, NC 27102

Step One, Inc.
545 North Trade Street
Winston-Salem, NC 27101

NORTH DAKOTA

BELCOURT

Turtle Mountain Counseling and Rehab Center
Belcourt, ND 58316

BISMARCK

Burleigh County Detoxification Center
514 East Thayer Avenue
Burleigh County Sheriff's Department
Bismarck, ND 58501

DE Counseling Service
626 North 6 Street
Bismarck, ND 58501

New Freedom Center
4205 State Street
Highway 83 North
Bismarck, ND 58501

North Dakota State Penitentiary Addiction Treatment Program
Bismarck, ND 58502

West Central Human Service Center Chemical Dependency Program
600 South 2 Street
Bismarck, ND 58504

Whole Person Recovery Center
919 South 7 Street
Suite 506
Bismarck, ND 58504

DEVILS LAKE

Lake Region Human Service Center Chemical Dependency Program
Highway 2 West
Devils Lake, ND 58301

DICKINSON

Badlands Human Service Center Chemical Dependency Program
Dickinson State University Campus
Pulver Hall
Dickinson, ND 58601

Heart River Alcohol/Drug Abuse Services
1260 West Villard Street
Dickinson, ND 58601

Saint Joseph's Hospital/Health Center Chemical Dependency Program
30 West 7 Street
Dickinson, ND 58601

FARGO

Addiction Counseling Services
213 9 Street South
Cass County Sheriff's Department
Fargo, ND 58107

Centre, Inc.
123 North 15 Street
Fargo, ND 58107

Consultation and Movement
2109 3 Street North
Fargo, ND 58102

Drake and Burau Counseling Services
1202 23 Street South
Suite 6
Fargo, ND 58103

Fargo Clinic Meritcare Chemical Dependency Program
700 South First Avenue
Fargo, ND 58103

Heartland Medical Center Mental Health Services
45 South 4 Street
Fargo, ND 58103

Lutheran Social Services Addiction Outreach for Recovery
1325 South 11 Street
Fargo, ND 58107

Margaret Volk Human Service Associates
806 North 6 Avenue
Fargo, ND 58102

Pathways Recovery Centers, Inc.
2579 Atlantic Drive
Fargo, ND 58103

Professional Resource Network Counseling Services
1323 23 Street South
Fargo, ND 58103

Share House
4227 9 Avenue SW
Fargo, ND 58103

Southeast Human Service Center Alcohol and Drug Abuse Unit
2624 9 Avenue South
Fargo, ND 58103

Veterans' Affairs Medical Center Substance Abuse Treatment Program
2101 Elm Street North
Fargo, ND 58102

GARRISON

Ron Stanley Counseling Service
36 3 Avenue NW
Garrison, ND 58540

GRAFTON

Michael Bryan Addiction Counseling
700 Manvel Avenue
Grafton, ND 58237

GRAND FORKS

Don Foley Counseling Service
711 North Washington Street
Grand Forks, ND 58203

Katy Wright Counseling
1407 South 24 Avenue
Suite 214
Grand Forks, ND 58201

Northeast Human Service Center Chemical Dependency Program
1407 24 Avenue South
Grand Forks, ND 58201

Northridge Counseling Centre, Inc.
215 North 3 Street
Suite 100
Grand Forks, ND 58203

Operation Transition Foundation
1407 10 Avenue South
Grand Forks, ND 58201

United Recovery Center Medical Park
Grand Forks, ND 58201

JAMESTOWN

Alcohol/Families and Children
419 2 Avenue NW
Jamestown, ND 58401

North Dakota State Hospital Chemical Dependency Unit
Jamestown, ND 58402

Northern Prairie Consultants
115 2 Street SW
Jamestown, ND 58401

South Central Human Service Center Chemical Dependency Program
520 3 Street NW
Jamestown, ND 58402

MANDAN

Heartview Foundation
1406 NW 2 Street
Mandan, ND 58554

North Dakota Industrial School Drug and Alcohol Program
701 16 Avenue SW
Mandan, ND 58554

MINOT

Dakota Boys' Ranch
RR 6
Minot, ND 58702

Gateway Drug and Alcohol Center
315 South Main Street
Suite 307-A
Minot, ND 58701

North Central Human Service Center Chemical Dependency Program
400 22 Avenue NW
Minot, ND 58701

Saint Joseph's Hospital Chemical Dependency Unit
3 Street SE and Burdick Expressway
Minot, ND 58701

Will Bachmeier LAC
1809 South Broadway
Lower Unit Suite FF
Minot, ND 58701

NEW TOWN

Circle of Life Alcohol Program
New Town, ND 58763

TRENTON

Trenton Alcohol and Drug Program
Trenton, ND 58853

WILLISTON

Mercy Recovery Center
1213 15 Avenue West
Williston, ND 58801

Northwest Human Service Center Chemical Dependency Program
316 2 Avenue West
Williston, ND 58801

OHIO

AKRON

Akron Health Department Alcoholism Division
177 South Broadway
Akron, OH 44308

Community Drug Board
725 East Market Street
Akron, OH 44305

Genesis Program
386 South Portage Path
Akron, OH 44320

Ramar Center
380 South Portage Path
Akron, OH 44320

Edwin Shaw Hospital Alcohol/ Drug Treatment Unit
1621 Flickinger Road
Akron, OH 44312

Employee Special Service Center
1717 Brittain Road
Suite 102
Akron, OH 44310

Family Services
212 East Exchange Street
Akron, OH 44304

Fairlawn Office
3050 West Market Street
Akron, OH 44333

First Step Alcohol/Drug Outreach Program
550 South Arlington Street
Akron, OH 44306

Interval Brotherhood Homes, Inc. Alcohol Rehabilitation Center
3445 South Main Street
Akron, OH 44319

North American Indian Cultural Center Cuyahoga Project
1062 Triplett Boulevard
Akron, OH 44306

Oriana House, Inc. DWI Program/Outpatient
40 East Glenwood Street
Akron, OH 44304

Recovery Services
665 West Market Street
Suite 1F
Akron, OH 44303

Senior Workers' Action Program Chemical Dependency Services
415 South Portage Path
Akron, OH 44320

Tri-County Employee Assistance Program
450 Grant Street
Suite 301
Akron, OH 44311

Urban Minority Alcoholism and Drug Abuse Outreach Program
762 Mallison Avenue
Akron, OH 44307

Urban Ounce of Prevention Services, Inc.
1544 South Hawkins Avenue
Akron, OH 44320

ALLIANCE

Quest Recovery Services, Inc. Alliance Division
724 South Union Street
Alliance, OH 44601

AMHERST

Alcohol Education and Assessment Program
Amherst, OH 44001

ASHLAND

Appleseed Counseling and Case Management
221 Church Street
Suites 4 5 6
Ashland, OH 44805

Ashland County Council on Alcoholism and Drug Abuse, Inc.
310 College Avenue
Ashland, OH 44805

ASHTABULA

Lake Area Recovery Center
Outpatient Drug Free Program
2801 C Court
Ashtabula, OH 44004

Turning Point
2711 Donohoe Drive
Ashtabula, OH 44004

ATHENS

Health Recovery Services, Inc.
Athens County Outpatient Clinic
406 Richland Avenue
Athens, OH 45701

Bassett House
154-B Bassett Road
Athens, OH 45701

Driver Intervention Program
28 North College Street
Athens, OH 45701

Rural Women's Recovery Program
9908 Bassett Road
Athens, OH 45701

BARBERTON

Barberton Citizens Hospital Adolescent Chemical Dependency Unit
155 5 Street NE
Barberton, OH 44203

Family Services of Summit County
480 West Tuscarawas Avenue
Suite 101
Barberton, OH 44203

BATAVIA

Clermont County Youth Services
2400 Clermont Center Drive
Medical/Social Services Building
Batavia, OH 45103

Clermont Recovery Center, Inc.
Camp Allyn DIP
2379 Clermont Center Drive
Batavia, OH 45103

Jail Program
4200 Filager Road
Batavia, OH 45103

Outpatient Services
2289 Bauer Road
Batavia, OH 45103

BEACHWOOD

Center for the Prevention of Domestic Violence
23875 Commerce Park Road
Beachwood, OH 44122

Jewish Family Services Assoc. of Cleveland Alcoholism/Chemical Dependency
24075 Commerce Park Road
Beachwood, OH 44122

BEDFORD

Community Hospital of Bedford
Drivers' Alternative Program
Substance Abuse Program
44 Blaine Avenue
Bedford, OH 44146

BELLAIRE

Bellaire Drug and Alcohol Office
First National Bank Building
Suite 210–211
Bellaire, OH 43906

Belmont County Student Services
Student Services Discovery
Program
3517 Guernsey Street
Bellaire, OH 43906

BELLEFONTAINE

Logan/Champaign Alcohol and
Drug Addiction Services
1513 TWP. Road 235
Bellefontaine, OH 43311

Mercy Memorial Hospital Mercy
Outreach
Memorial Hall
Lower Level
Bellefontaine, OH 43311

BEREA

United Methodist Alcohol and
Chemical Counseling, Inc.
(UMACC, Inc.)
298 Pineview Drive
Berea, OH 44017

BIDWELL

Family Addiction Community
Treatment Services
1770 Jackson Pike
Bidwell, OH 45614

BOWLING GREEN

Wood County Council on
Alcoholism/Drug Abuse, Inc.
320 West Gypsy Lane Road
Bowling Green, OH 43402

Wood County Impaired Driver
Program
320 West Gypsy Lane
Bowling Green, OH 43402

BRUNSWICK

Alcohol and Drug Dependency
Services of Medina County/
Brunswick Office
4274 Manhattan Circle Drive
Brunswick, OH 44212

BRYAN

Five County Alcohol/Drug
Program
820 East Edgerton Street
Bryan, OH 43506

BUCYRUS

Community Counseling Services,
Inc. Bucyrus Office
820 Plymouth Street
Bucyrus, OH 44820

Contact, Inc.
Bucyrus, OH 44820

BURTON

Awareness, Inc. Camp Burton
14282 Butternut Road
Burton, OH 44021

BYESVILLE

Guernsey County Alcohol and
Drug Addiction Treatment
Center, Inc.
Driver Intervention Program
Outpatient
60788 Southgate Road
Byesville, OH 43723

CADIZ

Belmont/Harrison/Monroe Drug
and Alcohol Council, Inc.
239 West Warren Street
Cadiz, OH 43907

CALDWELL

Noble Drug Abuse and
Alcoholism Council, Inc.
48 Olive Street
Caldwell, OH 43724

CAMBRIDGE

Residential Intervention Driver
Education (RIDE)
1251 Clark Street
Cambridge, OH 43725

CANTON

Canton Community Kidsummit
Against Drugs
1027 9 Street NE
Canton, OH 44704

Canton Urban League Alcohol
and Drug Program
1400 Sherrick Road SE
Canton, OH 44707

City of Canton Police Department
Alcohol and Drug Education
221 3 Street SW
Canton, OH 44702

Community Treatment and
Correction Center Inc./
Substance Abuse Program
1200 Market Avenue South
Canton, OH 44707

Crisis Intervention Center of
Stark County, Inc. Outpatient
2421 13 Street NW
Canton, OH 44708

Quest Driver Intervention
Program
1341 Market Avenue North
Canton, OH 44714

Quest Recovery Services, Inc.
1341 Market Avenue North
Canton, OH 44714

Quest Deliverance House/
Women's Residential Treatment
626 Walnut Avenue NE
Canton, OH 44702

Quest Recovery House
215 Newton Avenue NW
Canton, OH 44703

CARROLLTON

Carroll County
Alcohol and Drug Addiction Center
DUI Counterattack School
331 West Main Street
Carrollton, OH 44615

Self Help, Inc.
70 East Main Street
Carrollton, OH 44615

CELINA

Gateway Outreach Center
Nonresidential Alcohol Safety Program
Outpatient Services
441 East Market Street
Celina, OH 45822

CHARDON

Catholic Social Services of Geauga County
10771 Mayfield Drive
Chardon, OH 44024

Geauga County Board of Education Teen Inst. for Prevention of Drug and Alcohol Abuse
211 Main Street
3rd Floor
Chardon, OH 44024

CHILLICOTHE

Chillicothe Correctional Institute Recovery Services Department
15802 State Route 104
Chillicothe, OH 45601

Great Seal Family Care Center
425 Chestnut Street
Room 1
Chillicothe, OH 45601

Ross Correctional Institute Substance Abuse Program
16149 State Route 104
Chillicothe, OH 45601

Ross County Sheriff Drivers' Intervention Program
28 North Paint Street
Chillicothe, OH 45601

Scioto Paint Valley Mental Health Center Ross County Office
4449 State Route 159
Chillicothe, OH 45601

Veterans' Affairs Medical Center Substance Abuse Treatment Program
17273 State Route 104
Chillicothe, OH 45601

CINCINNATI

Alcoholism Council of Cincinnati Area
Alice Paul House
NCA/DIP
NCADD
118 East William Howard Taft Road
Cincinnati, OH 45219

Archdiocese of Cincinnati Board of Education Project Triple Outreach
100 East 8 Street
Drug Education Resource Center
Cincinnati, OH 45202

Bethesda Alcohol and Drug Treatment
Blueash Outpatient Family and Day Treatment
11305 Reed Hartman Highway
Cincinnati, OH 45241

Oak Program
619 Oak Street
Cincinnati, OH 45206

Careunit Hospital of Cincinnati
3156 Glenmore Avenue
Cincinnati, OH 45211

Center for Comp. Alcoholism Treatment
830 Ezzard Charles Drive
Cincinnati, OH 45214

Central Community Health Board Drug Services
3020 Vernon Place
Cincinnati, OH 45219

Christ Hospital Alcohol and Drug Center
2139 Auburn Avenue
Cincinnati, OH 45219

CPC/Alcohol and Substance Abuse, Inc. Outpatient Treatment Services
311 Martin Luther King Drive
C Building
Cincinnati, OH 45219

Drug and Poison Information Center and Ohio Prevention and Education Resource Center
231 Bethesda Avenue
Bridge ML 144 Room 7702
Cincinnati, OH 45267

Family Services of the Cincinnati Area/Central Unit
205 West 4 Street
Cincinnati, OH 45202

Prospect House
682 Hawthorne Avenue
Cincinnati, OH 45205

Reading Youth Service Bureau
1223 Jefferson Avenue
Cincinnati, OH 45215

Shaffer House, Inc.
583 Grand Avenue
Cincinnati, OH 45205

Shelterhouse Volunteer Group, Inc. Drop-In Center
217 West 12 Street
Cincinnati, OH 45210

Talbert House, Inc.
Alternatives
100 Shadybrook Drive
Cincinnati, OH 45216

Beekman Center
2438 Beekman Street
Cincinnati, OH 45214

Cornerstone for Men
2216 Vine Street
Cincinnati, OH 45219

Driver Intervention Program
1617 Reading Road
Cincinnati, OH 45202

Drug and Family Counseling
308 Reading Road
Cincinnati, OH 45202

Pathways
334 McGregor Avenue
Cincinnati, OH 45219

SA/MI Intensive Outpatient
SA/MI Case Management
3140 Harvey Avenue
Cincinnati, OH 45229

SA/MI Residential
2915 Highland Avenue
Cincinnati, OH 45219

Talbert House for Women
3123 Woodburn Avenue
Cincinnati, OH 45207

Talbert House Turning Point
2605 Woodburn Avenue
Cincinnati, OH 45206

Talbert House/Neighborhood Outreach
Drug and Family Counseling/Eastern
3806 Eastern Avenue
Cincinnati, OH 45214

Mt. Auburn
328 McGregor Avenue
Cincinnati, OH 45219

Madisonville
6100 Desmond Street
Madisonville YMCA
Cincinnati, OH 45227

Substance Abuse Services/Walnut Hills
2601 Melrose Avenue
Suite 106
Cincinnati, OH 45206

University of Cincinnati Medical Center Substance Abuse Services
234 Goodman Street
Cincinnati, OH 45269

Urban Appalachian Council Outpatient Substance Abuse Program
2115 West 8 Street
Cincinnati, OH 45204

Urban Minority Alcoholism/Drug Abuse Outreach Program of Cincinnati, Inc.
400 West 9 Street
Suite G
Cincinnati, OH 45203

CIRCLEVILLE

Pickaway Area Recovery Services
600 North Pickaway Street
Berger Hospital Room 205
Circleville, OH 43113

Scioto Paint Valley Mental Health Center Pickaway County Office
145 Morris Road
Circleville, OH 43113

CLEVELAND

Alcohol and Drug Recovery Center/P 47 Cleveland Clinic Foundation
9500 Euclid Avenue
Cleveland, OH 44195

Alcoholism Services of Cleveland
Homeless Project
2219 Payne Avenue
Cleveland, OH 44114

Residential Services for Homeless
2600 Carrol Avenue
Cleveland, OH 44118

East
2490 Lee Boulevard
Suite 300
Cleveland, OH 44118

West
14805 Detroit Avenue
Suite 320
Cleveland, OH 44107

Arts Educational Theatre Company Baggage on Alcohol and Other Drugs
140 Public Square
Suite 911
Cleveland, OH 44114

Bellefaire/Jewish Children's Bureau Pact Program
22001 Fairmount Boulevard
Cleveland, OH 44118

Catholic Social Services of Cuyahoga County
Chemical Dependency Services
3135 Euclid Avenue
Room 202
Cleveland, OH 44115

Hispanic Program
2012 West 25 Street
Suite 506
Cleveland, OH 44113

Center for Human Services The Rap Art Center
13429 Cedar Road
Cleveland, OH 44118

City of Cleveland Public Health Dept.
Focus II Westside Office/Mental Health and Substance Abuse
Student Assistance Program
4242 Lorain Avenue
Cleveland, OH 44113

Cleveland Hearing and Speech Center Alcohol/Drug Addiction Prevention Program
11206 Euclid Avenue
Cleveland, OH 44106

Cleveland Student Health Program East High Health Clinic
1349 East 79 Street
Cleveland, OH 44103

Cleveland Treatment Center, Inc.
1127 Carnegie Avenue
Cleveland, OH 44115

Cleveland Urban Minority Alcoholism/Drug Abuse Outreach Project
1215 East 79 Street
Cleveland, OH 44103

Community Action Against Addiction, Inc.
5209 Euclid Avenue
Cleveland, OH 44103

Community Assessment Program Weekend Intervention Program
5163 Broadway Avenue
Cleveland, OH 44127

Community Guidance, Inc.
Genesis House
1924 East 85 Street
Cleveland, OH 44106

Genesis/A New Beginning
3134 Euclid Avenue
Cleveland, OH 44115

Cuyahoga Community College Alcohol/Drug Awareness/ Prevention Team (ADAPT)
4520 Richmond Road
Eastern Campus Bldg East I Room 112
Cleveland, OH 44122

Cuyahoga County Hospital Metro Center/Alcoholism CD Treatment Service
2500 Metrohealth Drive
Hamann Building H-842
Cleveland, OH 44109

East Side Catholic Shelter Two San Program
9412 Heath Avenue
Cleveland, OH 44104

Exodus Program, Inc.
1809 East 89 Street
Cleveland, OH 44106

Family Health of Beechwood
3737 Lander Road
Cleveland, OH 44124

Free Medical Clinic of Greater Cleveland
12201 Euclid Avenue
Cleveland, OH 44106

Fresh Start, Inc. Cedar Avenue Unit
4807 Cedar Avenue
Cleveland, OH 44103

Greater Cleveland Schools Council Project Care
8001 Brecksville Road
Cleveland, OH 44141

Harbor Light Substance Abuse Division Outpatient/Detox Unit 1
1710 Prospect Avenue
Cleveland, OH 44115

Hispanic Urban Minority Alcoholism and Drug Abuse Outreach Program
2012 West 25 Street
Suite 517
Cleveland, OH 44113

Hitchcock Center for Women
1227 Ansel Road
Cleveland, OH 44108

Impaired Driver Education/ Awareness Project Idea
3865 Rocky River Drive
Suite 1
Cleveland, OH 44111

Inner City Renewal Society Ministerial Outreach Program
2230 Euclid Avenue
Cleveland, OH 44115

Lutheran Metropolitan Ministry
After-School Prevention Resources
6516 Detroit Avenue
Cleveland, OH 44102

Community Re-Entry Program
1468 West 25 Street
Cleveland, OH 44113

Maternity and Infant Health Care Saving Babies Through Drug Prevention
2500 Metrohealth Drive
Metrohealth Medical Center
Cleveland, OH 44109

Matt Talbot Inn Residential
2270 Professor Avenue
Cleveland, OH 44113

McIntyre Clinic, Inc. Driver Intervention Program
6285 Pearl Road
Cleveland, OH 44130

Metro Health Clement Center Alcohol and Other Drug Services
2500 East 79 Street
Cleveland, OH 44104

Metrohealth CMHA Outhwaite Homes Estate/Model Drug Treatment
4500 Kennard Court
Miracle Village
Cleveland, OH 44104

Metrohealth Saint Luke's Medical Center Adolescent Chemical Dependency Programs
11311 Shaker Boulevard
Cleveland, OH 44104

Neighborhood Centers Association Substance Abuse Prevention Program
3135 Euclid Avenue
Suite 103
Cleveland, OH 44115

New Directions
30800 Chagrin Boulevard
Cleveland, OH 44124

Northeast CMHC Turning Point
763 East 152 Street
Cleveland, OH 44110

Northeast Pre-Release Center Substance Abuse Services
2675 East 30 Street
Cleveland, OH 44101

Ohio Department of Youth Services Cuyahoga Hills Boys' School
4321 Green Road
Cleveland, OH 44128

Orca House, Inc.
Bradley House
1914 East 90 Street
Cleveland, OH 44106

Mary Gooden Center for Women
1909 East 89 Street
Cleveland, OH 44106

Men's Program
1905 East 89 Street
Cleveland, OH 44106

Project East DBA East Cleveland Straight Talk
12921 Euclid Avenue
Cleveland, OH 44112

Rocky River Community Challenge
19120 Detroit Road
Suite 10
Cleveland, OH 44116

Saint Vincent Charity Hospital Health Center
The Rosary Serenity Centers
2351 East 22 Street
Cleveland, OH 44115

Scarborough Hall
2430 West 10 Street
Cleveland, OH 44113

Southwest General Hospital Oak View Program
18697 East Bagley Road
Cleveland, OH 44130

Stella Maris Washington Avenue Unit
1320 Washington Avenue
Cleveland, OH 44113

Task Force on Violent Crime Substance Abuse Initiative of Greater Cleveland
614 Superior Avenue West
300 Rockefeller Building
Cleveland, OH 44113

Templum House
Cleveland, OH 44101

The Covenant West Side Ecumenical Ministry
1688 Fulton Road
3rd Floor
Cleveland, OH 44113

Transitional Housing, Inc.
1545 West 25 Street
Cleveland, OH 44113

United Labor Agency, Inc. Unemployed Women Substance Abuse Program
1800 Euclid Avenue
Cleveland, OH 44115

United Pastors in Mission
7510 Woodland Avenue
Cleveland, OH 44104

United Way Services First Call for Help Drug Info. Program
3113 Prospect Avenue
Cleveland, OH 44115

University Settlement Homeless Program
4800 Broadway
Cleveland, OH 44127

Veterans' Addiction Recovery Center Alcohol/Drug Dependence Treatment Unit
10701 East Boulevard
Cleveland, OH 44106

10000 Brecksville Road
Cleveland, OH 44141

West Side Women's Center
4207 Lorain Avenue
Cleveland, OH 44113

Westside Community Mental Health Center The Unbar
8301 Detroit Avenue
Cleveland, OH 44102

Women's Alliance for Recovery Services
2012 West 25 Street
Suite 620
Cleveland, OH 44113

CLEVELAND (PARMA)

Parmadale, Inc.
6753 State Road
Cleveland (Parma), OH 44134

CLEVELAND HEIGHTS

Institute for Creative Living
3630 Fairmount Boulevard
Cleveland Heights, OH 44118

COLUMBUS

Adolescent Drug Education Prevention as a Team Club Surrender
1181 East Main Street
Columbus, OH 43205

Africentric Personal Development Shop
1393 East Broad Street
Suite 105
Columbus, OH 43205

Amethyst, Inc.
92 Jefferson Avenue
Columbus, OH 43215

Pre-Release Center
1800 Harmon Avenue
Columbus, OH 43223

Buckeye Youth Center Substance Abuse Program
2280 West Broad Street
Columbus, OH 43223

Callvac Services
370 South 5 Street
Columbus, OH 43215

Choices for Victims of Domestic Violence, Inc.
Columbus, OH 43206

Columbus Area Council on Alcoholism
360 South 3 Street
Suite 306
Columbus, OH 43223

Columbus Community Hospital Chemical Dependency Unit
1430 South High Street
Columbus, OH 43207

Columbus Health Department
Alcoholism and Drug Abuse Programs
Franklin County Alcohol Safety Program
181 Washington Boulevard
Columbus, OH 43215

Compdrug Corporation
Senior and Youth Project
700 Bryden Road
Columbus, OH 43215

Vita Treatment Center/Methadone Services
156 Parsons Avenue
3rd Floor
Columbus, OH 43215

Crittenton Family Services Cedars Branch
1414 East Broad Street
Columbus, OH 43205

Deaf Hope
1500 West 3 Avenue
Columbus, OH 43212

Franklin County Juvenile Court High Risk Youth Project/New Direction
373 South High Street
6th Floor
Columbus, OH 43215

Franklin Pre-Release Center Residential Substance Abuse Program
1800 Harmon Avenue
Columbus, OH 43223

Freedom Center
Columbus, OH 43218

House of Hope for Alcoholics
825 Dennison Avenue
Columbus, OH 43215

Intervention and Assessment Program
1375 South Hamilton Road
Columbus, OH 43227

Maryhaven, Inc.
Maryhaven Exploring Sober Alternatives (MESA)
DIP
Alternatives to Crack Cocaine
1755 Alum Creek Drive
Columbus, OH 43207

Neighborhood House, Inc. Alcohol/Drug Counseling Program
1000 Atcheson Street
Columbus, OH 43203

North Central Calm Associates North Central Mental Health Sattellite Office
3620 North High Street
Columbus, OH 43214

North Central Mental Health Services
Drug and Alcohol Treatment Program
1301 North High Street
Columbus, OH 43201

Family Focus
162 West 5 Avenue
Columbus, OH 43201

Fowler House
422 East Lane Avenue
Columbus, OH 43201

North Community Counseling Centers, Inc. The Bridge
4897 Karl Road
Columbus, OH 43229

Ohio Employee Assistance Program
22 East Gay Street
7th Floor
Columbus, OH 43266

Orient Correctional Institution Residential Substance Abuse Program
Columbus, OH 43216

Parkside Lodge Driver Education Weekend
349 Ridenour Road
Columbus, OH 43230

Phoenix Human Services, Inc.
620 Alum Creek Drive
Suite 201
Columbus, OH 43205

Prevention Education Coalition
263 Carpenter Street
2nd Floor
Columbus, OH 43205

Project Linden
1500 East 17 Avenue
Columbus, OH 43219

Riverside Methodist Hospitals Alcohol and Drug Dependency Services
3535 Olentangy River Road
Columbus, OH 43214

Southeast CMHC Substance Abuse Services
1455 South 4 Street
Columbus, OH 43207

Southeast Counseling Services
1250 South High Street
Columbus, OH 43206

Special Achievement Council
373 South High Street
10th Floor
Columbus, OH 43215

Stephens House
1138 South High Street
Columbus, OH 43206

Talbot Hall at Park Medical Center
1492 East Broad Street
Columbus, OH 43205

Teen Challenge
47 East 12 Avenue
Columbus, OH 43216

Training Center for Youth
2280 West Broad Street
Columbus, OH 43266

Training Institute/Central Ohio
2130 West Broad Street
Columbus, OH 43266

Urban Minority Alcoholism/ Outreach of Franklin County, Inc.
98 Hamilton Park
Suite 1-F
Columbus, OH 43203

Volunteers of America Shelter for Homeless Men
379 West Broad Street
Columbus, OH 43215

Wait Program
Columbus, OH 43209

Women's Outreach for Women
1950-H North 4 Street
Columbus, OH 43201

COSHOCTON

Coshocton County
Drug and Alcohol Council, Inc.
Driver Intervention Program
140 1/2 South 6 Street
Coshocton, OH 43812

CUYAHOGA FALLS

Northeast Summit Family Service Chemical Abuse
2100 Front Street
Cuyahoga Falls, OH 44221

DAYTON

Afrikan/Amerikan Institute for Positive Living Inc. Web Dubois Academy
3011 Oakridge Drive
Dayton, OH 45417

Bergamo Center Greene County Outreach/Drug Prevention Project
4400 Shakertown Road
Dayton, OH 45430

Black Urban Minority Alcoholism and Drug Abuse Outreach Program
3807 West 3 Street
2nd Floor
Dayton, OH 45417

Combined Health District Center for Alcohol and Drug Addiction Services
4100 West 3 Street
VA Medical Center Building 410, 3rd Floor
Dayton, OH 45428

7 East 4 Street
Suite 900
Dayton, OH 45402

Daybreak, Inc. Peer/Adults Lending Support
819 Wayne Avenue
Dayton, OH 45410

Daymont West Outpatient Alcohol and Drug Treatment Program
1520 Germantown Street
Dayton, OH 45408

Dayton Boys' and Girls' Club Project Outreach/Substance Abuse Service
601 South Keowee Street
Dayton, OH 45410

Dayton Correctional Institute Project Rebound
4104 Germantown Street
Dayton, OH 45417

Eastway Corporation Pathways Residential Program
5300 Little Richmond Road
Dayton, OH 45426

Good Samaritan Hospital and Health Center Counseling and Treatment Centers
2222 Philadelphia Drive
Dayton, OH 45406

Greater Dayton Christian Council Substance Abuse Prevention Ministry
212 Belmonte Park East
Dayton, OH 45405

Miami Valley Hospital Born Free
1 Wyoming Street
Dayton, OH 45409

Montgomery County Court of Common Pleas Chemical Offender Program
810 North Main Street
Dayton, OH 45405

Montgomery County Sheriff's Department Project DARE
330 West 2 Street
Dayton, OH 45402

Nova House Association Treatment Program
732 Beckman Street
Dayton, OH 45410

Project Cure, Inc.
1800 North James H. McGee Boulevard
Dayton, OH 45427

Regional DWI Programs, Inc. Driver Intervention Program (DIP)
333 West First Street
Suite 515
Dayton, OH 45402

South Community, Inc. Alcohol/ Drug Treatment Program
8353 Yankee Street
Dayton, OH 45458

United Behavioral Systems, Inc. Positive Focus Professional Counseling
131 North Ludlow Street
Suite 258
Dayton, OH 45402

United Health Services of Dayton, Inc. Substance Abuse Prevention for Youth
184 Salem Avenue
Suite 210
Dayton, OH 45406

Wright State University School of Medicine Weekend Intervention Program
Dayton, OH 45401

DEFIANCE

Community Counseling of Northwest Ohio Driver Intervention Program
2603 Evansport Road
Yokefellow House
Defiance, OH 43512

Five County Alcohol/Drug Program First Timers' Retreat
418 Auglaize Street
Defiance, OH 43512

DELAWARE

Alcohol Education Program
100 Woodrow Avenue
Delaware, OH 43015

Delaware Area Recovery Resources, Inc.
540 U.S. Route 36 East
Delaware, OH 43015

Help Anonymous
11 East Central Avenue
Delaware, OH 43015

Riverview School for Boys
7790 Dublin Road
Delaware, OH 43266

DOVER

Eastern Alcohol Services and Training (EAST)
323 Lincoln Street
Dover, OH 44662

Tuscarawas County Alcohol and Addiction Center
897 East Iron Avenue Extended
Dover, OH 44622

Tuscarawas DUI Counterattack School
897 East Iron Avenue Extended
Dover, OH 44622

DUBLIN

Lifescapes, Inc. DIP/Chemical Assessment and Prevention
6047 Frantz Road
Suite 102
Dublin, OH 43017

EAST CLEVELAND

Center for Human Services Rap Art East
15040 Euclid Avenue
East Cleveland, OH 44112

East Cleveland Neighborhood Center, Inc.
CSCSU Prevention Program
Teen Turf/Teen Drop-In Center
15840 Euclid Avenue
East Cleveland, OH 44112

EAST LIVERPOOL

Columbiana County CMHC
East Liverpool Site Substance Abuse Services
Family Recovery Center
422 West 6 Street
East Liverpool, OH 43920

EATON

Mental Health And Recovery Board Absentee Prevention Program
121 North Barron Street
Eaton, OH 45320

Preble County Juvenile TASC
204 North Barron Street
Eaton, OH 45320

Preble County Recovery Center, Inc.
100 East Somers Street
Eaton, OH 45320

ELYRIA

Alcohol Education and Assessment Program
Elyria, OH 44035

Drinking Driver Intervention Program of Elyria Lorain County Site
143 Louisiana Avenue
Elyria, OH 44036

Lorain County Council on Alcoholism and Drug Abuse/ Outpatient
1131 East Broad Street
Elyria, OH 44035

230 4 Street
Elyria, OH 44035

Westside Intervention, Inc.
322 Broad Street
Elyria, OH 44035

Women's Renaissance Project
1131 East Broad Street
Elyria, OH 44035

FAIRBORN

Community Network, Inc.
919 South Central Street
Fairborn, OH 45324

FINDLAY

Alcohol Traffic Safety Institute, Inc.
1918 North Main Street
Findlay, OH 45840

Family Service of Hancock County
401 West Sandusky Street
Findlay, OH 45840

FOSTORIA

Fostoria Alcohol/Drug Center
150 Perry Street
Fostoria, OH 44830

Residential Alcohol Education Program
150 Perry Street
Fostoria, OH 44830

Sandusky Valley Center, Inc. Substance Abuse Services
304 North Main Street
Fostoria, OH 44830

FREMONT

Alcohol Education and Assessment Program
Fremont, OH 43420

Sandusky Valley Center, Inc. Substance Abuse Services
675 Bartson Road
Fremont, OH 43420

United Way First Call For Help
300 Croghan Street
Suite 203
Fremont, OH 43420

GALION

Community Counseling Services, Inc. Galion Office
1376 SR 598
Galion, OH 44833

GALLIPOLIS

Family Addiction Community Treatment Services Serenity House
Gallipolis, OH 45631

GENEVA

Lake Area Recovery Center Aware Program
Howard Johnson's Motel
Route 534 and Route 90
Geneva, OH 44041

GEORGETOWN

Brown and Adams Substance Abuse Center
200 Green Street
Room 5
Georgetown, OH 45121

Brown County Counseling
415 1/2 Home Street
Georgetown, OH 45121

GRAFTON

Lorain Correctional Institution Substance Abuse Services
2075 South Avon/Beldon Road
Grafton, OH 44044

GRAND RIVER

Lake County Mental Health Center Merrick Hutchinson Branch
512 River Road
Grand River, OH 44045

GRANVILLE

Quest International
537 Jones Road
Granville, OH 43023

GREENVILLE

Darke County Recovery Services
134 West 4 Street
Greenville, OH 45331

GROVE CITY

Buckeye Boys' Ranch, Inc.
5665 Hoover Road
Grove City, OH 43123

GROVEPORT

Southeast Counseling Services
4100 Venture Place
Groveport, OH 43125

HAMILTON

Alcohol and Chemical Abuse Council of Butler County Ohio, Inc.
Access/Driver Intervention Program
111 Buckeye Street
Hamilton, OH 45011

Drug Counseling Services of Butler County, Inc.
1475 Pleasant Avenue
Hamilton, OH 45015

Journey
202 South Monument Street
Hamilton, OH 45011

Family Service of Butler County Substance Abuse Treatment Program
111 Buckeye Street
Hamilton, OH 45011

Fort Hamilton Hughes Memorial Hospital Center Horizon Services
630 Eaton Avenue
Hamilton, OH 45013

Sojourner Home
449 North 3 Street
Hamilton, OH 45011

Herland Family Center
520 High Street
Hamilton, OH 45011

Southwestern Ohio Serenity Hall, Inc.
447 South 2 Street
Hamilton, OH 45011

HILLSBORO

Family Recovery Services for Alcohol and Drug Abuse, Inc.
Driver Intervention Program
972 West Main Street
Hillsboro, OH 45133

Scioto Paint Valley Mental Health Center Highland County Office
108 Erin Court
Hillsboro, OH 45133

HUDSON

Youth Development Center Genesis Program
996 Hines Hill Road
Hudson, OH 44236

IRONTON

Family Guidance Center
415 Rail Road Street
Ironton, OH 45638

Shawnee Mental Health Center, Inc. Lawrence County Program
225 Carlton Davidson Lane
Ironton, OH 45638

JACKSON

Family Addiction Community Treatment Services Facts/New Alternatives
731 East Main Street
Suite 12
Jackson, OH 45640

KENT

Addiction Intervention with the Disabled (AID)
Kent State University
Department of Sociology
Kent, OH 44242

Townhall II
123 South Water Street
Kent, OH 44240

KENTON

Hardin County Alcohol and Drug Abuse Center
213 West Columbus Street
Kenton, OH 43326

Northwest Center for Human Resources Recovery Services
718 East Franklin Street
Kenton, OH 43326

LANCASTER

Fairfield County Drug and Alcohol Recovery Center Inc./ Outpatient
904 East Main Street
Lancaster, OH 43130

Information and Crisis Service, Inc.
Lancaster, OH 43130

Southeastern Correctional Institution
5900 Bis Road
Lancaster, OH 43130

LEBANON

Center of Warren/Clinton Counties
107 Oregonia Road
Lebanon, OH 45036

Lebanon Correctional Institution Recovery Services Department
State Route 63
Lebanon, OH 45036

Warren Correctional Institution Substance Abuse Prevention Services
State Route 63
Lebanon, OH 45036

LIBERTY CENTER

Maumee Youth Center
RFD 2
Liberty Center, OH 43532

LIMA

Alcohol Traffic Safety Institute, Inc.
Nonresidential
Residential
1255 North Cole Street
Lima, OH 45801

Allen Correctional Facility Residential Program
2338 North West Street
Lima, OH 45802

Family Resource Center Project Inroads
799 South Main Street
Lima, OH 45804

Lima Correctional Institution Interface Residential Substance Abuse Treatment Program
2235 North West Street
Lima, OH 45802

Northwest Center for Human Resources Substance Abuse Program
529 South Elizabeth Street
Lima, OH 45804

Project Inroads Urban Minority Alcohol/Drug Outreach Program
1174 West North Street
Lima, OH 45804

Recovery Services of Northwest Center for Human Resources, Inc.
1255 North Cole Street
Lima, OH 45801

LISBON

Columbiana County Mental Health Center Substance Abuse Program
40722 State Route 154
Lisbon, OH 44432

Family Recovery Center
Outpatient Program
Steering Clear Driver Intervention Program
964 North Market Street
Lisbon, OH 44432

LOGAN

Health Recovery Services, Inc. Hocking County Outpatient Clinic
4 East Hunter Street
Logan, OH 43138

LONDON

Madison Correctional Institute Treatment Continuity Pilot Project
1851 State Route 56
London, OH 43140

Madison County Health Group, Inc. Unit 1/Alcohol Alternative Program
22 South Main Street
London, OH 43140

Madison County Hospital Substance Abuse Program/ Insight Program
210 North Main Street
London, OH 43140

LORAIN

Compass House
2130 East 36 Street
Lorain, OH 44055

Compass House II, Inc.
2926 Wood Avenue
Lorain, OH 44055

Urban Minority Alcoholism and Drug Abuse Outreach Program
517 East 28 Street
Lorain, OH 44055

LOUDONVILLE

Ohio Department of Youth Services Mohican Youth Center
Loudonville, OH 44842

LOUISVILLE

Molly Stark Hospital Alcohol and Drug Treatment Program
7900 Columbus Road NE
Louisville, OH 44641

LUCASVILLE

Southern Ohio Correctional Facility Substance Abuse Program
Lucasville-Minford Road
Lucasville, OH 45699

MANSFIELD

Center for Individual and Family Services Drug and Alcohol Program
741 Scholl Road
Mansfield, OH 44907

Community Action for Capable Youth
91 Park Avenue West
Mansfield, OH 44902

Mansfield Correctional Institution Innervisions
1150 North Main Street
Mansfield, OH 44901

New Beginnings Recovery House, Inc.
347 2 Avenue
Mansfield, OH 44905

Vocational Rehab
420 West 5 Street
Mansfield, OH 44903

Youth Home
416 West 5 Street
Mansfield, OH 44901

Richland Alternate Program
448 South Main Street
Mansfield, OH 44907

Richland Hospital
Serenity Hall
Substance Abuse Services
1451 Lucas Road
Mansfield, OH 44901

Urban Minority Alcoholism and Drug Abuse Outreach Program
400 Bowman Street
Mansfield, OH 44901

Volunteers of America Central Ohio, Inc.
290 North Main Street
Mansfield, OH 44901

MARIETTA

Eve, Inc.
Marietta, OH 45750

First City Recovery Center, Inc. Alcohol and Drug Abuse
520 Virginia Street
Marietta, OH 45750

Marietta Memorial Hospital Substance Abuse Services
401 Matthew Street
Marietta, OH 45750

MARION

Marion Area Counseling Center, Inc. Alcohol and Drug Program
320 Executive Drive
Marion, OH 43302

Marion County Halfway House
286 Patterson Street
Marion, OH 43302

Marion Driver Intervention Program
Marion, OH 43302

MARYSVILLE

Marysville Alcohol Education Program
125 West 6 Street
Marysville, OH 43040

MASON

Center of Warren/Clinton Counties
201 Reading Road
Mason, OH 45040

MASSILLON

City of Massillon Mayors Drug Task Force
1 James Duncan Plaza
Massillon Municipal Government Center
Massillon, OH 44646

Glenbeigh of Massillon Community Hospital
875 8 Street NE
Massillon, OH 44646

Massillon Division of Quest Recovery Services
111 Tremont Avenue SW
Massillon, OH 44646

Ohio Department of Youth Services Indian River School
2775 Erie Street SW
Massillon, OH 44646

Quest Driver Intervention Program Massillon Site
131 Tremont Street SE
SCE Massillon YMCA
Massillon, OH 44646

MAUMEE

Professional Systems Addiction Treatment Service
6010 Garden Street
Maumee, OH 43537

MCARTHUR

Health Recovery Services, Inc. Vinton County Outpatient Cinic
107 South Sugar Street
McArthur, OH 45651

MCCONNELSVILLE

Morgan County Drug/Alcohol and Substance Abuse Advisory Council, Inc.
State Route 372 South
Morgan County Prep Center
McConnelsville, OH 43756

MEDINA

Alcohol and Drug Dependency Services of Medina County, Inc.
246 Northland Drive
Suite 140
Medina, OH 44256

Alcohol and Drug Dependency Services Driver Intervention Program
246 Northland Drive
Medina, OH 44256

Alternative Paths, Inc.
246 Northland Drive
Suite 200A
Medina, OH 44256

City of Medina Medina Police Dept. Diversion Program
150 West Friendship Street
Medina, OH 44256

Sanderson and Associates, Inc.
445 West Libert Street
Suite 212
Medina, OH 44256

MENTOR

Lake County Mental Health Center North Coast Alternatives Program
6101 Heisley Road
Mentor, OH 44077

MIDDLETOWN

Alcohol and Chemical Abuse Council of Butler County Ohio, Inc.
29 City Centre Plaza
Middletown, OH 45042

Drug Counseling Services of Butler County, Inc.
1001 Reinatz Boulevard
Middletown, OH 45042

MILFORD

Cincinnati Teen Challenge, Inc.
1466 Route 60
Milford, OH 45150

MILLERSBURG

Holmes County Health Department
Addiction Recovery Program
Alcohol Offender Program
931 Wooster Road
Millersburg, OH 44654

Holmes County Office of Education Holmes County Schools/T1
110 South Clay Street
Millersburg, OH 44654

Human Resource Center of Wayne and Holmes Counties
127 West Jackson Street
Millersburg, OH 44654

MONTPELIER

Community Hospitals of Williams County Recovery Center
909 Snyder Avenue
Montpelier, OH 43543

MOUNT GILEAD

Hopeline Incorporated
Mount Gilead, OH 43338

Morrow County Council on Alcohol and Drugs, Inc.
950 Meadow Drive
Mount Gilead, OH 43338

MOUNT ORAB

Brown County Counseling Service Alcohol/Drug Program
13679 State Route 68
Mount Orab, OH 45154

MOUNT VERNON

Alcohol and Drug Freedom Center of Knox County
116 East High Street
Mount Vernon, OH 43050

NAPOLEON

Five County Alcohol/Drug Program
104 East Washington Street
Suite 204
Napoleon, OH 43545

Henry County Hospital Help Center
11-600 State Road 424
Napoleon, OH 43545

NELSONVILLE

Hocking Correctional Facility Substance Abuse Department
16759 Snake Hollow Road
Nelsonville, OH 45764

NEW ALEXANDRIA

Jefferson County Mental Health Center Vianney Woods
RD 1 School House Road
New Alexandria, OH 43938

NEW LEXINGTON

Perry County Alcohol and Drug Abuse Council, Inc.
203 North Main Street
New Lexington, OH 43764

NEW PHILADELPHIA

Harbor House, Inc. Shelter House
New Philadelphia, OH 44663

Self Help, Inc.
125 Fair Avenue NE
New Philadelphia, OH 44663

NEWARK

Center for Alternative Resources
35 South Park Place
Newark, OH 43055

Central Ohio Recovery Residences (CORR)
346 North 40 Street
Newark, OH 43055

Licking County Alcoholism Prevention Program
Driver Intervention Program
Outpatient Services
62 East Stevens Street
Newark, OH 43055

Shepherd Hill Hospital Substance Abuse Services
200 Messimer Drive
Newark, OH 43055

Spencer Halfway House, Inc.
69 Granville Street
Newark, OH 43055

NORTH JACKSON

Tri-County Drivers Intervention
9694 Mahoning Avenue
North Jackson, OH 44451

NORWALK

Alcohol/Drug Center of Huron County Driver Intervention Program
49 Benedict Avenue
Norwalk, OH 44857

NORWICH

Horizons Drug and Alcohol Intervention Program Norwich Residential DIP
8855 East Pike Street
Norwich, OH 43767

OBERLIN

Allen Memorial Hospital Recovery Center
200 West Lorain Street
Oberlin, OH 44074

ORIENT

Correctional Reception Center
11271 State Route 762
Orient, OH 43146

Pickaway Correctional Institution
Genesis Residential and Alcohol Unit
Oasis Therapeutic Comm. Prison Project
State Route 762
Orient, OH 43146

ORRVILLE

Education and Counseling Services, Inc.
Stark County United Way Residential Program
Wayne County Residential Program
1022 West High Street
Orrville, OH 44667

OTTAWA

Putnam County Alcoholism and Drug Abuse Center, Inc.
117 Court Street
Ohio Bank Building 2nd Floor
Ottawa, OH 45875

OXFORD

Alcohol and Chemical Abuse Council of Butler County Ohio, Inc.
23 East High Street
Oxford, OH 45056

Oxford Crisis and Referral Center
111 East Walnut Street
Oxford, OH 45056

PAINESVILLE

Awareness Incorporated Cedar Hills Driver Intervention Program
5811 Vrooman Road
Painesville, OH 44077

Catholic Service Bureau of Lake County
8 North State Street
Room 455
Painesville, OH 44077

Lake/Geauga Center on Alcohol/ Drug Abuse, Inc.
Lake House
42 East Jackson Street
Painesville, OH 44077

Oak House
796 Oak Street
Painesville, OH 44077

PAULDING

Paulding County Alcohol and Drug Services Council, Inc.
102 East Jackson Street
Paulding, OH 45879

PERRY

Teen Challenge of Greater Cleveland, Inc.
3032 Perry Park Road
Perry, OH 44081

PERRYSBURG

Wood County Council on Alcoholism/Drug Abuse, Inc.
1011 Sandusky Street
3 Meadows Office Building Suite S
Perrysburg, OH 43551

PICKERINGTON

Fairfield County Drug/Alcohol Recovery Center Pickerington Office
1251 Hill Road North
Suite 101
Pickerington, OH 43147

PIQUA

Miami County Alcoholism Program
Outpatient
Residential DWI Program
423 North Wayne Street
Piqua, OH 45356

Miami County CMHC Choices
1266 1/2 East Ash Street
Piqua, OH 45356

POMEROY

Health Recovery Services, Inc. Meigs County Outpatient Clinic
119 Butternut Street
Pomeroy, OH 45769

PORT CLINTON

Tri-County Addictions Center, Inc.
131 B Maple Street
Port Clinton, OH 43452

PORTSMOUTH

Scioto County Counseling Center, Inc. Weekend DWI Program
1321 2 Street
Portsmouth, OH 45662

Shawnee Mental Health Center, Inc. Scioto County Clinic
2203 25 Street
Portsmouth, OH 45662

Southern Ohio Drinking/Driving Awareness Program
1686 Charles Street
Portsmouth, OH 45662

PROCTORVILLE

Family Guidance Center Prevention for Rural Pre-Teen and Minority Youth
Route 243
Proctorville, OH 45669

Shawnee Mental Health Center, Inc. Substance Abuse Services
302 State Street
Proctorville, OH 45669

RAVENNA

Portage County
Alcohol and Drug Abuse Services/ Outpatient
127 East Main Street
Ravenna, OH 44266

Horizon Halfway House
147 East Spruce Street
Ravenna, OH 44266

Project Detour
127 East Main Street
Ravenna, OH 44266

Robinson Memorial Hospital First Step Unit
6847 North Chestnut Street
Ravenna, OH 44266

ROCK CREEK

Glenbeigh Health Sources
State Route 45
Rock Creek, OH 44084

ROOTSTOWN

Northeast Ohio Universities College of Medicine Office of Alcoholism and Substance Abuse
4209 State Route 44
Rootstown, OH 44272

SAINT CLAIRSVILLE

Belmont/Harrison/Monroe Drug and Alcohol Council, Inc.
255 West Main Street
Saint Clairsville, OH 43950

SANDUSKY

Alcohol Education and Assessment Program/Sandusky
209 East Water Street
Sandusky, OH 44870

Erie County School District Office of Education
2900 South Columbus Avenue
Sandusky, OH 44870

Firelands Community Hospital Firelands Center
2020 Hayes Avenue
Sandusky, OH 44870

North Coast Youth Services, Inc. Substance Abuse Prevention Program
1006 Grant Street
Sandusky, OH 44870

Tri-County Addictions Center, Inc. ATAC Program
334 East Washington Street
Sandusky, OH 44870

SHADYSIDE

Ohio Valley Learning Systems, Inc. Driver Intervention Program
3700 Central Avenue
Shadyside, OH 43947

SIDNEY

Shelby County Intervention Program
305 South Ohio Avenue
Sidney, OH 45365

SOLON

Cuyahoga County Reach Out
33995 Bainbridge Road
Solon, OH 44139

SOUTH POINT

Family Guidance Center River Valley Treatment Program
305 North 4 Street
South Point, OH 45680

SPRINGFIELD

Acosep Corp. Alternatives to Incarcerating Drunk Drivers (ACAIDD)
401 North Plum Street
Suite 2
Springfield, OH 45504

Alcohol/Drug Abuse Programs For Treatment (ADAPT)
825 East High Street
Springfield, OH 45501

Clark County Sherriff's Office Saturday's Child Program
120 North Fountain Avenue
Springfield, OH 45502

Community Hospital Recovery Center
2615 East High Street
Springfield, OH 45501

DARE Program
120 North Fountain Avenue
Springfield, OH 45502

McKinley Hall, Inc.
Intensive Day Treatment Program
1054 East High Street
Springfield, OH 44505

Outpatient Program
1101 East High Street
Springfield, OH 45505

Residential Treatment
1074 East High Street
Springfield, OH 44505

Springfield Metropolitan Housing Authority/Project Choice
437 East John Street
Springfield, OH 45505

The Community Network, Inc. Me Too Prevention/DEAF Youth
2100 East High Street
Suite 108
Springfield, OH 45505

STEUBENVILLE

Jefferson County Mental Health Center
Driver Intervention Program
Drug and Alcohol Unit
3200 Johnson Road
Steubenville, OH 43952

Saint John Medical Center Addiction Recovery Unit
Saint John Heights
Steubenville, OH 43952

TIFFIN

Sandusky Valley Center, Inc.
Impaired Driver Intervention Program
109 South Washington Avenue
Tiffin, OH 44883

Substance Abuse Services
67 Saint Francis Avenue
Tiffin, OH 44883

TOLEDO

Boysville of Michigan, Inc. Saint Anthony Villa
2740 West Central Avenue
Andre Hall
Toledo, OH 43606

Comprehensive Addiction Services Systems (COMPASS)
2465 Collingwood Boulevard
Toledo, OH 43620

Lucas County Women's Project TOTE UMADAOP
525 Hamilton Street
Suite 101B
Toledo, OH 43602

Nubia Neighborhood Youth Program
430 Nebraska Avenue
Toledo, OH 43602

Options, Inc.
151 North Michigan Street
Suite 329
Toledo, OH 43624

Rosary Cathedral Elementary School Athletes for Abstinence
2535 Collingwood Boulevard
Toledo, OH 43610

Saint Paul's Community Center Intervention Program
230 13 Street
Toledo, OH 43624

Saint Vincent Medical Center Tennyson Center
2213 Cherry Street
Toledo, OH 43608

Self Expression Teen Theater
1001 Indiana Avenue
Suite 203–204
Toledo, OH 43607

Substance Abuse Services, Inc. The Jameelah House
520 Buckeye Street
Toledo, OH 43611

Toledo Hospital Alcohol and Drug Treatment Center
2142 North Cove Boulevard
Toledo, OH 43606

Toledo Municipal Court DUI Intervention Program
555 North Erie Street
Toledo, OH 43624

Toledo/Lucas Count Cares High Risk Youth Project
301 Collingwood Boulevard
Toledo, OH 43602

Urban Minority Alcohol and Drug Abuse Outreach Program (UMADAOP)/Lucas County
525 Hamilton Street
Suite 101B
Toledo, OH 43602

TROY

Miami County Alcoholism Program
233 South Market Street
Troy, OH 45373

Miami County Mental Health Center Choices/Troy Satellite
1059 North Market Street
Troy, OH 45373

TWINSBURG

Family Services of Summit County North Summit Branch
2057 East Aurora Road
Twinsburg, OH 44087

UPPER SANDUSKY

Sandusky Valley Center, Inc.
127 South Sandusky Avenue
Upper Sandusky, OH 43351

URBANA

Logan/Champaign Alcohol and Drug Addiction Services
40 Monument Square
Suite 301
Urbana, OH 43078

Mercy Memorial Hospital Mercy Substance Abuse Program
803 Scioto Street
Urbana, OH 43078

VAN WERT

Alcohol Traffic Safety Institute, Inc. Van Wert Driver Intervention Program
120 West Main Street
Fountainview Center 2nd Floor
Van Wert, OH 45891

Fountainview Center
120 West Main Street
2nd Floor
Van Wert, OH 45891

VANDALIA

Vandalia Driver Intervention and Alcohol Awareness Program
333 James E. Bohanon Drive
Vandalia, OH 45377

WADSWORTH

Alcohol and Drug Dependency Services of Medina County/ Wadsworth Office
180 High Street
Wadsworth, OH 44281

WAPAKONETA

Northwest Center For Human Resources Recovery Services
15 Willipie Street
Wapakoneta, OH 45895

WARREN

Addiction Recovery Center of Hillside Hospital
8747 Squires Lane NE
Warren, OH 44484

Contact Community Connection
1569 Woodland Street
Suite 10
Warren, OH 44483

Mahoning Valley Substance Abuse Prevention Project
290 West Market Street
Warren, OH 44481

Pine Industries
820 Pine Avenue SE
Warren, OH 44483

Rebecca Williams Community House
760 Main Avenue SW
Warren, OH 44483

Urban Minority Alcohol and Drug Abuse Outreach Program
160 East Market Street
Suite 200
Warren, OH 44481

2 North Park, Inc.
347 North Park Avenue
Warren, OH 44481

WASHINGTON COURT HOUSE

Choice Recovery Center for Alcohol and Drug Services
133 South Main Street
Suite 109-12B
Washington Court House, OH 43160

Scioto Paint Valley Mental Health Center Fayette County Office
1300 East Paint Street
Washington Court House, OH 43160

WAUSEON

Five County Alcohol/Drug Program
125 North Fulton Street
Wauseon, OH 43567

WAVERLY

Pike County Recovery Council Drinking Driver Intervention Program
196 East Emmitt Avenue
Waverly, OH 45690

Scioto Paint Valley Mental Health Center Pike County Office
102 Dawn Lane
Waverly, OH 45690

WEST UNION

Brown and Adams Substance Abuse Center
214 East Main Street
West Union, OH 45693

Drunk Driver Intervention Program
214 East Main Street
West Union, OH 45693

WESTERVILLE

Concord Counseling Services, Inc.
924 Eastwind Drive
Westerville, OH 43081

The Campus
905 South Sunbury Road
Westerville, OH 43081

WESTLAKE

Saint John and West Shore Hospital Chemical Dependency Unit Serenity Hall
29000 Center Ridge Road
Westlake, OH 44145

WICKLIFFE

Alcohol Education and Assessment Program Cleveland/Wickliffe
Wickliffe, OH 44092

WILLOUGHBY

University Mednet New Leaf Recovery Program
34990 Melinz Parkway
Willoughby, OH 44095

WILMINGTON

Center of Warren/Clinton Counties
260 Charles Street
Wilmington, OH 45177

WOODSFIELD

Belmont/Harrison/Monroe Drug and Alcohol Council, Inc. Monroe County Health Clinic
37984 Airport Road
Woodsfield, OH 43793

WOOSTER

Douglas Pomeroy Pathway House
550 North Grant Street
Wooster, OH 44691

Human Resource Center of Wayne and Holmes Counties
2692 Akron Road
Wooster, OH 44691

Wayne County Alcoholism Services
149 East Liberty Street
Suite 211
Wooster, OH 44691

Beacon House
732–734 Spink Street
Wooster, OH 44691

WORTHINGTON

CAC, Inc. Worthington Site
6797 North High Street
Suite 203
Worthington, OH 43085

Harding Hospital Harding Addiction Recovery Center
445 East Granville Road
Worthington, OH 43085

XENIA

Community Network, Inc.
452 West Market Street
Xenia, OH 45385

Women's Recovery Center
515 Martin Drive
Xenia, OH 45385

YELLOW SPRINGS

Greene County Drug Free Schools Consortium
360 East Enon Road
Yellow Springs, OH 45387

YOUNGSTOWN

Alcoholic Clinic of Youngstown
2151 Rush Boulevard
Youngstown, OH 44507

Alcoholism Programs of Mahoning County, Inc.
Donofrio Alcoholism Rehabilitation Center
1161 McGuffey Road
Youngstown, OH 44505

Joseph P. Stephens 3/4 House
377 South Jackson Street
Youngstown, OH 44505

Churchill Counseling Services
310 Churchill/Hubbard Road
Suite A
Youngstown, OH 44505

Community Corrections Association, Inc.
Community Corrections Facility
1740 Market Street
Youngstown, OH 44507

Residential Treatment Center I
1764 Market Street
Youngstown, OH 44507

Residential Treatment Center II
1608 Market Street
Youngstown, OH 44507

Community Recovery Resource Centre
4214 Market Street
Youngstown, OH 44512

Drivers Educational Alternative Program (DEAP)
2516 Market Street
Youngstown, OH 44507

Family Service Agency
535 Marmion Avenue
Youngstown, OH 44502

Mahoning County Chemical Dependency Programs, Inc.
527 North Meridian Road
Youngstown, OH 44509

Ohio Network/Training and Assistance for School and Community, Inc.
212 Churchill Hubbard Road
Youngstown, OH 44505

Urban Minority Alcoholism and Drug Abuse Outreach Program
3119 Market Street
Suite 203
Youngstown, OH 44507

ZANESVILLE

Good Samaritan Medical Center Court Division Program Driver Intervention Program
800 Forest Avenue
Zanesville, OH 43701

Good Samaritan Medical Center Alcoholism and Drug Recovery Treatment Program
716 Adair Avenue
Zanesville, OH 43701

Muskingum County Drug/Alcohol and Substance Abuse Council, Inc.
575 Harding Road
Zanesville, OH 43701

OKLAHOMA

ADA

Ada Area Council on Alcoholism, Inc.
727 Arlington Street
Ada, OK 74820

Rolling Hills Hospital Substance Abuse Services
1000 Rolling Hills Lane
Ada, OK 74820

ALTUS AIR FORCE BASE

Air Force Base/Altus Social Actions Drug/Alcohol Abuse Control
97 Air Mobility Wing (AMC)
Altus Air Force Base, OK 73523

ALVA

Northwest Family Services
628 Flynn Street
Alva, OK 73717

ANADARKO

Consortium Against Substance Abuse
115 East Broadway
Anadarko, OK 73005

ARDMORE

Arbuckle Drug and Alcohol Information Center, Inc.
1219 K Street NW
Ardmore, OK 73401

Ardmore Valley Hope of Memorial Hospital of Southern Oklahoma
1011 14 Street NW
Ardmore, OK 73401

Community Children's Shelter, Inc. Prevention Program
15 Monroe Street NE
Ardmore, OK 73402

Mental Health Services of Southern Oklahoma Vantage Pointe
2530 South Commerce Street
Building C
Ardmore, OK 73401

Recovery Way, Inc. Ardmore Outpatient Program
Ardmore, OK 73402

BARTLESVILLE

Alcohol and Drug Center, Inc.
615 SE Frank Phillips Boulevard
Bartlesville, OK 74003

BROKEN ARROW

Country View/Saint John Medical Center Substance Abuse Services
12300 East 91 Street
Broken Arrow, OK 74012

CHICKASHA

Southwest Youth and Family Services, Inc.
198 East Almar Drive
Chickasha, OK 73023

CHOCTAW

Tri-City Youth and Family Center, Inc.
14625 NE 23 Street
Choctaw, OK 73020

CLAREMORE

Rogers County Drug Abuse Program, Inc.
118 West Will Rogers Boulevard
Claremore, OK 74018

CLINTON

Cheyenne Cultural Center
Route 4
Clinton, OK 73601

New Horizons Area Prevention and Resource Center
90 North 31 Street
Clinton, OK 73601

CUSHING

Valley Hope Alcoholism Treatment Center
100 South Jones Avenue
Cushing, OK 74023

EDMOND

Edmond Youth Council, Inc. Outpatient Drug/Alcohol Services
7 North Broadway
Suite E
Edmond, OK 73083

EL RENO

Federal Correctional Institution Growth Center Drug Abuse Program
West Highway 66
El Reno, OK 73036

ENID

Crossroads Counseling Center
309 West Cherokee Street
Enid, OK 73701

Enid Regional Hospital Vista Hill Treatment Center
401 South 3 Street
Enid, OK 73701

Wheatland Mental Health Center, Inc.
1900 West Willow Street
Enid, OK 73701

FORT SUPPLY

Fort Supply Alcohol and Drug Treatment Center
1 Mile East/Highway 270
Fort Supply, OK 73841

GROVE

House of Hope, Inc.
Route 1
Grove, OK 74344

GUTHRIE

Eagle Ridge Family Treatment Center
1916 East Perkins Street
Guthrie, OK 73044

Logan County Youth and Family Service
4710 South Division Street
Guthrie, OK 73044

GUYMON

Panhandle Treatment Center
1004 Highway 54 NE
Guymon, OK 73942

HOMINY

Hominy Health Services, Inc.
211 East 5 Street
Hominy, OK 74035

HUGO

Recovery Way, Inc.
Choctaw Memorial Hospital
1405 East Kirk Street
Hugo, OK 74743

IDABEL

Kiamichi Council on Alcoholism and Other Drug Abuse, Inc.
102 North East Avenue A
Idabel, OK 74745

McCurtain County Educational Coop. McCare
Court Plaza Building
Suite 4
Idabel, OK 74745

People Plus, Inc.
Highway 70 NW and 16 Street
Idabel, OK 74745

LAWTON

Area Prevention Resource Center
Region 8
116 NW 31 Street
Lawton, OK 73505

Cameron University Step Out
2800 West Gore Boulevard
Lawton, OK 73505

Comanche Tribe New Pathways
1401 Laurie Tatum Road
Lawton, OK 73501

Comanche Tribe Substance Abuse Program Substance Abuse Prevention Program
HC 32
Lawton, OK 73501

Jim Taliaferro Mental Health Center Substance Abuse Services
602 SW 38 Street
Lawton, OK 73505

Lawton Public Schools Employee/Student Assistance Program
753 Fort Sill Boulevard
Lawton, OK 73502

Marie Detty Youth Services, Inc.
811 SW 17 Street
Lawton, OK 73501

Roadback, Inc.
1502 SW D Street
Suite 4
Lawton, OK 73501

Specialized Treatment for Adolescent Recovery (STAR)
3401 West Gore Boulevard
Comanche County Memorial Hospital
Lawton, OK 73505

LONE WOLF

Southwestern Oklahoma Adolescent Addiction Rehab Ranch, Inc. (SOAARR)
Route 1
Lone Wolf, OK 73655

MANGUM

New Hope of Mangum Chemical Dependency Unit
2 Wickersham Drive
Route 2
Mangum, OK 73554

MARIETTA

Morning Star Adolescent Treatment Unit
Route 1
Marietta, OK 73448

MCALESTER

Carl Albert Community Mental Health Center Substance Abuse Program
1101 East Monroe Street
McAlester, OK 74502

The Oaks Rehabilitative Services Center
628 East Creek Street
McAlester, OK 74501

Youth Emergency Shelter
904 North 9 Street
McAlester, OK 74501

MIAMI

Inter Tribal Substance Abuse/ Prevention and Treatment Center
1 South Main Street
Miami, OK 74354

Northeastern Oklahoma Council on Alcoholism
316 Eastgate Boulevard
Miami, OK 74355

Willowcrest Hospital Drug/ Alcohol Prevention Program
130 A Street SW
Miami, OK 74354

MOORE

Moore Alcohol/Drug Center, Inc.
624 Northwest 5 Street
Moore, OK 73160

MUSKOGEE

Green Country Mental Health Services, Inc. Alcohol and Drug Abuse Services
619 North Main Street
Muskogee, OK 74401

Harbor House of Muskogee, Inc.
353 Callahan Street
Muskogee, OK 74401

Monarch Incorporated
501 Fredonia Street
Muskogee, OK 74403

Muskogee County Council of Youth Services
4409 Eufaula Avenue
Muskogee, OK 74401

NORMAN

Central Oklahoma CMHC Alcohol/Drug Program
909 East Alameda Street
Norman, OK 73071

Center for Child and Family Development
555 Constitution Street
Room 221
Norman, OK 73037

NAIC/Center for Oklahoma Alcohol and Drug Services, Inc.
225 West Duffy Street
Norman, OK 73069

Norman Alcohol and Drug Treatment Center
East Main Street and State Drive
Norman, OK 73071

NOWATA

Grand Lake Mental Health Center, Inc. Alcohol and Drug Abuse Services
114 West Delaware Street
Nowata, OK 74048

Recovery Way, Inc. Inpatient Program
Nowata, OK 74048

OKEMAH

Creek Nation Health System Creek Nation Alcoholism Program
309 North 14 Street
Okemah, OK 74859

OKLAHOMA CITY

A Chance to Change Foundation
5228 Classen Boulevard
Oklahoma City, OK 73118

AGAPE Recovery Services, Inc.
809 SW 89 Street
Suite G
Oklahoma City, OK 73139

Alcohol Training and Education, Inc.
2800 NW 36 Street
Suite 101
Oklahoma City, OK 73112

Carver Center
2801 SW 3 Street
Oklahoma City, OK 73108

Celebrations Educational Services, Inc.
431 SW 11 Street
Oklahoma City, OK 73109

Community Action Agency Turning Point
1900 NW 10 Street
Oklahoma City, OK 73106

Community Counseling Center
1140 North Hudson Street
Oklahoma City, OK 73103

Drug Recovery, Inc.
415 NW 7 Street
Oklahoma City, OK 73102

Hillcrest Health Center Behavioral Medicine Services
2129 SW 59 Street
Oklahoma City, OK 73119

Hope Community Services, Inc.
105 SE 45 Street
Oklahoma City, OK 73129

Horizon Lodge
3115 North Lincoln Boulevard
Oklahoma City, OK 73105

North Oklahoma County Mental Health Center Substance Abuse Services
6300 North Classen Boulevard
Building A
Oklahoma City, OK 73118

Oklahoma City Housing Authority OCHA/ERI/A Chance for Youth
1700 NE 4 Street
Oklahoma City, OK 73117

Oklahoma City Indian Clinic Family Services
1214 North Hudson Street
Suite 203
Oklahoma City, OK 73103

Oklahoma County Juvenile Bureau Detention Center
5905 North Classen Boulevard
Oklahoma City, OK 73118

Oklahoma Halfway House, Inc.
517 SW 2 Street
Oklahoma City, OK 73109

Orange Quarters, Inc. DBA The Life Improvement Center
3839 West Reno Street
Oklahoma City, OK 73107

Phoenix Recovery Institute Phoenix House
824 East Drive
Oklahoma City, OK 73105

Saint Anthony Recovery and Treatment (START)
1000 North Lee Street
Oklahoma City, OK 73101

Salvation Army Adult Rehabilitation Center
2041 NW 7 Street
Oklahoma City, OK 73106

The Referral Center for Alcohol and Drug Service of Central Oklahoma, Inc.
1215 NW 25 Street
Oklahoma City, OK 73106

Univ. of Oklahoma College of Nursing Focused Prevention Services
1100 North Stonewall Street
Oklahoma City, OK 73190

Veterans' Affairs Medical Center Chemical Dependency Unit
921 NE 13 Street
116C
Oklahoma City, OK 73104

OKMULGEE

Creoks Mental Health Services, Inc. Substance Abuse Services
206 South Grand Street
Suite 100
Okmulgee, OK 74447

Tri-County Area Prevention Resource Center
1801 East 4 Street
Okmulgee, OK 74447

PAWNEE

Community Alcoholism Services
6 and Denver Streets
Pawnee, OK 74058

United Community Action Program, Inc. Head Start/ Growing Up Strong
501 6 Street
Pawnee, OK 74058

PONCA CITY

Bridgeway, Inc.
610 West Grand Street
Ponca City, OK 74602

Edwin Fair Mental Health Center Alcohol and Drug Abuse Unit
1500 North 6 Street
Ponca City, OK 74601

Recovery Way, Inc. Ponca City Outpatient Program
422 West Grand Street
Ponca City, OK

Social Development Center
309 East Hartford Avenue
Ponca City, OK 74601

POTEAU

Le Flore County Youth Services, Inc.
404 Dewey Avenue
Poteau, OK 74953

PRYOR

Recovery Way, Inc. Pryor Outpatient Program
Pryor, OK 74361

RED ROCK

Otoe/Missouria Tribe Substance Abuse Program
Route 1
Red Rock, OK 74651

SEMINOLE

Seminole Public Schools High Risk Focus Program
617 North Timmons Street
Seminole, OK 74868

Tri-City Substance Abuse Center
214 East Oak Street
Seminole, OK 74868

SHAWNEE

Gateway to Prevention and Recovery
1010 East 45 Street
Shawnee, OK 74802

Indian Action Center
2041 South Gordon Cooper Drive
Shawnee, OK 74801

Mission Hill Memorial Hospital Substance Abuse Services
1900 Gordon Cooper Drive
Shawnee, OK 74801

STILLWATER

Panok Area Prevention Resource Center
1202 West Farm Road
Stillwater, OK 74078

Payne County Misdemeanant Program and Related Services, Inc.
801 South Main Street
Suite 5
Stillwater, OK 74074

Payne County Youth Services, Inc.
2224 West 12 Street
Stillwater, OK 74074

Starting Point II, Inc.
608 Highpoint Drive
Stillwater, OK 74075

STROUD

Sac and Fox Nation Office for Alcohol and Substance Abuse Program
Route 2
Stroud, OK 74079

TAHLEQUAH

Bill Willis CMHC Substance Abuse Services
1200 West 4 Street
Tahlequah, OK 74464

TALIHINA

Choctaw Nation Recovery Center
Off Hwy 271 1.5 Miles by Talihina Hosp
Talihina, OK 74571

TONKAWA

Alpha II, Inc.
1608 North Main Street
Tonkawa, OK 74653

TULSA

Associated Centers for Therapy, Inc.
7010 South Yale Avenue
Suite 215
Tulsa, OK 74136

DVIS/Mend Counseling Program
1419 East 15 Street
Tulsa, OK 74120

Family Mental Health Center Alcohol and Drug Abuse Services
2725 East Skelly Drive
Tulsa, OK 74105

Health Network, Inc. Assessment and Outpatient Services
2326 South Garnett Street
Suite C
Tulsa, OK 74129

How Foundation Rehabilitation Center of Oklahoma, Inc.
5649 South Garnett Road
Tulsa, OK 74146

Indian Health Care Resource Center of Tulsa
915 South Cincinnati Street
Tulsa, OK 74119

Margaret Hudson Program
1205 West Newton Street
Tulsa, OK 74127

Metro Tulsa Substance Abuse Services
1602 North Cincinnati Avenue
Tulsa, OK 74106

New Choice and Associates
4833 South Sheridan Road
Suite 408
Tulsa, OK 74135

Palmer Drug Abuse Program
711 South Sheridan Road
Tulsa, OK 74112

Parkside, Inc.
1620 East 12 Street
Tulsa, OK 74120

Salvation Army Adult Rehabilitation Center
601–611 North Main Street
Tulsa, OK 74106

Sobriety, Inc.
1014 South Detroit Avenue
Tulsa, OK 74120

Star Community Mental Health Center Substance Abuse Services
3604 North Cincinnati Street
Tulsa, OK 74106

Street School, Inc.
1135 South Yale Avenue
Tulsa, OK 74112

Tulsa Area Council on Alcoholism and Drug Abuse
4828 South Peoria Street
Suite 114
Tulsa, OK 74105

Tulsa Area Prevention Resource Center
1111 West 17 Street
Tulsa, OK 74107

Tulsa Regional Medical Center Chemical Dependency Unit
744 West 9 Street
Tulsa, OK 74127

Twelve and Twelve Transition House, Inc.
12 East 12 Street
Tulsa, OK 74119

Veterans' Affairs Medical Center Tulsa VA Substance Abuse Center
635 West 11th Street
VA Outpatient Clinic
Tulsa, OK 74127

Youth Services of Tulsa, Inc. Drug Abuse Prevention in Schools
302 South Cheyenne Street
Suite 114
Tulsa, OK 74103

VINITA

Vinita Alcohol and Drug Treatment Center
Vinita, OK 74301

WATONGA

Opportunities, Inc. Chemical Dependency Treatment Center
120 West First Street
Watonga, OK 73772

WETUMKA

Wetumka General Hospital Substance Abuse Services
325 South Washita Street
Wetumka, OK 74883

WEWOKA

Seminole Nation of Oklahoma Alcohol/Substance Abuse Program
506 South Mekusukey Street
Wewoka, OK 74884

WOODWARD

Northwest Oklahoma CMHC Substance Abuse Services
1222 10 Street
Suite 211
Woodward, OK 73801

YUKON

Chisholm Trail Counseling Services Substance Abuse Services
1501 Commerce Street
Yukon, OK 73085

OREGON

ALBANY

Addiction Counseling and Education Services, Inc. (ACES)
1856 Grand Prairie Road SE
Albany, OR 97321

Catherine Freer Wilderness Therapy Expeditions
1230 Pulver Lane NW
Albany, OR 97321

Linn County Alcohol and Drug Treatment Program
104 SW Lyons Street
Albany, OR 97321

Yes House
237 6 Avenue SE
Albany, OR 97321

ALOHA

Reconnections
3685 SW 170 Avenue
Aloha, OR 97007

ASTORIA

Clatsop County Mental Health Center Alcohol and Drug Programs
701 West Marine Drive
Astoria, OR 97103

Lobe and Associates
555 Bond Street
Astoria, OR 97103

BAKER

Baker County Council on Alcohol and Drug Problems, Inc.
2330 5 Street
Baker, OR 97814

Powder River Corrections Alcohol and Drug Treatment Program
3600 13 Street
Baker, OR 97814

BEAVERTON

Annand Counseling Center
9400 SW Beaverton/Hillsdale Highway
Suite 115
Beaverton, OR 97005

Counseling Intervention Programs, Inc. Project Stop
4576 SW 103 Street
Beaverton, OR 97005

Evans and Sullivan
9400 SW Beaverton-Hillsdale Highway
Suite 100
Beaverton, OR 97005

BEND

Central Oregon Alcohol and Drug Council
1725 NW Newport Avenue
Bend, OR 97701

Court Services Substance Abuse Prevention Services
2191 NE Kim Street
Bend, OR 97701

Deschutes County Mental Health Services Substance Abuse Services
409 NE Greenwood Avenue
Suite 2
Bend, OR 97701

BLACHLY

Ranch Care, Inc.
18624 Highway 36
Blachly, OR 97412

BURNS

Burns Indian Alcohol Program
100 Pasigo Street
Burns, OR 97720

Harney Counseling and Guidance Services
415 North Fairview Street
Burns, OR 97720

COOS BAY

Better Options to Corrections
320 Central Street
Suite 307
Coos Bay, OR 97420

Coos/Lower Umpqua/Siuslaw Alcohol and Drug Program
338 Wallace Street
Coos Bay, OR 97420

Recovery Plus, Inc.
455 South 4 Street
Suite 4
Coos Bay, OR 97420

CORVALLIS

Addiction Counseling and Education Services, Inc.
744 NW 4 Street
Corvallis, OR 97330

Benton County Alcohol Treatment Program
530 NW 27 Street
Public Service Building
Corvallis, OR 97330

Benton/Linn Council on Alcoholism Community Outreach, Inc.
128 SW 9 Street
Corvallis, OR 97333

Milestones Family Recovery Program
306 SW 8 Street
Corvallis, OR 97333

DALLAS

Bridgeway
128 Mill Street
Dallas, OR 97338

Polk County Mental Health Alcohol and Drug Treatment Program
182 SW Academy Street
Suite 304
Dallas, OR 97338

Recovery Counseling
289 East Ellendale Street
Suite 204
Dallas, OR 97338

ENTERPRISE

Wallowa County Mental Health Clinic Alcohol and Drug Program
616 West North Street
Enterprise, OR 97828

EUGENE

Addiction Counseling and Education Services, Inc. (ACES)
1639 Oak Street
Eugene, OR 97401

Buckley House Programs, Inc.
Carlton Substance Abuse Center
564 Lincoln Street
Eugene, OR 97401

Detoxification Services
605 West 4 Street
Eugene, OR 97402

Centro Latino Americana
944 West 5 Street
Eugene, OR 97402

Chemfree Outpatient Services
1902 Jefferson Street
Suite 3
Eugene, OR 97405

Coeur de Lane Program
1756 Williamette Street
Eugene, OR 97401

Lane County Alcohol/Drug/ Offender Program
135 East 6 Avenue
Eugene, OR 97401

Looking Glass Adolescent Recovery Program
1040 Oak Street
Eugene, OR 97401

Oregon Institute for Rational Recovery
317 West Broadway
Eugene, OR 97401

Passages
1075 Washington Street
Eugene, OR 97401

Pathways
2391 Centennial Boulevard
Eugene, OR 97401

Prevention and Recovery Northwest
1188 Olive Street
Eugene, OR 97401

Serenity Lane, Inc. New Hope
2133 Centennial Plaza
Eugene, OR 97401

Sunrise House
692 Jefferson Street
Eugene, OR 97402

White Bird Clinic Chrysalis Program
341 East 12 Street
Eugene, OR 97401

FLORENCE

West Lane Diversion
1495 West 8 Street
Florence, OR 97439

FOREST GROVE

Comm. Youth Services of Washington County Forest Grove Youth Services Center
2004 Main Street
Suite 200
Forest Grove, OR 97116

Tuality Chemical Dependency Services
1809 Maple Street
Forest Grove, OR 97116

GERVAIS

Red Willow Adolescent Chemical Dependency Treatment, Inc.
765 7 Street
Gervais, OR 97026

GOLD BEACH

Curry County Alcohol and Drug Treatment Program
145 East Moore Street
Curry County Courthouse Annex
Gold Beach, OR 97444

Curry County Community Corrections Substance Abuse Treatment Program
510 Colvin Street
Gold Beach, OR 97444

Curry County Council on Alcoholism
115 1/2 East 4 Street
Gold Beach, OR 97444

GRAND RONDE

Confederated Tribes of Grand Ronde Human Services Division Alcohol and Drug Program
9615 Grand Ronde Road
Grand Ronde, OR 97347

GRANTS PASS

Decisions Alcohol and Drug Counseling
245 SW G Street
Grants Pass, OR 97526

Downs Associates of Grants Pass
215 SE 6 Street
Suite 209
Grants Pass, OR 97526

Josephine Council on Alcohol/ Drug Abuse DBA Rogue Recovery Programs
707 NW A Street
Grants Pass, OR 97526

GRESHAM

Mount Hood Medical Center Chemical Dependency Treatment
24800 SE Stark Street
Gresham, OR 97030

Native American Rehab Association of the Northwest, Inc.
2022 NW Division Street
Gresham, OR 97030

Oregon Human Development Corporation Ayuda Community Services
439 West Powell Boulevard
Suite 5
Gresham, OR 97030

HEPPNER

Morrow County Mental Health Services South Office
150 Rock Street
Heppner, OR 97836

HILLSBORO

Oregon Human Development Corporation Ayuda Community Services
265 SE Oak Street
Suite E
Hillsboro, OR 97123

Washington County Alcohol Program
145 NE 2 Street
Washington County Courthouse
Hillsboro, OR 97123

Washington County Health and Human Services/Alcohol and Drug Treatment Program
155 North First Avenue
Hillsboro, OR 97124

Youth Contact
447 SE Baseline Street
Hillsboro, OR 97123

HOOD RIVER

Mount Hood Medical Center Chemical Dependency Treatment Services
216 Columbia Avenue
Union Building
Hood River, OR 97031

Pacific Northwest Counseling
1217 12 Street
Hood River, OR 97031

JOHN DAY

Grant County Center for Human Development
166 SW Brent Street
John Day, OR 97845

KLAMATH FALLS

Klamath Alcohol and Drug Abuse, Inc. (KADA)
310 South 5 Street
Klamath Falls, OR 97601

Klamath Tribal Health and Family Services Alcohol and Drug Program
3949 South 6 Street
Klamath Falls, OR 97603

Lutheran Family Services
2545 North Eldorado Avenue
Klamath Falls, OR 97601

Stepping Stones
5160 Summers Lane
Klamath Falls, OR 97603

Teen Challenge Women's Program
4825 Sunset Ridge Road
Klamath Falls, OR 97601

LA GRANDE

Union County Center for Human Development Alcohol and Drug Services
1100 K Avenue
La Grande, OR 97850

LAKEVIEW

Lake County Mental Health Center Substance Abuse Services
513 Center Street
Lakeview, OR 97630

LEBANON

Teen Challenge of Oregon, Inc. Men's Program
75 West Morton Street
Lebanon, OR 97355

MADRAS

Jefferson County Recovery Center
91 SE D Street
Madras, OR 97741

MCMINNVILLE

Recovery Counseling
1120 South Baker Street
McMinnville, OR 97128

Yamhill County Mental Health Chemical Dependency Program
626 North Ford Street
McMinnville, OR 97128

MEDFORD

Brook Counseling Services
345 North Bartlett Street
Suite 102
Medford, OR 97501

Jackson County Probation Department Substance Abuse Services
100 South Oakdale Street
Justice Building Room 105
Medford, OR 97501

Jackson County Substance Abuse Program
338 North Front Street
Medford, OR 97501

Ontrack, Inc.
221 West Main Street
Medford, OR 97501

Rogue Valley Addictions Recovery Center
841 South Riverside Drive
Medford, OR 97501

MILWAUKIE

Clackamas County Mental Health Alcohol and Drug Program
2100 SE Lake Road
Milwaukie, OR 97222

Providence Milwaukie Hospital Option
10150 SE 32 Avenue
Milwaukie, OR 97222

NEWBERG

Springbrook Northwest, Inc.
2001 Crestview Drive
Newberg, OR 97132

NEWPORT

Laurelhurst Coast Counseling
53 SW Lee Street
Newport, OR 97365

Lincoln County Alcohol and Drug Program
255 SW Coast Highway
Newport, OR 97365

Lincoln County Council on Alcohol and Drug Abuse
155 SW High Street
Newport, OR 97365

Reconnections
634 NE 3 Street
Newport, OR 97365

NORTH BEND

Alcohol/Drug Responsibility Program
1950 12 Street
North Bend, OR 97459

Center for Holistic Therapy
625 Oconnell Street
North Bend, OR 97459

Coos County Correctional Treatment Program
1975 McPherson Street
North Bend, OR 97459

Southwest Oregon Community Action AMBIT
1931 Meade Street
Phoenix Building Suite C
North Bend, OR 97459

ONTARIO

Malheur County Alcohol and Drug Authority Alcohol Recovery Center
686 NW 9 Street
Ontario, OR 97914

Malheur County Mental Health and Counseling Center
1108 SW 4 Street
Ontario, OR 97914

PENDLETON

Eastern Oregon Alcoholism Foundation
304 SW Hailey Avenue
Pendleton, OR 97801

Umatilla County Mental Health Program Substance Abuse Treatment Unit
721 SE 3 Street
Suite B
Pendleton, OR 97801

Umatilla Indian Alcohol and Drug Program
Pendleton, OR 97801

PHOENIX

Downs Associates, Inc. of Ashland
160 South Main Street
Phoenix, OR 97535

PORTLAND

A Minor Miracle
813 SW Alder Street
Suite 450
Portland, OR 97205

Addictions Recovery Association
3150 SE Belmont Street
Portland, OR 97214

Alcohol Treatment and Training Clinic
506 SW 6 Avenue
Wilcox Building 3rd Floor
Portland, OR 97204

Alder Associates for Recovery
808 SW Alder Street
2nd Floor
Portland, OR 97205

ASAP Treatment Services, Inc.
919 SW Taylor Street
7th Floor
Portland, OR 97205

CODA Drug Treatment Services
306 NE 20 Street
Portland, OR 97232

Columbia River Correctional Institution Turning Point
9111 NE Sunderland Avenue
Portland, OR 97211

Counseling Intervention Programs, Inc.
Project Stop
4413 SE 17 Avenue
Portland, OR 97202

De Paul Adult Treatment Center
1320 SW Washington Street
Portland, OR 97205

De Paul Youth Treatment Center
4411 NE Emerson Street
Portland, OR 97218

Diversion Associates
1949 SE 122 Avenue
Portland, OR 97233

Drug Abuse Rehab and Treatment (DART)
9747 SE Powell Boulevard
Portland, OR 97266

Emanuel Hospital Project Network
2801 North Gantenbein Avenue
Portland, OR 97227

General Health, Inc.
Delta Clinic/North
4037 NE Tillamook Street
Portland, OR 97212

Harbor Light Rehabilitation Center
134 West Burnside Street
Portland, OR 97209

Harmony House, Inc. Multnomah County
2270 SE 39 Avenue
Portland, OR 97214

Hooper Memorial Center
20 NE Martin Luther King Boulevard
Portland, OR 97232

Kaiser Permanente Recovery Resources
3414 North Kaiser Center Drive
East Interstate Medical Office
Portland, OR 97212

Adolescent Unit
2330 NE Siskiyou Street
Portland, OR 97212

Legacy Chemical Dependency Treatment Services Dual Diagnosis Program
1225 NE 2 Avenue
Holladay Park Hospital
Portland, OR 97232

Mainstream Youth Program, Inc.
4531 SE Belmont Street
Suite 300
Portland, OR 97215

Morrison Center Breakthrough/ Project Self Reliance
3390 SE Milwaukie Avenue
Portland, OR 97202

Northwest Treatment Services
9370 SW Greenburg Road
Suite 601
Portland, OR 97223

948 NE 102 Street
Suite 101
Portland, OR 97220

Portland Addictions/Acupuncture Center
727 NE 24 Street
Portland, OR 97232

Portland Adventist Medical Center New Day Center
6012 SE Yamhill Street
Portland, OR 97215

Project for Community Recovery
3525 NE Martin Luther King Jr. Blvd.
Portland, OR 97212

Providence Medical Center Addictions Treatment Services
5228 NE Hoyt Street
Portland, OR 97213

Rosemont Residential Treatment Center
597 North Dekum Street
Portland, OR 97217

Saint Vincent Hospital and Medical Center Alcohol and Chemical Dependency Program
9340 SW Barnes Road
Suite A
Portland, OR 97225

Serenity Lane
9221 SW Barbur Boulevard
Suite 205
Portland, OR 97219

TASC of Oregon, Inc.
1733 NE 7 Avenue
Portland, OR 97212

Transition Projects
1211 SW Main Street
Portland, OR 97205

Tualatin Valley Mental Health Center Substance Abuse Services
14600 NW Cornell Road
Portland, OR 97229

Volunteers of America Women's Residential Center
200 SE 7 Street
Portland, OR 97214

Western Health Clinics Southeast Portland Clinic
3777 SE Milwaukie Street
Portland, OR 97202

Woodland Park Hospital Pathways
10300 NE Hancock Street
Portland, OR 97220

PRINEVILLE

Lutheran Family Services Crook County Mental Health Program
203 North Court Street
Prineville, OR 97754

Rimrock Trails
1099 North Elm Street
Prineville, OR 97754

REEDSPORT

Douglas County Health and Social Services Dept./ Substance Abuse Program
680 Fir Avenue
Reedsport, OR 97467

ROSEBURG

Cow Creek Band Umpqua Tribe Drug and Alcohol Treatment Program
2400 Stewart Parkway
Suite 300
Roseburg, OR 97470

Douglas Community Hospital Gateways Program
738 West Harvard Boulevard
Roseburg, OR 97470

Douglas County Council on Alcohol and Drug Abuse Prevention and Treatment
621 West Madrone Street
Roseburg, OR 97470

First Step
1122 NW Garden Valley Boulevard
Suite 108
Roseburg, OR 97470

Roseburg Recovery Services
451 West Corey Court
Roseburg, OR 97470

Veterans' Affairs Medical Center Substance Abuse Treatment Program
913 NW Garden Valley Boulevard
Roseburg, OR 97470

SAINT HELENS

Columbia County Family Counseling Center Alcohol and Drug Program
161 Saint Helens Street
Saint Helens, OR 97051

SALEM

Bridgeway
2550 Coral Avenue NE
Salem, OR 97305

Chemawa Alcoholism Education Center
3760 Chemawa Road NE
Salem, OR 97305

Creekside Counseling, Inc.
868 Promontory Place
Salem, OR 97302

Eckles and Mauk Counseling Services, Inc.
756 Hawthorne Street NE
Salem, OR 97301

Hillcrest School of Oregon Hillcrest Alcohol and Drug Treatment Program
2450 Strong Road SE
Salem, OR 97310

Marion County Drug Treatment Program
3180 Center Street NE
Room 225
Salem, OR 97301

Nanitch Sahallie Treatment Center
5119 River Road NE
Salem, OR 97303

National Traffic Safety Institute Regional Office
1235 Woodrow Street NE
Salem, OR 97303

Oregon State Hospital
Cornerstone
2600 Center Street NE
Salem, OR 97310

Correctional Institution TRT Services
2600 Center Street NE
Cottage 21
Salem, OR 97310

Pacific Recovery
1655 Capitol Street NE
Suite 2
Salem, OR 97303

Rid/Jade
663 High Street NE
Salem, OR 97301

Seasons
3054 Lancaster Drive NE
Salem, OR 97305

Serenity Lane
755 Medical Center Drive NE
Salem, OR 97301

The Heart Center for Recovery and Healing
145 SE Wilson Street
Salem, OR 97302

SANDY

Sandy Family Services, Inc.
39332 Proctor Boulevard
Sandy, OR 97055

SEASIDE

Awakenings at Serenity by the Sea
321 South Prom and Avenue A
Seaside, OR 97138

Beginnings, Inc.
851 Broadway
Suite 2
Seaside, OR 97138

SHERIDAN

Federal Correctional Institution Substance Abuse Services
2707 Ballston Road
Sheridan, OR 97378

SILETZ

Siletz Tribal Council Alcohol and Drug Program
107 SE Swan Street
Siletz, OR 97380

STAYTON

Stayton Center for Improvement
223 Locust Street
Stayton, OR 97383

SWEET HOME

Eckles and Mauk Counseling
2200 Main Street
Sweet Home, OR 97386

THE DALLES

Mid-Columbia Center for Living
400 East 5 Street
Courthouse Annex A Room 106
The Dalles, OR 97058

The Counseling Center
511 Union Street
The Dalles, OR 97058

TIGARD

General Health, Inc. Delta Clinic
11945 SW Pacific Highway
Suite 113
Tigard, OR 97223

Harmony House, Inc./ Washington County
10362 SW McDonald Road
Tigard, OR 97224

Pacific Alcohol and Drug Counseling, Inc.
12950 SW Pacific Highway
Suite 6
Tigard, OR 97223

TILLAMOOK

Tillamook Counseling, Inc.
2405 5 Street
Tillamook, OR 97141

TUALATIN

Meridian Park Hospital Chemical Dependency Treatment Services
18770 SW Boones Ferry Road
Tualatin, OR 97062

WARM SPRINGS

Confederated Tribes of Warm Springs Alcohol and Drug Abuse Program
Warm Springs, OR 97761

WHITE CITY

Veterans' Affairs Domiciliary Alcohol/Drug Treatment Program (ADTP)
8495 Crater Lake Highway
White City, OR 97501

WOODBURN

Northwest Defensive Drivers, Inc.
158 Grant Street
Woodburn, OR 97071

Oregon Human Development Corporation Ayuda Community Services
1440 Newberg Highway
Woodburn, OR 97071

PENNSYLVANIA

ALIQUIPPA

Gateway Rehabilitation Center Moffett Run Road
RD 2
Aliquippa, PA 15001

ALLENTOWN

Allentown Osteopathic Medical Center (AOMC) Recovery Center
33 North Saint George Street
Allentown, PA 18104

Council on Alcohol and Drug Abuse Alcohol and Drug Abuse
126 North 9 Street
Allentown, PA 18102

Drug Treatment Program of Lehigh Valley and Reading
1810 Steelstone Road
Suite 101–102
Allentown, PA 18103

Florence Child Guidance Center
1812 Allen Street
Allentown, PA 18104

Lehigh County
Drug and Alcohol Intake Unit
139 North 8 Street
Allentown, PA 18101

Treatment Alternatives to Street Crime
521 Court Street
Allentown, PA 18101

Lehigh Valley Addictions Treatment Services Halfway Home of Lehigh Valley
117–121 North 8 Street
Allentown, PA 18101

Upward Drug and Alcohol Program
532 Saint John Street
Allentown, PA 18103

ALLENWOOD

White Deer Run Devitt Camp Road
Allenwood, PA 17810

ALTOONA

Altoona Hospital CMHC Community Crisis Center
620 Howard Avenue
Altoona, PA 16601

AMP/CEP Group Homes, Inc.
T/A Sobriety House
901 6 Avenue
Altoona, PA 16602

830 6 Avenue
Altoona, PA 16602

Blair County Driving Under the Influence (DUI) Program
222 Lakemont Park Boulevard
Altoona, PA 16602

Blair County Community Action Program Substance Abuse Services
5433 Industrial Avenue
Altoona, PA 16601

Home Nursing Agency Community Support Alternatives
201 Chestnut Avenue
Altoona, PA 16601

AMBLER

Turning Point, Inc.
22 North Main Street
Ambler, PA 19002

ARDMORE

Lower Merion Counseling Services
34 Rittenhouse Place
Ardmore, PA 19003

Womanspace/RHD
120 Ardmore Avenue
Ardmore, PA 19003

ASHLAND

Gaudenzia at Fountain Springs Women and Children Program
95 Broad Street
Ashland, PA 17921

BATH

Center for Humanistic Change Inc. (CHC)
7574 Beth-Bath Pike
Bath, PA 18014

BEAVER

The Medical Center, Inc. Alcohol and Drug Abuse Services
1000 Dutch Ridge Road
Beaver, PA 15009

BEDFORD

Twin Lakes Center For Drug and Alcohol Rehabilitation
130 West Penn Street
Bedford Professional Building
Bedford, PA 15522

BELLEFONTE

Centre Single County Authority Substance Abuse Education/ Prevention Unit
420 Holmes Street
Willowbank Building
Bellefonte, PA 16823

Counseling Services, Inc.
Drug and Alcohol Program
441 North Spring Street
Bellefonte, PA 16823

Howard House
139 West Howard Street
Bellefonte, PA 16823

BENSALEM

Libertae, Inc.
5245 Bensalem Boulevard
Bensalem, PA 19020

Mustard Seed, Inc. Street Road and Bristol Pike
Neshaminy Plaza 1 Suite 102
Bensalem, PA 19020

Renewal Centers Bensalem Office
1950 Street Road
Constitution Building Suite 301
Bensalem, PA 19020

BETHEL PARK

The Drug Connection
5171 Park Avenue
Bethel Park, PA 15102

BETHLEHEM

Alcohol Council/Lehigh Valley
520 East Broad Street
Bethlehem, PA 18018

Lehigh Valley Addictions Treatment Services, Inc.
Outpatient/Bethlehem
50 East Broad Street
Bethlehem, PA 18018

Intensive Outpatient Treatment Alternatives
800 Ostrum Street
Bethlehem, PA 18015

Valley Youth House Drug and Alcohol Prevention and Education
539 8 Avenue
Bethlehem, PA 18018

BIRDSBORO

The Drug and Alcohol Center
201 East Main Street
Birdsboro, PA 19508

BRADFORD

Alcohol and Drug Abuse Services Bradford Unit
36 South Avenue
Bradford, PA 16701

Federal Correctional Institution Substance Abuse Services
Bradford, PA 16731

Nelson Counseling Center
24 West Corydon Street
Bradford, PA 16701

BRIDGEPORT

Lincoln Center for Family and Youth
201 Union Avenue
Bridgeport, PA 19405

BRIDGEVILLE

Turtle Creek Valley MH/MH, Inc. Bridgeville Drug/Alcohol Outpatient
755 Washington Avenue
Bridgeville, PA 15017

BRISTOL

Family Service Assoc. of Bucks County
1200 New Rodgers Road
Suite D6
Bristol, PA 19007

BROOMALL

Changing Times Center
370 Reed Road
Broomall, PA 19008

BUTLER

Butler A Center, Inc.
165 Old Plank Road
Butler, PA 16001

Butler Alcohol Countermeasures Program
222 West Cunningham Street
2nd Floor
Butler, PA 16001

Butler County Council on Alcohol and Drug Dependency
227 South Chestnut Street
Butler, PA 16001

Irene Stacy CMHC Addictive Behaviors Unit
112 Hillvue Drive
Butler, PA 16001

Veterans' Affairs Medical Center Substance Abuse Treatment Unit (SATU)
325 New Castle Road
Butler, PA 16001

CAMP HILL

Camp Hill State Correctional Institution Chemical Abuse Department
Lisburn Road
Trailor Complex
Camp Hill, PA 17001

Guidance Associates
412 Erford Road
Camp Hill, PA 17011

Holy Spirit Hospital Drug and Alcohol Medical Service Unit
503 North 21 Street
Camp Hill, PA 17011

Mainstay Service, Inc.
423 North 21 Street
Camp Hill, PA 17011

CANONSBURG

The Care Center Crossroads Office
Canonsburg, PA 15317

CARLISLE

Cumberland/Perry Drug and Alcohol Commission
Cumberland County Court House
East Wing Room 206
Carlisle, PA 17013

New Insights
8 South Hanover Street
Carlisle, PA 17013

Stevens Center Drug and Alcohol Program
33 State Avenue
Carlisle, PA 17013

CENTRE HALL

The Meadows Psychiatric Center Psychiatric Chemical Dependency Program
Earlystown Road
RD 1
Centre Hall, PA 16828

CHAMBERSBURG

Contact Chambersburg
237 East Queen Street
Chambersburg, PA 17201

Mountain Manor Treatment Center
25 Penncraft Avenue
Professional Arts Building Room 302
Chambersburg, PA 17201

Twin Lakes Center Drug and Alcohol Rehabilitation
166 South Main Street
Kerrstown Square Suite 202
Chambersburg, PA 17201

Women In Need, Inc.
156 East Queen Street
Chambersburg, PA 17201

CHESTER

Crozer Chester Medical Center
CHS Methadone Program
CHS Outpatient Service
1 East 9 Street
Chester, PA 19013

CHS Positive Choice
Vedder House Residential Program
1 Medical Center Boulevard
Chester, PA 19013

UHS Keystone Center
2001 Providence Road
Chester, PA 19013

CLARION

Clarion County Counseling Center
Drug/Alcohol Administration
214 South 7 Avenue
Clarion, PA 16214

CLEARFIELD

Bi-County Treatment Center
214 North 2 Street
Clearfield, PA 16830

Gateway Institute and Clinic, Inc.
600 Leonard Street
Clearfield, PA 16830

COATESVILLE

Riverside Clinics, Inc. Riverside Brandywine
1825 East Lincoln Highway
Coatesville, PA 19320

CORRY

GECAC Drug and Alcohol Services/Corry
45 East Washington Street
Corry, PA 16407

COUDERSPORT

Alcohol and Drug Abuse Center
5 North Main Street
Office 5
Coudersport, PA 16915

CRESSON

Cedar Manor Treatment Center
109 Sumner Street
Cresson, PA 16630

DANVILLE

Columbia/Montour/Snyder/Union MH/MR Drug and Alcohol Program
Terrace Building
Danville, PA 17821

Geisinger Medical Center Substance Abuse Services
North Academy Avenue
Danville, PA 17822

DELTA

Adams Hanover Counseling Services Delta
RD 1
Delta, PA 17314

DOWNINGTOWN

Counseling Network of Brandywine Valley/Downingtown Clinic
351 East Lancaster Avenue
Downingtown, PA 19335

DOYLESTOWN

Aldie Counseling Center
228 North Main Street
Doylestown, PA 18901

Bucks County Correctional Facility Drug and Alcohol Unit
1730 South Easton Road
Doylestown, PA 18901

Bucks County Council on Alcoholism and Drug Dependence
Routes 313 and 611
252 West Swamp Road/Unit 33
Doylestown, PA 18901

Family Service Association of Bucks County
20 West Oakland Avenue
Doylestown, PA 18901

Youth Services of Bucks County, Inc.
Route 611 and Almshouse Road
Neshaminy Manor Center
Doylestown, PA 18901

DU BOIS

Bi-County Treatment Center
319 Daly Street
Du Bois, PA 15801

DUNCANSVILLE

The Encouragement Place
900 Sunbrook Drive
Duncansville, PA 16635

EAGLEVILLE

Eagleville Hospital Inpatient Program
100 Eagleville Road
Eagleville, PA 19408

EASTON

Lehigh Valley Addictions Treatment Services Easton Outpatient
905C Line Street
Easton, PA 18042

EBENSBURG

Cambria County DUI Counter Attack Program
201 North Julian Street
Ebensburg, PA 15931

Cambria County MH/MR Intake Center
Route 22
East Beth Energy Building
Ebensburg, PA 15931

EDINBORO

GECAC Drug and Alcohol Services/Edinboro
Edinboro University
White Hall Room 103
Edinboro, PA 16412

EMPORIUM

Alcohol and Drug Abuse Center
107 South Cherry Street
Emporium, PA 15834

EPHRATA

Addiction Recovery Corporation The Terraces
1170 South State Street
Ephrata, PA 17522

ERIE

Abraxas Foundation, Inc.
Abraxas Halfway House West
Abraxas II
502 West 6 Street
Erie, PA 16507

Community House, Inc. Women
521 West 7 Street
Erie, PA 16502

Crossroads Hall Facility of Serenity Hall, Inc.
414 West 5 Street
Erie, PA 16507

GECAC Drug and Alcohol Services Central Office
809 Peach Street
Erie, PA 16501

Hospitality House Services for Women, Inc.
Erie, PA 16507

Perseus House, Inc.
Andromeda House I
132 West 26 Street
Erie, PA 16508

Perseus House
516 West 7 Street
Erie, PA 16502

Transitional Living Program
527 West 8 Street
Erie, PA 16502

Saint Vincent Health Center Serenity Recovery Center for Substance Abuse Outpatient Program
232 West 25 Street
Erie, PA 16544

EVANS CITY

Irene Stacy CMHC Drug and Alcohol Unit
401 Smith Drive
Cranberry Professional Park Suite 100
Evans City, PA 16033

EXTON

Chester County Council on Addictive Diseases, Inc.
734 East Lancaster Avenue
Whiteland Business Park
Exton, PA 19341

FAIRVIEW

Saint Vincent Recovery Center
6816 West Lake Road
Fairview, PA 16415

FALLS CREEK

Clearfield/Jefferson Drug and Alcohol Commission
104 Main Street
Falls Creek, PA 15840

FARRELL

Shenango Valley Medical Center
Insights Inpatient
2200 Memorial Drive Extension
Farrell, PA 16121

Insights Outpatient
500 Darr Avenue
Southwest Gardens Memorial Building
Farrell, PA 16121

GETTYSBURG

Adams Hanover Counseling Services, Inc.
37 West Street
Gettysburg, PA 17325

Gettysburg YWCA Substance Abuse Services
909 Fairfield Road
Gettysburg, PA 17325

The Recovery Place
70 West Middle Street
Gettysburg, PA 17325

GIRARD

GECAC Drug and Alcohol Services/Girard
259 Main Street East
Girard, PA 16417

GLENMOORE

Vitae House, Inc.
Fairview Road
Glenmoore, PA 19343

GREENSBURG

Westmoreland Gateway Drug and Alcohol Abuse Center
532 West Pittsburgh Street
Greensburg, PA 15601

HANOVER

Adams Hanover Counseling Services, Inc.
625 West Elm Avenue
Hanover, PA 17331

HARRISBURG

Addictive Disease Clinic
1727 North 6 Street
Harrisburg, PA 17102

Capital Psychiatric and Psychological Associates
205 South Front Street
Brady Hall 5th Floor
Harrisburg, PA 17101

Dauphin County Department of Drugs and Alcohol Services/ Prevention Department
25 South Front Street
Harrisburg, PA 17101

Gaudenzia Concept 90
Spruce Road
Harrisburg State Hospital Building 21
Harrisburg, PA 17105

Greater Harrisburg Alcohol and Drug Counseling
3309 Spring Street
Progress Plaza Suite 204
Harrisburg, PA 17109

Teen Challenge
1419–21 North Front Street
Harrisburg, PA 17102

The First Step Program
650 North 12 Street
Harrisburg, PA 17105

HAVERTOWN

Mercy Haverford Hospital Substance Abuse Services
2000 Old West Chester Pike
Havertown, PA 19083

HAZLETON

Alcoholism and Drug Services of Lower Luzerne County/N. Cedar Unit 1
8 West Broad Street
Room 521
Hazleton, PA 18201

HOMESTEAD

Turtle Creek Valley MH/MR, Inc. Alternatives
201 East 18 Avenue
Homestead, PA 15120

HUMMELSTOWN

Mazzitti and Sullivan EAP Services
1305 Middletown Road
Suite 2
Hummelstown, PA 17036

INDIANA

The Open Door, Inc.
20 South 6 Street
Indiana, PA 15701

JOHNSTOWN

Cambria County Human Services Intake and Evaluation
417 Main Street
Johnstown, PA 15901

KANE

Alcohol and Drug Abuse Services Kane Unit
2 Thompson Park
Kane, PA 16735

KENNETT SQUARE

Bowling Green Inn Brandywine
495 Newark Road
Kennett Square, PA 19348

KITTANNING

ARC/Manor Annex
220 Garfield Street
Kittanning, PA 16201

Armstrong County Council On Alcohol and Other Drugs, Inc./ ARC Manor
301 Arthur Street
Kittanning, PA 16201

Ministries of Eden Christian Counseling Center
219 North McKean Street
Kittanning, PA 16201

LAKE ARIEL

The White House, Ltd.
Lake Ariel, PA 18436

LANCASTER

Catholic Charities Substance Abuse Services
925 North Duke Street
Lancaster, PA 17602

Drug and Alcohol Rehab Service, Inc. Manos Residential Therapeutic Community
121 South Prince Street
Lancaster, PA 17603

Family Service of Lancaster County
630 Janet Avenue
Lancaster, PA 17601

Lancaster Clinical Counseling Assoc.
131 East Orange Street
2nd Floor/Rear
Lancaster, PA 17602

Lancaster County Drug and Alcohol Program Prevention Unit
50 North Duke Street
Lancaster, PA 17603

Saint Joseph Hospital
Inpatient Chemical Dependency Services
Outpatient Chemical Dependency Services
250 College Avenue
Lancaster, PA 17604

Therapy Services
131 East Grant Street
Lancaster, PA 17602

Watson and Hogg Counseling Assoc., Inc.
202 Butler Avenue
Lancaster, PA 17601

LANGHORNE

Family Service Assoc. of Bucks County
1 Oxford Valley
Suite 717
Langhorne, PA 19047

LANSDALE

Help Line Center, Inc.
306 A Madison Avenue
Lansdale, PA 19446

LATROBE

Saint Vincent College Drug/ Alcohol Prevention Project
Wimmer Hall
Latrobe, PA 15650

LEBANON

Lebanon County Crisis Intervention/Information and Referral Service
4 and Walnut Streets
Good Samaritan Hospital
Lebanon, PA 17042

New Perspective/Roxbury Addiction Treatment Center
3030 Chestnut Street
Lebanon, PA 17042

Renaissance Counseling
701 Chestnut Street
Lebanon, PA 17042

Veterans' Affairs Medical Center Substance Abuse Treatment Unit (SATU)
1700 South Lincoln Avenue
Lebanon, PA 17042

LEESPORT

Council on Chemical Abuse (COCA) Prison Program
RD 1
Leesport, PA 19533

LEHIGHTON

Carbon/Monroe/Pike Drug/ Alcohol Commission, Inc.
128 South First Street
Lehighton, PA 18235

LEVITTOWN

Riverside Clinics, Inc. Riverside North
1609 Woodbourne Road
Levittown, PA 19057

LEWISBURG

United States Penitentiary Drug Abuse Program
Lewisburg, PA 17837

LEWISTOWN

Central Pennsylvania Individual and Family Counseling
48 Chestnut Street
Lewistown, PA 17044

LIMERICK

Creative Health Services, Inc. Drug and Alcohol Outpatient
168 West Ridge Pike
Limerick Office Court
Limerick, PA 19468

LOCK HAVEN

Green Ridge Counseling Center
Unit IV
350 East Main Street
Lock Haven, PA 17745

MALVERN

Malvern Institute
940 King Road
Malvern, PA 19355

MARIENVILLE

Abraxas Foundation, Inc. Abraxas I
Blue Jay Village
Marienville, PA 16239

MCKEES ROCKS

Mercy Center for Chemical Dependency Services McKees Rocks Center
710 Thompson Avenue
McKees Rocks, PA 15136

MCKEESPORT

Center for Substance Abuse
120 5 Avenue
McKeesport, PA 15132

MEADVILLE

Crawford County Drug and Alcohol Executive Commission
898 Park Avenue
Suite 12
Meadville, PA 16335

Meadville Medical Center Stepping Stones
1034 Grove Street
Meadville, PA 16335

MECHANICSBURG

Gaudenzia Foundation Inc. West Shore Outpatient Program
6 State Road
Suite 115
Mechanicsburg, PA 17055

MEDIA

Alcoholism and Addictions Council of Delaware County
115 West State Street
Suite 300
Media, PA 19063

Changing Times Center
204 South Avenue
Media, PA 19063

Delaware County Drug/Alcohol Commission Functional Prevention Unit
600 North Jackson Street
Media, PA 19063

Mirmont Treatment Center
100 Yearsley Mill Road
Media, PA 19063

Penn Recovery Systems, Inc.
100 West 6 Street
Media, PA 19063

Rainbow of Recovery
316–318 East Baltimore Pike
Media, PA 19063

Riverside Clinics, Inc. Media Adolescent Program
280 North Providence Road
Media, PA 19063

MIFFLINTOWN

Central Pennsylvania Individual and Family Counseling
104 North Main Street
Mifflintown, PA 17059

MILFORD

Carbon/Monroe/Pike Drug/Alcohol Commission Pike County Clinic
State Route 1
Milford, PA 18337

MILTON

Green Ridge Counseling Center Unit I
28 North Front Street
Milton, PA 17847

MOHNTON

Richard J. Caron Foundation Rosie Kearney Halfway House for Women
223–225 East Wyomissing Avenue
Mohnton, PA 19540

MONACA

Beaver County Alcohol Highway Safety Program
1260 North Brodhead Road
Beaver Valley Prof Center Suite 101
Monaca, PA 15061

MONONGAHELA

Freedom of Monongahela, Inc.
1290 Chess Street
Monongahela, PA 15063

Monongahela Valley Hospital Drug/Alcohol Detoxification Unit
Country Club Road
Route 88
Monongahela, PA 15063

MONROEVILLE

Saint Francis Medical Center Center for Chemical Dependency/Monroeville Outreach/Treatment
2550 Mosside Boulevard
Medical Arts Building Suite 212
Monroeville, PA 15146

MORRISVILLE

Good Friends, Inc.
Lincoln Highway
Morrisville, PA 19067

NEW CUMBERLAND

New Insights, Inc.
R 320 Bridge Street
Suite 96
New Cumberland, PA 17070

NEW KENSINGTON

Allegheny Valley Drug and Alcohol Prevention Program
310 Central City Plaza
New Kensington, PA 15068

Hope Haven
301 Greensburg Road
New Kensington, PA 15068

NORRISTOWN

Counseling Associates
1139 Markley Street
Norristown, PA 19401

Family House/Norristown
901 Dekalb Street
Norristown, PA 19401

Montgomery County Methadone Center
316 Dekalb Street
Norristown, PA 19401

Montgomery County MH/MR Emergency Service
50 Beech Drive
Norristown, PA 19401

Programs in Counseling
319 Swede Street
Norristown, PA 19401

Valley Forge Medical Center and Hospital
1033 West Germantown Pike
Norristown, PA 19403

NORTH EAST

GECAC Drug and Alcohol Services/North East
41 West Main Street
North East, PA 16428

PENNSBURG

Upper Perkiomen Valley Youth Service Bureau
One Walt Road
Pennsburg, PA 18073

PHILADELPHIA

AB Associates, Inc. America Beats Addiction
1523 West Erie Avenue
Philadelphia, PA 19140

Addiction Treatment Services Outpatient
111 North 49 Street
Philadelphia, PA 19139

Adept Alcohol and Drug Abuse Pregnant Treatment Program
11 and Tabor Streets
Sheerr Building
Philadelphia, PA 19141

Adult Probation Alcohol Highway Safety Program
121 North Broad Street
4th Floor
Philadelphia, PA 19107

Alcohol and Mental Health Associates
1200 Walnut Street
2nd Floor
Philadelphia, PA 19107

Asociacion de Puertorriquenos Proyecto Proyecto Nueva Vida
2143 North 6 Street
Philadelphia, PA 19122

Belmont Center for Comp. Treatment Woodside Hall Drug Treatment Program
4200 Monument Road
Philadelphia, PA 19131

Bridge Counseling Center
1912 Welsh Road
Philadelphia, PA 19115

Bridge Therapeutic Center at Fox Chase
8400 Pine Road
Philadelphia, PA 19111

Diagnostic and Rehabilitation Center
Main Clinic
229 Arch Street
Philadelphia, PA 19106

Residential Detox
100–110 North Bread Street
Philadelphia, PA 19106

Episcopal Hospital Substance Abuse Unit
100 East Lehigh Avenue
Philadelphia, PA 19125

Frankford Hospital
First Days Outpatient
4936 Griscom Street
Philadelphia, PA 19124

Frankford Campus
First Days Program
Frankford Avenue and Wakeling Street
Philadelphia, PA 19124

Genesis II, Inc.
1214 North Broad Street
Philadelphia, PA 19121

Caton House
3947 Lancaster Avenue
Philadelphia, PA 19104

Harbison Recovery Program
6737 Harbison Avenue
Philadelphia, PA 19149

Horizon House Residential Program
1714 Point Breeze Avenue
Philadelphia, PA 19145

Institute of Pennsylvania Hospital Strecker Partial Program
111 North 49 Street
Philadelphia, PA 19139

Intercommunity Action, Inc. (Interac) Alcohol/Education/ and Family Counseling Program
6122 Ridge Avenue
Philadelphia, PA 19128

Interim House, Inc.
333 West Upsal Street
Philadelphia, PA 19119

J. F. Kennedy MH/MR Center Centro de Servicios Para Hispanos
2742 North 5 Street
Philadelphia, PA 19133

Jefferson Intensive Cocaine Treatment Program
Jefferson Methadone Clinic
21 Street and Washington Avenue
Philadelphia, PA 19146

Jefferson Medical College My Sister's Place
5601 Kingsessing Avenue
Philadelphia, PA 19143

Jefferson Outreach Drug and Alcohol Program
Central District
1201 Chestnut Street
14th Floor
Philadelphia, PA 19107

Southern District
1715 McKean Street
Philadelphia, PA 19145

Jewish Family and Children's Service Project Pride
10125 Verree Road
Suite 200
Philadelphia, PA 19116

John F. Kennedy Comm. MH/MR Center Walk-In Clinic
112 Broad Street
Philadelphia, PA 19102

Kensington Hospital Alcohol and Drug Services
136 West Diamond Street
Philadelphia, PA 19122

Medical College of Pennsylvania Harbison Recovery Program/ Caring Together
3300 Henry Avenue
Philadelphia, PA 19129

Nazareth Hospital The Mustard Seed Program
2601 Holme Avenue
Philadelphia, PA 19152

New Journeys in Recovery
2927 North 5 Street
Philadelphia, PA 19133

North East Treatment Centers The Wharton Center/ Outpatient Services
2205 Bridge Street
Philadelphia, PA 19137

North Philadelphia Health System Girard Medical Center
Comprehensive Alcoholism Program/Outpatien
Goldman Clinic
Medical Rehab Services
Residential Recovery Program
Girard Avenue and 8 Street
Philadelphia, PA 19122

Detox Service
Girard Avenue and 16 Street
Philadelphia, PA 19122

Northeast Treatment Centers The Fire House Residential Center
2005 North 2 Street
Philadelphia, PA 19125

Parkside Human Services
Methadone Clinic
New Directions
4950 Parkside Avenue
Philadelphia, PA 19131

Penn Recovery Systems, Inc.
Mutch 4
Non Hospital Detox
39 and Market Streets
Philadelphia, PA 19104

Presbyterian Univ. of Pennsylvania Medical Center Alcohol and Drug Services
51 North 39 Street
Philadelphia, PA 19104

River's Bend Drug and Alcohol Unit
2401 Penrose Avenue
Philadelphia, PA 19145

School District of Philadelphia Student Substance Abuse Programs
734 Schuylkill Avenue
J. F. Kennedy Center Room 305
Philadelphia, PA 19146

Self Help Movement, Inc.
14000 Roosevelt Boulevard
Daniel Blaine Building
Philadelphia, PA 19154

Shalom, Inc.
311 South Juniper Street
Philadelphia, PA 19107

The Assistance Program
1225 Vine Street
2nd Floor
Philadelphia, PA 19107

The Northwest Center, Inc. Substance Abuse Treatment Program
21–23 East School House Lane
Philadelphia, PA 19144

Thomas Jefferson University Hospital Family Center Program
1201 Chestnut Street
11th Floor
Philadelphia, PA 19107

Veterans' Affairs Medical Center Alcohol/Drug Dependence Treatment Program
University and Woodland Avenues
Philadelphia, PA 19104

Washington House
1516–18 Washington Avenue
Philadelphia, PA 19146

West Philadelphia Comm. Mental Health Consortium Drug Abuse Rehabilitation Program
451 University Avenue
Philadelphia PA 19104

PHOENIXVILLE

Help Counseling Center
234 Church Street
Phoenixville, PA 19460

PITTSBURGH

Abraxas Foundation, Inc.
Abraxas Center for Adolescent Females
437 Turrett Street
Pittsburgh, PA 15206

Abraxas III
936 West North Avenue
Pittsburgh, PA 15233

Achates Mental Health Alcohol and Drug Abuse Services
7805 McKnight Road
Suite 103
Pittsburgh, PA 15237

Alcoholic Recovery Center
ARC House II
1216 Middle Street
Pittsburgh, PA 15212

800 East Ohio Street
Pittsburgh, PA 15212

Circle C Specialized Group Home for Chemically Dependent Adolescents
227 Seabright Street
Pittsburgh, PA 15214

Forbes Regional Health Center Mental Health Substance Abuse Services
2570 Haymaker Road
Pittsburgh, PA 15146

House of the Crossroads
2012 Centre Avenue
Pittsburgh, PA 15219

PBA, Inc. The Second Step Program
1425 Beaver Avenue
Pittsburgh, PA 15233

Saint Francis Medical Center
Adolescent Chemical Dependency Unit
45 and Pennsylvania Avenue
Pittsburgh, PA 15201

Adult Chemical Dependency Treatment Unit
400 45 Street
Pittsburgh, PA 15201

Center for Chemical Dependency Treatment
6714 Kelly Street
Pittsburgh, PA 15208

Salvation Army
Public Inebriate Program
44–54 South 9 Street
Pittsburgh, PA 15203

Public Inebriate Program
54 South 9 Street
Pittsburgh, PA 15203

Southwood Chemical Dependency Services
2575 Boyce Plaza Road
Pittsburgh, PA 15241

The Mercy Center for Chemical Dependency Services
1700 East Carson Street
Suite 4-C
Pittsburgh, PA 15203

Fleming Avenue Unit
3530 Fleming Avenue
Pittsburgh, PA 15212

The Whale's Tale
Penn Hills Office
12013 Frankstown Road
Pittsburgh, PA 15235

Family Treatment Center
844 Proctor Way
Pittsburgh, PA 15210

Shadyside Office
250 Shady Avenue
Pittsburgh, PA 15206

Turtle Creek Valley MH/MR, Inc. Alternatives
70 South 22 Street
Pittsburgh, PA 15203

University of Pittsburgh Maximizing Adolescent Potentials (MAPS)
5D21 Forbes Quadrangle
Pittsburgh, PA 15260

Veterans' Affairs Medical Center Substance Abuse Treatment Unit
Highland Drive
Unit 116A3/5
Pittsburgh, PA 15206

Western Psychiatric Institute & Clinic Center for Psychiatric and Chemical Dependency Services
3811 Ohara Street
Pittsburgh, PA 15213

Womanspace East, Inc.
Pittsburgh, PA 15230

PORT ALLEGANY

Maple Manor
120 Chestnut Street
Port Allegany, PA 16743

POTTSTOWN

CDM Counseling, Inc.
67 Walnut Street
Pottstown, PA 19464

Creative Health Services, Inc.
Drug and Alcohol Outpatient
365 High Street
Pottstown, PA 19464

Pottstown Area Drug Rehab Program
101–105 King Street
Pottstown, PA 19464

POTTSVILLE

Schuylkill County Organizing Project
619 Mahantongo Street
Cerebral Palsy Building
Pottsville, PA 17901

Turning Point
16 South Centre Street
Pottsville, PA 17901

QUAKERTOWN

Family Service Assoc. of Bucks County
515 South West End Boulevard
Second Floor
Quakertown, PA 18951

Renewal Centers, Inc. Quakertown Office
2705 Old Bethlehem Pike
Quakertown, PA 18951

READING

Are House
1010 Centre Avenue
Reading, PA 19601

Berks County Adult Probation Dept. Alcohol Safe Driving Program
Court and Reed Streets
Services Center
Reading PA 19601

Berks Youth Counseling Center
525 Franklin Street
Reading, PA 19602

Center for Mental Health Drug and Alcohol Center
6 and Spruce Streets
Building J
Reading, PA 19611

Council on Chemical Abuse Prevention Unit
220 North 5 Street
Reading, PA 19601

Drug Treatment Program of Reading
22 North 6 Avenue
Reading, PA 19611

Family Guidance Center Recovery Drug and Alcohol Program
631 Washington Street
Reading, PA 19603

Inroads Employee Assistance Program, Inc.
325 North 5 Street
Reading, PA 19601

RIDGWAY

Elk County General Hospital Substance Abuse Services
Ridgway, PA 15853

ROBESONIA

Richard J. Caron Foundation Chit Chat West
27 Freeman Street
Robesonia, PA 19551

ROSLYN

Growth Horizons, Inc.
1069 Easton Road
Roslyn, PA 19001

ROYERSFORD

Meridian Youth Services Drop-In and Prevention Program
201 North 4 Avenue
Royersford, PA 19468

RURAL RIDGE

Teen Challenge of Western Pennsylvania
Lefever Hill Road
Rural Ridge, PA 15075

SAINT MARYS

Alcohol and Drug Abuse Services Saint Marys Unit
20 North Michael Street
Saint Marys, PA 15857

Nelson Counseling Center
509 Arch Street Extension
Saint Marys, PA 15857

SCIOTA

Joan S. Wielgus
HRC 1
Sciota, PA 18354

SCRANTON

Community Intervention Center Drug and Alcohol Drop-In Center
614 Mulberry Street
Scranton, PA 18510

Lackawanna County SCA Drug and Alcohol Prevention Unit
200 Adams Avenue
Scranton, PA 18503

Voluntary Action Center of Northeastern Pennsylvania
225 North Washington Avenue
Park Plaza
Scranton, PA 18503

SELINSGROVE

Alternatives Counseling Service
720 North Market Street
Selinsgrove, PA 17870

SELLERSVILLE

Grand View Hospital Substance Abuse Services
700 Lawn Avenue
Sellersville, PA 18960

Penn Foundation, Inc. Recovery Center
807 Lawn Avenue
Sellersville, PA 18960

SHAMOKIN

Green Ridge Counseling Center Unit II
117 East Independence Street
Shamokin, PA 17872

SHARON HILL

Changing Times Center
800 Chester Pike
Sharon Hill, PA 19079

SHENANDOAH

Riverside Central
200 Pennsylvania Avenue
Shenandoah, PA 17976

SHICKSHINNY

Clear Brook Lodge
RD 2
Shickshinny, PA 18655

SHREWSBURY

Adams Hanover Counseling Services, Inc. Crossroads Counseling and Education Services
73 East Forest Avenue
Shrewsbury, PA 17361

SOMERSET

Twin Lakes Center for Drug/ Alcohol Rehabilitation
Byers Road
RD 7
Somerset, PA 15501

SPARTANSBURG

Perseus House, Inc. Andromeda House II
Mount Pleasant Road
Spartansburg, PA 16434

SPRING CITY

Creative Health Services, Inc.
1 Mennonite Church Road
Spring City, PA 19475

STAHLSTOWN

Ligonier Valley Treatment Center
RD 1
Stahlstown, PA 15687

STATE COLLEGE

Lawrence T. Clayton and Counseling Associates, Inc.
230 South Fraser Street
State College, PA 16801

On Drugs Incorporated
236 South Allen Street
State College, PA 16801

STROUDSBURG

Carbon/Monroe/Pike Drug/ Alcohol Commission Monroe County Clinic
14 North 6 Street
Suite 300
Stroudsburg, PA 18360

SUNBURY

Green Ridge Counseling Center Unit V
417B Market Street
Sunbury, PA 17801

Psychological Services Clinic
352 Arch Street
Sunbury, PA 17801

SWIFTWATER

Woodhaven in the Poconos
Route 611
Swiftwater, PA 18370

TUNKHANNOCK

Catholic Social Services
Falls
Route 92
Tunkhannock, PA 18657

TURTLE CREEK

Turtle Creek Valley MN/MR, Inc. Alternatives
519 Penn Avenue
Turtle Creek, PA 15145

TYRONE

Tyrone Hospital Chemical Dependency Program Detox Unit
1 Hospital Drive
Tyrone, PA 16686

UNION CITY

Union City Memorial Hospital Substance Abuse Services
130 North Main Street
Union City, PA 16438

UNIONTOWN

Fayette County Drug and Alcohol Commission, Inc.
100 New Salem Road
Suite 106
Uniontown, PA 15401

Uniontown Hospital Substance Abuse Services
500 West Berkeley Street
Uniontown, PA 15401

UPPER DARBY

Harwood House
9200 West Chester Pike
Upper Darby, PA 19082

Hipid Outpatient Drug Abuse Clinic
312 South 69 Street
Upper Darby, PA 19082

WARREN

Warren County Jail
Office of Counseling Services
407 Market Street
Warren, PA 16365

Warren General Hospital Chemical Dependency Center
204 Liberty Street
Warren, PA 16365

WASHINGTON

The Care Center
62 East Wheeling Street
Washington, PA 15301

Vision/Comprehensive Addiction Prevention Services
87 East Maiden Street
Washington, PA 15301

WAVERLY

Marworth
Lily Lake Road
Waverly, PA 18471

WAYNESBURG

The Care Center
63 South Washington Street
Waynesburg, PA 15370

WELLSBORO

Soldiers' and Sailors' Memorial Hospital Substance Abuse Detox Services
32–36 Central Avenue
Wellsboro, PA 16901

WERNERSVILLE

Richard J. Caron Foundation
Caron Adolescent Treatment Center
Chit Chat Farms
Galen Hall Road
Wernersville, PA 19565

WEST CHESTER

Chester County Hospital Substance Abuse Services
701 East Marshall Street
West Chester, PA 19380

Help Counseling Center/High Street
624 South High Street
West Chester, PA 19380

WEST GROVE

Southern Chester County Addiction Recovery Center
1011 West Baltimore Pike
Suite 101
West Grove, PA 19390

Southern Chester County Medical Center Medical Detoxification Program
1015 West Baltimore Pike
West Grove, PA 19390

WEST READING

National Council on Alcohol and Drug Dependency Berks County
529 Reading Avenue
Suite R
West Reading, PA 19611

WESTFIELD

Richard J. Caron Foundation Chit Chat Westfield
305 Church Street
Westfield, PA 16950

WEXFORD

The Mercy Center for Chemical Dependency Services North Hills Outpatient
Meadow Pointe Office Park
103 North Meadows Bldg 200 Suite 210
Wexford, PA 15090

WHITE HAVEN

Northeast Counseling Services Youth Forestry
Hickory Run State Park
White Haven, PA 18661

WILKES-BARRE

Choices at Nesbitt Memorial Hospital
518 Wyoming Avenue
Wilkes-Barre, PA 18704

Clear Brook Manor
10 East Northampton Street
Wilkes-Barre, PA 18702

Family Service Association of Wyoming Valley
31 West Market Street
Wilkes-Barre, PA 18701

Geisinger Wyoming Valley Medical Center Substance Abuse Services
1000 East Mountain Drive
Wilkes-Barre, PA 18711

Luzerne/Wyoming Counties Drug and Alcohol Primary Prevention Unit
111 North Pennsylvania Boulevard
Wilkes-Barre, PA 18701

Veterans' Affairs Medical Center Substance Abuse Treatment Unit
1111 East End Boulevard
Wilkes-Barre, PA 18711

WILLIAMSPORT

Crossroads Counseling, Inc.
460 Market Street
Room 218A
Williamsport, PA 17701

Green Ridge Counseling Center Unit III
829 West 4 Street
Williamsport, PA 17701

Helpline
815 West 4 Street
Williamsport, PA 17701

WYOMISSING

Richard J. Caron Counseling Services
845 Park Road
Wyomissing, PA 19610

YORK

Delphic Mental Health Associates and Addictions Services
1600 South George Street
York, PA 17403

New Insights, Inc.
707 Loucks Road
York, PA 17404

Stepping Stone Counseling and Education Services, Inc.
912 South George Street
York, PA 17043

211 South George Street
York, PA 17403

PUERTO RICO

AGUADILLA

Centro de Quimioterapia Anexo Hospital Regional de Aguadilla
Aguadilla, PR 00603

Centro Prevencion Aguadilla
Ave San Carlos Esq Betances
Altos
Aguadilla, PR 00603

Clinica Satelite de Tratamiento de Alcoholismo de Aguadilla
Calle Progreso 65
Aguadilla, PR 00603

Comunidad Terapeutica Guerrero Modulo 8
Carr 466 Bo Guerrero
Aguadilla, PR 00603

Modulo de Tratamiento Carcel Distrito Aguadilla
Calle Ruiz Belvis
Aguadilla, PR 00603

Teen Challenge de Aguadilla
Carretera 107 KM 3.5
Sector Playuela Barrio Borinquen
Aguadilla, PR 00603

Unidad Satelite Aguadilla
Ave Progreso 65
Aguadilla, PR 00603

ARECIBO

Arecibo Alcoholism Treatment Program
Ave Juan Rosado 160
Arecibo, PR 00612

Cede Arecibo
Carr 29 Antiguo Hospital de Distrito
Arecibo, PR 00612

Centro de Prevencion Arecibo
Avenida de Diego 358 Altos
Antiguo Teatro San Luis
Arecibo, PR 00612

Centro de Tratamiento Para Adultos de Arecibo
Antiguo Hospital de Distrito
Arecibo, PR 00612

Hogar Intermedio de Arecibo Carretera de Arecibo a Utuado
Bo Los Canos
Arecibo, PR 00612

Modulo de Tratamiento
Carcel Arecibo
Arecibo, PR 00613

BAYAMON

Bayamon Quimioterapia Antigua Central Juanita La Cambija
Bo Juan Sanchez
Bayamon, PR 00959

Bayamon Regional Metro Institution Alcoholism Treatment Module
Carr 167
Bayamon, PR 00956

Centro Prevencion Bayamon
Calle Barbosa 25 Esq Dr Veve Altos
Libreria Novedades
Bayamon, PR 00961

Centro Tratamiento Menores Bayamon
Calle Dr Veve
Esq Marti 51
Bayamon, PR 00956

Hogar Escuela Nuestra Senora de Fatima
Bo Cerro Gordo Camino Esteban Cruz
Ave Santa Juanita
Bayamon, PR 00956

Modulo de Tratamiento
Anexo 1072
Institucion Regional Metropolitana Bayamon Area Clinica
Institucion Regional Metropolitana Bayamon Sicoslipl
Carretera 167
Bayamon, PR 00956

Anexo 292/308
Institucion Regional Metropolitana Bayamon
Carretera 167
Bayamon, PR 00956

New Life for Girls de Puerto Rico
Carretera 830 KM 5.3
Bo Santaolaya
Bayamon, PR 00956

Renovados en Cristo
Carr 812 KM 6.4 Camino Los Ponos
Bo Guaraguao Sector La Pena
Bayamon, PR 00956

CAGUAS

Caguas Alcoholism Treatment Program
Monsenor Berrios 22 Altos
Caguas, PR 00725

Caguas Quimioterapia
Carr 796 KM 0.5 Sector La 25
Bo Bairoa
Caguas, PR 00725

Cede Caguas
Carretera 198 KM 2.6 Urbanizacion
Industrial Placidin Gonzalez Int
Caguas, PR 00725

Centro de Prevencion de Caguas
Calle Padial 2 3rd Piso
Caguas, PR 00725

Centro de Tratamiento Para Adultos Dec de Caguas/Casa Cesame
Carretera 189 KM 2.6 Urb Industrial
Placidin Gonzalez (Int)
Caguas, PR 00725

Centro Tratamiento Menores Caguas
Acosta 103
Caguas, PR 00725

Hogar Resurreccion
Carretera 175 KM 3 HM O
Bo San Antonio
Caguas, PR 00725

CAROLINA

Centro de Prevencion de Carolina Ave Roberto Clemente
Blq 27 Nuh 10
Carolina, PR 00985

Hogar El Buen Samaritano, Inc.
Unit 2
Carr 857 KM 9.5
Barrio Carruzo Sector Filipinas
Carolina, PR 00628

Proyecto Los Mirtos
Edificio 10 Apto 146
Carolina, PR 00987

CATANO

Centro Prevencion de Catano
Avenida Barbosa 91 Altos
Catano, PR 00962

FAJARDO

Centro Unidad Satelite de Fajardo Tratamiento Para Adultos
Centro Comunal Veve Calzada
Fajardo, PR 00738

Fajardo Satellite Alcoholism Treatment Clinic
Garrido Morales 2
Fajardo, PR 00738

GUAYAMA

Centro de Tratamiento a Menores Guayama
Calle Baldorioty/Final
Guayama, PR 00784

Centro de Tratamiento Para Adultos de Guayama
Calle Baldorioty Final Oeste
Guayama, PR 00784

Guayama Regional Detention Center
Alcoholism Treatment Module
Road 3
Melania Sector
Guayama, PR 00784

Guayama Satellite
Alcoholism Treatment Clinic
Hostos Street 46
Guayama, PR 00784

Modulo de Tratamiento Area Clinica
Centro Detencion Regional Guayama
Detox
Outpatient
Calle San Juan
Guayama, PR 00784

Proyecto Gipsy
Jardines de Guamani
Guayama, PR 00654

GUAYNABO

Unidad Evaluacion Diagostico
Carr 19 KM 5 HM 0
Guaynabo, PR 00970

GURABO

Gurabo Halfway House for Alcoholic Women
Carr 943 KM 2
Calle Santiago Final
Gurabo, PR 00778

Hogar El Buen Samaritano, Inc.
Carretera 941 KM 5 HM 0
Barrio Jaguas
Gurabo, PR 00778

HATO REY

Centro de San Juan Tratamiento Para Adultos Libre de Drogas
Ave Ponce de Leon 358
Hato Rey, PR 00918

Centro Hato Rey Tratamiento Residencial Para Varones
Hato Rey Ave Ponce de Leon 358
Hato Rey, PR 00918

Centro Tratamiento Menores San Juan
Ave Ponce de Leon 358
Hato Rey, PR 00918

Clinica Bodante San Juan
Bayomon/Cat
Avenida Barbosa 414
Hato Rey, PR 00917

Division de Educacion a la Comunidad Secretaria Auxiliar de Prevencion
Avenida Barbosa 414
Hato Rey, PR 00917

Residencial Mujeres
Avenida Ponce de Leon 358
Hato Rey, PR 00918

Secretaria Auxiliar de Prevencion Linea de Auxilio
Avenida Barbosa 414
Hato Rey, PR 00917

HUMACAO

CDE for Women Hospital Area Humacao
Hospt Area Humacao 6 Piso
Dr Victor Rincon Nunez Ave Tejas
Humacao, PR 00791

Centro de Prevencion de Humacao
Calle Dufresne 13 Bajos
Humacao, PR 00791

Corda
Centro Detox
Carr 924 KM 1.5
Bo Pitahaya
Humacao, PR 00791

Centro de Orientacion y Rehab Drogadictos y Alcoholicos
Carretera 3 Interseccion 910 Bo Catano
Antigua Escuela Alvira Alicea
Humacao, PR 00791

Centro de Prevencion y Servicios Multiples
Calle Antonio Lopez 52
Esquina Dr Vidal Altos
Humacao, PR 00791

Centro Pre Admision Orientacion y Referimiento
Calle Antonio Lopez 51
Esquina Dr. Vidal Altos
Humacao, PR 00791

Humacao Satellite Alcoholism Treatment Clinic
Ferrocarril Avenue
Humacao, PR 00791

Unidad Satelite Humacao Residencial Padre Rivera
Edificio 11
Humacao, PR 00791

ISABELA

Programa Pro Ayuda a Dictos and Familiares (PAAF) Isabela
Calle Corchado 81
Isabela, PR 00662

JUNCOS

Ebenezer
Carr 934 KM 0.5
Bo Ceiba Azul
Juncos, PR 00666

Hogar Nuevo Pacto
Carretera 31 KM 19
Bo Caimito 1
Juncos, PR 00777

MANATI

Unidad Satelite de Manati
Calle Obrero 15-A
Esquina Quinones
Manati, PR 00674

MAUNABO

Corda Prevencion Educacion y Ayuda Social
Calle Luis Munoz Rivera 16
Esquina Tierra Santa
Maunabo, PR 00707

Maunabo Satellite Alcoholism Treatment Clinic/CHC
Ave Kennedy Interior
Maunabo, PR 00707

MAYAGUEZ

Centro de Prevencion de Mayaguez
Calle McKinley 96 Bajos Oeste
Mayaguez, PR 00680

Centro de Tratamiento Para Adultos de Mayaguez
Casa de Salud 5T0 Piso
Mayaguez Medical Center Branch
Mayaguez, PR 00680

Centro Tratamiento a Menores Mayaguez
Avenue Duscombe 177
Mayaguez, PR 00680

Department of Addiction Services Mayaguez Alcoholism Program
Calle Mendez Vigo 155-E
Mayaguez, PR 00680

PONCE

Cede/Ponce
Centro Medico Carrt 14
Bo Machuelo
Ponce, PR 00731

Centro de Prevencion de Ponce
Calle Comercio 64
Ponce, PR 00731

Centro de Tratamiento Para Adultos de Ponce
Carretera Num 14 Barrio Machuelo
Facilidades de Centro Medico
Ponce, PR 00731

Centro Ponce Tratamiento Residencial Para Varones
Centro Medico
Carr 14 Bo Machuelo
Ponce, PR 00731

Centro Tratamiento a Menores Ponce
Carretera 14 Bo Machuelo
Centro Medico
Ponce, PR 00731

Centros Sor Isolina Ferre, Inc.
Del Dispensario San Antonio, Inc.
Calle E
Urb San Tomas
Ponce, PR 00731

Del Dispensario San Antonio Inc./2
Brisas Del Caribe El Tuque
Calle 13/654
Ponce, PR 00734

Del Dispensario San Antonio, Inc./3
Lirios Del Sur Bloque 29
Apartamentos 393–394–395 Playa
Ponce, PR 00731

Parcelas Amalia Marin Calle C Final 1
La Playa Sector Tabaiba
Ponce, PR 00731

Mision Refugio Incorporado
Bo Maraquez KM 4 HM 2
Ponce, PR 00731

Modulo de Tratamiento
Institucion Regional Del Sur
Detox
Outpatient
Bo El Tuque Sector Las Cucharas
Ponce, PR 00731

Ponce Alcoholism Treatment Program
Ponce Medical Center
Ponce, PR 00731

Quimioterapia Ponce
Area Hosp Distrito Carr 14
Bo Machuelos
Ponce, PR 00731

RIO PIEDRAS

Anexo 352 de Penitenciaria Estatal
Carretera 21 Penitenciaria Estatal
Barrio Monacillos
Rio Piedras, PR 00926

Cede San Juan
Pabellion B
Barrio Monacillos
Rio Piedras, PR 00925

Centro de Prevencion
Area Metropolitana
Quisqueya 176 Hato Rey
Urbanizacion Matienzo Cintron
Rio Piedras, PR 00019

Centro de Tratamiento Residencial A Menores Varones
Facilidades Centro Medico
Barrio Monacillos
Rio Piedras, PR 00935

Centro Quimioterapia San Juan Bo Monacillos/Facilidades Centro Medico
Rio Piedras, PR 00925

Centro San Juan Tratamiento Residential Varones
Pabellon J. Terrenos Centro Medico
Bo Monacillos
Rio Piedras, PR 00925

Dynamic Medical of Puerto Rico, Inc. Hospital Metropolitano
Road 21 Number 1785
Las Lomas
Rio Piedras, PR 00921

Emergency Detoxification and Hospitalization Unit
Casa de Salud Medical Center
Rio Piedras, PR 00928

Hogar Intermedio de Rio Piedras
Carretera 849 KM 1 HM 4
Bo Santo Domingo
Rio Piedras, PR 00929

La Montana Alcoholism Mobile Clinic
Ave Jose Celso Barbosa 414
Cuarto Piso
Rio Piedras, PR 00917

Modulo de Tratamiento Anexo Penitenciaria Estatal
Seguridad Maxima
Rio Piedras, PR 00926

Rio Piedras Psychiatric Hospital Rio Piedras Alcoholism Program
Building G
Centro Medico
Rio Piedras, PR 00925

State Penitentiary Annex Alcoholism Treatment Module
De Diego Avenue
In Front of Las Amapolas Public Housing
Rio Piedras, PR 00926

SAN JUAN

Casa La Providencia
Calle Norzagaray Street 200
Old San Juan
San Juan, PR 00902

Modulo de Tratamiento
Centro Detencion Parada 8
Avenida Fernandez Juncos
Parada 8
San Juan, PR 00903

Unidad Central de Admisiones
Esquina Calle San Andres 1063
Puerta de Tierra
San Juan, PR 00906

Veterans' Affairs Medical Center Drug Dependence Treatment Program
San Juan, PR 00927

SANTURCE

Modulo de Tratamiento Institucion Jovenes Adultos Miramar
Calle Villa Verde
Esquina Refugio Pda 10 Miramar
Santurce, PR 00907

TOA ALTA

Hogar Posada la Victoria, Inc.
C/Principal 165 KM 4 Hect 9 Parcela
52 Barrio Galateo Hoyo
Toa Alta, PR 00953

UTUADO

Centro de Prevencion de Utuado Centro Gubernamental Utuado
2 Do Piso
Utuado, PR 00641

VEGA ALTA

Escuela Industrial Para Mujeres Clinica de Desintoxicacion
Carretera 2 Int
Vega Alta, PR 00692

RHODE ISLAND

BARRINGTON

Comm. Organization for Drug Abuse Control Codac East
310 Maple Avenue
Suite 105
Barrington, RI 02806

Edward J. La Riviere Memorial Foundation IMP/ACT
3 Riverview Drive
Barrington, RI 02806

CENTRAL FALLS

Channel One/Central Falls
507 Broad Street
Central Falls, RI 02863

Progreso Latino
626 Broad Street
Central Falls, RI 02863

Proyecto Esperanza
400 Dexter Street
Central Falls, RI 02863

Ser Jobs for Progress, Inc.
626 Broad Street
Central Falls, RI 02863

CHARLESTOWN

South Shore Mental Health Center Alcohol and Drug Abuse Services
4705A Old Post Road
Charlestown, RI 02813

CRANSTON

Alcoholism Services of Cranston/ Johnston and Northwestern Rhode Island
311 Doric Avenue
Cranston, RI 02910

Comm. Organization for Drug Abuse Control
Codac I/Detoxification
1763 Broad Street
Cranston General Hospital
Cranston, RI 02905

Outpatient Drug Free
1052 Park Avenue
Cranston, RI 02910

Eastman House, Inc.
1545 Pontiac Avenue
Cranston, RI 02920

Human Ecology c/o Cranston Public Schools
845 Park Avenue
Cranston, RI 02910

Rhode Island Office of Substance Abuse
Division of Community Development
Cranston, RI 02920

Substance Abuse Detoxification Unit
Rhode Island Medical Center/ Howard Avenue
Benjamin Rush Building
Cranston, RI 02920

RISAP Treatment Alternative to Street Crime (TASC) Program/ Office of Substance Abuse
Rhode Island Medical Center
Louis Pasteur Building
Cranston, RI 02920

Senior High Prevention Network C/O Institute for Human Development
845 Park Avenue
Cranston, RI 02910

EAST PROVIDENCE

East Bay Mental Health Center Substance Abuse Services
610 Wampanoag Trail
East Providence, RI 02914

Edgehill East Providence Outpatient Clinic
850 Waterman Avenue
East Providence, RI 02914

ESMOND

Channel One of Smithfield
64 Farnum Pike
Esmond, RI 02917

EXETER

Marathon of Rhode Island, Inc. Residential
Exeter, RI 02822

JOHNSTON

Alcoholism Services of Cranston/ Johnston/Northwestern Rhode Island Tri-Town Community Health Center
190 Putnam Avenue
Johnston, RI 02919

Center for Behavioral Health
985 Plainfield Street
Johnston, RI 02919

Mental Health Services of Cranston/Johnston and Northwestern Rhode Island, Inc./Project Leap
1516 Atwood Avenue
Johnston, RI 02919

MIDDLETOWN

Child and Family Services of Newport
19 Valley Road
Middletown, RI 02840

Counseling for Independent Living, Inc.
26 Valley Road
Room 104
Middletown, RI 02840

NARRAGANSETT

Galilee Assistance Program
355 Great Island Road
Narragansett, RI 02882

Galilee Mission to Fishermen, Inc.
268 Kingstown Road
Narragansett, RI 02882

NEWPORT

Comm. Organization for Drug Abuse Control Codac III
93 Thames Street
Newport, RI 02840

Edgehill Newport
200 Harrison Avenue
Newport, RI 02840

The Good Hope Center
58 East Main Road
Newport, RI 02840

U.S. Naval Hospital Alcohol Rehabilitation Dept./Code 401
Newport, RI 02841

NORTH KINGSTOWN

The Center North Kingstown Office
1130 Ten Rod Road
North Kingstown, RI 02852

NORTH PROVIDENCE

Junction Human Service Corporation Junction North Providence
1910 Smith Street
North Providence, RI 02911

PASCOAG

Marathon, Inc. The Lodge at Wallum Lake
Route 100
Pascoag, RI 02859

Talbot Treatment Centers, Inc. Transitional Long-Term Care Program
2198 Wallum Lake Road
Pascoag, RI 02859

PAWTUCKET

Community Counseling Center, Inc. Substance Abuse Treatment Program
160 Beechwood Avenue
Pawtucket, RI 02860

Friends of Caritas House, Inc.
166 Pawtucket Avenue
Pawtucket, RI 02860

Pawtucket Addictions Counseling Services
104 Broad Street
Pawtucket, RI 02860

Rhode Island Council on Alcoholism and Other Drug Dependence
500 Prospect Street
Pawtucket, RI 02860

Rhode Island Youth Guidance Center Substance Abuse Prevention
82 Pond Street
Pawtucket, RI 02860

Robert J. Wilson House, Inc.
Outpatient Counseling Center Residential
80 Summit Street
Pawtucket, RI 02860

The Family Center
25 North Union Street
Pawtucket, RI 02860

PEACEDALE

South Shore Mental Health Center Community Support Program/Harbor House
27 North Road
Peacedale, RI 02883

PORTSMOUTH

The Center Middletown Office
West Main Road
Kings Grant Office Park
Portsmouth, RI 02871

PROVIDENCE

Alcohol and Drug Rehab Services, Inc.
Minority Alcohol Prog Outpatient Counseling
369 Broad Street
Providence, RI 02905

Minority Alcoholism Program/ Burnett
66 Burnett Street
Providence, RI 02907

Butler Hospital Alcohol and Drug Treatment Service
345 Blackstone Boulevard
Providence, RI 02906

Capitol Hill Interaction Council (CHIC) Alcohol Program
272 Smith Street
Providence, RI 02908

Comm. Organization for Drug Abuse Control Codac II
349 Huntington Avenue
Providence, RI 02909

Family Service Incorporated Substance Abuse Program
55 Hope Street
Providence, RI 02906

Genesis Center
620 Potters Avenue
Providence, RI 02907

Harvard Cocaine Recovery Project Rhode Island Division
265 Atwells Avenue
Suite 8
Providence, RI 02903

Hispanic Pro./Ed. Committee of Rhode Island Peer Counseling Partnership Program
741 Westminister Street
Room 203
Providence, RI 02903

Junction Human Service Corporation
16 Borinquen Street
Providence, RI 02905

Marathon House, Inc.
Outpatient
Inside the Walls
Outside the Walls
131 Wayland Avenue
Providence, RI 02906

Providence Academy Alternative Education Program/Daycare
662 Hartford Avenue
Providence, RI 02909

Providence Center for Counseling and Psychiatric Services
790 North Main Street
Providence, RI 02904

Providence Housing Authority Special Services
100 Broad Street
Providence, RI 02903

Rhode Island Hospital Emergency Department
593 Eddy Street
Providence, RI 02903

Rhode Island Department of Elementary and Secondary Education Drug Free Schools and Communities
22 Hayes Street
Providence, RI 02908

Roger Williams General Hospital Substance Abuse Treatment Center
825 Chalkstone Avenue
Providence, RI 02908

Saint Joseph Hospital Commitment to Change
21 Peace Street
Providence, RI 02907

Smith Hill Center
110 Ruggles Street
Providence, RI 02908

Socioeconomic Devel. Center for Southeast Asians, Inc. Substance Abuse Prevention Initiative for Cambodian Population
620 Potters Avenue
Providence, RI 02907

South Providence Addiction Center
1058 Broad Street
Providence, RI 02905

Talbot Treatment Centers, Inc.
Day Treatment Program for Women/Their Children
756 Eddy Street
Providence, RI 02903

Detoxification
Residential
265 Oxford Street
Providence, RI 02905

Outpatient Program
75 Oxford Street
Providence, RI 02905

The Center A Treatment Management Facility
1055 North Main Street
Providence, RI 02904

Theatre for Emily, Inc.
190 Mathewson Street
Providence, RI 02903

Urban League of Rhode Island Youth Substance Abuse Prevention Initiative
246 Prairie Avenue
Providence, RI 02905

Veterans' Affairs Medical Center Substance Abuse Treatment Program
830 Chalkstone Avenue
Unit 116A
Providence, RI 02908

RIVERSIDE

East Bay Human Resource Corporation
656 Bullocks Point Avenue
Riverside, RI 02915

WAKEFIELD

South Shore Mental Health Center Substance Abuse Services
Woodward Avenue
South Kingstown Office Park Bldg C4
Wakefield, RI 02879

Sympatico
57 Columbia Street
Wakefield, RI 02879

Women's Resource Center of South County Rejuvenations
61 Main Street
Wakefield, RI 02879

WARWICK

Channel One Warwick, Inc.
111 Greenwich Avenue
Warwick, RI 02886

Counseling and Intervention Services, Inc.
3649 Post Road
Warwick, RI 02886

Edgehill Warwick Outpatient Clinic
535 Centerville Road
The Landmark Center Suite 202
Warwick, RI 02886

Kent House, Inc.
2030 Elmwood Avenue
Warwick, RI 02888

Rhode Island Employee Assistance Program
120 Centerville Road
Center Point Office Park
Warwick, RI 02886

RIEAP, Inc. Student Assistance Services
120 Centerville Road
Center Point Office Park
Warwick, RI 02886

Warwick Community Action Program, Inc. Alcohol/Drug and Family Counseling
159 Winter Avenue
Warwick, RI 02889

WEST GREENWICH

Good Hope Center
Administrative
Day Treatment/Outpatient
John Potter Road
West Greenwich, RI 02816

WEST WARWICK

Directions
1071 Main Street
West Warwick, RI 02893

Warwick Community Action Program Alcohol/Drug and Family Counseling Services
328 Cowessett Avenue
Suite 2
West Warwick, RI 02893

WESTERLY

Good Hope Center Outpatient
28 High Street
Westerly, RI 02891

WOONSOCKET

Family Resources, Inc.
460 South Main Street
Woonsocket, RI 02895

Good Hope Center Outpatient
63 Eddy Dowling Highway
Park Square Medical Center Bldg., Room 9
Woonsocket, RI 02895

Northern Rhode Island Community Mental Health Center, Inc. Substance Abuse Services/Dually Diagnosed
181 Cumberland Street
Woonsocket, RI 02895

Road Counseling Program, Inc.
8 Court Street
Woonsocket, RI 02895

The Center Woonsocket Office
1 Cumberland Plaza
3rd Floor
Woonsocket, RI 02895

Tri-Hab Counseling
282 South Main Street
Woonsocket, RI 02895

Tri-Hab House, Inc.
79 Asylum Street
Woonsocket, RI 02895

King House
80 Hamlet Avenue
Woonsocket, RI 02895

SOUTH CAROLINA

ABBEVILLE

Abbeville County Commission on Alcohol and Drug Abuse
111 South Main Street
Abbeville, SC 29620

AIKEN

Aiden County Commission on Alcohol and Drug Abuse
214 Newberry Street SW
Aiken, SC 29801

Aurora Pavilion
655 Medical Park Drive
Aiken, SC 29801

ANDERSON

Anderson/Oconee Counties Alcohol and Drug Abuse Commission
212 South Main Street
Anderson, SC 29624

Anderson/Oconee/Pickens Mental Health Center/Alcohol and Drug Service
200 McGee Road
Anderson, SC 29625

Patrick B. Harris Hospital Substance Abuse Services
Anderson, SC 29621

BARNWELL

Barnwell County Commission on Alcohol and Drug Abuse
Barnwell, SC 29812

BEAUFORT

Beaufort County Alcohol and Drug Abuse Department
1905 Duke Street
Beaufort Cnty Human Services Bldg., 2nd fl
Beaufort, SC 29902

BENNETTSVILLE

Marlboro County Commission on Alcohol and Drug Abuse
100 West Main Street
Bennettsville, SC 29512

BISHOPVILLE

Lee County Commission on Alcohol and Drug Abuse
Lee County Courthouse
Room 300
Bishopville, SC 29010

CAMDEN

Kershaw County Commission on Alcohol and Drug Abuse
416 Rutledge Street
Camden, SC 29020

CHARLESTON

Charleston Air Force Base Social Actions
101 North Graves Avenue
437 MSSQ/MSL Suite B
Charleston, SC 29404

Charleston County Substance Abuse Commission
25 Courtenay Drive
Charleston, SC 29401

Medical University of South Carolina Dept. Psych. Outpatient Alcohol Substance Abuse Program
171 Ashley Avenue
Charleston, SC 29425

Veterans' Affairs Medical Center Substance Abuse Treatment Center
109 Bee Street
Charleston, SC 29401

CHERAW

Tri-County Mental Health Center Chesterfield County Alcohol and Drug Services
Cheraw, SC 29520

CHESTER

Chester County Commission on Alcohol and Drug Abuse
130 Hudson Street
Chester, SC 29706

COLUMBIA

Baptist Medical Center Columbia Alcohol and Drug Treatment Services
1410 Blanding Street
Suite 204
Columbia, SC 29202

Chaps Recovery Programs
7020 Two Notch Road
Columbia, SC 29223

Crafts Farrow State Hospital
7901 Farrow Road
Columbia, SC 29203

Earle E. Morris, Jr. Alcohol and Drug Addiction Treatment Center
610 Faison Drive
Columbia, SC 29203

Lexington/Richland Alcohol and Drug Abuse Council
1325 Harden Street
Columbia, SC 29204

CONWAY

Coastal Carolina Hospital
152 Waccamaw Medical Park Drive
Conway, SC 29526

Horry County Commission on Alcohol and Drug Abuse
1004 Bell Street
Conway, SC 29526

Waccamaw Center for Mental Health
1804 North Main Street
Conway, SC 29526

DILLON

Dillon County Commission on Alcohol and Drug Abuse
104 East Harrison Street
Dillon, SC 29536

FLORENCE

Florence County Commission on Alcohol and Drug Abuse
Bruce Hall
601 Gregg Avenue
Florence, SC 29501

Palmetto Center
Florence, SC 29502

FORT JACKSON

Alcohol and Drug Abuse Prevention Program
Hill and Sumter Streets
Building 4250 Bay A
Fort Jackson, SC 29207

GAFFNEY

Cherokee County Commission on Alcohol and Drug Abuse
201 West Montgomery Street
Gaffney, SC 29341

GEORGETOWN

Georgetown County Alcohol and Drug Abuse Commission
1423 Winyah Street
Georgetown, SC 29440

GREENVILLE

Greenville County Commission on Alcohol and Drug Abuse
3336 Buncombe Road
Greenville, SC 29609

Holsmesview Center
Old Easley Bridge Road
Greenville, SC 29610

Resource I, Inc.
803 East North Street
Greenville, SC 29601

GREENWOOD

Faith Home Christian Alcohol and Drug Rehab
Buck Level Road
Greenwood, SC 29646

Greenwood/Edgefield/McCormick Commission on Alcohol and Drug Abuse
1420 Spring Street
Greenwood, SC 29646

Self Memorial Recovery Center at Self Memorial
115 Academy Avenue
Greenwood, SC 29646

GREER

Charter Hospital of Greenville Adult Addictive Disease Program
2700 East Phillips Road
Greer, SC 29650

HAMPTON

Low Country Commission on Alcohol and Drug Abuse
Courthouse Annex
Room 202
Hampton, SC 29924

HARTSVILLE

Rubicon, Inc.
510 East Carolina Avenue
Hartsville, SC 29550

JOHNS ISLAND

Fenwick Hall Hospital
1709 River Road
Johns Island, SC 29457

KINGSTREE

Williamsburg County Department on Alcohol and Drug Abuse
115 Short Street
Kingstree, SC 29556

LANCASTER

Elliott White Springs Memorial Hospital Lancaster Recovery Center
800 West Meeting Street
Lancaster, SC 29720

Lancaster County Commission on Alcohol and Drug Abuse
114 South Main Street
Lancaster, SC 29720

LAURENS

Laurens Commission on Alcohol and Drug Abuse
Industrial Park Road
Laurens, SC 29360

MANNING

Clarendon County Commission on Alcohol and Drug Abuse
14 North Church Street
Manning, SC 29102

MARION

Marion County Alcohol and Drug Abuse Program
103 Court Street
Marion, SC 29571

MONCKS CORNER

Berkeley County Commission on Alcohol and Drug Abuse
109 West Main Street
Moncks Corner, SC 29461

NEWBERRY

Newberry Commission on Alcohol and Drug Abuse
909 College Street
Newberry, SC 29108

NORTH CHARLESTON

Baker Hospital Chaps Baker Treatment Center
2750 Speissegger Drive
North Charleston, SC 29405

Morningstar Recovery
3125 Ashley Phosphate Road
Suite 126
North Charleston, SC 29418

ORANGEBURG

Tri-County Commission on Alcohol and Drug Abuse
897 Russell Street
Orangeburg, SC 29115

PICKENS

Pickens County Commission on Alcohol and Drug Abuse
309 East Main Street
Pickens, SC 29671

ROCK HILL

York County Council on Alcohol and Drug Abuse/Keystone
199 South Herlong Avenue
Rock Hill, SC 29730

RUBY

Good Samaritan Colony Substance Abuse Center
Highway 9
Ruby, SC 29741

SALUDA

Saluda County Commission on Alcohol and Drug Abuse
204 Ramage Street
Saluda, SC 29138

SPARTANBURG

Flynn Christian Fellowship Home
332 South Church Street
Spartanburg, SC 29306

Spartanburg Area Mental Health Center Substance Abuse Services
149 East Wood Street
Spartanburg, SC 29303

Spartanburg County Commission on Alcohol and Drug Abuse
131 North Spring Street
Spartanburg, SC 29306

SUMMERVILLE

Dorchester County Commission on Alcohol and Drug Abuse
535 North Cedar Street
Summerville, SC 29483

SUMTER

Sumter County Commission on Alcohol and Drug Abuse
115 North Harvin Street
Sumter, SC 29150

TRAVELERS REST

North Greenville Hospital Addlife Addiction Services
807 North Main Street
Travelers Rest, SC 29690

UNION

Union County Commission on Alcohol and Drug Abuse
201 South Herndon Street
Union, SC 29379

WALTERBORO

Colleton County Commission on Alcohol and Drug Abuse
561 Recold Road
Walterboro, SC 29488

WEST COLUMBIA

Charter Rivers Hospital Alcohol and Drug Abuse Services
2900 Sunset Boulevard
West Columbia, SC 29169

WILLIAMSTON

Anderson Memorial Hospital Wellspring Chemical Dependency Treatment
313 William Street
Williamston, SC 29697

WINNSBORO

Chaps Fairfield
321 By Pass
Winnsboro, SC 29180

Fairfield County Substance Abuse Commission
200 Calhoun Street
Winnsboro, SC 29180

SOUTH DAKOTA

ABERDEEN

Health Path Counseling Services
12 2 Avenue SW
Aberdeen, SD 57401

Northern Alcohol/Drug Referral and Information Center (NADRIC)
221 South First Street
Aberdeen, SD 57402

Saint Luke's Midland Regional Medical Center Worthmore Treatment Center
1400 15 Avenue NW
Aberdeen, SD 57401

BROOKINGS

East Central Mental Health Chemical Dependency Center
211 4 Street
Brookings, SD 57006

CANTON

Keystone Treatment Center
1010 East 2 Street
Canton, SD 57013

CHAMBERLAIN

Missouri River Adolescent Development Center Alcohol/ Drug Program
211 West 16 Avenue
Chamberlain, SD 57325

CUSTER

Youth Forestry Camp
HC 83
Custer, SD 57730

DEADWOOD

Northern Hills General Hospital New Horizons Addictions Services/Behavioral Health
61 Saint Charles Street
Deadwood, SD 57732

EAGLE BUTTE

Cheyenne River Sioux Tribe (CRST) Alcohol/Drug Prevention Program
Eagle Butte, SD 57625

FORT MEADE

Veterans' Affairs Medical Center Substance Abuse Unit (SAU)
Building 148 Ward E
Fort Meade, SD 57741

FORT THOMPSON

Swift Horse Lodge Crow Creek Sioux Tribe Alcoholism Program
Fort Thompson, SD 57339

HOT SPRINGS

Southern Hills Alcohol/Drug Referral Center
330 1/2 South Chicago Street
Hot Springs, SD 57747

Veterans' Affairs Medical Center Substance Abuse Program
500 North 5 Street
Hot Springs, SD 57747

HURON

Community Counseling Services Alcohol and Drug Unit
1552 Dakota Street South
Huron, SD 57350

Our Home, Inc.
Rediscovery
510 Nebraska Avenue SW
Huron, SD 57350

Rediscovery Inhalant Abuse Program
East Centennial Road
Huron, SD 57350

KYLE

Oglala Sioux Tribe/Project Phoenix Adolescent Alcohol Treatment Center
North Flesh Road
7 Miles SE of Kyle South Dakota
Kyle, SD 57752

LAKE ANDES

Yankton Sioux Alcohol Program Halfway House
East Highway 18
Lake Andes, SD 57356

LEMMON

Professional Consultation Services, Inc.
402 West First Avenue
Lemmon, SD 57638

MADISON

Community Counseling Services
914 NE 3 Street
Madison, SD 57042

MISSION

White Buffalo Calf Woman Society
North Main Street
Mission, SD 57555

MITCHELL

Community Alcohol/Drug Center, Inc.
124 West First Street
Mitchell, SD 57301

Saint Luke's Midland Regional Medical Center Worthmore Treatment Center
728 North Kimball Street
Mitchell, SD 57301

PARKSTON

Our Home, Inc. Group Care Center
501 West Main Street
Parkston, SD 57366

PHILIP

Addiction Recovery Center Outreach Office
Philip, SD 57567

PIERRE

Capital Area Counseling Service, Inc. Drug and Alcohol Unit
804 North Euclid Street
Pierre, SD 57501

PLANKINTON

State Training School Alcohol and Drug Program
Plankinton, SD 57368

RAPID CITY

Big Brothers and Big Sisters of the Black Hills
2100 South 7 Street
Suite 260
Rapid City, SD 57701

City/County Receiving and Referral Center
725 North Lacrosse Street
Rapid City, SD 57701

Focus, Inc.
114 Kinney Avenue
Rapid City, SD 57709

Friendship House, Inc.
211 West Boulevard North
Rapid City, SD 57709

Rapid City Regional Hospital, Inc. Addiction Recovery Center (ARC)
915 Mount View Road
Rapid City, SD 57702

Saint James Street House Adolescent Transitional Care Facility
1205 East Saint James Street
Rapid City, SD 57701

Western Prevention Resource Center
924 North Maple Street
Rapid City, SD 57701

Youth and Family Counseling Services
924 North Maple Street
Rapid City, SD 57709

Youth and Family Services Group
910 Wood Avenue
Rapid City, SD 57709

REDFIELD

Lamont Youth Development Center Alcohol and Drug Unit
North West 3 Avenue
Redfield, SD 57469

SIOUX FALLS

Carroll Institute
2nd Street Manor
826 West 2 Street
Sioux Falls, SD 57104

Outpatient Alcohol and Drug Center
Project Awareness
231 South Phillips Avenue
Suite 450
Sioux Falls, SD 57102

Arch Halfway House
Sioux Falls Detoxification Center
333 South Spring Avenue
Sioux Falls, SD 57104

Charter Hospital of Sioux Falls Substance Abuse Services
2812 South Louise Avenue
Sioux Falls, SD 57106

First Step Counseling Services
5201 West 41 Street
Sioux Falls, SD 57106

Glory House of Sioux Falls
4000 South West Avenue
Sioux Falls, SD 57105

Home Based Services
801 North Sycamore Drive
Sioux Falls, SD 57103

Keystone Outreach Program
1908 West 42 Street
Sioux Falls, SD 57105

South Dakota State Penitentiary Alcohol and Drug Program
1600 North Drive
Sioux Falls, SD 57117

South Dakota Urban Indian Health, Inc. Youth Alcohol/ Drug Program
100 West 6 Street
Sioux Falls, SD 57102

Southeast Alcohol and Drug Abuse Prevention Resource Center
1401 West 51 Street
Sioux Falls, SD 57105

Threshold Youth Services
1401 Valley Drive
Sioux Falls, SD 57105

Veterans' Affairs Medical Center Psychiatry Service/Substance Abuse Program
2501 West 22 Street
Sioux Falls, SD 57717

Youth Enrichment Services, Inc.
824 East 14 Street
Sioux Falls, SD 57104

SISSETON

Tetakwitha Adolescent Treatment Center
RR2
Sisseton, SD 57262

SPRINGFIELD

Springfield Correctional Facility Chemical Dependency Program
Springfield, SD 57062

STURGIS

Northern Hills Alcohol and Drug Service
950 Main Street
Sturgis, SD 57785

VALE

New Dawn Center
Rural Route 1
Vale, SD 57788

VERMILLION

University of South Dakota Student Counseling Center
414 East Clark Street
Vermillion, SD 57069

WATERTOWN

Human Service Agency
Alcohol/Drug Referral/Treatment Center
Northeastern Drug/Alcohol Abuse Prevention Residential Center
123 19 Street NE
Watertown, SD 57201

Keystone Outreach Program
525 5 Street SE
Suite 6
Watertown, SD 57201

WINNER

Southern Plains Mental Health Center Decision 1
500 East 9 Street
Winner, SD 57580

Winner Alcohol/Drug Counseling Service
223 South Main Street
Winner, SD 57580

YANKTON

Federal Prison Camp Substance Abuse Services
1201 Douglas Street
Yankton, SD 57078

Lewis and Clark Mental Health Center Alcohol/Drug Unit
1028 Walnut Street
Yankton, SD 57078

South Dakota Human Services Center
Gateway Chemical Dependency Treatment Center
Steps Beyond Chemical Dependency Treatment Program
North Highway 81
Yankton, SD 57078

TENNESSEE

BOLIVAR

Quinco Mental Health Center Alcohol and Drug Services
Route 1
Bolivar, TN 38008

BRISTOL

Bristol Regional Counseling Center, Inc. Chemical Dependency Program
26 Midway Street
Bristol, TN 37620

BURNS

New Life Lodge
999 Girl Scout Road
Burns, TN 37029

CASTALIAN SPRINGS

Pathfinders, Inc.
875 Highway 231 South
Castalian Springs, TN 37031

CENTERVILLE

Hickman County Board of Education Substance Abuse Services
115 Murphree Avenue
Centerville, TN 37033

CHATTANOOGA

Council for Alcohol and Drug Abuse Services
207 Spears Avenue
Chattanooga, TN 37405

Fortwood Center, Inc.
1028 East 3 Street
Chattanooga, TN 37403

Joseph W. Johnson Mental Health Center Alcohol and Drug Program
Moccasin Bend Road
Chattanooga, TN 37405

CLARKSVILLE

Harriett Cohn Mental Health Center Alcohol and Drug Services
511 8 Street
Clarksville, TN 37040

CLEVELAND

Hiwassee Mental Health Center Turning Point Alcohol and Drug Division
2700 Executive Park Place
Cleveland, TN 37364

CLINTON

Anderson County Health Council Prevention/Education
141 East Broad Street
Clinton, TN 37716

COLUMBIA

Columbia Area Comp. Mental Health Center Alcohol and Drug Abuse Program
1219 Trotwood Avenue
Columbia, TN 38401

COOKEVILLE

Plateau Mental Health Center Drug Abuse Program
Burgess Falls Road
Cookeville, TN 38501

COVINGTON

Tri-County Mental Health Center Alcohol and Drug Abuse Services
1997 Highway 51 South
Covington, TN 38019

ELIZABETHTON

Crossroads Alcohol Drug Abuse Services Prevention/Education
413 Elk Avenue
Elizabethton, TN 37643

GREENEVILLE

Nolachuckey/Holston Mental Health Center Alcohol and Drug Abuse Outpatient Program
Holston Drive
Greeneville, TN 37744

HOHENWALD

Buffalo Valley, Inc.
221 South Maple Street
Hohenwald, TN 38462

JACKSON

Aspell Manor
331 North Highland Avenue
Jackson, TN 38301

Jackson Area Council on Alcoholism and Drug Dependency
900 East Chester Street
Jackson, TN 38301

West. Tennessee Behavioral Center Alcohol and Drug Program
238 Summar Drive
Jackson, TN 38301

JOHNSON CITY

Comprehensive Community Services
323 West Walnut Street
Johnson City, TN 37604

Recovery North Side Hospital Chemical Dependency Unit
401 Princeton Road
Johnson City, TN 37601

Watauga Area Mental Health Center Alcohol and Drug Abuse Program
109 West Watauga Avenue
Johnson City, TN 37601

KINGSPORT

Holston Mental Health Services Substance Abuse Services
1570 Waverly Road
Kingsport, TN 37664

MCC Behavioral Care, Inc.
108 East Main Street
Suite 204
Kingsport, TN 37660

KNOXVILLE

Agape, Inc. Halfway House
205–211–215 East Scott Avenue
Knoxville, TN 37917

Child and Family Services
114 Dameron Avenue
Knoxville, TN 37917

Detoxification Rehab Institute (DRI)
6400 Papermill Road
Suite 100
Knoxville, TN 37919

DRD Knoxville Medical Clinic
1501 Cline Street
Knoxville, TN 37921

E. M. Jellinek Center
130 Hinton Street
Knoxville, TN 37917

Helen Ross McNabb Center, Inc. Alcohol and Drug Program
1520 Cherokee Trail
Knoxville, TN 37920

Knoxville Knox County Community Action Com. Counseling and Recovery Services
2247 Western Avenue
Knoxville, TN 37921

Midway Rehabilitation Center
1715 Magnolia Avenue
Knoxville, TN 37927

Overlook Center, Inc. Alcohol and Drug Abuse Services
3001 Lake Brook Boulevard
Knoxville, TN 37909

Saint Mary's Medical Center Alcoholism Treatment Unit
900 East Oak Hill Avenue
Knoxville, TN 37917

University of Tennessee Medical Center Alcohol and Drug Recovery Center
1924 Alcoa Highway
Knoxville, TN 37920

LEBANON

Cumberland Mental Health Services Drug and Alcohol Program
1404 Winter Drive
Lebanon, TN 37087

LOUISVILLE

Peninsula Hospital Chemical Dependency Program
Louisville, TN 37777

MADISON

Dede Wallace Center Alcohol and Drug Program
620 Gallatin Road South
Madison, TN 37115

Tennessee Christian Medical Center Center for Addictions
500 Hospital Drive
Madison, TN 37115

MARTIN

Northwest Counseling Center Substance Abuse Treatment Division
457 Hannings Lane
Martin, TN 38237

MARYVILLE

Blount Memorial Hospital Mountain View Recovery Center
907 East Lamar Alexander Parkway
Maryville, TN 37801

MEMPHIS

Federal Correctional Institution Drug Abuse Program
1101 John A. Denie Road
Memphis, TN 38134

Frayser/Millington Mental Health Center Alcohol and Drug Abuse Services
2150 Whitney Avenue
Memphis, TN 38127

Goodwill Homes Community Services, Inc. Early Intervention
4590 Goodwill Road
Memphis, TN 38186

Grace House, Inc. Women's Program
329 North Bellevue Street
Memphis, TN 38105

Harbor House, Inc. Alcoholic Rehabilitation Center
1979 Alcy Road
Memphis, TN 38114

MCC Managed Behavioral Care
6401 Poplar Avenue
Suite 290
Memphis, TN 38119

Memphis Alcohol and Drug Council Prevention/Education
1450 Poplar Street
Memphis, TN 38104

Memphis City Schools Mental Health Center Substance Abuse Services
2597 Avery Avenue
Room 102
Memphis, TN 38112

Memphis Recovery Centers, Inc.
219 North Montgomery Street
Memphis, TN 38104

Methodist Outreach, Inc.
2009 Lamar Avenue
Memphis, TN 38114

Mid-Town Mental Health Center Alcohol and Drug Abuse Services
427 Linden Avenue
Memphis, TN 38126

New Directions, Inc.
642 Semmes Street
Memphis, TN 38111

Northeast Community Mental Health Center Alcohol and Drug Abuse Program
5515 Shelby Oaks Drive
Memphis, TN 38134

Raleigh Professional Associates
2960-B Austin Peay Highway
Memphis, TN 38128

Serenity Houses of Memphis, Inc.
1094 Poplar Avenue
Memphis, TN 38105

Southeast Mental Health Center, Inc. Alcohol/Drug Abuse Program
3810 Winchester Road
Memphis, TN 38118

Veterans' Affairs Medical Center Psychiatry Services/Alcohol/ Drug Dependency Treatment Program
1030 Jefferson Avenue
Memphis, TN 38104

Whitehaven/Southwest Mental Health Center Alcohol and Drug Abuse Program
3127 Stonebrook Circle
Memphis, TN 38116

MILLINGTON

Naval Air Station Memphis Counseling and Assistance Center
Naval Air Station Memphis
Building S-52
Millington, TN 38054

MORRISTOWN

Cherokee Mental Health Center Substance Abuse Treatment Program
815 West 5 North Street
Morristown, TN 37814

MOUNTAIN HOME

Veterans' Affairs Medical Center Alcohol/Drug Dependence Treatment Program
Johnson City
Mountain Home, TN 37684

MURFREESBORO

Middle Tennessee Medical Center, Inc. Chemical Dependency Services
400 North Highland Avenue
Murfreesboro, TN 37130

Rutherford County Guidance Center and Williamson County Counseling Center
118 North Church Street
Murfreesboro, TN 37130

Veterans' Affairs Alvin C. York Medical Center Substance Abuse Rehabilitation Program
3400 Lebanon Road
Building 7A
Murfreesboro, TN 37129

NASHVILLE

Alcohol and Drug Council of Middle Tennessee, Inc./ Prevention/Education
2612 Westwood Drive
Nashville, TN 37204

Baptist Hospital Drug and Alcohol Recovery Center
2000 Church Street
Nashville, TN 37236

Bethlehem Center Prevention/ Education
1417 Charlotte Avenue
Nashville, TN 37203

Cumberland Heights Alcohol and Drug Treatment
River Road
Route 2
Nashville, TN 37209

Davidson County Sheriff's Office New Avenues/Straight Time
506 2 Avenue North
Nashville, TN 37201

Downtown Service Center Chemical Dependency Program
526 8 Avenue South
Nashville, TN 37203

Edgehill Center
935 Edgehill Avenue
Nashville, TN 37203

Human Growth Corporation
3931-B Gallatin Road
Nashville, TN 37216

Life Challenge of Nashville Women's Residence
1017 Burchwood Avenue
Nashville, TN 37216

Lloyd C. Elam Mental Health Center Meharry Alcohol and Drug Abuse Program
1005 Dr. David B. Todd, Jr., Boulevard
Nashville, TN 37208

Luton Community Mental Health Center Alcohol and Drug Abuse Program
5240 Harding Place
Nashville, TN 37217

Mayor's Office of Drug Policy
211 Union Street
Suite 601
Nashville, TN 37201

MCC Managed Behavioral Care, Inc.
2416 21 Avenue South
Suite 202A
Nashville, TN 37212

Oasis Center Prevention/ Education
1221 16 Avenue South
Nashville, TN 37212

Parthenon Pavilion CMC Dual Treatment Program
2401 Murphy Avenue
Nashville, TN 37203

Payne's Chapel Prevention/ Education
212 Neill Avenue
Nashville, TN 37206

Samaritan Recovery Community, Inc.
319 South 4 Street
Nashville, TN 37206

Vanderbilt Community Mental Health Center Alcohol and Drug Services
1500 21 Avenue South
VAV2200
Nashville, TN 37212

OAK RIDGE

Hope of East Tennessee, Inc.
176 Northwestern Avenue
Suite C
Oak Ridge, TN 37830

Methodist Medical Center Turning Point Recovery Center
990 Oak Ridge Turnpike
Oak Ridge, TN 37830

Ridgeview Psychiatric Hospital and Center Alcohol/Drug Abuse Program
240 West Tyrone Road
Oak Ridge, TN 37830

PARIS

Carey Counseling Center Alcohol and Drug Abuse Program
408 Virginia Avenue
Paris, TN 38242

SAVANNAH

Care of Savannah, Inc. Jack Gean Shelter for Women
Route 3
Savannah, TN 38372

SHELBYVILLE

Tony Rice Center, Inc.
Shelbyville, TN 37160

TULLAHOMA

Multi-County Comprehensive Mental Health Center Substance Abuse Program
1803 North Jackson Street
Tullahoma, TN 37388

TEXAS

ABILENE

Abilene Regional MH/MR Center Substance Abuse Services
190 Woodlawn Drive
Abilene, TX 79603

Hendrick Medical Center Care Center
1242 North 19 Street
Abilene, TX 79601

Serenity Foundation of Texas
Detox Unit
1502 North 2 Street
Abilene, TX 79601

Residential and Outpatient
1546 North 2 Street
Abilene, TX 79601

Serenity House of Abilene, Inc. Reclaiming Adolescents for Tomorrow
317 North 6 Street
Abilene, TX 79601

ALICE

Alice Counseling Center Adolescent Supportive Outpatient
63 South Wright Street
Alice, TX 78333

ALVIN

Alvin Recovery Center
301 Medic Lane
Alvin Community Hospital
Alvin, TX 77511

AMARILLO

Alcoholic Area Recovery Home
5407 East Amarillo Boulevard
Amarillo, TX 79107

Amarillo Alcoholic Women's Recovery Center The Haven
1308 South Buchanan Street
Amarillo, TX 79102

Panhandle Alcoholic Recovery Center
1505 West 10 Street
Amarillo, TX 79101

Veterans' Affairs Medical Center Substance Abuse Treatment Program
6010 Amarillo Boulevard West
Ward 2-A
Amarillo, TX 79106

ANGLETON

Gulf Coast Center Substance Abuse Recovery Program
2512 North Velasco Street
Angleton, TX 77515

ARCHER CITY

Archer County Outreach Clinic Archer County Courthouse
Archer City, TX 76351

ARGYLE

The Argyle Center, Inc. Specialty Hospital
914 Country Club Road
Argyle, TX 76226

ARLINGTON

Counterpoint Center of CPC
1011 North Cooper Street
CPC Millwood Hospital
Arlington, TX 76011

Family Service, Inc. Substance Abuse Treatment
601 West Sanford Street
Suite 209
Arlington, TX 76011

Metropolitan Clinic of Counseling, Inc. Evening Recovery Program
2001 East Lamar Boulevard
Suite 100
Arlington, TX 76001

Viewpoint Recovery and Learning Centers, Inc. Adult Intensive Outpatient
2212 Arlington Downs Center
Suite 200
Arlington, TX 76011

AUSTIN

Aeschbach and Associates Substance Abuse Services
2011-D East Riverside Drive
Austin, TX 78741

Association for Retarded Citizens
2818 San Gabriel Street
Austin, TX 78705

Austin Drug and Alcohol Abuse Program
13377 Pond Springs Road
Suite 104
Austin, TX 78729

Austin Family House, Inc.
2604 Paramount Avenue
Austin, TX 78704

Austin Rehabilitation Center, Inc.
Adolescent Outpatient Program
3809 South 2 Street
Building B Suite B400
Austin, TX 78704

Adult Outpatient
3921 Steck Street
Suite A105
Austin, TX 78759

Female Adolescent Program
1900 Pearl Street
Austin, TX 78705

Gardner/Betts Juvenile Justice System
2515 South Congress Street
Austin, TX 78704

Lavender Center for Adolescents
3207 Slaughter Lane
Austin, TX 78748

Treatment Rehab Center
1808 West Avenue
Austin, TX 78701

Austin/Travis County MH/MR
Collier Outpatient Clinic
1430 Collier Street
Austin, TX 78704

East 2nd Street Outpatient Clinic
1633 East 2 Street
Austin, TX 78702

Gateway House
1210 Rosewood Avenue
Austin, TX 78702

Oak Springs Detox Treatment Center
4110 Guadalupe Street
Building 785 Ward 8
Austin, TX 78764

Oak Springs Treatment Center
3000 Oak Springs Drive
Austin, TX 78702

Rosewood Outpatient Clinic
2800 Webberville Road
Suite 107
Austin, TX 78702

Methadone Maintenance
1631-A East 2 Street
Austin, TX 78702

Bergstrom Air Force Base Social Actions Substance Abuse Control
Austin, TX 78743

Charter Hospital of Austin Intensive Outpatient Program
1812 Centre Creek Drive
Austin, TX 78754

CPC Capital Hospital Counterpoint
12151 Hunters Chase Drive
Austin, TX 78729

ESH, Inc. The Cottage
403 East 33 Street
Austin, TX 78705

Faulkner Center
1900 Rio Grande Street
Austin, TX 78705

Greater Austin Council on Alcoholism and Drug Abuse, Inc.
1609 Shoal Creek Street
Suite 303
Austin, TX 78701

Human Affairs International, Inc.
505 East Huntland Street
Suite 500
Austin, TX 78752

McCabe Center
1915 East Martin Luther King Boulevard
Austin, TX 78702

Middle Earth Unlimited, Inc.
3816 South First Street
Austin, TX 78704

Oaks Treatment Center Amethyst Program
1407 West Stassney Lane
Austin, TX 78745

Renaissance Outpatient Program for Chemical Dependency
1600 West 38 Street
Suite 340
Austin, TX 78731

Resolve Program
2520 Longview Street
Suite 412
Austin, TX 78705

Saint David's Pavilion Chemical Dependency Partial Program
1000 East 32 Street
Suite 7
Austin, TX 78705

Salvation Army Adult Rehabilitation Center
6510 Berkman Drive
Austin, TX 78723

Shoal Creek Hospital Inpatient Substance Abuse Services
3501 Mills Avenue
Austin, TX 78731

Solutions Counseling and Treatment Center
3809 South 2 Street
Suite C-200
Austin, TX 78704

Teen and Family Counseling Center
3536 Bee Caves Road
Suite 100
Austin, TX 78746

Texas Youth Commission Central Texas Regional Office
44 East Avenue
Suite 200
Austin, TX 78701

The Bridge Recovery Center of Austin, Inc.
2579 Western Trails Boulevard
Suite 220
Austin, TX 78745

BASTROP

Federal Correctional Institution Drug Abuse Program
Highway 95
Bastrop, TX 78602

Greater Austin Council on Alcohol and Drug Abuse/ Bastrop Outpatient
1001-B Tahitian Drive
Bastrop, TX 78602

BAY CITY

Riceland Regional Mental Health Authority Substance Abuse Service System
400 Avenue F
Bay City, TX 77414

BAYTOWN

Baytown Medical Center, Inc. Adult/Adolescent/Outpatient CDC
1700 James Bowie Drive
Baytown, TX 77520

BEAUMONT

Baptist Hospital of Southeast Texas Parkside Recovery Unit
3450 Stagg Drive
Beaumont, TX 77701

Charter Counseling Center of Beaumont
85 IH-10 North
Suite 207
Beaumont, TX 77707

HCA Beaumont Neurological Hospital
Pinebrook Inpatient Program
Pinebrook Outpatient Center
3250 Fannin Street
Beaumont, TX 77701

Jefferson County COADA
Bridge Program
Pride Outpatient Program
390 Elizabeth Street
Beaumont, TX 77701

Drug/Alcohol Abuse Recovery Center (DAARC)
2235 South Street
Beaumont, TX 77701

Jefferson County Juvenile Probation Dept. Project Intercept
215 Franklin Street
Beaumont, TX 77701

Land Manor, Inc.
Adams House/Adolescent Residential
1490 Avenue F
Beaumont, TX 77701

Franklin House Recovery Center
1990 Franklin Street
Beaumont, TX 77701

Melton Center
1785 Washington Boulevard
Beaumont, TX 77705

Life Resource/A CMHC Addiction Treatment and Recovery (ATAR)
2750 South 8 Street
Beaumont, TX 77701

BEDFORD

Harris Methodist Springwood Hospital Addiction Treatment Center
1608 Hospital Parkway
Bedford, TX 76022

Treatment Center at Bedford, Inc. DBA The Residential Treatment Center
2904 Bedford Road
Bedford, TX 76021

BELLAIRE

Care Centers of Texas, Inc. Bellaire Unit
4500 Bissonnet Street
Suite 333
Bellaire, TX 77401

BELLVILLE

Riceland Regional Mental Health Authority Substance Abuse Service System
1412 South Front Street
Bellville, TX 77418

BIG SPRING

Federal Correctional Institution Substance Abuse Services
1900 Simler Avenue
Big Spring, TX 79720

Veterans' Affairs Medical Center Substance Abuse Treatment Program
2400 South Gregg Street
116
Big Spring, TX 79720

BONHAM

Fannin County Mental Health Center Substance Abuse Program
1221 East 6 Street
Bonham, TX 75418

BOWIE

Montague County Outreach Clinic
507 Pelham Road
Bowie, TX 76230

BRACKETTVILLE

United Medical Center Health Counseling Services
201 James Street
Brackettville, TX 78832

BRECKENRIDGE

Stephens/Shackelford Counties Mental Health Clinic
607 South Rose Street
Breckenridge, TX 76024

BRENHAM

Washington County Outreach Center Substance Abuse Treatment Program
307 North Baylor Street
Brenham, TX 77833

BROOKS AIR FORCE BASE

Brooks Air Force Base Substance Abuse Control Center for Social Action
Building 511
Brooks Air Force Base, TX 78235

BROWNSVILLE

Cameron County Housing Authority Jovenes Unidos
833 East Price Road
Brownsville, TX 78520

Rio Grande Valley Midway House Brownsville Outpatient Program
1150 East Madison Street
Brownsville, TX 78520

Tropical Texas Center for MH/MR Brownsville Outpatient Substance Abuse Services
5 Boca Chica
Commercial Plaza West Suite 5
Brownsville, TX 78520

BROWNWOOD

Brownwood Reception Center
Brownwood, TX 76804

Brownwood State School Substance Abuse Services
Bangs Road
Brownwood, TX 76804

BRYAN

Brazos Valley Council on Alcohol Substance Abuse Adolescent Treatment
1713 Broadmoor Drive
Suite 204
Bryan, TX 77805

Federal Prison Camp Substance Abuse Services
1100 Ursuline Street
College Park
Bryan, TX 77803

MH/MR Authority of Brazos Valley
Outpatient Substance Abuse Program
623 Mary Lake Drive
Bryan, TX 77801

Substance Abuse Treatment Program
1835 Sandy Point Road
Brazos County Minimum Security Jail
Bryan, TX 77801

Substance Abuse Treatment Program
300 East 26 Street
Brazos County Jail Suite 105
Bryan, TX 77801

BUDA

Austin Rehabilitation Center, Inc. REAP Intensive Residential
1888 Wright Road
Buda, TX 78610

CALDWELL

Burleson County Outreach Center Substance Abuse Treatment Program
201 Fawn Street
Burleson County Jail
Caldwell, TX 77836

MH/MR Authority of Brazos Valley Substance Abuse Treatment Program
206 South Echols Street
Caldwell, TX 77836

CANTON

Sundown Ranch, Inc.
Route 4
Canton, TX 75103

CANYON

Careunit at Palo Duro Hospital
2 Hospital Drive
Canyon, TX 79015

CARROLLTON

National Medical Enterprises, Inc. DBA Trinity Medical Center Addictions Service
4343 North Josey Lane
Carrollton, TX 75010

North Dallas Drug Rehabilitation Center
1606 South I-35
Suite 101
Carrollton, TX 75006

CARTHAGE

Sabine Valley Center Substance Abuse Services
1701-A South Adams Street
Carthage, TX 75633

CENTERVILLE

MH/MR Authority of Brazos Valley Leon County Outreach Center
Highway 75 South
Centerville, TX 75833

CHAPPELL HILL

Daytop Chappell Hill Residential
Route 1
Chappell Hill, TX 77426

CHILDRESS

Childress Outreach Service Center
100 3 Street NW
Childress, TX 79201

CHILTON

Texas P J J, Inc. Chilton House
4006 Street
Chilton, TX 76632

CLEBURNE

Johnson County MH/MR Center Outpatient Substance Abuse Services
1601 North Anglin Street
Cleburne, TX 76031

COLLEGE STATION

K W Center Outpatient Program
2911 Texas Avenue South
Suite 202B
College Station, TX 77845

Scott and White Regional Clinic Alcohol and Drug Dependency Outpatient Treatment Program
511 University Drive
Suite 202
College Station, TX 77840

COLUMBUS

Riceland Regional Mental Health Authority Substance Abuse Service System
624 Preston Street
Columbus, TX 78934

CONROE

Tri-County MH/MR Services
Crisis Resolution Unit
301 South First Street
Conroe, TX 77301

Life Sync
200 River Pointe Street
Suite 310
Conroe, TX 77304

CORPUS CHRISTI

Charter Hospital of Corpus Christi, Inc. Addictive Disease Program
3126 Rodd Field Road
Corpus Christi, TX 78414

Coastal Bend Alcohol/Drug Rehab Center
Detox and Evaluation
25 North Country Club Place
Corpus Christi, TX 78407

Henderson House
36 North Country Club Place
Corpus Christi, TX 78407

Ivy House for Women
41 North Country Club Place
Corpus Christi, TX 78407

Coastal Bend Council on Alcohol/ Drug Abuse Alpha Project
3154 Reid Drive
Suite 2
Corpus Christi, TX 78404

Corpus Christi Drug Abuse Council
Drug Free
Methadone Clinic
405 John Sartain Street
Corpus Christi, TX 78401

Counseling and Assistance Center
Naval Air Station
11001 D Street Suite 143
Corpus Christi, TX 78419

Nueces County MH/MR Center
Positive Life Skills
Client and Family Support Services
Youth Residential
1546 South Brownlee Street
Corpus Christi, TX 78404

Residential Unit
615 Oliver Court
Corpus Christi, TX 78408

Toxicology Associates, Inc.
3822 Leopard Street
Corpus Christi, TX 78408

CORSICANA

Corsicana State Home
West 2 Avenue
Corsicana, TX 75110

CROCKETT

Crockett State School Substance Abuse Services
Crockett, TX 75835

DALLAS

Alternatives for Recovery
10 Medical Parkway
Suite 303
Dallas, TX 75234

Baylor University Medical Center Baylor Center for Addictive Diseases
3500 Gaston Avenue
Collias Hospital
Dallas, TX 75246

Bomar Epstein and Associates, Inc. The Cliffs
2303 West Ledbetter Street
Suite 400
Dallas, TX 75224

Catholic Charities Saint Joseph Youth Center
901 South Madison Street
Dallas, TX 75208

Charlton Methodist Hospital Alcohol and Drug Treatment Center
3500 Wheatland Road
Dallas, TX 75237

Dallas County Juvenile Department Drug Intervention Unit
4711 Harry Hines Street
Dallas, TX 75235

Dallas House
7929 Military Parkway
Dallas, TX 75227

Darco Drug Services, Inc.
2608 Inwood Road
Dallas, TX 75235

Day Treatment Center of Dallas Drug and Alcohol Program
3450 Wheatland Road
Suite 110
Dallas, TX 75237

East Dallas Counseling Center
4306 Bryan Street
Dallas, TX 75204

Ethel Daniels Foundation, Inc.
Connie Nelson House
1810 North Prairie Street
Dallas, TX 75206

Intensive Outpatient Services
1910 Henderson Street
Dallas, TX 75206

First Step Counseling Center
13612 Midway Road
Suite 108
Dallas, TX 75244

Green Oaks Ambulatory Care Corp. Substance Abuse Services
7808 Clodus Fields Drive
Dallas, TX 75251

Help Is Possible, Inc. (HIP)
723 South Peak Street
Dallas, TX 75223

Holmes Street Foundation, Inc. Holmes Street Adolescent Residential
2719 Holmes Street
Dallas, TX 75209

Homeward Bound, Inc. Trinity Recovery Center
1015 Browder Street
Dallas, TX 75215

Kelly Coutee and Associates Chemical Dependency Program
8350 Meadow Road
Suite 268
Dallas, TX 75231

La Posada Servicios Internacionales Outpatient Program
921 North Peak Street
Dallas, TX 75204

Metropolitan Clinic of Counseling, Inc. Evening Recovery Program
12720 Hillcrest Road
Suite 400
Dallas, TX 75230

Nexus, Inc.
8733 La Prada Drive
Dallas, TX 75228

Oak Lawn Community Services
3434 Fairmont Street
Dallas, TX 75201

Our Brothers Keeper/NDUGU
4200 South Fitzhugh Street
Dallas, TX 75210

Permanente Medical Association of Texas Kaiser Permanente Chemical Dependency Treatment Program
9250 Amberton Parkway
Dallas, TX 75243

Psychological and Addiction Services
9550 Forest Lane
Building 1 Room 118
Dallas, TX 75228

Saint Paul Medical Center Saint Paul Substance Abuse Program
5909 Harry Hines Boulevard
Dallas, TX 75235

Salvation Army Social Service Center Substance Abuse Services Program
5302 Harry Hines Boulevard
Dallas, TX 75235

The New Place
4301 Bryan Street
Suite 120
Dallas, TX 75204

Timberlawn Psychiatric Hospital, Inc. Substance Abuse Services
4600 Samuell Boulevard
Dallas, TX 75227

Turtle Creek Manor, Inc.
2707 Routh Street
Dallas, TX 75201

Outpatient Services
2021 Cedar Springs Street
Dallas, TX 75201

Veterans' Affairs Medical Center Drug/Alcohol Dependence Treatment Programs
4500 South Lancaster Road
Unit 116A5
Dallas, TX 75216

West Dallas Community Centers Outpatient Youth Treatment Program
8200 Brookriver Drive
Suite N-704
Dallas, TX 75247

DECATUR

Wise County Outreach Center
300 North Trinity Street
Decatur, TX 76234

DEL RIO

United Medical Centers Health Counseling Services
1102 Bedell Street
Del Rio, TX 78840

DEL VALLE

Travis County Jail System
Correctional Center/Del Valle
Buildings CCE and CCD and 4
Units B and C
Del Valle, TX 78617

Unit CCB Bay 1 and 2
3614 Bill Price Road
Del Valle, TX 78617

Unit 1H
Del Valle, TX 78617

DENISON

Texoma Medical Center Texoma Parkside Treatment Center
401 East Hull Street
Denison, TX 75020

DENTON

Denton County MH/MR Center Intensive Outpatient Substance Abuse Treatment Program
515 South Locust Street
Denton, TX 76201

Parkside Lodge/Westgate Inpatient/Outpatient Services
4601 North Interstate Highway 35
Denton, TX 76207

DIMMITT

Central Plains MH/MR and Substance Abuse Dimmitt Counseling Center
109 NE 2 Street
Dimmitt, TX 79027

EAGLE PASS

Kickapoo Traditional Tribe of Texas Prevention Intervention Program
1109 Ferry Street
Eagle Pass, TX 78852

United Medical Centers Health Counseling Services
2315 El Indio Road
Eagle Pass, TX 78853

EDINBURG

Tropical Texas Center for MH/MR Methadone/Edinburg Outpatient Substance Abuse Services
1901 South 24 Street
Edinburg, TX 78539

EL PASO

Aliviane NO/AD, Inc.
Inner Resources Women's/Children's Residential Center
11960 Golden Gate Road
El Paso, TX 79936

Inner Resources Recovery Center
10690 Socorro Road
El Paso, TX 79927

Outpatient Clinic
5160B El Paso Drive
El Paso, TX 79905

Alternative House, Inc.
4910 Alameda Avenue
El Paso, TX 79905

Casa Blanca Therapeutic Communities
917 North Ochoa Street
El Paso, TX 79902

Communities in Schools/El Paso
2244 Trawood Street
Suite 206
El Paso, TX 79935

Community Assistance Program Fort Bliss
Building 7124
El Paso, TX 79924

El Paso County Alcohol/Drug Abuse Treatment Services
Arcoiris Program
Administration and Treatment
1014 North Stanton Street
El Paso, TX 79902

Paisano Street Unit
702 East Paisano Street
El Paso, TX 79901

El Paso Methadone Maintenance and Detox Treatment Center
5004 Alameda Avenue
El Paso, TX 79905

El Paso State Center for MH/MR Dual Diagnosis Treatment Program
6700 Delta Drive
Cottage 509
El Paso, TX 79905

Genesis Recovery Center Gateway East Unit
6070 East Gateway Boulevard
Suite 305
El Paso, TX 79905

Life Management Center for MH/MR Services
Adolescent Substance Abuse Outpatient Treatment Program
Helping Kids Cope Prevention Program
1014 North Stanton Street
El Paso, TX 79902

Drug Abuse Services
Substance Abuse Services
5304 El Paso Drive
El Paso, TX 79905

Schaeffer House
8716 Independence Avenue
El Paso, TX 79907

Serenity Outpatient Services, Inc.
4625 Alabama Street
El Paso, TX 79930

Sunrise
1300 Murchison Street
Suite 380
El Paso, TX 79902

Veterans' Affairs Substance Abuse Treatment Program (VASAT) 122
5919 Brook Hollow Drive
El Paso, TX 79925

William Beaumont Army Medical Center Residential Treatment Facility
Fort Bliss
Building 7115
El Paso, TX 79920

ELGIN

Greater Austin Council on Alcohol/Drug Abuse/Bastrop County Restit. Center
122 Fisher Street
Elgin, TX 78621

FLOYDADA

Central Plains MH/MR Substance Abuse Center Floydada Counseling Center
206 Floyd County Courthouse
Floydada, TX 79235

FORT WORTH

Careunit Hospital of Dallas/Fort Worth Careunit Outpatient Services
1066 West Magnolia Street
Fort Worth, TX 76104

Cenikor Foundation, Inc. North Texas Facility
2209 South Main Street
Fort Worth, TX 76110

Cherry Lane Hospital and Treatment Center The Pavilion at Cherry Lane
701 South Cherry Lane
Fort Worth, TX 76108

Family Service, Inc. Substance Abuse Treatment
1424 Hemphill Street
Fort Worth, TX 76104

Federal Correctional Institution Drug Abuse Program
3150 Horton Road
Fort Worth, TX 76119

Goodrich Center for the Deaf Alcohol and Drug Abuse Program
2500 Lipscomb Street
Fort Worth, TX 76110

Holistic Addictions Recovery and Treatment of Texas, Inc. (HART)
4809 Brentwood Stair Road
Suite 403
Fort Worth, TX 76103

New Day of Fort Worth Chemical Dependency Program
508 South Adams Street
Fort Worth, TX 76104

Oak Bend Hospital Counter Point Center
7800 Oakmont Boulevard
Fort Worth, TX 76134

Permanente Medical Association of Texas Kaiser Permanente Chemical Dependency Treatment Program
1001 12 Avenue
Fort Worth, TX 76104

Polytechnic Community Clinic Mom and Baby DF for the Health of It
1518 Vaughn Street
Fort Worth, TX 76104

Saint Joseph Hospital Mental Health Services Chemical Dependency Program
1401 South Main Street
Fort Worth, TX 76104

Salvation Army
Community Corrections Treatment Center
1855 East Lancaster Street
Fort Worth, TX 76103

First Choice
2110 Hemphill Street
Fort Worth, TX 76110

Schick Shadel Hospital of Dallas/Fort Worth, Inc.
4101 Frawley Drive
Fort Worth, TX 76180

Society of Mary Stars, Inc. Star House
2740 Avenue K
Fort Worth, TX 76105

Tarrant Council on Alcoholism and Drug Abuse
1200 Summit Avenue
Suite 320
Fort Worth, TX 76102

Tarrant County Medical Education and Research Foundation
Outpatient
904 Southland Avenue
Fort Worth, TX 76104

Addiction Services
3840 Hulen Street
Fort Worth, TX 76107

Alcohol and Drug Abuse Program
1501 El Paso Street
Fort Worth, TX 76102

Dual Diagnosis Unit
902 Pennsylvania Street
Fort Worth, TX 76105

Spanish Alcohol Outpatient
2400 NW 24 Street
Fort Worth, TX 76106

Substance Abuse Outpatient Unit 2
1500 South Main Street
Fort Worth, TX 76104

Volunteers of America Northern Texas, Inc.
Gemini House
2841 Avenue H
Fort Worth, TX 76105

Rosies House
2837 Avenue H
Fort Worth, TX 76105

Western Clinical Health Services Pennsylvania Avenue Clinic
514 Pennsylvania Avenue
Fort Worth, TX 76104

Willoughby House
8100 West Elizabeth Lane
Fort Worth, TX 76116

FREEPORT

Brazoria County Alcohol Recovery Center Brazos Place
11034 North Avenue H
Freeport, TX 77541

FRIONA

Central Plains Center for MH/MR and Substance Abuse Parmer County Counseling Center
715 Main Street
Friona, TX 79035

GAINESVILLE

Gainesville State School Substance Abuse Services
4701 East Farm Road
Gainesville, TX 76240

GALVESTON

Bay Area Council on Drugs and Alcohol, Inc. Know More R and R/22 Street
415 22 Street
Suite 400
Galveston, TX 77550

Gulf Coast Center Administrative Unit
123 Rosenberg Street
Suite 6
Galveston, TX 77553

Island Counseling Center Forgive
1923 Sealy Avenue
Galveston, TX 77550

GARLAND

D. Gonzalez and Associate
2848 West Kingsley Road
Suite B
Garland, TX 75041

Horizon Recovery Center at Garland Community Hospital
2696 West Walnut Street
Garland, TX 75042

GEORGETOWN

Georgetown Hospital Authority
Fred M. Carter Treatment Unit
2000 Scenic Drive
Georgetown, TX 78626

San Gabriel Treatment Center
605 East University Avenue
Georgetown, TX 78626

Williamson County Council on Alcohol and Drug Abuse
707 1/2 Main Street
Georgetown, TX 78627

GIDDINGS

Giddings State School Substance Abuse Services
Giddings, TX 78942

GILMER

Sabine Valley Center Gilmer Outpatient Clinic
103 Madison Street
Gilmer, TX 75644

GLEN ROSE

Pecan Valley MH/MR Region Alcohol/Drug Rehabilitation Program
209B Barnard Street
Glen Rose, TX 76043

GRAHAM

Young County Mental Health Center Graham Outreach Center
806 Cherry Street
Graham, TX 76046

GRANBURY

Pecan Valley MH/MR Region Alcohol and Drug Abuse Services
104 Charles Street
Granbury, TX 76048

GRAND PRAIRIE

American Indian Center of Dallas, Inc. Rehabilitation Program
818 East Davis Street
Grand Prairie, TX 75050

GREENVILLE

Crossroads Council on Alcohol and Drug Abuse/Outpatient
2612 Jordan Street
Greenville, TX 75401

Glen Oaks Hospital New Frontier Unit
301 East Division Street
Greenville, TX 75401

Green Villa
1014 Walnut Street
Greenville, TX 75401

A Transitional Living Center
Route 2
Greenville, TX 75401

Hunt County Family Services Center, Inc. Chemical Dependency Service
3900 Joe Ramsey Boulevard
Building G
Greenville, TX 75401

HARLINGEN

Rio Grande State Center Chemical Dependency Unit
1401 Rangerville Road
Harlingen, TX 78550

Rio Grande Valley
Midway House for Men
1605 North 7 Street
Harlingen, TX 78550

Midway House/Women
1617 North 7 Street
Harlingen, TX 78550

South Texas Hospital Drug and Alcohol Counseling Program
1/2 Mile South Rangerville Road
Harlingen, TX 78550

Tropical Texas Center for MH/ MR Harlingen Outpatient Substance Abuse Services
913 Grimes Street
Harlingen, TX 78550

Valley House
1438 North 77 Sunshine Strip
Harlingen, TX 78550

HASKELL

Headstream Mental Health Center Substance Abuse Services
1301 North First Street
Haskell, TX 79521

HEARNE

Robertson County Outreach Center
217 4 Street
Hearne, TX 77859

HEMPSTEAD

Riceland Regional Mental Health Authority Adolescent Alternatives
926 6 Street
Hempstead, TX 77445

HENDERSON

Sabine Valley Center Henderson Outpatient Clinic
116 South Marshall Street
Henderson, TX 75652

HENRIETTA

Clay County Mental Health Center Substance Abuse Services
310 West South Street
Clay County Memorial Hospital
Henrietta, TX 76365

HOCKLEY

Friendship Ranch
Route 1
Hockley, TX 77447

HOUSTON

AAMA/CASA Phoenix
4039 Gramercy Street
Houston, TX 77025

Addiction Treatment Centers Addiction Treatment Center of Fannin Clinic
7225 Fannin Street
Houston, TX 77036

Adult Rehabilitation Services
6624 Hornwood Street
Houston, TX 77074

Advance Treatment Center Ruth E. Winn Methadone Treatment
1320 McGowen Street
Houston, TX 77004

Assoc. for the Adv. of Mexican Americans (AAMA) Comp. Inhalant Drug Abuse Program
204 Clifton Street
Houston, TX 77011

Bay Area Council on Drugs and Alcohol, Inc. Know More R and R/Egret Bay
18201-A Egret Bay Boulevard
Houston, TX 77058

Ben Taub General Hospital Nursery Followup Developmental Clinics
1504 Taub Loop
Houston, TX 77030

Campus New Spirit Volunteers of America
4515 Lyons Street
Houston, TX 77020

Communities in Schools/Houston Youth Empowerment Against Substance Abuse
1100 Milam Street
Suite 3590
Houston, TX 77002

Extended Aftercare, Inc.
5115 Del Sur
Houston, TX 77018

Family Service Center El Centro Familiar
7151 Office City Drive
Suite 101
Houston, TX 77087

Five Oaks Residential Treatment Center
1122 Bissonnet Street
Houston, TX 77005

Group Plan Clinic, Inc.
DBA New Spirit
1775 Saint James Corp. Office
Suite 200
Houston, TX 77056

New Spirit
2411 Fountain View Drive
Suite 175
Houston, TX 77056

Gulf Coast Community Services Assoc. Substance Abuse Services Program
6300 Bowling Green Street
Suite 147
Houston, TX 77021

Harris County Community Supervision and Corrections
49 San Jacinto Street
Houston, TX 77002

Harris County Hospital District Alcoholism/Drug Abuse Services
1502 Taub Loop
Houston, TX 77030

Harris County Psychiatric Center Substance Abuse Unit
2800 South MacGregor Way
Houston, TX 77021

HCA Belle Park Hospital Chemical Dependency Treatment Program
4427 Belle Park Drive
Houston, TX 77072

HCA Gulf Pines Hospital Substance Abuse Services
205 Hollow Tree Lane
Houston, TX 77090

Houston Child Guidance Center Substance Abuse Program
3214 Austin Street
Houston, TX 77004

Houston Council on Alcohol and Drug Abuse
Community Unit Probation Services/Cups I
4605 Wilmington Street
Houston, TX 77051

Community Unit Probation Services/Cups 2
4014 Market Street
Houston, TX 77020

Community Unit Probation Services/Cups 4
170 Heights Street
Houston, TX 77007

Community Unit Probation Services/Cups 5
6610 Harwin Street
Houston, TX 77036

Informal Adjustment
1927 West Lamar Street
Houston, TX 77019

Houston Northwest Medical Center Psychiatric Services
710 Farm Market Road
1960 West
Houston, TX 77090

Houston Substance Abuse Clinic
7428 Park Place Boulevard
Houston, TX 77087

Innerquest Counseling Center, Inc. Outpatient Chemical Dependency Program
17030 Nanes Street
Suite 211
Houston, TX 77090

Kelsey/Seybold Clinic PA Alcoholism Consulting and Treatment
5757 Woodway Drive
Suite 180
Houston, TX 77057

Maintenance Clinic
4900 Fannin Street
Houston, TX 77004

Martin Luther King Community Center Drug Abuse Prevention Program
2720 Sampson Street
Houston, TX 77004

Methodist Hospital and Baylor College of Medicine Chemical Dependency Program
6565 Fannin Street
Houston, TX 77030

Metropolitan Clinic of Counseling, Inc. First Step Recovery Program
1900 West Loop South
Suite 675
Houston, TX 77027

Montrose Counseling Center Chemical Dependency Program
900 Lovett Boulevard
Suite 203
Houston, TX 77006

Murillo Enterprises Counseling and Training Services
2510 Broad Street
Suite 200
Houston, TX 77087

Narcotics Withdrawal Center
4949 West 34 Street
Suite 111
Houston, TX 77092

South
1050 Edgebrook Drive
Suite 2
Houston, TX 77034

Nightingale Adult Day Center
5802 Holly Street
Houston, TX 77074

Odyssey House Texas, Inc.
5629 Grapevine Street
Houston, TX 77085

Over the Hill
3402 Dowling Street
Houston, TX 77004

Oxford Counseling Center
4625 North Freeway
Suite 101
Houston, TX 77022

Passages, Inc.
7722 Westview Drive
Houston, TX 77055

Phoenix Recovery Center, Inc.
9610 Long Point Road
Suite 206
Houston, TX 77055

Riverside General Hospital
Houston Recovery Campus
4514 Lyons Street
Houston, TX 77020

Total Care/Stress Unit
3204 Ennis Street
Houston, TX 77004

Sam Houston Memorial Hospital Alcohol and Drug Help Unit
1615 Hillendahl Street
Houston, TX 77055

Santa Maria Hostel, Inc.
807 Paschall Street
Houston, TX 77009

Sunrise Recovery Program
530 Wells Fargo Drive
Suite 119
Houston, TX 77090

Texas Alcoholism Foundation, Inc.
Texas House
10950 Beaumont Highway
Houston, TX 77078

Texas House Outpatient Treatment Program
2208 West 34 Street
Houston, TX 77213

Texas Clinic/Fulton Street
6311 Fulton Street
Houston, TX 77022

Westview Drive
9320 Westview Drive
Suite 10
Houston, TX 77055

The Houston Clinic PA Outpatient Chemical Dependency
7505 Fannin Street
Suite 680
Houston, TX 77054

The Recovery Center, Inc.
Blue Bonnett House
1420 Missouri Street
Houston, TX 77006

Jasmine House
1920 West Clay Street
Houston, TX 77019

Magnolia House
4917 Jackson Street
Houston, TX 77004

The Shoulder, Inc.
7655 Bellfort Avenue
Houston, TX 77061

Toxicology Associates
530 North Belt Street
Suite 311
Houston, TX 77060

Texas Recovering Addictions Foundation The Serenity Center
2511 Gregg Street
Houston, TX 77026

Univ. of Texas Health Science Center Substance Abuse Research Center
1300 Moursund Street
Houston, TX 77030

Veterans' Affairs
Alcohol Inpatient Treatment Program
Drug Inpatient Program
2002 Holcombe Boulevard
Houston, TX 77030

Vocational Guidance Service, Inc. (VGS)
Alternative Drug Abuse Treatment Program
2525 San Jacinto Street
Houston, TX 77002

Beaumont Highway Unit
10950 Beaumont Highway
Houston, TX 77078

North Belt Unit
255 North Point
Suite 2
Houston, TX 77060

Parole Satellite Program
4949 West 34 Street
Houston, TX 77092

Reveille Road Clinic
3716 Reveille Road
Houston, TX 77087

Volunteers of America
McGovern House/Transitional Services
312 East Rogers Street
Houston, TX 77022

Rogers Street Recovery Center
308 East Rogers Street
Houston, TX 77022

Wesley Community Center
1410 Lee Street
Houston, TX 77009

West Oaks/The Psychiatric Institute Hoskins Achievement Place
2329 Hoskins Street
Houston, TX 77080

HUNT

La Hacienda Treatment Center La Hacienda Treatment Services
FM 1340
Hunt, TX 78024

HUNTSVILLE

Hunstville Alcohol/Drug Abuse Program
115 North Highway 75
Huntsville, TX 77340

Hunstville Clinic, Inc.
3010 Old Houston Road
Huntsville, TX 77340

Tri-County MH/MR Services Life Sync
1 Financial Plaza
Suite 250-D
Huntsville, TX 77340

Texas Dept. of Criminal Justice/ Instit. Division Substance Abuse Treatment Program
815 11 Street
Huntsville, TX 77342

HURST

Tarrant County MH/MR Services Alcohol Outpatient Unit
129 Harmon Road
Hurst, TX 76053

IRVING

Planned Performance
2311 Texas Drive
Suite 102
Irving, TX 75062

Texas Youth Commission North Texas Regional Office
800 West Airport Freeway
Suite 503
Irving, TX 75062

JACKSBORO

Jack County Mental Health Center Substance Abuse Services
115 South Main Street
Jacksboro, TX 76458

KERRVILLE

Kerrville State Hospital Substance Abuse Unit
721 Thompson Drive
Kerrville, TX 78028

Veterans' Affairs Medical Center Alcohol and Drug Dependence Treatment Program
3600 Memorial Boulevard
Kerrville, TX 78028

KILLEEN

Central Counties Center for MH/MR Services Killeen Substance Abuse Program
100 East Avenue A
Killeen, TX 76541

Christian Farms/Treehouse, Inc.
Farm Road 2410
Killeen, TX 76513

KINGWOOD

Charter Hospital of Kingwood Inpatient Unit
2001 Ladbrook Drive
Kingwood, TX 77339

Tri-County MH/MR Services Life Sync
900 Rockmead Drive
Suite 147
Kingwood, TX 77339

KYLE

Wackenhut Corrections Corporation Kyle Facility Treatment Program
701 South IH 35
Cell Block A
Kyle, TX 78640

LA GRANGE

Colorado Valley Council on Alcohol and Drug Abuse, Inc./ Outpatient Treatment
258 West Pearl Street
La Grange, TX 78945

LA MARQUE

Toxicology Associates
2411 Franklin Street
La Marque, TX 77568

LAKE JACKSON

Brazosport Memorial Hospital Alpha Center
100 Medical Drive
Lake Jackson, TX 77566

LAREDO

BASTA/DEAR/PASA
1205 East Hillside Street
Suite 1
Laredo, TX 78041

Laredo State Center La Familia Inhalant Abuse Program
413 Cherry Hill Drive
Laredo, TX 78041

LEAGUE CITY

Gulf Coast Center
Circuit Breakers CSA Program
Circuit Breakers Parents' Program
Family Innovation
4444 West Main Street
League City, TX 77573

LEWISVILLE

Denton County MH/MR Center Adolescent Substance Abuse Program
1310 South Stemmons Freeway
Suite A-4
Lewisville, TX 75067

LIBERTY

Tri-County MH/MR Substance Abuse Services
612 Highway 90
Liberty, TX 77575

LITTLEFIELD

Central Plains Center for MH/MR and Substance Abuse Littlefield Counseling Center
100 West 4 Street
Littlefield, TX 79339

LONGVIEW

East Texas Council on Alcoholism and Drug Abuse
450 East Loop 281
Suite B
Longview, TX 75601

Sabine Valley Center Woodbine Treatment Center
103 Woodbine Place
Longview, TX 75608

LUBBOCK

Charter Plains Hospital Addictive Disease Treatment Program
801 North Quaker Avenue
Lubbock, TX 79416

Lubbock Faith Center, Inc.
Center Recovery Program
2809 Clovis Road
Lubbock, TX 79415

Lubbock Regional MH/MR Center
Alcoholism Program
Lubbock Employee Assistance Network
1202 Main Street
Lubbock, TX 79401

Billy Meeks Addiction Center
1601 Vanda Avenue
Lubbock, TX 79401

Drug Free and Methadone Treatment Program
1210 Texas Avenue
Lubbock, TX 79401

Lubbock Regional MH/MR Center Project Hope
1202 Main Street
Lubbock, TX 79401

Texas Tech. Univ. Health Sciences Center Southwest Institute for Addictive Diseases
Department of Psychiatry
4000 24 Street/Saint Mary Plaza
Lubbock, TX 79410

Walker House
1614 Avenue K
Lubbock, TX 79401

LUFKIN

Alcohol and Drug Abuse Council of Deep East Texas/First Street Unit
304 North Raguet Street
Lufkin, TX 75901

Deep East Texas Regional MH/MR Services
Choices
1601 B South Chestnut Street
Lufkin, TX 75901

Peavy Switch Recovery Center
FM 2497
Lufkin, TX 75901

Recover
1406 Turtle Creek Street
Lufkin, TX 75901

Memorial Medical Center of East Texas Lifecare
1201 Frank Street
Lufkin, TX 75901

MADISONVILLE

MH/MR Authority of Brazos Valley
Madison County Outreach Center
203 South Madison Street
Madisonville, TX 77864

Substance Abuse Treatment Program/Madisonville
101 West Main Street
Madison County Jail
Madisonville, TX 77864

MARBLE FALLS

Greater Austin Council on Alcohol and Drug Abuse/ Highland Lakes
1119 Highway 1431 West
Marble Falls, TX 78654

MARSHALL

Sabine Valley Center
Grove Moore Center
401 North Grove Street
Marshall, TX 75670

Kirkpatrick Family Center
3427 South Garrett Street
Marshall, TX 75670

Oak Haven Recovery Center
Highway 154
Marshall, TX 75670

MCALLEN

BETO House
4112 North 22 Street
McAllen, TX 78501

Charter Palms Hospital Substance Abuse Services
1421 East Jackson Avenue
McAllen, TX 78503

Universal Health Services of McAllen McAllen Medical Recovery Center
301 West Expressway 83
McAllen, TX 78503

MCKINNEY

Collin County MH/MR Center Substance Abuse Services
825 North McDonald Street
McKinney, TX 75069

MESQUITE

Garland Community Hospital Youth Crossing of Horizon Recovery Center
3500 Interstate 30
Suite 300
Mesquite, TX 75150

MIDLAND

Glenwood Hospital Glenwood Chemical Dependency Program
3300 South FM 1788
Midland, TX 79703

Permian Basin Community Centers for MH/MR
Country Mesa
5803 County Road 1200 South
Midland, TX 79703

Johnson Center
502 North Carver Street
Midland, TX 79701

Midland Outpatient Substance Abuse Services
Texas Place
3701 North Big Spring Street
Midland, TX 79701

MIDLOTHIAN

Reach Midlothian, Inc.
1351 North Highway 67
Midlothian, TX 76065

MINEOLA

Andrews Center Substance Abuse Services
703 West Patton Street
Mineola, TX 75773

MINERAL WELLS

Pecan Valley MH/MR Region Substance Abuse Services
Route 3
Building 244
Mineral Wells, TX 76067

MULESHOE

Central Plains Center for MH/MR and Substance Abuse Muleshoe Counseling Center
623 West 2 Street
Muleshoe, TX 79347

NACOGDOCHES

Alcohol and Drug Abuse Council of Deep East Texas/Pilar Street Unit
118 East Hospital Street
Suite 307
Nacogdoches, TX 75961

Stephen F. Austin State University Campus Assistance Program
Nacogdoches, TX 75961

NAVASOTA

Grimes County Outreach Center
702 La Salle Street
Navasota, TX 77868

ODESSA

Permian Basin Community Centers for MH/MR
Turning Point
Odessa Supportive Outpatient Services
2000 Maurice Road
Odessa, TX 79763

Faces
701 East 7 Street
Odessa, TX 79760

Permian Basin Regional Council on Alcohol and Drug Abuse
3641 North Dixie Street
Odessa, TX 79762

ORANGE

Baptist Hospital/Orange Substance Abuse Recovery Center
608 Strickland Drive
Orange, TX 77630

Life Resource/ A CMHC Substance Abuse Services
4303 North Tejas Parkway
Orange, TX 77630

PADUCAH

Paducah Outreach Service Center
922 Gober Street
Paducah, TX 79248

PALESTINE

Daytop Pine Mountain Residential
Route 3
Palestine, TX 75801

PAMPA

Genesis House, Inc.
Administrative Unit
615 West Buckler Street
Pampa, TX 79066

Genesis House for Boys
612 West Browning Street
Pampa, TX 79065

Genesis House for Girls
420 North Ward Street
Pampa, TX 79066

Pampa Counseling Services for Adolescents
1224 North Hobart Street
NBC Plaza Unit 1
Pampa, TX 79065

PARIS

Lamar County Alcohol/Drug Center, Inc.
Adult Outpatient
Youth Outpatient
50 North Main Street
Paris, TX 75460

PASADENA

Bayshore Medical Center Chemical Dependency Center
4000 Spencer Highway
Pasadena, TX 77504

Houston Council on Alcohol and Drug Abuse Community Unit Probation Services/Cups 3
302 East Shaw Street
Pasadena, TX 77506

Pasadena Substance Abuse Clinic
1803 Strawberry Road
Pasadena, TX 77502

Sand Dollar, Inc. Sand Dollar Emergency Shelter
527 Spring Drive
Pasadena, TX 77504

PLAINVIEW

Central Plains Center for MH/MR and Substance Abuse
Plainview Counseling Center
620 West 7 Street
Plainview, TX 79072

Institute for Adolescent Addictions
404 Floydada Street
Plainview, TX 79072

Plainview Women's Center
405 Ennis Street
Plainview, TX 79072

W. W. Allen Treatment Center
715 Houston Street
Plainview, TX 79072

Methodist Hospital Plainview Lonetree Recovery Center
2601 Dimmitt Road
Plainview, TX 79072

Serenity Center, Inc.
806 El Paso Street
Plainview, TX 79072

PLANO

Collin County MH/MR Center Plano Clinic
3920 Alma Drive
Plano, TX 75023

Presbyterian and Children's Seay Behavioral Healthcare Center
6110 West Parker Road
Plano, TX 75093

PORT ARTHUR

Organization of Christians Assisting People, Inc.
600 Foley Avenue
Port Arthur, TX 77640

Port Arthur Drug Abuse Clinic Methadone Program
530 Waco Avenue
Port Arthur, TX 77640

PYOTE

West Texas Children's Home Substance Abuse Services
Pyote, TX 79777

QUANAH

Foard/Hardeman Counties Outreach Center
510 King Street
Quanah, TX 79252

RICHARDSON

Daytop Dallas/Richardson Outreach
902 Saint Paul Drive
Richardson, TX 75080

New Place Program
1730 North Greenville Avenue
Richardson, TX 75081

ROANOKE

Lena Pope Home, Inc. Unhooked/Roanoke
195 Dunham Road
Roanoke, TX 76262

ROMA

Roma Independent School District Student Assistance Program
700 Gladiator Drive
Roma, TX 78584

ROSENBERG

Riceland Regional Mental Health Authority Substance Abuse Service System
1110 Avenue G
Rosenberg, TX 77471

RUSK

Rusk State Hospital Substance Abuse Program
Highway 69 North
Rusk, TX 75785

SAN ANGELO

Concho Valley Regional Hospital First Step
2018 Pulliam Street
San Angelo, TX 76902

San Angelo Council on Alcohol/ Drug Abuse
Adolescent Outpatient Clinic
1017 Caddo Street
San Angelo, TX 76901

Outpatient
1021 Caddo Street
San Angelo, TX 76901

Williams House
134 West College Avenue
San Angelo, TX 76903

San Angelo Detox Center
609 North Park Drive
San Angelo, TX 76901

Women's Halfway House of San Angelo, Inc. Women and Children Halfway House
401 West Twohig Street
San Angelo, TX 76901

SAN ANTONIO

Alamo Mental Health Group, Inc. Spectrum
4242 Medical Drive
Suite 7250
San Antonio, TX 78229

Alcoholic Rehab Center of Bexar County
Women and Children Residential Program
Women and Children's Center
1616 North Saint Marys Street
San Antonio, TX 78212

Alpha Home, Inc.
300 East Mulberry Avenue
San Antonio, TX 78212

Baptist Medical Center Baptist Recovery Center
111 Dallas Street
San Antonio, TX 78205

Center for Health Care Services
Cloudhaven Unit
127 Cloudhaven Street
San Antonio, TX 78209

Detoxification Unit
711 East Josephine Street
San Antonio, TX 78208

IH 10 West Unit
3031 IH 10 West
San Antonio, TX 78201

Outpatient Unit
402 Austin Street
San Antonio, TX 78215

Charter Real Hospital Substance Abuse Services
8550 Huebner Road
San Antonio, TX 78240

Community Counseling Center Alcohol/Drug Abuse Prevention Program
Fort Sam Houston
Building 1123
San Antonio, TX 78234

CPC Afton Oaks Hospital Counterpoint Center
620 East Afton Oaks Boulevard
San Antonio, TX 78232

Healthcare San Antonio, Inc. Variable Intensity Chemical Dependency Treatment Program
17720 Corporate Woods Drive
San Antonio, TX 78270

Integrated Therapy Program
14855 Blanco Road
Suite 220
San Antonio, TX 78216

Mexican American Unity Council, Inc.
Amigos Project
1921 Buena Vista Street
Rear
San Antonio, TX 78207

Casa Del Sol/Casa Adelante
2303 West Commerce Street
San Antonio, TX 78207

San Antonio Youth Res. Project/Girls
2415 West Commerce Street
San Antonio, TX 78207

San Antonio Youth Res. Project/Post
248 Post Street
San Antonio, TX 78215

New Day of San Antonio Substance Abuse Services
7220 Louis Pasteur Street
Suite 100
San Antonio, TX 78229

River City Rehabilitation Center, Inc.
1422 Nogalitos Street
San Antonio, TX 78204

San Antonio State Chest Hospital Alcoholism Treatment Program
2303 SE Military Drive
San Antonio, TX 78223

San Antonio State Hospital Substance Abuse Unit
5900 South Presa Street
San Antonio, TX 78223

South Texas Counseling Centers, Inc. Lifeway
11330 IH 10 West
Suite 2800
San Antonio, TX 78249

Teen Challenge of South Texas, Inc.
3850 South Loop 1604
San Antonio, TX 78221

Texas Biodyne, Inc. San Antonio Northeast
1919 Oakwell Farms Parkway
Suite 125
San Antonio, TX 78218

The Patrician Movement
Site 1/Residential
222 East Mitchell Street
San Antonio, TX 78210

Site 2/Outpatient Treatment Program
1249 South Saint Mary's Street
San Antonio, TX 78210

Site 3/Outpatient Treatment Program
215 Claudia Street
San Antonio, TX 78210

Site 4/Outpatient Treatment Program
278 East Mitchell Street
San Antonio, TX 78210

Wilford Hall U.S. Air Force Medical Center Alcoholism Rehabilitation Center
2289 McChord Street
Lackland Air Force Base Building 1355
San Antonio, TX 78236

SCHERTZ

Schertz/Cibolo Universal City Independent School Dist. Substance Abuse Services
1060 Elbel Road
Schertz, TX 78154

SEAGOVILLE

Federal Correctional Institution Drug Abuse Program
Psychology Department
Seagoville, TX 75159

SEGUIN

Guadalupe Valley Hospital Teddy Buerger Center for Alcohol/ Drug Abuse
1215 East Court Street
Seguin, TX 78155

SEYMOUR

Baylor/Throckmorton Counties Outreach Service Center
300 South Stratton Street
Seymour, TX 76380

SHERMAN

Alcoholic Services of Texoma, Inc.
House of Hope
225 West Brockett Street
Sherman, TX 75090

House of Hope Outpatient Program
400 North Crockett Street
Suite B
Sherman, TX 75090

Texoma Council on Alcoholism and Drug Abuse
103 South Travis Street
Sherman, TX 75090

SPRING

Practical Recovery
20807 Sunshine Lane
Spring, TX 77388

SPUR

White River Retreat HCR 2
White River Lake
Spur, TX 79370

STEPHENVILLE

Summer Sky, Inc. Chemical Dependency Treatment Center
1100 McCart Street
Stephenville, TX 76401

SUGAR LAND

Charter Hospital of Sugar Land
1550 First Colony Boulevard
Sugar Land, TX 77479

SULPHUR SPRINGS

Red River Council on Alcohol and Drug Abuse Hopkins County Substance Abuse Program
468 Shannon Square
Suite 9
Sulphur Springs, TX 75483

TAFT

Shoreline, Inc.
1220 Gregory Street
Taft, TX 78390

TEMPLE

CEN/TEX Alcoholic Rehabilitation Center
2500 South General Bruce Drive
Temple, TX 76504

Central Counties Center for MH/ MR Services Temple Substance Abuse Program
304 South 22 Street
Temple, TX 76501

Christian Farms/Treehouse, Inc. The Treehouse
613 South 9 Street
Temple, TX 76504

Olin E. Teague Veterans' Center Psychiatry Service SATP/20AE
1901 South First Street
Temple, TX 76504

Scott and White Santa Fe Center Alcohol and Drug Dependence Treatment Program
600 South 25 Street
Temple, TX 76503

TERRELL

Terrell State Hospital Substance Abuse Unit
East Brin Street
Terrell, TX 75160

TEXARKANA

Care Unit Hospital of Fort Worth Care Unit Program of Texarkana
4520 Summerhill Road
Texarkana, TX 75503

Choices
301 Westlawn Drive
Texarkana, TX 75501

Federal Correctional Institution Drug Abuse Program
Psychology Department
Texarkana, TX 75505

Liberty/Eylau Independent School District School of Success
Route 11
Texarkana, TX 75501

Red River Council on Alcohol and Drug Abuse
Administrative Unit
Respond Outpatient
222 West 5 Street
Texarkana, TX 75501

Dowd House
2101 Dudley Avenue
Texarkana, TX 75502

TEXAS CITY

Alcohol Drug Abuse Women's Center, Inc. 5th Avenue Unit
712 5 Avenue North
Texas City, TX 77590

Gulf Coast Center Substance Abuse Recovery Program
6510 Memorial Drive
Texas City, TX 77591

Texas City Men's Center Intensive Residential Unit 1
124 6 Street North
Texas City, TX 77592

THE WOODLANDS

Innerquest Counseling Center, Inc. Outpatient Chemical Dependency Program
25211 Grogans Mill Street
Suite 211
The Woodlands, TX 77381

TULIA

Central Plains Center for MH/MR and Substance Abuse Tulia Counseling Center
310 West Broadway
Tulia, TX 79088

Driskill Halfway House
Highway 87 North
Tulia, TX 79088

TYLER

Andrews Center The Beginning
1010 Timberwild Street
Suite 100
Tyler, TX 75703

County Rehabilitation Center, Inc.
313 Ferrell Place
Tyler, TX 75702

Smith County Council on Alcohol/Drug Abuse/Adult Counseling/Education
3027 Loop 323 South SE
Tyler, TX 75701

University Park Hospital Chemical Dependency Program
4101 University Boulevard
Tyler, TX 75701

VERNON

Haney House, Inc.
1908 Wilbarger Street
Vernon, TX 76384

Vernon State Hospital South Drug Dependent Youth Program
FM 433 and SH 283
Vernon, TX 76384

Wilbarger County Outreach Alcohol and Drug Abuse Services
2500 Wilbarger Street
Vernon, TX 76384

VICTORIA

Citizens' Medical Center Alcohol and Drug Recovery Program
2701 Hospital Drive
Victoria, TX 77901

Devereux Foundation Devereux Psychiatric Residential Treatment Center of Texas
120 David Wade Drive
Victoria, TX 77902

Gulf Bend MH/MR Talbot House/ Residential and Inpatient
210 East Juan Linn Street
Victoria, TX 77901

WACO

Freeman Center
Dear Unit
1619 Washington Avenue
Waco, TX 76703

Men's Residential
1401 Columbus Avenue
Waco, TX 76701

Women's Residential
1425 Columbus Avenue
Waco, TX 76703

Heart of Texas MH/MR Substance Abuse Program
110 South 12 Street
Waco, TX 76703

Lake Shore Center for Psychological Services PC Better Way Chemical Dependency Treatment Program
4555 Lake Shore Drive
Waco, TX 76714

Providence Health Center Substance Abuse Services
6901 Medical Parkway
Waco, TX 76712

WEBSTER

Baywood Hospital Baywood Partial Hospitalization Services
709 Medical Center Boulevard
Webster, TX 77598

Humana Hospital Corporation, Inc. Recovery Source
500 Medical Center Boulevard
Webster, TX 77598

WESLACO

Tropical Texas Center for MH/ MR Weslaco Outpatient Substance Abuse Services
601 West 6 Street
Weslaco, TX 78596

WHARTON

Riceland Regional Mental Health Authority Substance Abuse Service System
3007 North Richmond Road
Wharton, TX 77488

WHITE OAK

Sabine Valley Center Dear Recovery Center
2000 U.S. Highway 80
White Oak, TX 75693

WICHITA FALLS

HCA Red River Hospital Crossings
1505 8 Street
Wichita Falls, TX 76301

Helen Farabee Center
Substance Abuse Outpatient Treatment Program
500 South Broad Street
Wichita Falls, TX 76301

Substance Abuse Triage Program
1501 7 Street
Wichita Falls, TX 76301

Hopecrest Lodge, Inc.
2001 Harriett Street
Wichita Falls, TX 76309

Wichita Falls State Hospital Substance Abuse Recovery Program
Lake Road
Highway 79
Wichita Falls, TX 76307

396th Medical Group Alcohol Rehabilitation Center/SGHMA
Sheppard Air Force Base/Texas
Wichita Falls, TX 76311

WYLIE

The Country Place Adolescent Residential Treatment Center, Inc.
2708 FM 1378
Wylie, TX 75098

UTAH

BLANDING

San Juan Mental Health Substance Abuse Services
522 North 100 East
Blanding, UT 84511

BRIGHAM CITY

Family Preservation Institute
693 South 400 East
Brigham City, UT 84302

Office of Social Services Substance Abuse Program
1050 South 500 West
Brigham City, UT 84302

DRAPER

Utah State Prison Substance Abuse Treatment Program
Utah State Prison
Draper, UT 84020

DUGWAY

Community Counseling Center
Dugway Proving Ground
Building 5236 Room 1226
Dugway, UT 84022

FARMINGTON

Davis County Comp. Mental Health Center Administrative Unit
291 South 200 West
Farmington, UT 84025

HEBER CITY

Wasatch County Alcohol and Drug Treatment and Prevention Program
805 West 100 South
Heber City, UT 84032

HILL AIR FORCE BASE

Hill Air Force Base Social Actions Drug/Alcohol Section
11 Street
Building 555
Hill Air Force Base, UT 84056

IBAPAH

Goshute Tribal Alcoholism Program
Ibapah, UT 84034

KOOSHAREM

Sorenson's Ranch School, Inc.
Koosharem, UT 84744

LOGAN

Logan Regional Hospital Dayspring
1400 North 500 East
Logan, UT 84321

MOUNT PLEASANT

Central Utah Mental Health/ Substance Abuse Center Substance Abuse Services/ Mount Pleasant
255 West Main Street
Mount Pleasant, UT 84647

OGDEN

McKay/Dee Hospital Dayspring Chemical Dependency Unit
5030 Harrison Boulevard
Ogden, UT 84403

Problems Anonymous Action Group (PAAG)
2522 Wall Avenue
Ogden, UT 84401

Professional Services Corporation
533 26 Street
Suite 100
Ogden, UT 84401

Saint Benedict's Hospital/ACT Alcohol and Chemical Dependency Treatment Center
5475 South 500 East
Ogden, UT 84405

Salvation Army Ogden Center
2615 Grant Avenue
Ogden, UT 84401

Weber Dept. of Alcohol and Drug Abuse
2650 Lincoln Avenue
Ogden, UT 84401

OREM

Charter Canyon Hospital Chemical Dependency Disease Program
1350 East 750 North
Orem, UT 84057

Gathering Place
555 South State Street
Suite 203
Orem, UT 84058

Heritage Center
1426 East 800 North
Orem, UT 84404

Utah County Council on Drug Abuse Rehabilitation (UCCODAR)
555 South State Street
Suite 203
Orem, UT 84058

PARK CITY

Valley Mental Health Summit County Unit
1753 Sidewinder Drive
Park City, UT 84060

PAYSON

Mountain View Hospital Step One/Pavilion
1000 East Highway 6
Payson, UT 84651

PRICE

Four Corners Mental Health Center Price Clinic
Price, UT 84501

PROVO

Charter Counseling Center of Provo Outpatient Chemical Dependency Program
2474 North University Avenue
Country Club Court Suite 100
Provo, UT 84604

Provo Canyon School Substance Abuse Services
4501 North University Avenue
Provo, UT 84603

Rocky Mountain Consultants
420 North Freedom Boulevard
Provo, UT 84601

Utah County Division of Substance Abuse
100 East Center
Suite 3300
Provo, UT 84606

Wasatch Mental Health Substance Abuse Services
Provo, UT 84603

ROOSEVELT

Uintah Basin Counseling, Inc. Roosevelt Office
510 West 200 North
Roosevelt, UT 84066

SAINT GEORGE

Southwest Utah Mental Health Alcohol and Drug Center
354 East 600 South
Suite 202
Saint George, UT 84770

SALT LAKE CITY

Asian Association of Utah
28 East 2100 South
Suite 102
Salt Lake City, UT 84115

Community Counseling Center
660 South 200 East
Suite 308
Salt Lake City, UT 84111

First Step House
411 North Grant Street
Salt Lake City, UT 84116

Great Basin Family Support
9730 South 700 East
Suite 208
Salt Lake City, UT 84070

Highland Ridge Hospital Substance Abuse Services
4578 Highland Drive
Salt Lake City, UT 84117

Latter-Day Saints Hospital Intermountain Health Care Dayspring Program
C Street and 8 Avenue
Salt Lake City, UT 84143

Neo Genesis
744 South 500 East
Salt Lake City, UT 84102

Northwest Passage, Inc.
432 North 300 West
Salt Lake City, UT 84103

Odyssey House, Inc. Adult Treatment Program
68 South 600 East
Salt Lake City, UT 84102

Olympus View Hospital Substance Abuse Services
1430 East 4500 South
Salt Lake City, UT 84117

Pioneer Valley Hospital Chemical Dependency Program
3460 South Pioneer Parkway
Salt Lake City, UT 84120

Professional Services Corporation Substance Abuse Services
4525 South 2300 East
Suite 201B
Salt Lake City, UT 84117

Project Reality
150 East 700 South
Salt Lake City, UT 84111

Saint Mary's Home for Men
1206 West 200 South
Salt Lake City, UT 84104

Salt Lake County Alcohol Counseling and Education Center
231 East 400 South
2nd Floor
Salt Lake City, UT 84111

Salvation Army Alcohol Rehabilitation Program
252 South 500 East
Salt Lake City, UT 84102

The Cottage Program International, Inc.
57 West South Temple
Suite 420
Salt Lake City, UT 84101

The Haven
974 East South Temple
Salt Lake City, UT 84102

Turnabout
2738 South 2000 East
Salt Lake City, UT 84109

University of Utah Alcohol and Drug Abuse Clinic
50 North Medical Drive
Room 1R52
Salt Lake City, UT 84132

Utah Alcoholism Foundation Combined Facilities
2880 South Main Street
Suite 210
Salt Lake City, UT 84115

Valley Mental Health Alcohol and Drug Treatment Unit
404 East 4500 South
Suite 22B and Suite 22A
Salt Lake City, UT 84107

Veterans' Affairs Medical Center Substance Abuse Treatment Units
500 Foothill Boulevard
Salt Lake City, UT 84148

Volunteers of America Alcohol and Drug Detoxification Center
249 West 700 South
Salt Lake City, UT 84101

Wasatch Canyons Hospital Dayspring
5770 South 1500 West
Salt Lake City, UT 84123

Wasatch Youth Support Systems
3540 South 4000 West
Suite 550
Salt Lake City, UT 84120

Western Institute of Neuropsychiatry Recovery Center/Substance Abuse Services
501 Chipeta Way
Salt Lake City, UT 84108

TOOELE

Tooele Army Depot Community Counseling Program
Commander SDSTE-PSO
Tooele, UT 84074

Tooele Mental Health/Substance Abuse Counseling Center
305 North Main Street
Room 240
Tooele, UT 84074

VERNAL

Uintah Basin Counseling Vernal Office
559 North 1700 West
Vernal, UT 84078

WOODS CROSS

Benchmark Regional Hospital Substance Abuse Services
592 West 1350 South
Woods Cross, UT 84087

VERMONT

BENNINGTON

United Counseling Service of Bennington County, Inc.
Ledge Hill Drive
Bennington, VT 05201

BRATTLEBORO

Brattleboro Retreat Adult Alcohol and Substance Abuse Program
75 Linden Street
Brattleboro, VT 05301

Marathon of Brattleboro, Inc.
101 Western Avenue
Brattleboro, VT 05301

Youth Services Incorporated
11 Walnut Street
Brattleboro, VT 05301

BURLINGTON

Champlain Drug and Alcohol Services
45 Clarke Street
Burlington, VT 05401

Howard Mental Health Services Substance Abuse Unit
300 Flynn Avenue
Burlington, VT 05401

Medical Center Hospital of Vermont Day One
200 Twin Oaks Terrace
Suite 6
Burlington, VT 05403

HUNTINGTON

Huntington Lodge
Delfrate Road
RR 1
Huntington, VT 05462

MIDDLEBURY

Counseling Service of Addison County Drug Abuse Treatment Unit
89 Main Street
Middlebury, VT 05753

MONTPELIER

Washington County Youth Service Bureau
38 Elm Street
Montpelier, VT 05602

MORRISVILLE

Lamoille County Mental Health Services Substance Abuse Treatment Unit
Washington Highway
Morrisville, VT 05661

NEWPORT

Northeast Kingdom Community Action Alcohol and Drug Services
30 Coventry Street
Newport, VT 05855

RANDOLPH

Orange County Mental Health Service Substance Abuse Treatment Unit
4 Highland Avenue
Randolph, VT 05060

RUTLAND

Rutland Mental Health Service Evergreen Center for Alcohol/ Drug Services
230 West Street
Rutland, VT 05701

SAINT ALBANS

Franklin/North Grand Isle Mental Health Services, Inc. Drug Abuse Prevention Unit
8 Ferris Street
Saint Albans, VT 05478

SAINT JOHNSBURY

Northeastern Vermont Regional Hospital
Founders' Hall Non CDS Unit
Outpatient Recovery Center
Hospital Drive
Saint Johnsbury, VT 05819

SPRINGFIELD

Canterbury Counseling Services
374 North River Road
Springfield, VT 05156

UNDERHILL

Maple Leaf Farm Associates, Inc.
Stevensville Road
RR 1
Underhill, VT 05489

WALLINGFORD

Recovery House
Wallingford, VT 05773

WATERBURY

Vermont Office of Alcohol and Drug Abuse Programs Administrative Unit
103 South Main Street
State Office Building
Waterbury, VT 05671

VIRGIN ISLANDS

CHARLOTTE AMALIE

Council on Alcoholism Saint Thomas/Saint John (Coast)
7 Bred Gade
Charlotte Amalie, VI 00802

CHRISTIANSTED

Mental Health Alcoholism and Drug Dependency Services/ Unit 1
3500 Richmond Street
Charles Harwood Complex
Christiansted, VI 00820

The Village/Virgin Islands Partners in Recovery
2012 Queen Street
Lot 38
Christiansted, VI 00820

SAINT THOMAS

Mental Health Alcoholism and Drug Dependency Services/ Unit 2
C Street
Oswald Harris Court
Saint Thomas, VI 00802

VIRGINIA

ACCOMAC

Eastern Shore Alcohol Safety Action Project
23386 Front Street
Accomac, VA 23301

ALEXANDRIA

Alexandria Alcohol Safety Action Program (ASAP)
421 King Street
Suite 210
Alexandria, VA 22314

Alexandria Community Services Board Substance Abuse Services
2355-A Mill Road
Alexandria Health Department
Alexandria, VA 22314

Living Free Alcohol and Chemical Dependence Program
6391 Little River Turnpike
Alexandria, VA 22312

Second Genesis, Inc. King Street Clinic
1001 King Street
Alexandria, VA 22314

ARLINGTON

Arlington Alcohol Safety Action Program
1400 North Courthouse Road
Suite 230
Arlington, VA 22201

Arlington Hospital
Addiction Treatment Program
1701 North George Mason Drive
Arlington, VA 22205

Counseling Center
1916 Wilson Boulevard
Suite 100
Arlington, VA 22201

Drewry Center Day Treatment
1725 North George Mason Drive
Arlington, VA 22205

Northern Virginia Doctors Hospital Star Program
601 South Carlin Springs Road
Arlington, VA 22204

The Women's Home, Inc.
1628 North George Mason Drive
Arlington, VA 22205

Vanguard Services, Ltd. Phoenix Program
506 North Pollard Street
Arlington, VA 22203

BUENA VISTA

Rockbridge Alcohol Safety Action Program
2044 Sycamore Avenue
Buena Vista, VA 24416

Rockbridge Area Community Services Board Alcohol and Drug Services
2224 Magnolia Avenue
Buena Vista, VA 24416

CARTERSVILLE

Human Resources, Inc. Willow Oaks
2123 Cartersville Road
Cartersville, VA 23027

CEDAR BLUFF

Cumberland Mental Health Center Substance Abuse Services
Route 19
Cedar Bluff, VA 24609

CHARLOTTESVILLE

James River Alcohol Safety Action Program
104 4 Street NE
Suite 201
Charlottesville, VA 22902

Region 10 Community Services Board Alcohol and Drug Abuse Services
413 East Market Street
Suite 103
Charlottesville, VA 22901

CHESAPEAKE

Chesapeake Substance Abuse Program
524 Albermarle Drive
Chesapeake, VA 23320

CHESTERFIELD

John Tyler Alcohol Safety Action Project
9520 Iron Bridge Road
Chesterfield, VA 23832

CHRISTIANSBURG

New River Valley Alcohol Safety Action Program
100 Arbor Street
Suite 6
Christiansburg, VA 24073

CLIFTON FORGE

Alleghany Highlands Comm. Services Board Substance Abuse Services
601 Main Street
Clifton Forge, VA 24422

CLINTWOOD

Dickenson County Community Services Substance Abuse Services
Highway 83
Kendrick Building
Clintwood, VA 24228

CULPEPER

Rapidan Community Mental Health Services Board Substance Abuse Services
401 South Main Street
Culpeper, VA 22701

DANVILLE

Alcoholic Counseling Center, Inc. Hope Harbor
1021 Main Street
Danville, VA 24541

Dan River ASAP
530 Main Street
Signet Bank Building 2nd Floor
Danville, VA 24541

Danville/Pittsylvania Mental Health Services Board Substance Abuse Services Division
245 Hairston Street
Danville, VA 24540

Interventions Counseling and Consulting Services
105 South Union Street
Suite 800
Danville, VA 24541

Memorial Hospital Substance Abuse Center First Step Program
142 South Main Street
Danville, VA 24541

FAIRFAX

Fairfax/Falls Church Comm. Services Board Alcohol and Drug Services
3900 Jermantown Road
Suite 200
Fairfax, VA 22030

Metropolitan Alcoholism Center
3863 Plaza Drive
Fairfax, VA 22030

Thesis 96, Inc. Life Line Addictions Program
10565 Lee Highway
Suite 100
Fairfax, VA 22030

FALLS CHURCH

Comprehensive Addiction Treatment Services (CATS)
3300 Gallows Road
Falls Church, VA 22046

Ethos Foundation, Inc.
3 Skyline Place
Suite 100
Falls Church, VA 22041

Skyline Psychiatric Associates, Inc. Addictive Disorders Program
5113 Leesburg Pike
Falls Church, VA 22041

FARMVILLE

Crossroads
Farmville, VA 23901

FORT BELVOIR

Fort Belvoir Counseling Center Alcohol/Drug Abuse Prevention and Control Program
Building 1826
North Post
Fort Belvoir, VA 22060

FORT LEE

Fort Lee Alcohol and Drug Community Counseling Center
Building 12000
Fort Lee, VA 23801

FREDERICKSBURG

Rappahannock Area Community Services Board Alcohol and Drug Outpatient Services
600 Jackson Street
Fredericksburg, VA 22401

Serenity Home, Inc. Substance Abuse ICF and Halfway Treatment Services
521–523 Sophia Street
Fredericksburg, VA 22401

GALAX

Galax Treatment Center, Inc. Life Center of Galax/Adult Residency Program
112 Painter Street
Galax, VA 24333

GLEN ALLEN

Henrico Area MH/MR Services Substance Abuse Services
10299 Woodman Road
Glen Allen, VA 23060

GLOUCESTER

Middle Peninsula/Northern Neck Counseling Center
Gloucester, VA 23061

HAMPTON

Alternatives
2021B Cunningham Drive
Suite 5
Hampton, VA 23666

Hampton/Newport News Community Services Board Substance Abuse Services
1520 Aberdeen Road
Suite 202
Hampton, VA 23666

Peninsula Drug Rehabilitation Services
1834 Todds Lane
Hampton, VA 23666

Peninsula Hospital Adult Chemical Dependency Unit
2244 Executive Drive
Hampton, VA 23666

Perspectives Health Program
22 Enterprise Parkway
Suite 200
Hampton, VA 23666

HARRISONBURG

Arlington Treatment Center Outpatient Program
Highway 11 North
Harrisonburg, VA 22801

Harrisonburg/Rockingham Comm. Services Board Outpatient Substance Abuse Services
1241 North Main Street
Harrisonburg, VA 22801

Rockingham/Harrisonburg Alcohol Safety Action Program
44 East Market Street
Harrisonburg, VA 22801

HERNDON

Reston Recovery Center Substance Abuse Services
171 Elden Street
Suite 204
Herndon, VA 22071

LANGLEY AIR FORCE BASE

Langley Air Force Base Substance Abuse Control Program
Building 147
Langley Air Force Base, VA 23665

LAWRENCEVILLE

Brunswick Correctional Center
Route 1
Lawrenceville, VA 23686

LEBANON

Southwest Virginia Alcohol Safety Action Program
Route 19 Russell County
Shopping Center/2nd Floor
Lebanon, VA 24266

LEESBURG

Springwood Psychiatric Institute Chemical Dependency Program
Route 4
Leesburg, VA 22075

LYNCHBURG

Central Virginia Alcohol Safety Action Program
2316 Atherholt Road
Lynchburg, VA 24502

Community Services Board of Central Virginia Substance Abuse Services
2235 Landover Place
Lynchburg, VA 24501

Virginia Baptist Hospital Pathways Treatment Center
3300 Rivermont Avenue
Lynchburg, VA 24503

MANASSAS

Prince William Addiction and Psychiatric Treatment Centers
8700 Sudley Road
Manassas, VA 22110

Upstairs Counseling and Chemical Dependency Center
8409 Dorsey Circle
Suite 201
Manassas, VA 22110

MARION

Southwestern Virginia Mental Health Institute Medical Detox Unit
502 East Main Street
Marion, VA 24354

MARTINSVILLE

Passages
817 Starling Avenue
Martinsville, VA 24112

Patrick Henry Drug and Alcohol Council Crossroads/Intensive Outpatient Program
29 Jones Street
Martinsville, VA 24114

NASSAWADOX

Eastern Shore CMHC Substance Abuse Services
10129 Rogers Drive
Nassawadox, VA 23413

NEWPORT NEWS

CAPO Center Detox
1003 28 Street
Newport News, VA 23607

New Foundations
610 Thimble Shoals Boulevard
Newport News, VA 23606

Peninsula Alcohol Safety Action Program
760 J. Clyde Morris Boulevard
Newport News, VA 23601

Peninsula Alcoholism Services Intake and Outpatient
732 Thimble Shoals Boulevard
Suite 701-J
Newport News, VA 23606

Riverside Regional Medical Center New Foundations
500 J. Clyde Morris Boulevard
Newport News, VA 23601

Serenity House Substance Abuse Recovery Program, Inc.
926 H J. Clyde Morris Boulevard
Unit 4
Newport News, VA 23606

NORFOLK

Naval Amphibious Base Little Creek Counseling and Assistance Center
Building 3007
Norfolk, VA 23521

Rehabilitation Services, Inc.
300 West 20 Street
Norfolk, VA 23517

Tidewater Psychiatric Institute Substance Abuse Treatment Program
860 Kempsville Road
Norfolk, VA 23502

U.S. Naval Station
Alcohol Rehabilitation Center
Building J-50
Norfolk, VA 23511

Counseling and Assistance Center
Concord Hall
Building IE
Norfolk, VA 23511

PETERSBURG

District 19 Substance Abuse Services
116 South Adams Street
Petersburg, VA 23803

Federal Correctional Institution Drug Abuse Program
River Road
Petersburg, VA 23804

Hiram Davis Medical Center Substance Abuse Services
Petersburg, VA 23803

Poplar Springs Hospital Chemical Dependency Program
350 Poplar Drive
Petersburg, VA 23805

PILOT

Serenity House, Inc.
Route 1
Pilot, VA 24138

PORTSMOUTH

Portsmouth Community Service Board Substance Abuse Services
500 Crawford Street
Suite 400
Portsmouth, VA 23704

Southeastern Virginia Alcohol Safety Action Program
505 Washington Street
Suite 710
Portsmouth, VA 23704

T. W. Neumann and Associates
720 Rodman Avenue
Portsmouth, VA 23707

POWHATAN

Goochland/Powhatan Community Service Board Substance Abuse Services
Powhatan, VA 23139

QUANTICO

Substance Abuse Counseling Center Marine Corps Combat Development Center
Building 3035
Quantico, VA 22134

RICHMOND

Capital Area Alcohol Safety Action Project
5407 Patterson Avenue
Suite 200
Richmond, VA 23226

Charter Westbrook Hospital Chemical Dependency Program
1500 Westbrook Avenue
Westbrook Psychiatric Hospital
Richmond, VA 23227

Human Resources' Inc.
Division of Addiction Services/Drug Free Unit
2926 West Marshall Street
Richmond, VA 23230

Drug Residential
919 West Grace Street
Richmond, VA 23220

Outpatient Methadone Program
15 West Cary Street
Richmond, VA 23220

Medical College of Virginia Division of Substance Abuse Medicine
1200 East Broad Street
11th Floor South Wing
Richmond, VA 23298

Richmond Aftercare, Inc. Men's and Women's Program
1109 Bainbridge Street
Richmond, VA 23224

Rubicon, Inc.
1300 Mactavish Avenue
Richmond, VA 23230

Veterans' Affairs Medical Center Substance Abuse Treatment Program
1201 Broad Rock Boulevard
Richmond, VA 23249

Virginia Health Center Adolescent Medicine Program Substance Abuse
2203 East Broad Street
Richmond, VA 23223

ROANOKE

Bethany Hall Women's Recovery Home Chemical Dependency Treatment
1109 Franklin Road SW
Roanoke, VA 24016

Mental Health Services of Roanoke Valley Substance Abuse Services
301 Elm Avenue
Roanoke, VA 24016

Total Action Against Poverty, Inc. TAP/MHS High Risk Prevention Program
145 West Campbell Avenue
Roanoke, VA 24011

SALEM

Mount Regis Center
405 Kimball Avenue
Salem, VA 24153

Roanoke Valley Alcohol Safety Action Program
220 East Main Street
Salem Bank and Trust Bldg., Suite 218
Salem, VA 24153

SOUTH BOSTON

Southside Community Services Board Substance Abuse Treatment Services
424 Hamilton Boulevard
South Boston, VA 24592

STAUNTON

Valley Alcohol Safety Action Program Holiday Court
Suite B
Staunton, VA 24401

Valley Community Services Board Substance Abuse Services
110 West Johnson Street
Staunton, VA 24401

STEPHENSON

Shalom Et Benedictus, Inc.
Route 664
Stephenson, VA 22656

SUFFOLK

Western Tidewater Mental Health Center Substance Abuse Department
1218 Holland Road
Suffolk, VA 23434

VIRGINIA BEACH

Tidewater Virginia Alcohol Safety Action Program
5163 Cleveland Street
Virginia Beach, VA 23462

Virginia Beach Substance Abuse Services
Pembroke Six
Suite 126
Virginia Beach, VA 23462

Virginia Beach Psychiatric Center Chemical Dependency Program
1100 First Colonial Road
Virginia Beach, VA 23454

WILLIAMSBURG

Bacon Street, Inc.
247 McLaws Circle
Suite 100
Williamsburg, VA 23185

Colonial Services Board Substance Abuse Services
1657 Merrimac Trail
Williamsburg, VA 23185

Williamsburg Community Hospital Substance Abuse Services
1238 Mount Vernon Avenue
Williamsburg, VA 23185

WINCHESTER

Council on Alcohol Lord Fairfax Comm., Inc. Lord Fairfax House
512 South Braddock Street
Winchester, VA 22601

New Life Center, Inc.
315 East Cork Street
Winchester, VA 22601

New Life Counseling Services
19 North Washington Street
Winchester, VA 22601

Northwestern Community Services Board Substance Abuse Service
1855 Front Royal Road
Winchester, VA 22601

Old Dominion Alcohol Safety Action Project
317 South Cameron Street
Winchester, VA 22601

Winchester Medical Center Choices
1840 Amherst Street
Winchester, VA 22601

WYTHEVILLE

Mount Rogers Mental Health Services Substance Abuse Services
770 West Ridge Road
Wytheville, VA 24382

YORKTOWN

U.S. Coast Guard RTC Drug and Alcohol Representative
Yorktown, VA 23690

WASHINGTON

ABERDEEN

Grays Harbor Community Hospital Eastcenter Recovery
1006 H Street
Aberdeen, WA 98520

ARLINGTON

Stillaguamish Tribe Substance Abuse Services
3439 Stoluckguamish Lane
Arlington, WA 98223

AUBURN

Federal Way Clinic Western Clinical Health Services
34507 Pacific Highway South
Suite 3
Auburn, WA 98003

BELLEVUE

Bellevue District Court Probation Division DWI Assessments
475 112 Avenue SE
Bellevue, WA 98009

Eastside Alcohol/Drug Center
606 120 Avenue NE
Park 120 Suite D-204
Bellevue, WA 98005

Group Health Cooperative Adapt Branch Facility
2661 Bel-Red Road
Suite 100
Bellevue, WA 98008

Professional Alcohol and Drug Services
555 116 Street NE
Suite 115
Bellevue, WA 98004

Starting Over Adolescent Recovery Services
515 116 Avenue NE
Suite 285
Bellevue, WA 98004

BELLINGHAM

Belair Clinic
1130 North State Street
Bellingham, WA 98225

Combined Treatment
1000 North Forest Street
Bellingham, WA 98225

Lummi Care Program
1790 Bayon Road
Bellingham, WA 98226

Olympic Center/Bellingham
1603 East Illinois Street
Bellingham, WA 98226

Unitycare
202 Unity Street
Lower Level Suite 100
Bellingham, WA 98225

BOTHELL

Alpha Center for Treatment, Inc.
10614 Beardslee Boulevard
Suite D
Bothell, WA 98011

Lakeside Recovery Centers, Inc.
14500 Juanita Drive NE
Bothell, WA 98011

Residence XII North/Bothell
14506 Juanita Drive NE
Bothell, WA 98011

BREMERTON

Group Health Adapt/Bremerton
5002 Kitsap Way
Suite 202
Bremerton, WA 98312

Kitsap Mental Health Services Substance Abuse Services
500 Union Avenue
Bremerton, WA 98312

Kitsap Recovery Center
1975 NE Fuson Road
Bremerton, WA 98310

Right Choice Counseling Service
1740 Northeast Riddell Road
Suite 314
Bremerton, WA 98310

BURLINGTON

Follman Agency
127 South Spruce Street
Burlington, WA 98233

CAMAS

Day One Treatment Center
2045 SW 6 Avenue
Camas, WA 98607

CENTRALIA

Maple Lane School
20311 Old Highway 9 Southwest
Centralia, WA 98531

South Sound Advocates for Disabled Citizens/Chemical Dependency Services
114 West Pine Street
Centralia, WA 98531

CHELAN

Lake Chelan Community Hospital Chemical Dependency Unit
503 East Highland Avenue
Chelan, WA 98816

CLARKSTON

Rogers Counseling Center
900 7 Street
Clarkston, WA 99403

COLVILLE

Stevens County Counseling Services
1707 East Birch Avenue
Colville, WA 99114

DARRINGTON

Sauk/Suiattle Indian Tribe Substance Abuse Services
5318 Chief Brown Lane
Darrington, WA 98241

DAVENPORT

Lincoln County Alcohol/Drug Center
407 Morgan Street
Davenport, WA 99122

DAYTON

Columbia County Services Substance Abuse Program
221 East Washington Street
Dayton, WA 99328

ELLENSBURG

Alcohol/Drug Dependency Service
507 Nanum Street
Room 111
Ellensburg, WA 98926

EVERETT

A Solution Recovery Program
7207 Evergreen Way
Suite M
Everett, WA 98203

Catholic Community Services Lifeline Recovery Program
1918 Everett Avenue
Everett, WA 98201

Community Alcohol and Drug Services DBA The Phoenix Center
2808 Hoyt Avenue
Everett, WA 98201

Evergreen Manor, Inc.
Outpatient Services
Recovery House/Detox Services
2601 Summit Avenue
Everett, WA 98201

Focus
909 SE Everett Mall Way
Suite C-364
Everett, WA 98204

Pacific Treatment Alternatives
1114 Pacific Avenue
Everett, WA 98201

FAIRCHILD AIR FORCE BASE

Fairchild AFB Social Actions
92 MSSQ/MSS
Building 3509
Fairchild Air Force Base, WA 99011

FEDERAL WAY

Intercept Associates
30620 Pacific Highway South
Suites 108–109
Federal Way, WA 98003

FERNDALE

Avalon Counseling and Treatment Services
1980 Main Street
Ferndale, WA 98248

FORKS

West End Outreach Services Forks Community Hospital
RR 3
Forks, WA 98331

FORT LEWIS

Alcohol and Drug Abuse Prevention and Control Program
HQ I Corps/Fort Lewis
Building 4290 Room 207
Fort Lewis, WA 98433

FRIDAY HARBOR

San Juan Community Alcoholism Center
955 Guard Street
Friday Harbor, WA 98250

GIG HARBOR

Gig Harbor Counseling and Recovery Center
3620 Grandview Street
Gig Harbor, WA 98335

GRANDVIEW

Phoenix Addiction Counseling Services
242 Division Street
Grandview, WA 98930

HOQUIAM

Evergreen Chemical Dependency Program
510 1/2 8 Street
Hoquiam, WA 98550

KELSO

Drug Abuse Prevention Center
2112 South Kelso Drive
Kelso, WA 98626

KENT

Careunit Clinics of Washington/ Kent
25400 74 Avenue South
Kent, WA 98032

Comprehensive Alcohol Services
1609 South Central Avenue
Suite 1
Kent, WA 98032

Kent Youth and Family Services
232 South 2 Avenue
Suite 201
Kent, WA 98032

Southeast Community Alcohol/ Drug Center
213 4 Avenue South
Kent, WA 98032

KIRKLAND

Youth Eastside Services Lake Washington
13009 85 Street
Kirkland, WA 98033

LONGVIEW

Recovery Northwest/Longview
Inpatient Center
Intensive Outpatient
600 Broadway
Longview, WA 98632

LYNNWOOD

Crosby Enterprises, Inc.
3924 204 Street SW
Lynnwood, WA 98036

Northwest Alternatives, Inc. Alcohol/Drug Outpatient Services
15332 Highway 99
Suite 2
Lynnwood, WA 98037

Pacific Treatment Alternatives
19324 40 Avenue West
Suite A
Lynnwood, WA 98036

MAPLE VALLEY

Cedar Hills Treatment Center
15900 227 Avenue SE
Maple Valley, WA 98038

MCCHORD AIR FORCE BASE

McChord AFB Social Actions Drug/Alcohol Abuse Control Branch
Building 100
Room 2057
McChord Air Force Base, WA 98438

MONROE

Valley General Hospital Alcohol and Drug Recovery Center
14701 179 Street SE
Monroe, WA 98272

MOSES LAKE

Grant County Alcohol and Drug Center
510 West Broadway
Moses Lake, WA 98837

MOUNT VERNON

Skagit Community Mental Health Center Chemical Dependency Treatment Program
108 Broadway
Mount Vernon, WA 98273

Skagit Recovery Center John King Recovery House
1905 Continental Place
Mount Vernon, WA 98273

NEWPORT

Pend Oreille Community Alcoholism Center
South 230 Garden Avenue
Newport, WA 99156

OAK HARBOR

Whidbey Island Naval Air Station Counseling and Assistance Center
Building 103
Oak Harbor, WA 98278

OLALLA

Olalla Guest Lodge, Inc.
12851 Lala Cove Lane SE
Olalla, WA 98359

OLYMPIA

C. G. Campbell Associates, Inc.
2633A Parkmont Lane SW
Suite A-1
Olympia, WA 98502

Community Mental Health Center Chemical Dependency Unit
112 East State Street
Olympia, WA 98501

Damon Counseling Services
4412 Pacific Avenue SE
Suite B
Olympia, WA 98503

Northwest Resources
115 McCormick Street
Olympia, WA 98506

Saint Peter Chemical Dependency Program
4800 College Street SE
Olympia, WA 98503

Thurston/Mason Addictions Recovery Council (TAMARC)
1625 Mottman Road SW
Olympia, WA 98502

OMAK

Okanogan County Counseling Services Chemical Dependency Programs
307 South Main Street
Omak, WA 98841

OTHELLO

Adams County Community Counseling Service Alcohol/ Drug Abuse Program
165 North First Street
Suite 120
Othello, WA 99334

PASCO

Benton Franklin Alcohol/Drug Services
720 West Court Street
Sheridan Building Office 4
Pasco, WA 99301

Discovery Substance Abuse Services Detoxification Center
1020 South 7 Avenue
Pasco, WA 99301

Our Lady of Lourdes Health Center Alcohol/Drug Treatment Unit
520 4 Avenue
Pasco, WA 99302

Unity Counseling Services
303 North 20 Street
Pasco, WA 99301

POMEROY

Rogers Counseling Center
856 Main Street
Pomeroy, WA 99347

PORT ANGELES

Lower Elwha Chemical Dependency Program
1666 Lower Elwha Road
Port Angeles, WA 98362

North Olympic Alcohol and Drug Center
315 East 8 Street
Port Angeles, WA 98362

Peninsula Community Mental Health Center Substance Abuse Services
118 East 8 Street
Port Angeles, WA 98362

PORT TOWNSEND

Jefferson County Community Alcoholism/Drug Abuse Center
802 Sheridan
Mailstop 115
Port Townsend, WA 98368

PULLMAN

Whitman County Alcohol Center
NE 340 Maple Street
Room 2
Pullman, WA 99163

PUYALLUP

Counselor
315 39 Avenue SW
Suite 11
Puyallup, WA 98373

REDMOND

Group Health Cooperative Alcohol and Drug Abuse Unit
2700 152 Avenue NE
Redmond, WA 98052

RENTON

Drug Free Systems
817 South 3 Street
Porcello Building Suite 8
Renton, WA 98055

Renton Area Youth Services (RAYS)
1025 South 3 Street
Renton, WA 98055

Valley Medical Recovery Center
400 South 43 Street
Renton, WA 98055

RICHLAND

Psychological Consultants Outpatient Drug Free Program
846 Stevens Drive
Richland, WA 99352

SEATTLE

Addiction Recovery Center at Fifth Avenue Hospital
10560 5 Avenue NE
Seattle, WA 98125

Alcohol/Drug 24 Hour Help Line
3700 Rainier Avenue South
Suite B
Seattle, WA 98144

Alternatives
1818 Westlake Avenue North
Suite 106
Seattle, WA 98109

BLAADE
8806 6 Avenue NW
Seattle, WA 98102

Cairn/Justice Associates, Inc.
1207 North 200 Street
Suite 217
Seattle, WA 98133

Careunit Clinics of Washington/ Seattle
12345 15 Avenue NE
Seattle, WA 98125

Cascade Recovery Center Bainbridge Island
600 Winslow Way East
Suite 237
Seattle, WA 98110

Center for Human Services
17011 Meridian Avenue North
Seattle, WA 98133

10501 Meridian Avenue North
Suite E
Seattle, WA 98133

Central Seattle Recovery Center Jefferson Street Unit
1401 East Jefferson Street
Jefferson Ctr. Prof. Building, Suite 300
Seattle, WA 98122

Chinook Center
220 Queen Anne Avenue North
Seattle, WA 98109

Cornerstone Treatment Centers, Inc.
610 44 Street NW
Seattle, WA 98107

Dell Craig Therapists, Inc. Marina Professional Center
22030 7 Avenue South
Seattle, WA 98198

Drug Free Systems/TASC
811 First Avenue
Colman Building Suite 610
Seattle, WA 98104

Evergreen Treatment Services
1250 First Avenue South
Seattle, WA 98134

Genesis House
621 34 Avenue
Seattle, WA 98122

Group Health Cooperative Alcohol and Drug Abuse Unit
1703 Minor Avenue
Metropolitan Park II Suite 1500
Seattle, WA 98101

Guardian Recovery Program
2608 3 Avenue
Seattle, WA 98121

Hope Recovery Services
974 Industry Drive
Seattle, WA 98188

King County Alcoholism Treatment Facility Detoxification Center
1421 Minor Avenue
Seattle, WA 98101

King County Perinatal Treatment Program
1005 East Jefferson Street
Seattle, WA 98122

Milam Recovery Program
12845 Ambaum Boulevard SW
Seattle, WA 98146

North Seattle Treatment Services
4106 Stone Way North
Seattle, WA 98103

Northwest Treatment Center
130 Nickerson Street
Suite 210
Seattle, WA 98109

Praxis
3245 Fairview Avenue East
Suite 200
Seattle, WA 98102

Rainbow Treatment Services
14900 Interurban Avenue South
Suite 255
Seattle, WA 98168

Ruth Dykeman Youth and Family Service
15001 8 Avenue SW
Seattle, WA 98166

Schick Shadel Hospital Substance Abuse Program
12101 Ambaum Boulevard SW
Seattle, WA 98146

Seadrunar
Phase I/Georgetown
976 South Harney Street
Seattle, WA 98108

Phase II Capitol Hill
809 15 Avenue East
Seattle, WA 98112

Queenanne
200 West Comstock Street
Seattle, WA 98119

Seattle Indian Health Board Alcohol/Drug Outpatient
611 12 Avenue South
Seattle, WA 98144

Shamrock Group, Inc.
8535 Phinney Avenue North
Seattle, WA 98103

Shared Health Services
6300 South Center Boulevard
Suite 205
Seattle, WA 98188

South King County Drug and Alcohol Recovery Centers
15025 4 Avenue SW
Seattle, WA 98166

Stonewall Recovery Services
430 Broadway East
Seattle, WA 98102

Swedish Hospital Medical Center
Addiction Recovery Program
747 Summit Avenue
Seattle, WA 98104

Ballard
NW Market and Barnes Avenues
Seattle, WA 98107

Therapeutic Health Services, Inc. Midvale Treatment Center
17962 Midvale Avenue North
Suite 150
Seattle, WA 98133

Thunderbird Treatment Center
9236 Renton Avenue South
Seattle, WA 98118

Veterans' Affairs Medical Center Addiction Treatment Center
1660 South Columbian Way
Seattle, WA 98108

Virginia Mason Chemical Dependency Program
1100 Olive Way
Metro Park West Tower Suite 1000
Seattle, WA 98101

Women's Recovery Center
6532 Phinney Avenue North
Building B Room 16A
Seattle, WA 98103

SEDRO WOOLLEY

Pioneer Center North
2268 Hub Drive
Sedro Woolley, WA 98284

United General Hospital United Recovery Center
1971 Highway 20
Sedro Woolley, WA 98284

SEQUIM

Family Center
237 North Sequim Avenue
Sequim, WA 98382

Jamestown S. Klallam Chemical Dependency Program
305 Old Blyn Highway
Sequim, WA 98382

SHELTON

Opportunities Counseling
2201 Olympic Highway North
Shelton, WA 98584

SILVERDALE

Cascade Recovery Center Silverdale
9481 Silverdale Way
Silverdale, WA 98383

SNOQUALMIE

Echo Glen Children's Center Exodus
33010 SE 99 Street
Snoqualmie, WA 98065

SPOKANE

Addiction Outpatient Services of Spokane
East 901 2 Avenue
Suite 100
Spokane, WA 99202

Valley Office Branch
11704 East Montgomery Drive
Suite 5
Spokane, WA 99206

Addiction Recovery Systems, Inc.
West 601 Francis Avenue
Spokane, WA 99205

Alcohol/Drug Network
1101 West College Avenue
Spokane, WA 99201

Center for Drug Treatment
East 115 Indiana Street
Spokane, WA 99207

Colonial Clinic
West 315 9 Avenue
Suite 210
Spokane, WA 99204

Daybreak of Spokane
Intensive Inpatient Program for Youth
South 4611 Dyer Road
Spokane, WA 99223

Outpatient Treatment
East 918 Mission Avenue
Spokane, WA 99202

Genesis Counseling Service, Inc.
North 10103 Division Street
Suite 100
Spokane, WA 99218

Group Health Northwest Chemical Dependency Program
316 West Boone Avenue
Rock Pointe Corporate Center Suite 650
Spokane, WA 99201

Inland Tribal Consortium Youth Treatment Center
North 1617 Calispel
Spokane, WA 99205

Isabella House
West 2308 3 Avenue
Spokane, WA 99204

Mountainview Outpatient
517 South Division Street
Suite 9
Spokane, WA 99202

New Horizon Counseling Services
West 2317 3 Avenue
Spokane, WA 99204

River City Counseling Services
North 7307 Division
Suite 300
Spokane, WA 99208

Salvation Army
Alcohol and Drug Outpatient Services
North 2020 Division Street
Spokane, WA 99204

Booth Care Center
West 3400 Garland Avenue
Spokane, WA 99205

Spokane Addiction Recovery Centers (SPARC)
Inpatient Services
West 1403 7 Avenue
Spokane, WA 99210

Outpatient Services
South 812 Walnut Street
Spokane, WA 99204

Recovery House
1509 West 8 Avenue
Spokane, WA 99204

Spokane Care Service
South 165 Howard Street
Spokane, WA 99204

Stepps/Youth Help Association
West 522 Riverside
Suite 600
Spokane, WA 99201

TASC of Spokane County
1324 North Ash Street
Spokane, WA 99201

Veterans' Affairs Medical Center Substance Abuse Treatment Program
North 4815 Assembly Street
Spokane, WA 99205

SUNNYSIDE

Merit Resource Services
702 East Franklin Street
Sunnyside, WA 98944

TACOMA

Affirmation Counseling Services
4301 South Pine Street
Suite 30
Tacoma, WA 98409

Family Recovery Resources MOMS
2367 Tacoma Avenue South
Tacoma, WA 98402

Griffin and Griffin EAP, Inc.
4218 South Steele Street
Suite 304
Tacoma, WA 98409

Group Health Cooperative Adapt
4301 South Pine Street
Suite 219
Tacoma, WA 98409

Metro Dev. Council Chemical Dependency Division Detox Center and Outpatient Treatment
622 Tacoma Avenue South
Suite 6
Tacoma, WA 98402

Passages Professional Counseling and Educational Services
10510 Gravelly Lake Drive SW
Suite 200
Tacoma, WA 98499

Pierce County Alliance
Outpatient Treatment Program
710 South Fawcett Avenue
Tacoma, WA 98402

YOP/405 Program/Youth
1110 South 12 Street
Tacoma, WA 98405

Plaza Hall
1415 Center Street
Tacoma, WA 98409

Puyallup Tribal Treatment Center
2209 East 32 Street
Tacoma, WA 98402

Tacoma Detoxification Center
721 Fawcett Avenue
Room 100
Tacoma, WA 98402

Tacoma Pierce County Methadone Maintenance Program
3629 South D Street
CHD-049
Tacoma, WA 98408

Transitions Limited
10116 Portland Avenue
Suite B
Tacoma, WA 98445

Veterans' Affairs Medical Center Alcohol/Drug Dependence Treatment Program
American Lake
Unit 116C/D
Tacoma, WA 98493

Western Washington Alcohol Center, Inc.
3049 South 36 Street
Suite 214
Tacoma, WA 98409

TAHOLAH

Quinault Indian Nation Alcoholism Treatment Program
Taholah, WA 98587

TOPPENISH

Mwerit Resource Services
307 Asotin Street
Toppenish, WA 98948

VANCOUVER

Clark County Council on Alcohol and Drugs 8th Street Branch
509 West 8 Street
Vancouver, WA 98660

Recovery Northwest Alcohol/ Drug Outpatient Center
210 West 11 Street
Vancouver, WA 98660

Southwest Washington Medical Center Turn Around at Vancouver
400 NE Mother Joseph Way
Vancouver, WA 98664

Starting Point
2703 East Mill Plain Boulevard
Vancouver, WA 98661

Veterans' Affairs Medical Center Chemical Addictions Rehab Section
O Street and 4 Plain Street
Ward 30
Vancouver, WA 98661

WALLA WALLA

Veterans' Affairs Medical Center Chemical Dependency Treatment Unit
77 Wainwright Drive
Walla Walla, WA 99362

WENATCHEE

Center for Alcohol and Drug Treatment Casa, Inc.
327 Okanogan Avenue
Wenatchee, WA 98801

WHITE SALMON

Counseling and Resource Center Chemical Dependency Program
40 Skyline Hospital
White Salmon, WA 98672

WOODINVILLE

Motivations
17311 135 Avenue NE
Suite C-400
Woodinville, WA 98072

YAKIMA

A J Alcohol and Drug Services
32 North 3 Street
Room 310
Yakima, WA 98901

Barth Clinic
105 East E Street
Suite 2
Yakima, WA 98901

Bollman Counseling Center
412 South 3 Street
Suite A
Yakima, WA 98901

Central Washington Comprehensive Mental Health Drug Program
321 East Yakima Avenue
Yakima, WA 98901

Dependency Health Services
401 South 5 Avenue
Yakima, WA 98902

Riel House
1408 West Yakima Avenue
Yakima, WA 98902

Sundown M Ranch
2280 SR 821
Yakima, WA 98901

Yakima Community Alcohol and Drug Center
102 South Naches Avenue
Yakima, WA 98901

DBA Dependency Health Services
315 Holton Avenue
Suite B-1
Yakima, WA 98902

YELM

Resolution A Counseling Service
806 Yelm Avenue East
Suite 7
Yelm, WA 98597

WEST VIRGINIA

ALDERSON

Federal Correctional Institution Substance Abuse Services
Alderson, WV 24910

BECKLEY

Fayette/Monroe/Raleigh/ Summers (EMRS) Mental Health Council, Inc.
101 South Eisenhower Drive
Beckley, WV 25801

Southern West Virginia Fellowship Home, Inc.
Beckley, WV 25801

Veterans' Affairs Medical Center Substance Abuse Treatment Program
200 Veterans Avenue
Beckley, WV 25801

BLUEFIELD

Mercer County Fellowship Home
421 Scott Street
Bluefield, WV 24701

CHARLESTON

AFL/CIO Appalachian Council, Inc. Occupational Alcoholism Program
501 Broad Street
Charleston, WV 25301

Charleston Area Medical Center, Inc.
Brooks and Washington Streets
Charleston, WV 25325

Comprehensive Alcohol Rehab and Evaluation Services (CARES)
1716 7 Avenue
Charleston, WV 25312

Kanawha Valley Fellowship Home, Inc.
1107 Virginia Street East
Charleston, WV 25301

Shawnee Hill Substance Abuse Peers
1591 Washington Street East
Charleston, WV 25311

Shawnee Hills Community MH/ MR Center, Inc.
Brookside Campus Substance Abuse Services
Extended Care Service for Women
705 South Park Road
Charleston, WV 25325

Southway Treatment Center
4605 Maccorkle Avenue SW
Charleston, WV 25309

Threshold at Brookside
705 South Park Road
Charleston, WV 25304

CLARKSBURG

Summit Center for Human Development
6 Hospital Plaza
Clarksburg, WV 26301

Veterans' Affairs Medical Center Substance Abuse Services
Route 98
Clarksburg, WV 26301

CROSS LANES

Phoenix House
5405 Alpine Drive
Cross Lanes, WV 25313

ELKINS

Appalachian Mental Health Center Alcoholism and Drug Abuse Program
725 Yokum Street
Elkins, WV 26241

FAIRMONT

Fairmont General Hospital Addiction Recovery Unit
1325 Locust Avenue
Fairmont, WV 26554

GYPSY

Rainbow House
158 Main Street
Gypsy, WV 26361

HOPEMONT

Shawnee Hills Rehabilitation Unit
Hopemont, WV 26764

HUNTINGTON

HCA River Park Hospital Bridges Chemical Dependency Program
1230 6 Avenue
Huntington, WV 25701

Prestera Center for Mental Health Services, Inc.
3375 U.S. Route 60 East
Huntington, WV 25705

New Directions
1530 Norway Avenue
Huntington State Hospital
Huntington, WV 25709

Saint Mary's Hospital Substance Abuse Unit
2900 First Avenue
Huntington, WV 25701

Transition House
432 6 Avenue
Huntington, WV 25705

West Virginia Department of Health Substance Abuse Unit
1530 Norway Avenue
Huntington, WV 25709

KINGWOOD

Olympic Center/Preston Adolescent Treatment Program
Route 7
Kingwood, WV 26537

Preston Addiction Treatment Center
300 South Price Street
Kingwood, WV 26537

LEWISBURG

Hunter House
100 Church Street
Lewisburg, WV 24901

LOGAN

Logan/Mingo Area Mental Health Substance Abuse Services
Route 10
3 Mile Curve
Logan, WV 25601

MARTINSBURG

Eastern Panhandle CMHC, Inc. Main Unit
235 South Water Street
Martinsburg, WV 25401

MORGANTOWN

Chestnut Ridge Hospital WVU Dept. Behavioral Medicine/ Psych. Addiction Recovery Unit
930 Chestnut Ridge Road
Morgantown, WV 26505

Federal Correctional Institution Chemical Abuse Program Drug Abuse Unit/Gerard
Morgantown, WV 26505

Valley Comprehensive CMHC Main Unit
301 Scott Avenue
Morgantown, WV 26505

PARKERSBURG

Mid-Ohio Valley Fellowship Home, Inc.
1030 George Street
Parkersburg, WV 26101

Western District Guidance Center, Inc. Substance Abuse Program
2121 7 Street
Parkersburg, WV 26101

PETERSBURG

Potomac Highlands Guild, Inc.
1 Virginia Avenue
Petersburg, WV 26847

PRINCETON

Southern Highlands Community Mental Health Center, Inc.
200 12 Street Extension
Princeton, WV 24740

RIPLEY

Jackson General Hospital Jackson Treatment Center
Pinnell Street
Ripley, WV 25271

SUMMERSVILLE

Seneca MH/MR Council, Inc.
1305 Webster Road
Summersville, WV 26651

SUTTON

Braxton County Fellowship Home
11 Stonewall Street
Sutton, WV 26601

WEIRTON

Hancock/Brooke Mental Health Service
501 Colliers Way
Weirton, WV 26062

WHEELING

Northwood Health Systems
2121 Eoff Street
Wheeling, WV 26003

New Hope
52 15 Street
Wheeling, WV 26003

Ohio Valley Medical Center Alcoholism Services
2000 Eoff Street
Wheeling, WV 26003

WISCONSIN

ALGOMA

Kewaunee County Community Programs Alcohol and Drug Abuse Treatment Program
522 4 Street
Algoma, WI 54201

ALMA

Buffalo County Dept. of Human Services
Courthouse
Alma, WI 54610

Lutheran Hospital Counseling Services
Courthouse
Alma, WI 54610

ANTIGO

Langlade Health Care Center
1225 Langlade Road
Antigo, WI 54409

Langlade Memorial Hospital Substance Abuse Emergency Services
112 East 5 Avenue
Antigo, WI 54409

APPLETON

Casi Programs, Inc.
217 South Badger Avenue
Appleton, WI 54914

Lutheran Social Services Waples Teen Recovery Services
1412 North Rankin Street
Appleton, WI 54913

Meridian House
1308 North Leona Street
Appleton, WI 54913

Saint Elizabeth Hospital
Alcohol/Drug Program (ADP)
1506 South Oneida Street
Appleton, WI 54915

Residential Living Facility/Women
1310 South Madison Street
Appleton, WI 54915

The Mooring Halfway House, Inc.
607 West 7 Street
Appleton, WI 54911

ARCADIA

Saint Joseph's Hospital Emergency Room Substance Abuse Services
464 South Saint Joseph Avenue
Arcadia, WI 54612

ASHLAND

Ashland County Information and Referral Center
206 6 Avenue West
Room 213
Ashland, WI 54806

Memorial Medical Center, Inc. Memorial Medical Treatment Center
1635 Maple Lane
Ashland, WI 54806

BARABOO

Saint Clare Hospital Saint Clare Center
707 14 Street
Baraboo, WI 53913

BAYFIELD

Red Cliff Tribal Council Alcohol and Drug Program
Route 1
Bayfield, WI 54814

BEAVER DAM

Beaver Dam Community Hospital AODA Detox/AODA Inpatient Emergency Care
707 South University Avenue
Beaver Dam, WI 53916

Lutheran Social Services
1010 De Clark Street
Beaver Dam, WI 53916

BELOIT

Addiction Treatment and Education Program
2091 Shopiere Road
Beloit, WI 53511

Beloit Inner City Council Substance Abuse Services
934 Alice Street
Beloit, WI 53511

Intoxicated Driver Program
136 West Grand Avenue
Beloit, WI 53511

Parkside Lodge of Wisconsin
2185 Shopiere Road
Beloit, WI 53511

Rock Valley Correctional Programs, Inc. Treatment Alternative Program
431 Olympian Boulevard
Beloit, WI 53511

BERLIN

Berlin Memorial Hospital Emergency Detoxification
225 Memorial Drive
Berlin, WI 54923

BLACK RIVER FALLS

Lutheran Hospital Counseling Services of Black River Memorial Hospital/Substance Abuse Services
610 West Adams Street
Black River Falls, WI 54615

BOSCOBEL

Memorial Hospital of Boscobel Substance Abuse Services
205 Parker Street
Boscobel, WI 53805

BOWLER

Stockbridge/Munsee Health Center Tribal Alcoholism Treatment Program
Route 1 Moheconnock Road
Bowler, WI 54416

BROOKFIELD

Elmbrook Memorial Hospital Alcohol and Drug Treatment Center
19333 West North AVenue
Brookfield, WI 53045

CHILTON

Calumet County Human Service Dept. Alcohol and Other Drug Abuse Unit
206 Court Street
Courthouse
Chilton, WI 53014

Calumet Medical Center Substance Abuse Detox Services
614 Memorial Drive
Chilton, WI 54014

CHIPPEWA FALLS

Council on Alcohol and Other Drug Abuse, Inc.
404 1/2 North Bridge Street
Box 4
Chippewa Falls, WI 54729

L. E. Phillips Libertas Center for the Chemically Dependent
2661 County Road I
Chippewa Falls, WI 54729

Saint Joseph's Hospital Emergency Room
2661 County Road I
Chippewa Falls, WI 54729

Serenity House, Inc. Transitional Living Program
205 East Grand Avenue
Chippewa Falls, WI 54729

Transitus House
1830 Wheaton Street
Chippewa Falls, WI 54729

CLINTONVILLE

Clintonville Community Hospital Emergency Detoxification
35 North Anne Street
Clintonville, WI 54929

CRANDON

Forest County Potawatomi Alcohol and Drug Program
Crandon, WI 54520

Lakeland Council on AODA, Inc.
211 East Madison Street
Crandon, WI 54520

Sokaogon Community Center Office of Alcohol and Other Drug Abuse
Route 1
Crandon, WI 54520

CUMBERLAND

Cumberland Memorial Hospital Emergency Detoxification
1110 7 Avenue
Cumberland, WI 54829

Northern Pines Unified Service Center Board Chemical Dependency Service
1066 8 Avenue
Cumberland, WI 54829

DARLINGTON

Lafayette County Department of Human Services/AODA Program
700 North Main Street
Darlington, WI 53530

Memorial Hospital of Lafayette County
800 Clay Street
Darlington, WI 53530

DODGEVILLE

Unified Counseling Services Dodgeville Outpatient Clinic
410 North Union Street
Dodgeville, WI 53533

DURAND

Pepin County Department of Human Services
740 7 Avenue West
Durand, WI 54736

EAGLE RIVER

Eagle River Memorial Hospital Emergency Room Substance Abuse Services
Eagle River, WI 54521

Lakeland Council on AODA, Inc.
Highway 45 North
Eagle River, WI 54521

EAU CLAIRE

Eau Claire Academy
550 North Dewey Street
Eau Claire, WI 54701

Fahrman Center
3136 Craig Road
Eau Claire, WI 54701

Guidance Clinic, Inc.
202 Graham Avenue
Eau Claire, WI 54701

Luther Hospital Genesis Adolescent Chemical Dependency Program
1221 Whipple Street
Eau Claire, WI 54702

Lutheran Social Services
3136 Craig Road
Eau Claire, WI 54701

Midelfort Clinic Journey
733 West Clairemont Avenue
Eau Claire, WI 54701

Triniteam, Inc. Treatment Alternative Program
515 South Barstow Street
Suite 114
Eau Claire, WI 54701

EDGERTON

Memorial Community Hospital Parkside Lodge/Edgerton
313 Stoughton Road
Edgerton, WI 53534

ELKHORN

Lakeland Medical Center Emergency Room Substance Abuse Services
Elkhorn, WI 53121

Walworth County Department of Human Services Center
County Highway NN
Elkhorn, WI 53121

ELLSWORTH

Pierce County Dept. of Human Services Alcohol and Other Drug Abuse Services
412 West Kinne Street
Ellsworth, WI 54011

ELROY

Saint Francis Community Programs Pinecrest Center Extended Care II
1510 Academy Street
Elroy, WI 53929

FOND DU LAC

Blandine House, Inc.
25 North Park Avenue
Fond du Lac, WI 54935

Fond du Lac County Alcohol and Other Drug Abuse Counseling Center
459 East First Street
Fond du Lac, WI 54935

Robert E. Berry Halfway House
178 6 Street
Fond du Lac, WI 54935

Saint Agnes Hospital Alcohol Rehabilitation Unit
430 East Division Street
Fond du Lac, WI 54935

FORT ATKINSON

Fort Atkinson Memorial Hospital Substance Abuse Emergency Medical Detox
611 East Sherman Avenue
Fort Atkinson, WI 53538

FRIENDSHIP

Adams County Dept. of Community Programs
108 East North Street
Friendship, WI 53934

Adams County Memorial Hospital Substance Abuse Emergency Services
402 West Lake Street
Friendship, WI 53934

GREEN BAY

Bellin Psychiatric Center Addictive Services
725 South Webster Avenue
Green Bay, WI 54305

Brown County Mental Health Center Alcohol and Other Drug Abuse Services
2900 Saint Anthony Drive
Green Bay, WI 54311

Family Service Association Outpatient AODA Program
1546 Dousman Street
Green Bay, WI 54303

Genesis II
2900 Saint Anthony Drive
Green Bay, WI 54311

Our Lady of Charity Center, Inc. Family Program
2640 West Point Road
Green Bay, WI 54307

Saint Mary's Hospital Medical Center Substance Abuse Emergency Detox Services
1726 Shawano Avenue
Green Bay, WI 54303

Samaritan House
630 Cherry Street
Green Bay, WI 54301

GREEN LAKE

Green Lake County Human Services Dept. Community Services Unit
500 Lake Steel Street
Green Lake, WI 54941

GRESHAM

Maehnowesekiyah Treatment Program
N 4587 County C
Gresham, WI 54128

HALES CORNERS

Family Social and Psychotherapy
5300 South 108 Street
Hales Corners, WI 53130

HARTFORD

Kettle Moraine Treatment Center Hartford Clinic
1567 Sumner Street
Hartford, WI 53027

HAYWARD

Hayward Area Memorial Hospital Substance Abuse Services
Route 3
Hayward, WI 54843

Lac Courte Oreilles Alcohol/Drug and Mental Health Program
Route 2
Hayward, WI 54843

Sawyer County Council on AODA Hill House
County Hill Road
Hayward, WI 54843

Sawyer County Info. and Referral Center on Alcohol and Other Drug Abuse
315 West 5 Street
Hayward, WI 54843

HERTEL

Saint Croix Tribal Alcoholism Program
Hertel, WI 54845

HILLSBORO

Saint Joseph's Hospital Emergency Detoxification
400 Water Avenue
Hillsboro, WI 54634

HUDSON

Burkwood Residence
615 Old Mill Road
Hudson, WI 54016

Hudson Medical Center Chemical Health Recovery Center
400 Wisconsin Street
Hudson, WI 54016

HURLEY

Iron County Council on Alcohol and Drug Abuse
408 Silver Street
Hurley, WI 54534

JANESVILLE

Alcohab, Inc.
New Horizons Halfway House
170 Linn Street
Janesville, WI 53545

River Commons Halfway House
786 South Main Street
Janesville, WI 53545

Crossroads Counseling Center
301 East Milwaukee Street
Janesville, WI 53545

Lutheran Social Services/Rock County Alcohol and Drug Treatment Unit
205 North Main Street
Suite 102
Janesville, WI 53545

Parkside Lodge of Wisconsin
320 Lincoln Street
Janesville, WI 53547

Rock County Psychiatric Hospital Substance Abuse Services
3530 North Cty Trunk F
Janesville, WI 53545

JEFFERSON

Jefferson County Human Services Dept. Hillside Chemical Dependency Unit
N3995 Annex Road
Jefferson, WI 53549

JUNEAU

Dodge County Dept. of Human Services Chemical Dependency Services
199 Home Road
Juneau, WI 53039

KENOSHA

Alcohol and Drug Consultants
7543 17 Avenue
Kenosha, WI 53143

Alcohol and Other Drugs Council of Kenosha County, Inc.
1115 56 Street
Kenosha, WI 53140

Interconnections
920 60 Street
Kenosha, WI 53140

Interventions
6755 14 Avenue
Kenosha, WI 53140

Kenosha Hospital and Medical Center
3 West
6308 8 Avenue
Kenosha, WI 53140

Kenosha Mental Health and Family Counseling Center
1605 Birch Road
Kenosha, WI 53140

Kenosha Youth Development Services (KYDS) Substance Abuse Program
5407 8 Avenue
Kenosha, WI 53140

Lakeshore Counseling Associates
3618 8 Avenue
Lakeshore Medical Building
Kenosha, WI 53140

Lincoln Neighborhood Community Center, Inc. Inner City Project on Prevention of Substance Abuse
1607 65 Street
Kenosha, WI 53143

Oakwood Clinical Associates, Ltd.
3734 7 Avenue
Suite 3
Kenosha, WI 53140

Saint Catherine's Hospital Behavioral Services/3E
3556 7 Avenue
Kenosha, WI 53140

KESHENA

Menominee County Human Services Dept. Alcohol and Other Drug Abuse Program
Keshena, WI 54135

KEWASKUM

Exodus Transitional Care Facility, Inc.
1421 Fond Du Lac Avenue
Kewaskum, WI 53040

LA CROSSE

Coulee Council on Alcohol and Other Chemical Abuse
921 West Avenue South
La Crosse, WI 54601

La Crosse County Human Service Dept. Clinical Services Section
300 North 4 Street
La Crosse, WI 54601

La Crosse Lutheran Hospital Residential Treatment Center CBRE
1312 5 Avenue South
La Crosse, WI 54601

Lutheran Hospital/La Crosse Regional Center for Chemical Dependency Treatment
1910 South Avenue
La Crosse, WI 54601

Unity House I
1920 Miller Street
La Crosse, WI 54601

Unity House II
1924 Miller Street
La Crosse, WI 54601

Lutheran Social Services of Wisconsin and Upper Michigan Inc./Southwest Office Area
2350 South Avenue
Suite 5
La Crosse, WI 54601

Saint Francis Community Programs, Inc.
Laar House
1022 Division Street
La Crosse, WI 54601

Scarseth House
535 South 17 Street
La Crosse, WI 54601

Saint Francis Medical Center Chemical Dependency Services
700 West Avenue South
La Crosse, WI 54601

LAC DU FLAMBEAU

Family Resource Center Chippewa Health Center
450 Old Abe Road
Lac du Flambeau, WI 54538

LADYSMITH

Rusk County Memorial Hospital Substance Abuse Services
900 College Avenue West
Ladysmith, WI 54848

LANCASTER

Unified Counseling Services Lancaster Outpatient Clinic
230 West Cherry Street
Lancaster, WI 53813

MADISON

ARC Community Services, Inc. Arc House
202 North Paterson Street
Madison, WI 53703

Hope Haven, Inc.
Colvin Manor
425 West Johnson Street
Madison, WI 53703

Hope Haven
425 West Johnson Street
Madison, WI 53703

North Bay Lodge
3602 Memorial Drive
Madison, WI 53704

Lutheran Social Services
101 Nob Hill Road
Suite 200
Madison, WI 53713

Madison Inner City Council on Substance Abuse, Inc.
1244 South Park Street
Lower Level
Madison, WI 53715

Matkom Outpatient Clinic
1 South Park Street
Suite 420
Madison, WI 53715

Mendota Mental Health Institute Substance Abuse Services
301 Troy Drive
Madison, WI 53704

Mental Health Center of Dane County Alcohol and Drug Treatment Unit
625 West Washington Avenue
Madison, WI 53703

Meriter Hospital/New Start
New Start East
1310 Mendota Street
Suite 110
Madison, WI 53714

New Start West
1015 Gammon Lane
Madison, WI 53719

Raymond Road Unit
8221 Raymond Road
Madison, WI 53719

Washington Avenue Unit
309 West Washington Avenue
Madison, WI 53703

Prevention and Intervention Center for Alcohol and Other Drug Abuse (PICADA)
2000 Fordem Avenue
Madison, WI 53704

Rebos House of Wisconsin I
Women
1903 University Avenue
Madison, WI 53705

House of Wisconsin II
Men
810 West Olin Avenue
Madison, WI 53715

Tellurian Community, Inc.
Tellurian Apogee
Adult Residential Program (ARP)
Day Treatment Program
Network AM/PM
300 Femrite Drive
Madison, WI 53716

Tellurian Ucan, Inc.
Detox Unit
2914 Industrial Drive
Madison, WI 53713

Thoreau House
1102 Spaight Street
Madison, WI 53703

Univ. of Wisconsin Hospital and Clinics Adolescent Alcohol/ Drug Abuse Intervention Program
122 East Olin Avenue
Suite 275
Madison, WI 53705

MANITOWOC

Counseling and Development Center
3618 Calumet Avenue
Manitowoc, WI 54220

Holy Family Memorial Medical Center Chemical Dependency Unit
333 Reed Avenue
Manitowoc, WI 54220

Manitowoc County Human Services Dept. Counseling Center
339 Reed Avenue
Manitowoc, WI 54220

Marco
1114 South 11 Street
Manitowoc, WI 54220

MARINETTE

Bay Area Medical Center Emergency Drug Abuse Programs
3100 Shore Drive
Marinette, WI 54143

Goodwill Industries, Inc. Marinette County
1428 Main Street
Marinette, WI 54743

Marinette County Human Services Adapt
400 Wells Street
Marinette, WI 54143

MAUSTON

Hess Memorial Hospital Emergency Room Substance Abuse Services
1050 Division Street
Mauston, WI 53948

Juneau County Human Service Center
220 East La Crosse Street
Mauston, WI 53948

Wisconsin Winnebago Health Authority Alcohol/Drug Outpatient Program
104 West State Street
Mauston, WI 53948

MEDFORD

Taylor County Human Services Department
219 South Wisconsin Avenue
Medford, WI 54451

MENASHA

Family Service Association of Fox Valley
1488 Kenwood Center
Menasha, WI 54952

New Horizons Counseling Services Menasha Office
222 South Washington Street
Menasha, WI 54952

Theda Clark Center for Recovery
324 Nicolet Boulevard
Menasha, WI 54952

MENOMONIE

Dunn County Association on Alcohol and Other Drug Abuse
320 21 Street North
Menomonie, WI 54751

MERRILL

Good Samaritan Medical Center Emergency Detoxification
601 South Center Avenue
Merrill, WI 54452

Lincoln Health Care Center Merrill Office
503 South Center Avenue
Merrill, WI 54452

Sacred Heart Outpatient Clinic Oasis Recovery Program
402 West Main Street
Merrill, WI 54452

MILWAUKEE

Addiction Fighters, Inc.
1840 North Farwell Avenue
Suite 304
Milwaukee, WI 53202

American Indian Council on Alcoholism
2240 West National Avenue
Milwaukee, WI 53204

Children and Family Service, Inc.
6249 West Fond du Lac Avenue
Milwaukee, WI 53218

Council for the Spanish Speaking Salud de la Familia
614 West National Street
Milwaukee, WI 53204

Crossroads Recovery Center, Inc.
2436 North 50 Street
Milwaukee, WI 53210

De Paul Belleview Extended Care
1904 East Belleview Place
Milwaukee, WI 53211

De Paul Hospital, Inc. Addiction and Mental Health Center
4143 South 13 Street
Milwaukee, WI 53221

Eastside Youth and Family Clinic
1840 North Farwell Avenue
Suite 304
Milwaukee, WI 53202

Harambee Ombudsman Project, Inc. Imani II
3614 North 39th Street
Milwaukee, WI 53216

Horizon House
2511 West Vine Street
Milwaukee, WI 53205

Inner City Council on Alcoholism (ICCA)
4365 North 27 Street
Milwaukee, WI 53216

Family Program and Driver Improvement Program
3660 North Teutonia Avenue
Milwaukee, WI 53206

Southside Outpatient Center
1675 South 8 Street
Milwaukee, WI 53204

Ivanhoe Treatment, Inc.
2203 East Ivanhoe Place
Milwaukee, WI 53202

Kettle Moraine Milwaukee County Alcohol Detoxification
1218 West Highland Boulevard
Milwaukee, WI 53233

Lutheran Social Services of Wisconsin and Upper Michigan, Inc.
3200 West Highland Boulevard
Milwaukee, WI 53208

Matt Talbot Lodge
2613 West North Avenue
Milwaukee, WI 53205

META for Women and Children
2571–2579 North Farwell Avenue
Milwaukee, WI 53211

Milwaukee Council on Alcoholism
2266 North Prospect Avenue
Suite 324
Milwaukee, WI 53202

Milwaukee Council on Drug Abuse
1442 North Farwell Avenue
Milwaukee, WI 53202

Milwaukee County Dept. of Human Services Adult Services Division Admin. Office
235 West Galena Street
Suite 270
Milwaukee, WI 53212

Milwaukee Health Department
841 North Broadway
Milwaukee, WI 53202

Milwaukee Health Services System
4383 North 27 Street
Milwaukee, WI 53216

Milwaukee Indian Health Center Substance Abuse Services
930 North 27 Street
Milwaukee, WI 53208

Milwaukee Psychiatric Hospital Division of Chemical Dependency
1220 Dewey Avenue
Milwaukee, WI 53213

Milwaukee Women's Center Alcohol and Drug Abuse Program
611 North Broadway
Suite 230
Milwaukee, WI 53202

Multi-Cultural Counseling Services DBA Renew Counseling Services
1225 West Mitchell Street
Suite 223
Milwaukee, WI 53204

2014 West North Avenue
Milwaukee, WI 53233

Next Door Foundation Genesis
726 North 31 Street
Milwaukee, WI 53208

Northwest General Hospital Substance Abuse Services
5310 West Capitol Drive
Milwaukee, WI 53216

Psychological Addiction Consultants
5401 North 76 Street
Milwaukee, WI 53218

Reach, Inc. Comprehensive Mental Health Clinic Substance Abuse
6001 West Center Street
Suite 97LL
Milwaukee, WI 53210

S A F E Group Services, Inc.
3500 North Sherman Boulevard
Suite 302
Milwaukee, WI 53216

Saint Mary's Hill Hospital Alcohol and Drug Abuse Services
2350 North Lake Drive
Milwaukee, WI 53211

Saint Michael's Hospital Mental Health Center Alcohol and Other Drug Abuse Outpatient Treatment Program
2400 West Villard Avenue
Milwaukee, WI 53209

Sinai Samaritan Medical Center Substance Abuse Services
2000 West Kilbourn Avenue
Milwaukee, WI 53233

Social Development Commission Outpatient AODA Services
2707 North 54 Street
Milwaukee, WI 53210

United Community Center New Beginning Clinic
1028 South 9 Street
Milwaukee, WI 53204

Veterans' Affairs Medical Center Substance Abuse Treatment Program
5000 West National Avenue
Routing Symbol 116E
Milwaukee, WI 53295

Wisconsin Cipe, Inc.
1925 West Hampton Avenue
Milwaukee, WI 53206

Wisconsin Correctional Service (WCS)
Outpatient Substance Abuse Program
436 West Wisconsin Avenue
Milwaukee, WI 53203

Residential Drug Abuse Treatment Program
2105 North Booth Street
Milwaukee, WI 53212

MINOCQUA

Lakeland Council on AODA, Inc. Outpatient Counseling
415 Menominee Street
Minocqua, WI 54548

MONROE

Green County Human Services Alcohol/Other Drug Abuse Services
Pleasant View Complex
Box 216
Monroe, WI 53566

MONTELLO

Marquette Chemical Dependency Service
Highway 22 South
Montello, WI 53949

NEENAH

Freedom House
135 Curtis Street
Neenah, WI 54956

Haukeness Counseling Services
307 South Commercial Street
Old Post Office Building
Neenah, WI 54956

NEILLSVILLE

Clark County Community Services Alcohol and Other Drug Abuse Program
517 Court Street
Neillsville, WI 54456

NEW FRANKEN

Alpine Country House, Inc.
5628 Sturgeon Bay Road
New Franken, WI 54229

NEW LONDON

New London Family Medical Center Emergency Room Substance Abuse Services
1405 Mill Street
New London, WI 54961

Rawhide
E7475 Weiland Road
New London, WI 54961

NEW RICHMOND

Saint Croix County Human Services Alcohol and Drug Abuse Services
1246 185 Avenue
New Richmond, WI 54017

OCONOMOWOC

Kettle Moraine Outpatient Clinic
612 East Summit Avenue
Oconomowoc, WI 53066

OCONTO FALLS

Community Memorial Hospital Emergency Room Substance Abuse Services
855 South Main Street
Oconto Falls, WI 54154

Oconto County Dept. of Human Services Clinical Services Division/Substance Abuse Unit
835 South Main Street
Oconto Falls, WI 54154

ODANAH

Bad River Alcohol/Drug Program
Odanah, WI 54861

ONEIDA

Oneida Social Services
Kahni Kuhliyo Family Center/ Oneida AODA
3000 Seminary Road
Oneida, WI 54155

Luthahi Yosta Group Home
4370 Meadow Drive
Oneida, WI 54155

OSHKOSH

Mercy Medical Center Family Renewal Center
631 Hazel Street
Oshkosh, WI 54901

New Horizons Counseling Services, Inc.
429 Algoma Boulevard
Oshkosh, WI 54901

Summit House
2501 Harrison Street
Oshkosh, WI 54901

Terra Programs, Inc.
Horizon House
968 Butler Avenue
Oshkosh, WI 54901

Nova Treatment Center
111 Josslyn Street
Oshkosh, WI 54901

Terra House
105 Josslyn Street
Oshkosh, WI 54901

Winnebago County Department of Community Programs
471 High Avenue
Oshkosh, WI 54901

OXFORD

Federal Correctional Institution Substance Abuse Services
Oxford, WI 53952

PARK FALLS

Flambeau Medical Center Emergency Detox Services
98 Sherry Avenue
Park Falls, WI 54552

PHILLIPS

Alcohol and Drug Abuse Recovery Center
124 North Avon Avenue
Phillips, WI 54555

Counseling and Personal Development AODA Treatment Program
171 Chestnut Street
Phillips, WI 54555

PLATTEVILLE

Unified Counseling Services Platteville Outpatient Clinic
6057 South Chestnut Street
Platteville, WI 53818

PLOVER

Washington House
1608 Washington Avenue
Plover, WI 54467

PORT WASHINGTON

Ozaukee Council, Inc. Alcohol and Other Drug Abuse
125 North Franklin Street
Port Washington, WI 53074

Ozaukee County Dept. of Community Programs Ozaukee County Counseling Center
121 West Main Street
Port Washington, WI 53074

Saint Mary's Hospital/Ozaukee Saint Mary's Chemical Dependency Program
743 North Montgomery Street
Port Washington, WI 53074

PORTAGE

Divine Savior Hospital Emergency Room Substance Abuse Detox Services
1015 West Pleasant Street
Portage, WI 53901

Pathfinder
711 East Cook Street
Portage, WI 53901

Pathfinder House
108 East Pleasant Street
Portage, WI 53901

PRAIRIE DU CHIEN

Saint Francis Community Programs, Inc. Villa Succes
121 South Prairie Street
Prairie du Chien, WI 53831

RACINE

Crisis Center of Racine, Inc. Goodwill Industries of Southeastern Wisconsin
209 8 Street
Racine, WI 53403

Crossroads Consultants, Ltd.
3308 Washington Avenue
Racine, WI 53405

Kettle Moraine Hospital
Durand Home
4606 Durand Avenue
Racine, WI 53406

Clinic
1254 West Boulevard
Racine, WI 53405

Lighthouse Counseling Associates
5605 Washington Avenue
Racine, WI 53406

Professional Services Group
3105 Lathrop Avenue
Racine, WI 53405

Racine Psychological Services
840 Lake Avenue
Racine, WI 53403

Saint Luke's Hospital Emergency Room
1320 Wisconsin Avenue
Racine, WI 53403

Southeastern Wisconsin Medical and Social Services Racine Office
1055 Prairie Drive
Racine, WI 53406

Urban League Racine/Kenosha Youth Drug Prevention Intervention Program
718 North Memorial Drive
Racine, WI 53404

Waybridge Manor
1633 South Memorial Drive
Racine, WI 53403

Westwind Treatment Center
5625 Washington Avenue
Racine, WI 53406

REEDSBURG

Sauk County Dept. of Human Services Alcoholism and Drug Abuse Outpatient Services
425 6 Street
Reedsburg, WI 53959

RHINELANDER

Human Service Center DWI Assessment
705 East Timber Drive
Rhinelander, WI 54501

Koinonia
1991 Winnebago Drive
Rhinelander, WI 54501

Lakeland Council on AODA, Inc.
17A West Davenport Street
Rhinelander, WI 54501

RICE LAKE

Lutheran Social Services
40 West Newton Street
Rice Lake, WI 54868

Parkview Center
24 East Douglas Street
Rice Lake, WI 54868

RICHLAND CENTER

Richland County Community Programs
1000 Highway 14 West
Richland Center, WI 53581

RIVER FALLS

Kinnic Falls Alcohol and Drug Abuse Services
900 South Orange Street
River Falls, WI 54022

SAINT CROIX FALLS

Saint Croix Valley Memorial Hospital Chemical Dependency Center
204 South Adams Street
Saint Croix Falls, WI 54024

SHAWANO

Department of Community Programs
504 Lakeland Road
Shawano, WI 54166

Shawano Community Hospital Emergency Room
309 North Bartlette Street
Shawano, WI 54166

Shawano County Community Programs Professional Services Center
125 North Main Street
Shawano, WI 54166

SHEBOYGAN

Counseling and Development Center
2205 Erie Avenue
Sheboygan, WI 53081

Kettle Moraine Hospital Kettle Moraine Sheboygan Center
503 Wisconsin Avenue
Sheboygan, WI 53081

Rebos Manor
908 Jefferson Street
Sheboygan, WI 53081

Sheboygan County Human Services Outpatient Services
1011 North 8 Street
Sheboygan, WI 53081

Sheboygan Memorial Medical Center Chemical Dependency Services
2629 North 7 Street
Sheboygan, WI 53083

SHELL LAKE

Ain Dah Ing, Inc.
108 West 6 Avenue
Shell Lake, WI 54871

Indianhead Residential Care Facility Sunshadows, Inc.
230 West 5 Avenue
Shell Lake, WI 54871

SPARTA

Monroe County Guidance Clinic
Route 2
County Trunk B
Sparta, WI 54656

Saint Mary's Hospital Inpatient Medical and Social Detox
West Main and K Streets
Sparta, WI 54656

SPOONER

Pines II, Inc. Detox
1885 North Rice Lake Road
Spooner, WI 54801

Spooner Community Memorial Hospital Substance Abuse Detox
819 Ash Street
Spooner, WI 54801

STEVENS POINT

Community Alcohol and Drug Abuse Center
209 Prentice Street North
Stevens Point, WI 54481

Oakside Residential Living Facility
201 North Prentice Street
Stevens Point, WI 54481

Family Crisis Center/CAP Services
1616 West River Street
Stevens Point, WI 54481

Saint Michael's Hospital Emergency Room Substance Abuse Services
900 Illinois Avenue
Stevens Point, WI 54481

Soma House
2201 Julia Street
Stevens Point, WI 54481

STOUGHTON

Lutheran Social Services HOMME Programs/Serenity Unit
209 North Division Street
Stoughton, WI 53589

Stoughton Hospital Association Share Program/Inpatient/Outpatient
900 Ridge Street
Stoughton, WI 53589

STURGEON BAY

Door County Memorial Hospital Emergency Room Substance Abuse Services
330 South 16 Place
Sturgeon Bay, WI 54235

Door County Unified Board
421 Nebraska Street
Sturgeon Bay, WI 54235

SULLIVAN

Kettle House
4756 North Indian Point Road
Sullivan, WI 53066

SUPERIOR

Lutheran Social Services Recovery Center, Inc.
2231 Catlin Avenue
Suite 2 East
Superior, WI 54880

TOMAH

Tomah Memorial Hospital Substance Abuse Detox Services
321 Butts Avenue
Tomah, WI 54660

Veterans' Affairs Medical Center Alcohol/Drug Dependence Treatment Program
Veterans Road
116C/4
Tomah, WI 54660

TOMAHAWK

Lincoln Health Care Center Tomahawk Office/Substance Abuse Services
310 West Wisconsin Avenue
Tomahawk, WI 54487

Sacred Heart Saint Mary's Hospital Oasis Recovery Program
216 North 7 Street
Tomahawk, WI 54487

VIROQUA

Douglas Mental Health Services Substance Abuse Services
Highway 14 North
Viroqua, WI 54665

Vernon Memorial Hospital Emergency Detoxification
507 South Main Street
Viroqua, WI 54665

WASHBURN

Lutheran Social Services
320 Superior Avenue
Washburn, WI 54891

WATERTOWN

Directions Counseling Center
129 Hospital Drive
Watertown, WI 53094

WAUKESHA

Century House
N1 W24940 Northview Road
Waukesha, WI 53188

Genesis House
1002 Motor Avenue
Waukesha, WI 53188

Kettle Moraine Outpatient Clinic
414 West Moreland Boulevard
Waukesha, WI 53186

La Casa de Esperanza AODA Prevention/Education
410 Arcadian Avenue
Waukesha, WI 53186

Lutheran Social Services House of Hope
325 Sentinel Drive
Waukesha, WI 53186

Southeastern Wisconsin Homes Noah House
West 222 South 3210 Racine Avenue
Waukesha, WI 53186

Waukesha County Community Human Services Alcohol and Drug Abuse Clinic
500 Riverview Avenue
Waukesha, WI 53188

Waukesha County Council on Alcoholism and Other Drug Abuse
310 South Street
Waukesha, WI 53186

Waukesha Memorial Hospital Chemical Dependency Treatment and Education Center
725 American Avenue
Waukesha, WI 53188

WAUPACA

Community Alcohol and Drug Abuse Center
1035B East Royalton Street
Waupaca, WI 54981

Riverside Medical Center Emergency Room Substance Abuse Services
800 Riverside Drive
Waupaca, WI 54981

Waupaca County Dept. of Human Services Outpatient Treatment Services
811 Harding Street
Waupaca, WI 54981

WAUSAU

Crossroads Mental Health Services
529 McClellan Street
Wausau, WI 54401

Lutheran Social Services Outpatient Services
725 Gilbert Street
Wausau, WI 54401

North Central Health Care Facilities
1100 Lake View Drive
Wausau, WI 54401

WAUTOMA

Alcoholism and Drug Abuse Services of Waushara County
310 South Scott Street
Wautoma, WI 54982

WEBSTER

Northwest Passage
Webster, WI 54893

WEST BEND

Council on Alcohol and Other Drug Abuse of Washington County
279 South 17 Avenue
Suite 4
West Bend, WI 53095

Kettle Moraine Treatment Center West Bend Clinic
344 South 6 Avenue
West Bend, WI 53095

WILD ROSE

Wild Rose Community Hospital Emergency Inpatient Detox
601 Grove Street
Wild Rose, WI 54984

WINNEBAGO

Anchorage
Winnebago, WI 54985

Winnebago Mental Health Institute Gemini
Winnebago, WI 54985

WISCONSIN RAPIDS

Riverview Hospital Emergency Inpatient Detox
410 Dewey Street
Wisconsin Rapids, WI 54494

Wood County Unified Services
2611 South 12 Street
Wisconsin Rapids, WI 54494

WOODRUFF

Howard Young Medical Center Substance Abuse Emergency Detox Services
Woodruff, WI 54568

Marshfield Clinic Lakeland Center
519 Hemlock Street
Woodruff, WI 54568

WYOMING

AFTON

Lincoln County Mental Health Assoc. Substance Abuse Services
Hospital Lane
Afton, WY 83110

BASIN

Big Horn County Counseling
220 South 4 Street
Basin, WY 82410

BUFFALO

Northern Wyoming Mental Health Center Substance Abuse Services
521 West Lott Street
Buffalo, WY 82834

CASPER

Central Wyoming Counseling Center Substance Abuse Program
1200 East 3 Street
Suite 330
Casper, WY 82601

Crest View Hospital Lifeworks Chemical Dependency Program
2521 East 15 Street
Casper, WY 82609

Mercer House, Inc.
425 Cy Avenue
Casper, WY 82601

New Directions
104 South Lowell Street
Casper, WY 82601

New Horizons
837 East C Street
Casper, WY 82601

The Prairie Institute, Inc.
1236 South Elm Street
Casper, WY 82601

Wyoming Medical Center
High Plains
1233 East 2 Street
Casper, WY 82601

CHEYENNE

Cheyenne Community Drug Abuse Treatment Council, Inc./ Pathfinder
803 West 21 Street
Cheyenne, WY 82001

Southeast Wyoming Mental Health Center
Chemical Health Services
1609 East 19 Street
Cheyenne, WY 82001

Cheyenne Alcohol Receiving Center
Cheyenne Halfway House
1623 East Lincolnway
Cheyenne, WY 82001

CODY

Cedar Mountain Center at West Park Hospital
707 Sheridan Avenue
Cody, WY 82414

DOUGLAS

Eastern Wyoming Mental Health Center Substance Abuse Services
1841 Madora Avenue
Douglas, WY 82633

DUBOIS

Fremont Counseling Service
706 Meckem Street
Dubois, WY 82513

EVANSTON

Southwestern Wyoming Alcohol Rehab Association Westland Services
1235 Uinta Street
Evanston, WY 82930

Uinta County Rehabilitation Center Alcohol and Drug Services
350 City View Drive
Suite 303
Evanston, WY 82930

Wyoming State Hospital Forensic Treatment Unit
Highway 150
Evanston, WY 82930

GILLETTE

Powder River Chemical Dependency, Inc.
400 South Kendrick Avenue
Suite 101
Gillette, WY 82716

Wyoming Regional Counseling Center
900 West 6 Street
Gillette, WY 82716

JACKSON

Curran/Seeley Foundation
610 West Broadway
Suite L-1
Jackson, WY 83001

KEMMERER

Lincoln County Mental Health Assoc. Substance Abuse Services
230 Highway 233
Kemmerer, WY 83101

LANDER

Fremont Counseling Service
748 Main Street
Lander, WY 82520

LARAMIE

Southeast Wyoming Mental Health Center Substance Abuse Services
710 Garfield Street
Suite 320
Laramie, WY 82070

LOVELL

Big Horn County Counseling
441 Montana Avenue
Lovell, WY 82431

LUSK

Eastern Wyoming Mental Health Center Substance Abuse Services
521 East 10 Street
Lusk, WY 82225

NEWCASTLE

Northern Wyoming Mental Health Center Substance Abuse Services
18 Stempede Street
Newcastle, WY 82701

PINEDALE

Sublette Community Counseling Services
41 1/2 South Franklin Street
Pinedale, WY 82941

RAWLINS

Carbon County Counseling Center
721 West Maple Street
Rawlins, WY 82301

RIVERTON

Fremont Counseling Service
322 North 8 Street West
Riverton, WY 82501

ROCK SPRINGS

Rosen Recovery Center
1414 9 Street
Rock Springs, WY 82901

Southwest Counseling Service
1124 College Road
Rock Springs, WY 82901

SHERIDAN

Northern Wyoming Mental Health Center Substance Abuse Services
1221 West 5 Street
Sheridan, WY 82801

Sheridan House, Inc.
1003 Saberton Street
Sheridan, WY 82801

Thunder Child Sheridan Veterans' Admin. Hospital
Building 24
Sheridan, WY 82801

Veterans' Affairs Medical Center Substance Abuse Treatment Program
Sheridan, WY 82801

SUNDANCE

Northern Wyoming Mental Health Center Substance Abuse Services
209 Cleveland Street
Sundance, WY 82729

THERMOPOLIS

Hot Springs County Counseling Service
121 South 4 Street
Thermopolis, WY 82443

TORRINGTON

Southeast Wyoming Mental Health Center Substance Abuse Services
1942 East D Street
Torrington, WY 82240

WHEATLAND

Southeast Wyoming Mental Health Center Substance Abuse Services
103 Park Avenue
Wheatland, WY 82201

WORLAND

Washakie County Mental Health Services
509 Big Horn Avenue
Worland, WY 82401

APPENDIX IV

Bureau of Justice Statistics

INTRODUCTION

The Bureau of Justice Statistics is an agency of the U.S. Department of Justice, Washington, DC. In a 1992 report entitled *Drugs, Crime, and the Justice System*, the bureau presented an overview of how the U.S. justice system attempts to combat illegal drugs.

Many areas of society are included in the overview. Here we present summarized data in easy to review format. Much of the information offered here is fully discussed throughout the alphabetical entries of the encyclopedia—in Volumes 1, 2, and 3. Consult the Index at the end of this volume for references to items of further interest.

POLICIES, STRATEGIES, AND TACTICS USED TO CONTROL THE ILLEGAL DRUG PROBLEM

POLICIES

Prohibition is the ban on the distribution, possession, and use of specified substances made illegal by legislative or administrative order and the application of criminal penalties to violators.

Regulation is control over the distribution, possession, and use of specified substances. Regulations specify the circumstances under which substances can be legally distributed and used. Prescription medications and alcohol are the substances most commonly regulated in the U.S.

STRATEGIES

Demand reduction strategies attempt to decrease individuals' tendency to use drugs. Efforts provide information and education to potential and casual users about the risks and adverse consequences of drug use, and treatment to drug users who have developed problems from using drugs.

Supply reduction focuses diplomatic, law enforcement, military, and other resources on eliminating or reducing the supply of drugs. Efforts focus on foreign countries, smuggling routes outside the country, border interdiction, and distribution within the U.S.

User accountability emphasizes that all users of illegal substances, regardless of the type of drug they use or the frequency of that use, are violating criminal laws and should be subject to penalties. It is closely associated with zero tolerance.

Zero tolerance holds that drug distributors, buyers, and users should be held fully accountable for their offenses under the law. This is an alternative to policies that focus only on some violators such as sellers of drugs or users of cocaine and heroin while ignoring other violators.

TACTICS

Criminal justice activities include enforcement, prosecution, and sentencing activities to apprehend, convict, and punish drug offenders. Although thought of primarily as having supply reduction goals, criminal sanctions also have demand reduction effects by discouraging drug use.

Prevention activities are educational efforts to inform potential drug users about the health, legal, and other risks associated with drug use. Their goal is to limit the number of new drug users and dissuade casual users from continuing drug use as part of a demand reduction strategy.

Taxation requires those who produce, distribute, or possess drugs to pay a fee based on the volume or value of the drugs. Failure to pay subjects violators to penalties for this violation, not for the drug activities themselves.

Testing individuals for the presence of drugs is a tool in drug control that is used for safety and monitoring purposes and as an adjunct to therapeutic interventions. It is in widespread use for employees in certain jobs such as those in the transportation industry and criminal justice agencies. New arrestees and convicted offenders may be tested. Individuals in treatment are often tested to monitor their progress and provide them an incentive to remain drug free.

Treatment (therapeutic interventions) focus on individuals whose drug use has caused medical, psychological, economic, and social problems for them. The interventions may include medication, counseling, and other support services delivered in an inpatient setting or on an outpatient basis. These are demand reduction activities to eliminate or reduce individuals' drug use.

HISTORIC MILESTONES IN EARLY U.S. DRUG CONTROL EFFORTS

Drugs of abuse have changed since the 1800s—most rapidly over the past quarter century. Problems with opiate addiction date from widespread use of patent medicines in the 1800s. The range of drugs included opium, morphine, laudanum, cocaine, and, by the turn of the century, heroin. The tonics, nostrums, and alleged cures that contained or used such drugs were sold by itinerant peddlers, mail order houses, retail grocers, and pharmacists. There also was unrestricted access to opium in opium-smoking dens and to morphine through retailers.

When morphine was discovered in 1806, it was thought to be a wonder drug. Its use was so extensive during the Civil War that morphine addiction was termed the "army disease." The availability of the hypodermic syringe allowed nonmedicinal use of morphine to gain popularity among veterans and other civilians. After 1898, heroin was used to treat respiratory illness and morphine addiction in the belief that it was nonaddicting.

In the 1880s coca became widely available in the U.S. as a health tonic and remedy for many ills. Its use was supported first by the European medical community and later by American medical authorities. In the absence of restrictive national legislation, its use spread. Initially cocaine was offered as a cure for opiate addiction, an asthma remedy (the official remedy of the American Hay Fever Association), and an antidote for toothaches.

By 1900, in the face of an estimated quarter of a million addicts, State laws were enacted to curb drug addiction. The major drugs of abuse at the time were cocaine and morphine.

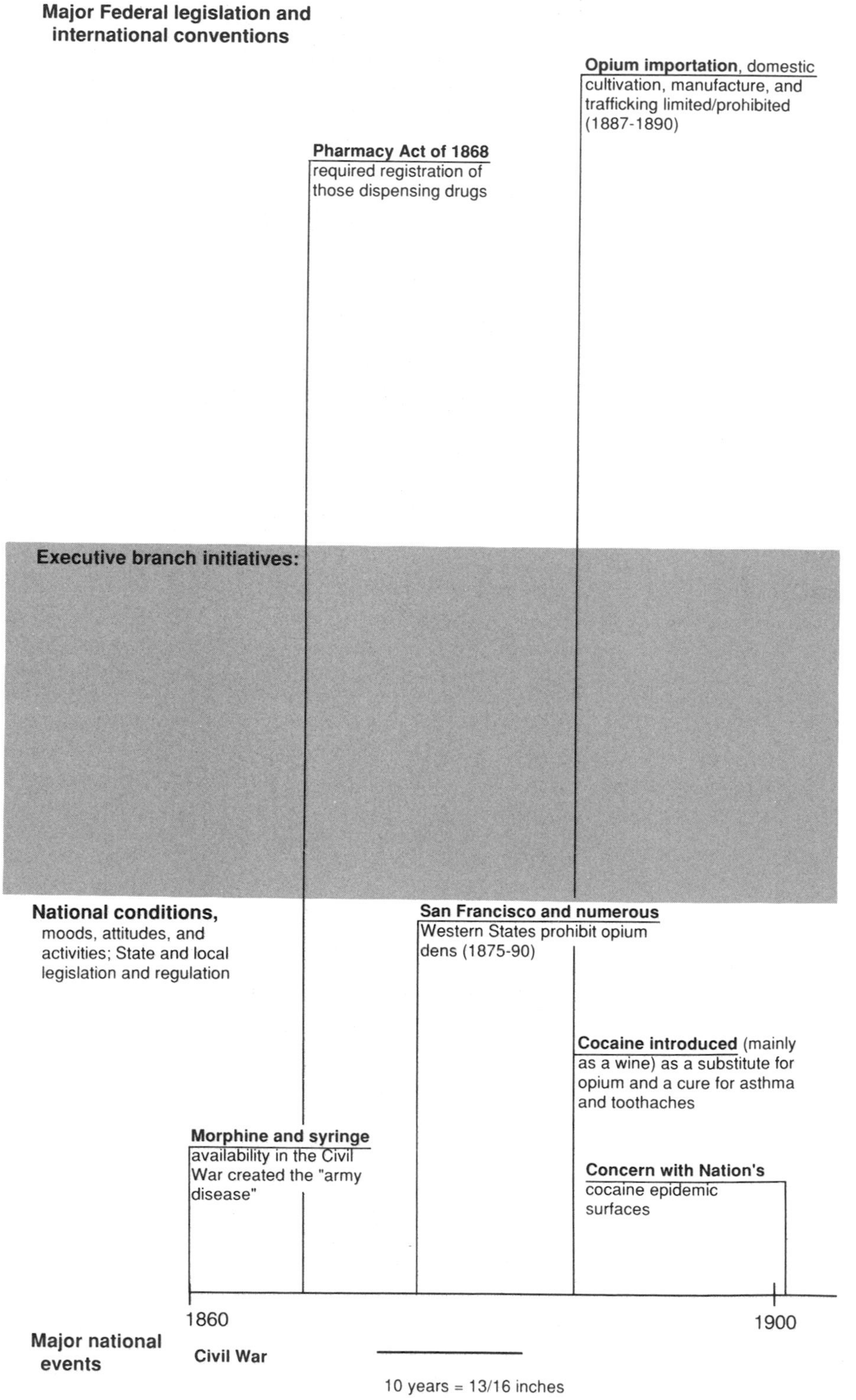

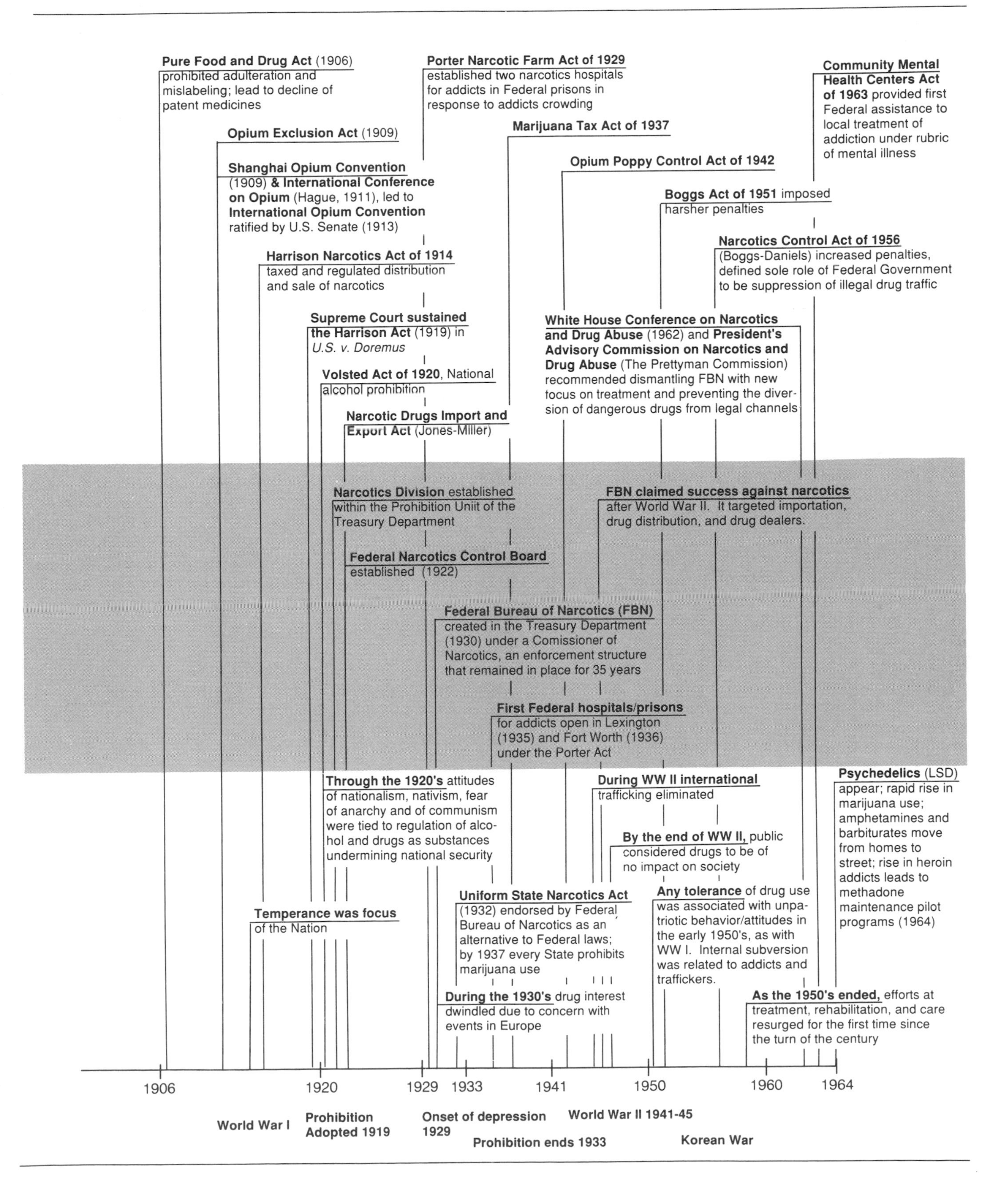

Pure Food and Drug Act (1906) prohibited adulteration and mislabeling; lead to decline of patent medicines
Opium Exclusion Act (1909)
Shanghai Opium Convention (1909) **& International Conference on Opium** (Hague, 1911), led to **International Opium Convention** ratified by U.S. Senate (1913)
Harrison Narcotics Act of 1914 taxed and regulated distribution and sale of narcotics
Supreme Court sustained the Harrison Act (1919) in *U.S. v. Doremus*
Volsted Act of 1920, National alcohol prohibition
Narcotic Drugs Import and Export Act (Jones-Miller)
Porter Narcotic Farm Act of 1929 established two narcotics hospitals for addicts in Federal prisons in response to addicts crowding
Marijuana Tax Act of 1937
Opium Poppy Control Act of 1942
Boggs Act of 1951 imposed harsher penalties
Community Mental Health Centers Act of 1963 provided first Federal assistance to local treatment of addiction under rubric of mental illness
Narcotics Control Act of 1956 (Boggs-Daniels) increased penalties, defined sole role of Federal Government to be suppression of illegal drug traffic
White House Conference on Narcotics and Drug Abuse (1962) and **President's Advisory Commission on Narcotics and Drug Abuse** (The Prettyman Commission) recommended dismantling FBN with new focus on treatment and preventing the diversion of dangerous drugs from legal channels
Narcotics Division established within the Prohibition Uniit of the Treasury Department
FBN claimed success against narcotics after World War II. It targeted importation, drug distribution, and drug dealers.
Federal Narcotics Control Board established (1922)
Federal Bureau of Narcotics (FBN) created in the Treasury Department (1930) under a Comissioner of Narcotics, an enforcement structure that remained in place for 35 years
First Federal hospitals/prisons for addicts open in Lexington (1935) and Fort Worth (1936) under the Porter Act
Through the 1920's attitudes of nationalism, nativism, fear of anarchy and of communism were tied to regulation of alcohol and drugs as substances undermining national security
During WW II international trafficking eliminated
By the end of WW II, public considered drugs to be of no impact on society
Psychedelics (LSD) appear; rapid rise in marijuana use; amphetamines and barbiturates move from homes to street; rise in heroin addicts leads to methadone maintenance pilot programs (1964)
Temperance was focus of the Nation
Uniform State Narcotics Act (1932) endorsed by Federal Bureau of Narcotics as an alternative to Federal laws; by 1937 every State prohibits marijuana use
Any tolerance of drug use was associated with unpatriotic behavior/attitudes in the early 1950's, as with WW I. Internal subversion was related to addicts and traffickers.
During the 1930's drug interest dwindled due to concern with events in Europe
As the 1950's ended, efforts at treatment, rehabilitation, and care resurged for the first time since the turn of the century
1906
1920
1929
1933
1941
1950
1960
1964
World War I
Prohibition Adopted 1919
Onset of depression 1929
Prohibition ends 1933
World War II 1941-45
Korean War

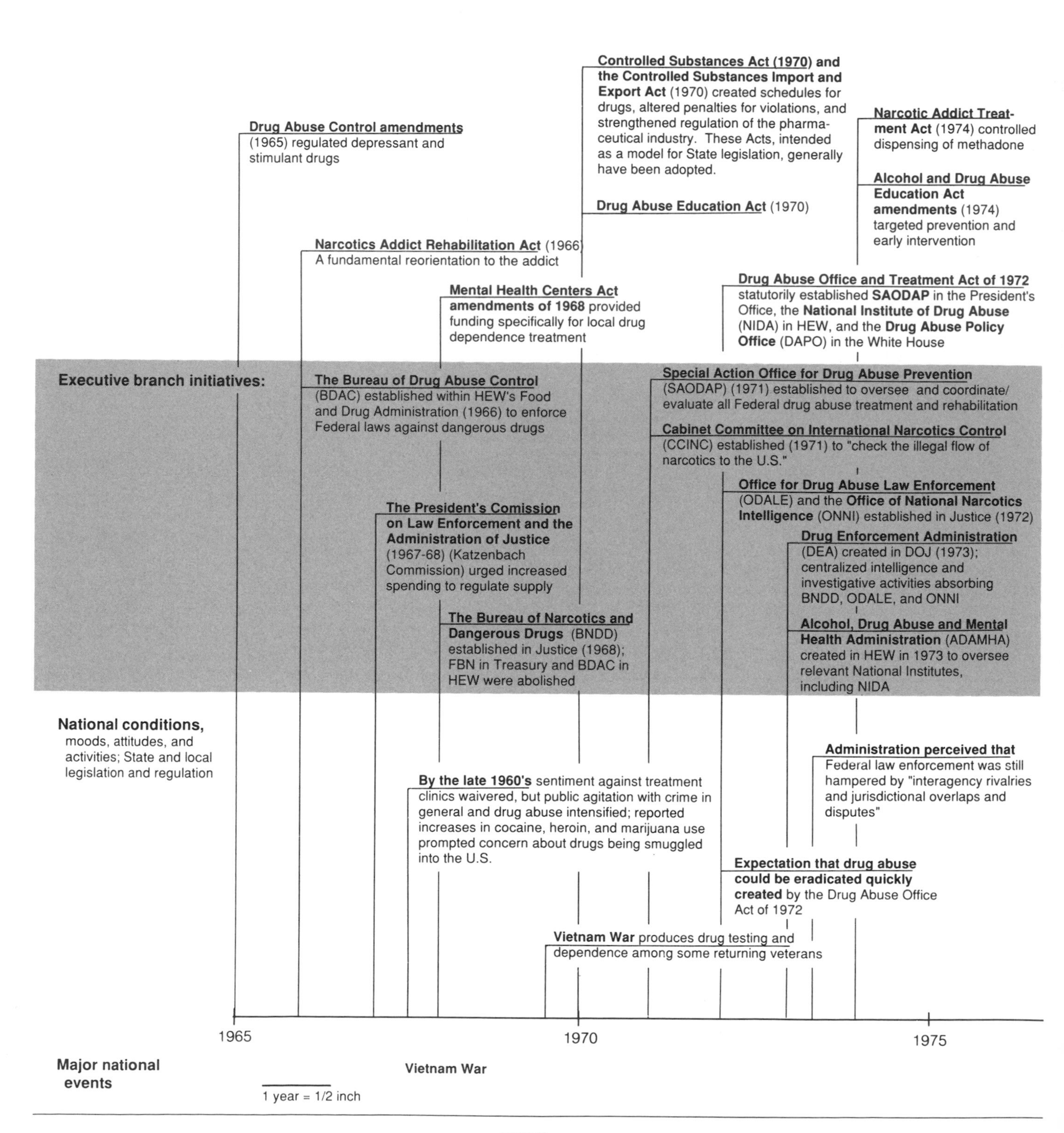
Major Federal legislation and international conventions
Drug Abuse Control amendments (1965) regulated depressant and stimulant drugs
Narcotics Addict Rehabilitation Act (1966) A fundamental reorientation to the addict
Mental Health Centers Act amendments of 1968 provided funding specifically for local drug dependence treatment
Controlled Substances Act (1970) and the Controlled Substances Import and Export Act (1970) created schedules for drugs, altered penalties for violations, and strengthened regulation of the pharmaceutical industry. These Acts, intended as a model for State legislation, generally have been adopted.
Drug Abuse Education Act (1970)
Narcotic Addict Treatment Act (1974) controlled dispensing of methadone
Alcohol and Drug Abuse Education Act amendments (1974) targeted prevention and early intervention
Drug Abuse Office and Treatment Act of 1972 statutorily established SAODAP in the President's Office, the National Institute of Drug Abuse (NIDA) in HEW, and the Drug Abuse Policy Office (DAPO) in the White House
Executive branch initiatives:
The Bureau of Drug Abuse Control (BDAC) established within HEW's Food and Drug Administration (1966) to enforce Federal laws against dangerous drugs
The President's Comission on Law Enforcement and the Administration of Justice (1967-68) (Katzenbach Commission) urged increased spending to regulate supply
The Bureau of Narcotics and Dangerous Drugs (BNDD) established in Justice (1968); FBN in Treasury and BDAC in HEW were abolished
Special Action Office for Drug Abuse Prevention (SAODAP) (1971) established to oversee and coordinate/evaluate all Federal drug abuse treatment and rehabilitation
Cabinet Committee on International Narcotics Control (CCINC) established (1971) to "check the illegal flow of narcotics to the U.S."
Office for Drug Abuse Law Enforcement (ODALE) and the Office of National Narcotics Intelligence (ONNI) established in Justice (1972)
Drug Enforcement Administration (DEA) created in DOJ (1973); centralized intelligence and investigative activities absorbing BNDD, ODALE, and ONNI
Alcohol, Drug Abuse and Mental Health Administration (ADAMHA) created in HEW in 1973 to oversee relevant National Institutes, including NIDA
National conditions, moods, attitudes, and activities; State and local legislation and regulation
By the late 1960's sentiment against treatment clinics waivered, but public agitation with crime in general and drug abuse intensified; reported increases in cocaine, heroin, and marijuana use prompted concern about drugs being smuggled into the U.S.
Administration perceived that Federal law enforcement was still hampered by "interagency rivalries and jurisdictional overlaps and disputes"
Expectation that drug abuse could be eradicated quickly created by the Drug Abuse Office Act of 1972
Vietnam War produces drug testing and dependence among some returning veterans
1965
1970
1975
Major national events
Vietnam War
1 year = 1/2 inch

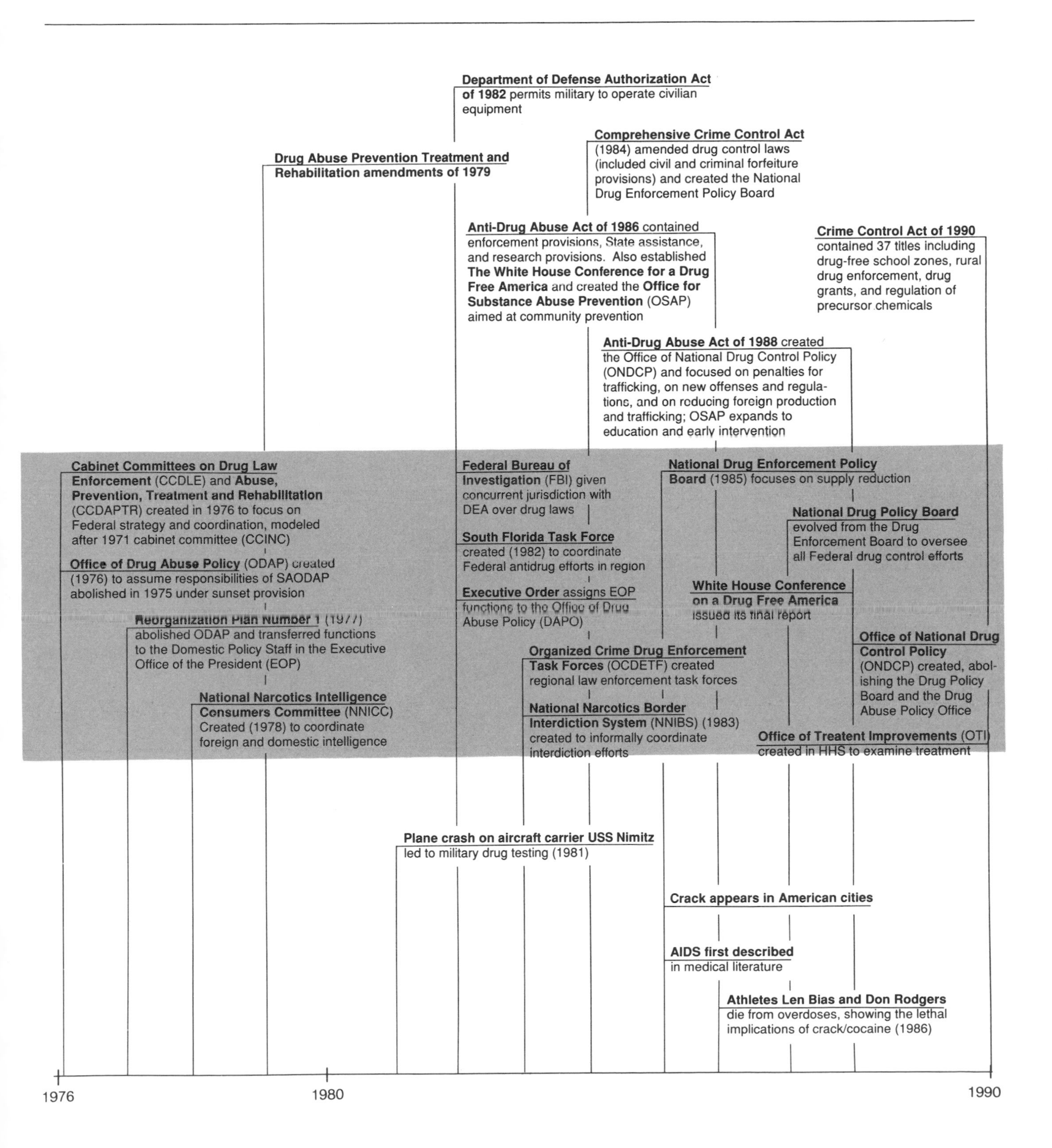
Department of Defense Authorization Act of 1982 permits military to operate civilian equipment
Comprehensive Crime Control Act (1984) amended drug control laws (included civil and criminal forfeiture provisions) and created the National Drug Enforcement Policy Board
Drug Abuse Prevention Treatment and Rehabilitation amendments of 1979
Anti-Drug Abuse Act of 1986 contained enforcement provisions, State assistance, and research provisions. Also established The White House Conference for a Drug Free America and created the Office for Substance Abuse Prevention (OSAP) aimed at community prevention
Crime Control Act of 1990 contained 37 titles including drug-free school zones, rural drug enforcement, drug grants, and regulation of precursor chemicals
Anti-Drug Abuse Act of 1988 created the Office of National Drug Control Policy (ONDCP) and focused on penalties for trafficking, on new offenses and regulations, and on reducing foreign production and trafficking; OSAP expands to education and early intervention
Cabinet Committees on Drug Law Enforcement (CCDLE) and Abuse, Prevention, Treatment and Rehabilitation (CCDAPTR) created in 1976 to focus on Federal strategy and coordination, modeled after 1971 cabinet committee (CCINC)
Office of Drug Abuse Policy (ODAP) created (1976) to assume responsibilities of SAODAP abolished in 1975 under sunset provision
Reorganization Plan Number 1 (1977) abolished ODAP and transferred functions to the Domestic Policy Staff in the Executive Office of the President (EOP)
National Narcotics Intelligence Consumers Committee (NNICC) Created (1978) to coordinate foreign and domestic intelligence
Federal Bureau of Investigation (FBI) given concurrent jurisdiction with DEA over drug laws
South Florida Task Force created (1982) to coordinate Federal antidrug efforts in region
Executive Order assigns EOP functions to the Office of Drug Abuse Policy (DAPO)
Organized Crime Drug Enforcement Task Forces (OCDETF) created regional law enforcement task forces
National Narcotics Border Interdiction System (NNIBS) (1983) created to informally coordinate interdiction efforts
National Drug Enforcement Policy Board (1985) focuses on supply reduction
National Drug Policy Board evolved from the Drug Enforcement Board to oversee all Federal drug control efforts
White House Conference on a Drug Free America issued its final report
Office of National Drug Control Policy (ONDCP) created, abolishing the Drug Policy Board and the Drug Abuse Policy Office
Office of Treatent Improvements (OTI) created in HHS to examine treatment
Plane crash on aircraft carrier USS Nimitz led to military drug testing (1981)
Crack appears in American cities
AIDS first described in medical literature
Athletes Len Bias and Don Rodgers die from overdoses, showing the lethal implications of crack/cocaine (1986)
1976
1980
1990

FOUR MAJOR FEDERAL ANTIDRUG BILLS, ENACTED 1984–1990

The 1984 Crime Control Act—
- expanded criminal and civil asset forfeiture laws
- amended the Bail Reform Act to target pretrial detention of defendants accused of serious drug offenses
- established a determinate sentencing system
- increased Federal criminal penalties for drug offenses

The 1986 Anti-Drug Abuse Act—
- budgeted money for prevention and treatment programs, giving the programs a larger share of Federal drug control funds than previous laws
- restored mandatory prison sentences for large-scale distribution of marijuana
- imposed new sanctions on money laundering
- added controlled substances' analogs (designer drugs) to the drug schedule
- created a drug law enforcement grant program to assist State and local efforts
- contained various provisions designed to strengthen international drug control efforts.

The 1988 Anti-Drug Abuse Act—
- increased penalties for offenses related to drug trafficking, created new Federal offenses and regulatory requirements, and changed criminal procedures
- altered the organization and coordination of Federal antidrug efforts
- increased treatment and prevention efforts aimed at reduction of drug demand
- endorsed the use of sanctions aimed at drug users to reduce the demand for drugs
- targeted for reduction drug production abroad and international trafficking in drugs

The Crime Control Act of 1990—
- doubled the appropriations authorized for drug law enforcement grants to States and localities
- expanded drug control and education programs aimed at the Nation's schools
- expanded specific drug enforcement assistance to rural States
- expanded regulation of precursor chemicals used in the manufacture of illegal drugs
- provided additional measures aimed at seizure and forfeiture of drug trafficker assets
- sanctioned anabolic steroids under the Controlled Substances Act
- included provisions on international money laundering, rural drug enforcement, drug-free school zones, drug paraphernalia, and drug enforcement grants.

APPENDIX V

Illicit and Licit Drugs of Abuse—Schedules of Controlled Substances

INTRODUCTION

U.S. legislation called the Controlled Substances Act of 1970 has ranked and categorized drugs according to their effects, medical use, and potential for abuse. Ongoing research may reclassify drugs from one category to another, as has happened in the past.

At the federal level, Schedule I is the most strictly controlled—with the highest abuse potential; Schedule V is the least strictly controlled—drugs sold with or without prescription by mail and in shops, with instructions for use, dosages, and warnings about effects and side effects printed on the packaging of over-the-counter (OTC) medications. The schedules shown below in simplified form are followed by extensive schedules (which are discussed fully in Volume 1, in the article entitled Controls: Scheduled Drugs/Drug Schedules, U.S.). A discussion of the Controlled Substances Act of 1970 precedes it.

Drugs are scheduled under federal law according to their effects, medical use, and potential for abuse

DEA Schedule	*Abuse Potential*	*Examples of Drugs Covered*	*Some of the Effects*	*Medical Use*
I	highest	heroin, LSD, hashish, marijuana, methaqualone, designer drugs	unpredictable effects, severe psychological or physical dependence, or death	no accepted use; some are legal for limited research use only
II	high	morphine, PCP, codeine, cocaine, methadone, Demerol®, benzedrine, dexedrine	may lead to severe psychological or physical dependence	accepted use with restrictions
III	medium	codeine with aspirin or Tylenol®, some amphetamines, anabolic steroids	may lead to moderate or low physical dependence or high psychological dependence	accepted use
IV	low	Darvon®, Talwin®, phenobarbital, Equanil®, Miltown®, Librium®, diazepam	may lead to limited physical or psychological dependence	accepted use
V	lowest	over-the-counter or prescription compounds with codeine, Lomotil®, Robitussin A-C®	may lead to limited physical or psychological dependence	accepted use

SOURCE: Adapted from Drug Enforcement Administration, *Drugs of abuse: 1989.*

SCHEDULES OF U.S. CONTROLLED DRUGS

CRITERIA FOR U.S. DRUG SCHEDULING

	Potential for:		*Medical Use*
Schedule	*Abuse*	*Dependence*	*& Safety*
I	+ + + +	+ + + +	No
II	+ + + +	+ + + +	Yes
III	+ + +	+ + +	Yes
IV	+ +	+ +	Yes
V	+	+	Yes

LIST OF CONTROLLED DRUGS

SCHEDULE I

Opiates

Acetyl-alpha-methylfentanyl
Acetylmethadol
Allylprodine
Alphameprodine
Alphamethadol
Alpha-methylfentanyl
Alpha-methylthiofentanyl
Benzethidine
Betacetylmethadol
Beta-hydroxyfentanyl
Beta-hydroxy-3-methylfentanyl
Betameprodine
Betamethadol
Betaprodine
Clonitazene
Dextromoramide
Diampromide
Diethylthiambutene
Difenoxin
Dimenoxadol
Dimepheptanol
Dimethylthiambutene
Dioxaphetyl butyrate
Dipipanone
Ethylmethylthiambutene
Etonitazene
Etoxeridine
Furethidine
Hydroxpethidine
Ketobemidone
Levomoramide
Levophenacylmorphan
3-methylfentanyl
3-methylthiofentanyl
Morpheridine
MPPP
Noracymethadol
Norlevorphanol
Normethadone
Norpipanone
Para-fluorofentanyl
PEPAP
Phenadoxone
Phenampromide
Phenomorphan
Phenoperidine
Piritramide
Proheptazine
Properidine
Propiram
Racemoramide
Thiofentanyl
Tilidine
Trimeperidine

Opium Derivatives

Actorphine
Acetyldihydrocodeine
Benzylmorphine
Codeine methylbromide
Codeine-N-Oxide
Cyprenorphine
Desomorphine
Dihydropmorphine
Drotebanol
Etorphine (except HCl salt)
Heroin
Hydromorphinol
Methyldesorphine
Methyldihydromorphine
Morphine methylbromide
Morphine methylsulfonate
Morphine-N-Oxide
Myrophine
Nicocodeine
Nicomorphine
Normorphine
Pholcodine
Thebacon

Hallucinogens

4-bromo-2,5-DMA
2,5-DMA
PMA
MMDA
DOM, STP
MDA
MDMA
MDEA
N-hydroxy MDA
3,4,5-trimethoxy amphetamine
Bufotenine
DET
DMT
Ibogaine
LSD
Marihuana
Mescaline
N-ethyl-3-piperidyl benzilate
N-methyl-3-piperidyl benzilate
Peyote
Phencyclidine analogs PCE, PCPy, TCP, TCPy
Psilocybin
Psilocyn
Synhexyl
Tetrahydrocannabinols

Depressants

Mecloqualone
Methaqualone

Stimulants

Fenethylline
(±) cis-4-methylaminorex
N-ethylamphetamine
N,N-dimethyl-amphetamine

LIST OF CONTROLLED DRUGS

SCHEDULE II

Opiates	Opium & Derivatives	Hallucinogens	Depressants	Stimulants	Others
Alfentanil	Raw opium	Dronabinol	Amobarbital	Amphetamine	Opium poppy
Alphaprodine	Opium extracts	Nabilone	Glutethimide	Methamphetamine	Poppy straw
Anileridine	Opium fluid		Pentobarbital	Phenmetrazine	Coca leaves
Bezitramide	Powdered opium		Phencyclidine	Methylphenidate	Intermediate precursors
Bulk dextropropoxy-phene	Granulated opium		Secobarbital		Amphetamine
Carfentanil	Tincture of opium				Methamphetamine
Dihydrocodeine	Codeine				Phencyclidine
Diphenoxylate	Ethylmorphine				
Fentanyl	Etorphine hydrochloride				
Isomethadone	Hydrocodone				
Levomethorphan	Hydromorphone				
Levorphanol	Metopon				
Metazocine	Morphine				
Methadone	Oxycodone				
Metadone-Intermediate	Oxymorphone				
Moramide-Intermediate	Thebaine				
Pethidine					
Pethidine-Intermediate-A					
Pethidine-Intermediate-B					
Pethidine-Intermediate-C					
Phenazocine					
Piminodine					
Racemorphan					
Racemthrophan					
Sufentanil					

LIST OF CONTROLLED DRUGS

SCHEDULE III

Narcotics	Depressants	Stimulants	Anabolic Steroids		Others
Limited quantities of:	Mixtures of	Limited mixtures of Schedule II amphetamines	Boldenone	Methandrostenolone	Nalorphine
Codeine,	Amobarbital		Chlorotestosterone	Methenolone	
Dihydrocodeinone,	Secobarbital		Clostebol	Methyltestosterone	
Dihydrocodeine,	Pentobarbital	Benzphetamine	Dehydrochlor-methyltestosterone	Mibolerone	
Ethylmorphine,	Derivative of barbituric acid	Chlorphentermine		Nandrolone	
Opium, and		Clortermine	Dihydrotestosterone	Norethandrolone	
Morphine	Chlorhexadol	Phendimetrazine	Drostanolone	Oxandrolone	
in combination	Lysergic acid		Ethylestrenol	Oxymesterone	
with nonnarcotics.	Lysergic acid amide		Fluoxymesterone	Oxymetholone	
	Methyprylon		Formebulone	Stanolone	
	Sulfondiethylmethane		Mesterolone	Stanozolol	
	Sulfonethylmethane		Methandienone	Testolacton	
	Sulfonmethane		Methandranone	Testosterone	
	Tiletamine		Methandriol	Trenbolone	
	Zolazepam				

LIST OF CONTROLLED DRUGS

SCHEDULE IV

Narcotics	Depressants		Stimulants	Others
Limited quantity of diferoin in combination with atropine sulfate.	Alprazolam	Loprazolam	(+)-norpseudoephedrine	Fenfluramine
	Barbital	Lorazepam	Diethylpropion	
	Bromazepam	Lormetazepam	Fencamfamin	Pentazocine
Dextropropoxyphene	Camazepam	Mebutamate	Fenproporex	
	Chloral betaine	Medazepam	Mazindol	
	Chloral hydrate	Meprobamate	Mefenorex	
	Chlordiazepoxide	Methohexital	Pemoline	
	Clobazam	Methylphenobarbital	Phentermine	
	Clonazepam	Midazolam	Pipradrol	
	Clorazepate	Nimetazepam	SPA	
	Clotiazepam	Nitrazepam		
	Cloxazolam	Nordiazepam		
	Delorazepam	Oxazepam		
	Diazepam	Oxazolam		
	Estazolam	Paraldehyde		
	Ethchlorvynol	Petrichloral		
	Ethinamate	Phenobarbital		
	Ethyl loflazepate	Pinazepam		
	Fludiazepam	Prazepam		
	Flunitrazepam	Quazepam		
	Flurazepam	Temazepam		
	Halazepam	Tetrazepam		
	Haloxazolam	Triazolam		
	Ketazolam			

LIST OF CONTROLLED DRUGS

SCHEDULE V

Narcotics	Stimulants
Buprenorphine Limited quantities (less than Schedules III & IV) of: Codeine, Dihydrocodeine, Ethylmorphine, Diphenoxylate, Opium, and Difenoxin in combination with nonnarcotics.	Pyrovalerone

Index

Numbers in **boldface** *refer to the main entry on the subject.*
Numbers in *italic* refer to illustrations.